Paul B. Zbar

Basic Electricity

A Text-Lab Manual

Fifth Edition

Gregg Division
McGraw-Hill Book Company

New York Atlanta Dallas St. Louis
San Francisco Auckland Bogotá
Guatemala Hamburg Johannesburg
Lisbon London Madrid Mexico
Montreal New Delhi Panama
Paris San Juan São Paulo Singapore
Sydney Tokyo Toronto

Other Books by Paul B. Zbar

Basic Electronics: A Text-Lab Manual, Fifth Edition (with A. P. Malvino)
Basic Radio: Theory and Servicing, A Text-Lab Manual, Third Edition
Basic Television: Theory and Servicing, A Text-Lab Manual, Third Edition (with P. Orne)
Electricity–Electronics Fundamentals: A Text-Lab Manual, Second Edition (with Joseph Sloop)
Electronic Instruments and Measurements: A Text-Lab Manual
Industrial Electronics: A Text-Lab Manual, Third Edition

Sponsoring Editor: Paul Berk
Editing Supervisor: Marcos E. Ricardo
Design and Art Supervisor: Caryl Valerie Spinka
Production Supervisor: Laurence Charnow

Cover Designer: Jorgé Hernandez
Technical Studio: Fine Line, Inc.

Library of Congress Cataloging in Publication Data

Zbar, Paul B (date)
Basic electricity.

(The basic electricity—electronics series)
1. Electronics—Laboratory manuals. I. Title.
II. Series.
TK7818.Z18 1983 621.31′028 82-17155
ISBN 0-07-072801-1

BASIC ELECTRICITY: A Text-Lab Manual, Fifth Edition

1 2 3 4 5 6 7 8 9 0 SEM SEM 8 9 8 7 6 5 4 3

ISBN 0-07-072801-1

CONTENTS

NOTE ON EXPERIMENT CONTENT

Each of the experiments described below is set up in the following manner:

OBJECTIVES The objectives are enumerated and clearly stated.
INTRODUCTORY INFORMATION The theory and basic principles involved in the experiment are clearly stated.
SUMMARY A summary of the salient points is given.
SELF-TEST A self-test, based on the material included in Introductory Information, helps the student evaluate his understanding of the principles covered, prior to the experiment proper. The self-test should be taken before the experiment is undertaken. Answers to the self-test questions are given at the end of each experiment.
MATERIALS REQUIRED All the materials required to do the experiment—including test equipment and components—are enumerated.
PROCEDURE A detailed step-by-step procedure is given for performing the experiment.
QUESTIONS The conclusions reached by the student are brought out by a series of pertinent questions.
EXTRA CREDIT Design problems and questions are included for the more advanced student.

EXPERIMENTS

SERIES PREFACE

Electronics is at the core of a wide variety of specialized technologies which have been developing over several decades. Challenged by a rapidly expanding technology and the need for increasing numbers of technicians, the Consumer Electronics Group Service Committee of the Electronic Industries Association (EIA), in association with the Voorhees Technical Institute, Oklahoma State University, and various publishers, has been active in creating and developing educational materials to meet these challenges.

In recent years, a great many consumer electronic products have been introduced and the traditional radio and television receivers have become more complex. As a result, the pressing need for training programs to permit students of various backgrounds and abilities to enter this growing industry has induced EIA to sponsor the preparation of an expanding range of materials. Three branches of study have been developed in two specific formats. The tables list the books in each category; the paragraphs following them explain these materials and suggest how best to use them to achieve the desired results.

The Basic Electricity-Electronics Series

Title	Author	Publisher
Electricity-Electonics Fundamentals	Zbar/Sloop	Gregg/McGraw-Hill Book Company
Basic Electricity	Zbar	Gregg/McGraw-Hill Book Company
Basic Electronics	Zbar/Malvino	Gregg/McGraw-Hill Book Company

The laboratory text-manuals in the Basic Electricity-Electronics Series provide in-depth, detailed, completely up-to-date technical material by combining a comprehensive discussion of the objectives, theory, and underlying principles with a closely coordinated program of experiments. *Electricity-Electronics Fundamentals* provides an introductory course especially suitable for preparing service technicians; it can also be used for other broad-based courses. *Basic Electricity* and *Basic Electronics* are planned as 270-hour courses, one to follow the other, providing a more thorough background for all levels of technician training. A related instructor's guide is available for each course.

The Radio-Television-Audio Servicing Series

Title	Author	Publisher
Television Symptom Diagnosis—An Entry into TV Servicing	Tinnell	Howard W. Sams & Co., Inc.
Television Symptom Diagnosis audiovisual materials	Tinnell	Howard W. Sams & Co., Inc.
Television Servicing with Basic Electronics	Sloop	Howard W. Sams & Co., Inc.
Advanced Color Television Servicing	Sloop	Howard W. Sams & Co., Inc.
Audio Servicing — Theory and Practice	Wells	Gregg/McGraw-Hill Book Company
Audio Servicing—Text-Lab Manual	Wells	Gregg/McGraw-Hill Book Company
Basic Radio: Theory and Servicing	Zbar	Gregg/McGraw-Hill Book Company
Basic Television: Theory and Servicing	Zbar and Orne	Gregg/McGraw-Hill Book Company

The Radio-Television-Audio Servicing Series includes materials in two catgories: those designed to prepare apprentice technicians to perform in-home servicing and other apprenticeship functions, and those designed to prepare technicians to perform more sophisticated and complicated servicing such as bench-type servicing in the shop.

The two titles in the apprenticeship servicing course are *Television Symptom Diagnosis—An Entry into TV Servicing* (text, student workbook, instructor's guide) and *Television Symptom Diagnosis*, a series of 33 film loops. The first is a set consisting of a well-illustrated text, a student response manual, and an instructor's guide—all designed for people with no previous electronics training—to provide them with job-entry troubleshooting skills for servicing in the home and shop. The text utilizes the "cue-response" concept of diagnosis, concentrates on identifying abnormal circuit operation and symptom analysis, and develops skills in troubleshooting. In using the response manual, students are exposed to hundreds of television trouble symptoms through color photos and illustrated problems. The instructor's guide is a complete and essential professional course of study that also contains the answers to questions and problems in the text and lab manual.

Television Symptom Diagnosis consists of a series of color-sound motion picture film loops or slides, a student workbook, and an instructor's guide. These audiovisual materials provide integrated learning systems for color-television adjustment and setup procedures, trouble-symptom diagnosis, and the ability to isolate troubles to a given stage in the receiver, concentrating on the requirements for servicing in the customer's home. But these audiovisual materials can also be used to supplement all levels of television courses. This medium is especially suitable for students who may have reading and, in turn, learning difficulties.

Since only a minimum of electronics theory is presented in the two courses described above, it is expected that apprentices completing these programs will be motivated to progress to more comprehensive programs in order to deepen their understanding of electronics, to really know what makes the radio and television receiver work, and to become proficient in servicing all consumer electronics entertainment items. These in-depth studies are provided in the following materials.

The intermediate *Television Servicing with Basic Electronics* (text, student workbook, instructor's guide) goes beyond the basics and expands on the math and use of test equipment introduced in the beginning text. The book continues the diagnostic troubleshooting method.

Advanced bench-type diagnosis servicing techniques are covered in *Advanced Color Television Servicing* (text, student workbook, instructor's guide). Written primarily for color television servicing courses in schools and in industry, this set follows the logical diagnostic troubleshooting approach consistent with the manufacturers' approach to bench servicing.

Audio Servicing (theory and practice, text-lab manual, and instructor's guide) covers each component of a modern home stereo with an easy-to-follow block diagram and a diagnosis approach consistent with the latest industry techniques.

The bench-type service technician courses consist of *Basic Radio: Theory and Servicing* and *Basic Television: Theory and Servicing*. These books provide a series of experiments, with preparatory theory, designed to provide the in-depth, detailed training necessary to produce skilled radio-television service technicians for both home and bench servicing of all types of radio and television. A related instructor's guide for these books is also available.

The Industrial Electronics Series

Title	Author	Publisher
Industrial Electronics	Zbar	Gregg/McGraw-Hill Book Company
Electronic Instruments and Measurements	Zbar	Gregg/McGraw-Hill Book Company

Basic laboratory courses in industrial control and computer circuits and laboratory standard measuring equipment are provided by the Industrial Electronics Series and their related instructor's guides. *Industrial Electronics* is concerned with the fundamental building blocks in industrial electronics technology, giving the student an understanding of the basic circuits and their applications. *Electronic Instruments and Measurements* fills the need for basic training in the complex field of industrial instrumentation. Prerequisites for both courses are *Basic Electricity* and *Basic Electronics*.

The foreword to the first edition of the EIA cosponsored basic series states: "The aim of this basic instructional series is to supply schools with a well-integrated, standardized training program, fashioned to produce a technician tailored to industry's needs." This is still the objective of the varied training program that has been developed through joint industry-educator-publisher cooperation.

Peter McCloskey, President
Electronic Industries Association

PREFACE

Basic Electricity: A Text-Lab Manual, has been updated for this fifth edition in response to changing electrical technology and the growth in the electricity/electronics industry. Incorporating the suggestions of instructors who have used the fourth edition, and written, as were all its predecessors, under the guidance of an Industrial Advisory Committee, the fifth edition combines the recommendations of the industry with sound educational practice. By emphasizing learning by doing, the student experiences the theoretical and practical sides of the dynamic world of electricity. The textbook-lab manual is intended for use in technical-vocational schools, community colleges, and military and industrial training programs.

The manual provides a complete and tested laboratory program in electric circuits for both dc and ac. As in previous editions, emphasis is on scientific method, analysis, and deduction.

The 69 experiments in the fifth edition provide students with intensive experience in the use of test instruments that they will find in industry.

The experiments which have been added treat the following topics: the Superposition Theorem; dc relays; internal resistance of a battery; peak, average, and rms values of alternating current; measurement of frequency and phase with an oscilloscope whose sweep is time calibrated; measurement of power in ac circuits. In addition, an Instructor's Guide is available for instructors using this book. The guide contains tables with representative data for each experiment, together *with answers to all the questions at the end of each experiment*.

Each experiment follows the format established in previous editions: Title, Objectives, Introductory Information, Summary, Self-test, Materials Required, Procedure, Questions, Answers to Self-test. Objectives are stated in measurable behavioral terms. At the end of each experiment, student and teacher can evaluate the results of that learning experience in terms of the stated objectives.

A detailed summary follows the introductory information in which the principles of the experiment are discussed. Immediately after the summary comes a self-test (questions with answers) to help students determine how well they have understood the principles of the experiment. The results of the self-test will indicate to students their readiness to undertake the experimental procedure which follows or possibly a need to review the basic principles. The presentation of objectives, introductory information, self-test, and answers to the self-test will therefore pace student's progress and prepare them for a meaningful hands-on experience in the procedural steps of the experiment.

As in past editions, students are stimulated to discover for themselves the principles and laws of electricity. Scientific method, analysis, and logical deduction mark this aspect of their learning. As in previous editions also, discussions of circuits and of testing and measurement techniques are presented clearly and simply. Solutions of sample problems are included, as are sample graphs and interpretations of these graphs. Stimulating questions, following the experimental procedure and based on the data obtained, oblige students to evaluate their data and draw the necessary conclusions. Extra credit procedures and questions are provided to stimulate student interest and resourcefulness.

The author wishes to thank the members of the Service Education Sub-committee of the Consumer Electronics Group (CEG) of the Electronic Industries Association (EIA) for their guidance and encouragement, and particularly the following, whose reviews, comments, and suggestions were so helpful: Jerry Surprise (General Electric), William Dugger (Virginia Polytechnic Institute), Gene O. Jadwin (Sharp), Mark Kirschner (Sharp), and Richard G. Majer (General Electric).

The author also wishes to acknowledge with thanks the help of: John E. Gardner (Central Piedmont Community College, Charlotte, NC), Don Cooke (Carborundum Co.), James Farrell (B&K Dynascan), Gerald N. Goldberger (Simpson Electric Co.), and Jack Moore (Hickok Teaching Systems).

Acknowledgment is also gratefully made for permission to use equipment and component photographs to: Aerovox Corporation; Centralab (Division of Globe Union Inc.); Chicago Telephone Company; Heath/Schlumberger Instruments; Hewlett-Packard; Motorola Semiconductor Products; Ohmite Manufacturing Company; Radio Corporation of America; F. W. Sickles Division of General Instrument Corp.; Simpson Electric Company (Division of American Gage and Machine Company); Triplett Electrical Instruments Company; and Western Electric Company.

And last but not least, the author wishes to thank his wife May for her encouragement, inspiration, and many, many hours of proofreading.

Paul B. Zbar

SAFETY

Electronics technicians work with electricity, electronic devices, motors, and other rotating machinery. They are often required to use hand and power tools in constructing prototypes of new devices or in setting up experiments. They use test instruments to measure the electrical characteristics of components, devices, and electronic systems. They are involved in any of a dozen different tasks.

These tasks are interesting and challenging, but they may also involve certain hazards if the technician is careless in his or her work habits. It is therefore essential that student technicians learn the principles of safety at the very start of their career and then practice these principles throughout their busy and exciting lives.

Safe work requires a careful and deliberate approach to each task. Before undertaking a job, the technician must understand what he or she is to do and how to do it. The technician must plan the job, setting out on the work bench in a neat and orderly fashion the tools, equipment, and instruments which will be needed. Extraneous items shoud be removed and cables should be secured as far as possible.

When working on or near rotating machinery, loose clothing should be anchored, neckwear firmly tucked away.

Line (power) voltages should be isolated from ground by means of an isolation transformer. Power-line voltages can kill, so these should *not* be contacted with the hands or body. Line cords should be checked before use. If the insulation on line cords is brittle or cracked, these cords must *not* be used. TO THE STUDENT: Avoid direct contact with any voltage source. Measure voltages with one hand in your pocket. Wear rubber-soled shoes or stand on a rubber mat when working at your experiment bench. Be certain that your hands are dry and that you are not standing on a wet floor when making tests and measurements in a live circuit. Shut off power before connecting test instruments in a live circuit.

Be certain that line cords of power tools and non-isolated equipment use safety plugs (polarized 3-post plugs). Do not defeat the safety feature of these plugs by using ungrounded adapters. Do not defeat any safety device, such as fuse or circuit breaker, by shorting across it or by using a higher amperage fuse than that specified by the manufacturer. Safety devices are intended to protect you and your equipment.

Handle tools properly and with care. Don't indulge in horseplay or play practical jokes in the laboratory. When using power tools, secure your work in a vise or jig. Wear gloves and goggles when required.

Exercise good judgment and common sense and your life in the laboratory will be safe, interesting, and rewarding.

FIRST AID

If an accident should occur, shut off power immediately. Report the accident at once to your instructor. It may be necessary for you to render first aid before a physician can come, so you should know the principles of First Aid. A proper knowledge of these may be acquired by attendance at a Red Cross First Aid course.

Some First Aid suggestions are set forth here as a simple guide.

An injured person should be kept lying down until medical help arrives and should be kept warm to prevent shock. Do not attempt to give the victim water or other liquids if he or she is unconscious. Be sure nothing is done to cause further injury. Keep the injured person comfortable and cheerful until medical help arrives.

ARTIFICIAL RESPIRATION

Severe electrical shock may cause stoppage of breathing. Be prepared to start artificial respiration at once if breathing has stopped. The two recommended techniques are:

1. Mouth-to-mouth breathing, considered the most effective
2. Schaeffer method

These techniques are described in First Aid books. You should master one or the other so that if the need arises you will be able to save a life by applying artificial respiration.

These instructions are not intended to frighten you but to make you aware that there are hazards attendant upon the work of an electronics technician. But then there are hazards in every job. Therefore you must exercise common sense, good judgment, and safe work habits in this, as in every job.

1

EXPERIMENT

EVM—ELECTRONIC VOLTMETER—FAMILIARIZATION

OBJECTIVES

1. To identify the operating controls and write down the function of each control of an electronic voltmeter (EVM)
2. To identify each of the leads of an electronic voltmeter
3. To read the voltage value at a specified point on each of the voltage ranges of an EVM
4. To set the zero and ohms adjust on the meter

INTRODUCTORY INFORMATION

Electronic Voltmeter

Voltage measurements are made primarily with an instrument called a *voltmeter*. There are other instruments which may be used to measure voltage, for example the oscilloscope, but voltage measurement is not the primary purpose of these devices.

Two types of voltmeters are used today, (1) the analog and (2) the digital. A pointer on the analog meter indicates the voltage on a calibrated scale. The voltage measured by a digital meter appears as a number on a numerical (digital) display, so that any one who reads numbers can read the voltage.

The reading errors associated with analog meters (these are discussed in the experiment) are eliminated by the digital voltmeter (DVM). Moreover, DVMs are more accurate than analog meters. Despite the fact that digital meters are more costly than analog meters, their advantages have led to their widespread adoption.

Another way of identifying voltmeters is to specify them as (1) electronic voltmeters or (2) electromechanical meters. The vacuum-tube voltmeter (VTVM), the transistorized voltmeter (TVM), and the digital voltmeter (DVM) are examples of electronic voltmeters (EVMs). The volt-ohm-milliammeter (VOM) is an example of an electromechanical multimeter which is used to measure voltage.

Meters may be single-function devices, such as the voltmeter, which can measure only voltage. Or meters may be multipurpose devices, such as the multimeter, which is used to measure a variety of electrical quantities, such as voltage, current, and resistance. The VTVM, the TVM, the DMM (digital multimeter) or the digital volt-ohm-milliammeter (DVOM), and the VOM are examples of multimeters.

In this experiment we will be concerned with electronic multimeters, which we will designate simply as EVMs. When an EVM is specified in this experiment or in those which follow, any one of the three types of electronic multimeters (TVM, DVOM, or VTVM) may be used.

Prior to the invention of transistors, the VTVM (vacuum-tube voltmeter) was the electronic meter used to measure dc and ac voltage and resistance. On rare occasions it also included a current-measuring function. VTVMs are still in use in industry and in school laboratories, although they are no longer being manufactured. In new installations, they have been replaced by the TVM and the DVM. The TVM and DVM include all the measuring functions of a VTVM plus current-measuring capability.

Electronic voltmeters sometimes have other specialized scales, such as the decibel (dB) scale for sound level measurement.

It is evident that the transistor part of the name TVM stems from the use of transistors and other solid-state devices, while in the VTVM, vacuum tubes are employed to perform similar functions. The electromechanical VOM uses neither vacuum tubes nor transistors. The general-purpose VTVM is usually a line-operated instrument—it is plugged into an electrical outlet and derives its operating power from the line. TVMs and DVMs are either battery or line operated, and frequently they contain facilities for both types of power operation.

Operating Controls (VTVM)

Figure 1-1*a* is an illustration of a typical vacuum-tube voltmeter. Note that the operating controls and meter scales are on the front panel. Of the four controls, two are labeled "Zero" and "Ohms." Though the other two are not identified by name, their functions are apparent. The lower left-hand knob might well be called the function switch. The instrument is turned on when it is switched from off to any of its four other settings. The − Volts and + Volts positions are for measurement of positive and negative dc volts. The AC Volts position is for measuring ac voltage. The Ohms position is for measuring resistance.

The lower right-hand control is a range switch. It works in conjunction with the function switch to select the proper operating range for a specific measurement. For example, a

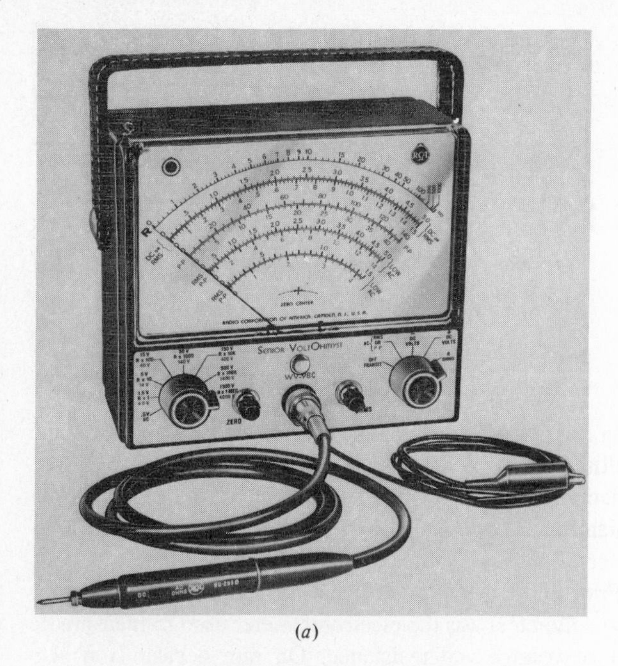

(*a*)

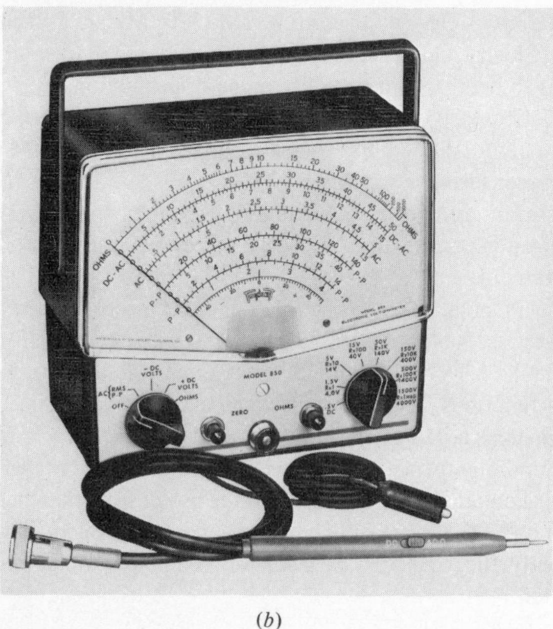

(*b*)

Fig. 1-1. (*a*) VTVM (*RCA*); (*b*) VTVM (*The Triplett Electrical Instrument Co.*).

The purpose of the Zero control is to set the meter pointer on zero (at the left side) of the meter scales. When the meter is first turned to any voltage position, the pointer may go off zero during the warmup period. After the meter has been on for about five minutes, it should stabilize and the pointer should return to its zero setting. If it does not, it can be brought to zero by the Zero adj.

When the function switch is in the Ohms position, the meter pointer will move all the way to the right on the *R* scale. The pointer should line up on the last calibration mark on the *R* (top) scale. If it does not, the Ohms adj. will bring the pointer into position. It should be noted that the Zero control must be properly set *before* Ohms control is adjusted.

There are many variations among manufacturers both as to appearance of the instrument and as to operating controls. In the VTVM of Fig. 1-1*b*, for example, the function switch is on the right and the range selector on the left. The two controls labeled "Zero" and "Ohms" are comparable to those on the previously discussed VTVM.

Operating Controls (TVM)

The operating controls on a TVM are very similar to those on a VTVM. In Fig. 1-2, for example, a function switch is on the lower left, a range switch is on the lower right, and there are a Zero ("0") control and an Ohms (Ω) control. The Zero and Ohms controls are adjusted like the same controls on the VTVM, and the range switch is similar in operation to that of a VTVM. It should be noted, however, that the range switch has 12 positions, including a low 0.05 dc volts position and a 0.005 ac volts position, contrasted with the eight positions on the VTVMs in Fig 1-1. The function switch is also more complex, reflecting the greater versatility of the TVM. Thus,

Fig. 1-2. TVM—solid-state VOM (*The Triplett Electrical Instrument Co.*).

technician who wishes to measure positive dc volts, and who knows that the voltage to be measured lies between 5 and 15 V, will turn the function switch to +Volts and the range switch to 15. Note that the range number represents the maximum voltage which can be read on that range. The meter in Fig. 1-1*a* has eight voltage ranges, 0.5 to 1500 V.

The range switch shown in Fig. 1-1*a* acts also as a resistance-scale multiplier. Thus, the 15-V position is also the $R \times 100$ position. When measuring resistance the function switch is set on ohms. Resistance readings on the *R* scale must be multiplied by 100 on the $R \times 100$ position. On the $R \times 1000$ position, they must be multiplied by 1000, etc.

in addition to the off, + and − DCV, ACV, and conventional ohms (Ω) positions, there are LPΩ (low-power ohms), DC-MA, AC-MA, and Batt. positions. The low-power ohms function is used for resistance measurements in solid-state circuits. DC and AC-MA positions are for measuring, respectively, direct and alternating current, and the Batt. position is used to test the battery which powers this instrument. This is *not* a line-operated TVM.

In TVMs, also, there are variations among manufacturers both as to appearance of the instrument and operating controls.

Fig. 1-3. Digital multimeter (*Hickok Electrical Instrument Co.*).

Operating Controls for a Digital Multimeter

Figure 1-3 shows a digital multimeter which uses push buttons for function and range selection. The function selector push buttons (on the right) are labeled OHMS, AC, DC, and SPECIAL FUNCTION. Except for Special Function their names indicate the use for which they are intended. The six push buttons on the left are for range selection. The scale markings above these controls indicate the maximum value of ohms, volts, or amps which can be measured on each range. Not shown, are jacks on the side of the meter into which the meter test leads may be inserted. There is a Common (−) jack and separate jacks for volts, ohms and amperes.

The Special Function selector makes it possible to make "logic" checks in digital circuits. Also, when both the Special Function and 20-Ω buttons are pushed in, the meter becomes a low-range ohmmeter, capable of measuring up to 20 Ω of resistance. On this range the meter must be nulled. This is accomplished by the 20-Ω NULL adjust, on the right, located above the POWER push button.

Digital Multimeter Display

The display on a digital multimeter shows the numerical value of the function which is being measured. Thus, in Fig. 1-3, the display reads 1.980. Note that the range selector is set on 2 and the function selector is set on DC. The reading is therefore 1.980 V or mA, depending on which jack the "hot" instrument lead is plugged into.

It should be noted that the display will show the segmented digits 0 through 9, together with a decimal point whose position will be automatically determined by the setting of the range selector. So it is fairly evident that this meter provides a direct, numerical (digital) readout.

Meter Scales for the TVM and VTVM

In contrast with the DVOM, which provides a direct numerical readout, the TVM and VTVM give *analog* measurements, which must be interpreted on a meter scale. In general, the *voltage* scales on a TVM and VTVM are linear, while the *ohms* scale is nonlinear.

The scale on a ruler is the most familiar example of a linear scale. On a ruler equal distances are represented by equally spaced calibrations. Thus, a 1-in distance is represented by the same calibration interval whether it is 1 in between 1 and 2 or between 9 and 10. This same relationship holds true for any linear scale, that is, the distance between consecutive corresponding calibration marks is everywhere the same on the scale.

The voltmeter scale (Fig. 1-4) is another example of a linear scale. Here the meter pointer travels equal distances along the scale *arc* for equal voltage changes. Thus, the *arc* distance between 0 and 10 V is the same as between 10 and 20 V or between 90 and 100 V.

In reading the scale in Fig. 1-4, it is necessary to supply numbers for the unnumbered calibrations. This is easily accomplished. For example, there are five equally spaced intervals between numbered voltage calibrations. Each of these distances represents one-fifth of the voltage difference between the numbered calibrations. In the case of Fig. 1-4, every calibration marker would therefore represent a 2-V interval. Thus, in Fig. 1-5, the meter pointer reads 44 V, that is, 40 V plus 2 intervals × 2 V per interval.

When the pointer falls between scale markers, the reading is approximated. For example, in Fig. 1-6 the reading is 45 V.

The voltage scales of a voltmeter are not always calibrated in the same units as the range setting. For example, the scale of Fig. 1-4 would serve admirably for a 0- to 100-V range. However, it could serve as well for a 0- to 10-V range or a 0- to 500-V range. In the case of a 10-V range, every scale reading must be divided by 10. For the 500-V range, every scale reading must be multiplied by 5.

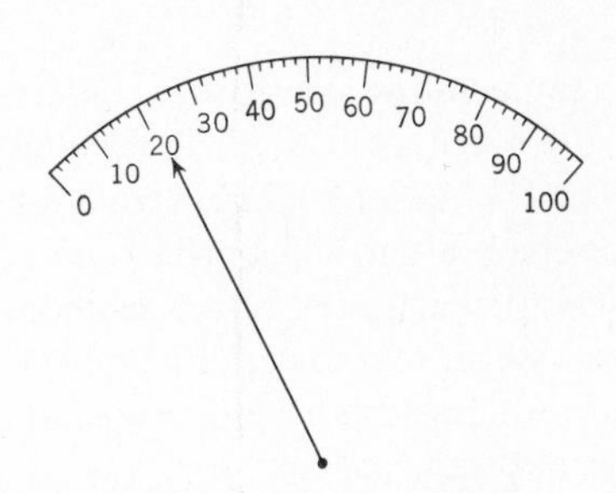

Fig. 1-4. Linear meter scale.

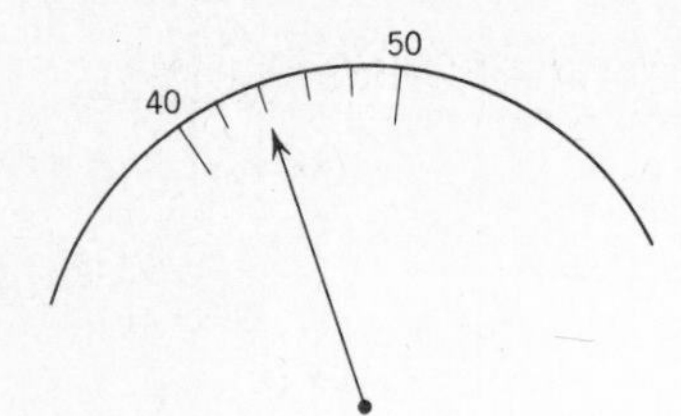

Fig. 1-5. Meter pointer reads 44.

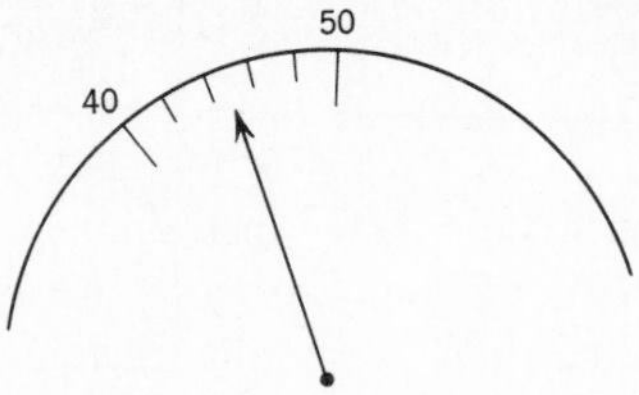

Fig. 1-6. Meter pointer reads 45.

EVM Leads

A TVM, DVM, and a VTVM usually require two or three leads. The three-lead arrangement normally consists of a flexible common ground (black), a shielded coaxial lead for direct current, and a shielded coaxial lead for alternating current and ohms. The ground lead is terminated in an alligator clip which is clipped onto the common return or ground of the circuit. The two coaxial leads are terminated in probes, completely insulated except for the metallic tip which extends from the end of the probe. The technician holds the insulated probe in measuring and does not come into contact with the conductive tip, which is applied to the measurement point. Since there are other variations of the three-lead arrangement, the technician should consult the instruction manual for a particular meter being used until he or she is entirely familiar with the instrument.

The two-lead arrangement shown in Fig. 1-1 includes a flexible common "ground" lead and a coaxial cable. The coaxial cable is terminated in a probe. There is a two-position switch on the probe. In one position the cable serves as a dc voltage lead. In the other it is an ac voltage or Ohms lead.

The meter end of the common ground lead is terminated in a pin tip or banana plug, which plugs into an appropriate jack on the meter. The coaxial cable is connected to the meter by means of a microphone-type connector. Leads should be left on the meter when it is not in use.

SUMMARY

1. The TVM (transistorized voltmeter), DVM (digital voltmeter), and VTVM (vacuum-tube voltmeter) are classified as electronic voltmeters, EVMs.
2. Electronic voltmeters are used to measure dc and ac voltages and resistance. The TVM and DVM also have current-measuring ranges.
3. Electronic voltmeters are line-powered or battery-powered.
4. The meter operating controls include: (*a*) function switch, (*b*) range switch, (*c*) zero adjust, (*d*) ohms adjust. The DVM does not have zero or ohms controls, but it does have a zero adjustment.
5. The function switch selects the desired measurement function: voltage, resistance, or current.
6. The range switch selects the proper operating range for a particular measurement.
7. The zero adjust on a TVM or VTVM sets the meter pointer at the zero calibration marker on the left side of the meter scales.
8. The ohms adjust on a TVM or VTVM is used to line up the meter pointer with the last calibration marker (infinity) on the ohms scale.
9. The voltage and current scales of an electronic voltmeter are linear.
10. The ohms scale is nonlinear.
11. The electronic voltmeter uses two or three leads. In both cases a flexible black lead serves as the common or ground lead.
12. In a three-lead meter, the other two leads are coaxial (shielded) leads, one for dc measurements and the other for ac and resistance readings.
13. In a two-lead meter, the single coaxial lead is terminated in a probe containing a two-position switch. One switch position is for dc measurements, the other position for ac and resistance measurements.
14. The DVM has a numerical (digital) readout.

SELF-TEST

Check your understanding by answering these questions:

1. The two types of instruments used for measurement of electrical quantities are the electronic voltmeter and the ____________ ____________.
2. The meter which employs transistors in its circuitry is called a ____________.
3. VTVMs are normally ____________-powered devices.
4. Electronic voltmeters usually have four operating controls. These are: (*a*) ____________; (*b*) ____________; (*c*) ____________; and (*d*) ____________.
5. The purpose of each control listed in the answer to question 4 is: (*a*) ____________; (*b*) ____________; (*c*) ____________; and (*d*) ____________.
6. An EVM has two or three leads. These are for: (*a*) ____________; (*b*) ____________; (*c*) ____________ and ____________.
7. The voltage and current scales of an electronic voltmeter are linear. ____________ (true/false)
8. Refer to Fig. 1-4. The range switch is set to the 10-V range. What is the meter reading? ____________ V

MATERIALS REQUIRED

■ Equipment: EVM—both a TVM and a DVOM.

PROCEDURE

TVM

1. Examine the meter assigned to you. Draw a panel view of the meter, showing the operating controls and the functions and ranges associated with the switches. Draw also the voltage scales and the ohm scale.
2. Read the instruction manual and learn how to operate the meter.
3. Turn the EVM **on**. Set the function switch on +dc volts. Permit the meter to warm up if required. Then vary the Zero adj. control. Observe its effect on the pointer. Set the pointer on zero. Now turn the range switch through every setting and note whether the meter pointer remains on zero.
4. Check the zero setting of the meter on each range of −dc volts and on each range of ac volts.

NOTE: A well-designed, properly adjusted meter should maintain its zero setting on every voltage function and range.

5. Set the function switch on ohms and the range switch on $R \times 1$. Be sure the lead tips do not touch each other or a metal object. Vary Ohms adj. and observe its effect on the pointer. Set this control so that the pointer rests on the maximum calibration on the resistance scale. In this position the resistance circuit is open and the meter registers infinite resistance. Check the setting of the pointer on every position of the range switch ($R \times 10$, $R \times 100$, etc.).
6. Short the ohms and ground leads (that is, touch the metal tips of these leads). The pointer should swing to zero. If it does not come to rest on zero with the leads shorted, set it on zero with the Zero adj. Now open the leads and check the position of the pointer. If it is no longer on the maximum-resistance marker (infinity), set it there with the Ohms adj.
7. Recheck the zero and infinity positions.

NOTE: Do not leave meter leads shorted together on Ohms for any length of time. In the Ohms position, an internal battery is connected in the circuit. The battery voltage will be greatly reduced in a short period of time if the leads remain shorted together.

DVOM

You will now become familiar with the operation of a DVOM.

8. Repeat steps 1 and 2.
9. Turn the meter **on**. Set the function switch on +dc volts. Permit the meter to warm up for several minutes. Short the meter leads together. Now turn the range switch to the lowest dc range. The display should read 000. If it does not, advise your instructor. Repeat for each setting of the +dc volt ranges.
10. Repeat step 9 for each setting of the −dc volt ranges.
11. Repeat step 9 for ohms (set function switch on ohms).
12. Repeat step 9 for ac volts.

QUESTIONS

1. List the controls on the panel of your EVM and state the purpose of each.
2. Is it possible to use the same scale on the 3-V range as on the 300-V range of your meter? Explain.
3. Draw a linear scale with number calibrations 0, 1, 2, etc., through 10. Set off each major subdivision into ten minor subdivisions. Show where 8.7 would be on your scale.
4. Is the ohm scale on your meter linear or nonlinear? Justify your answer by referring specifically to resistance calibrations.
5. Explain in detail how you would zero the meter for dc voltage readings. Identify the control or controls you would use.
6. Explain in detail how you would check to determine whether the meter is properly zeroed on the $R \times 1$ range. Identify the controls you would use.
7. If in question 6 you find the meter is not properly zeroed, explain in detail how you would zero it.
8. What is the difference between the readout on a TVM and a DVOM?

Answers to Self-Test

1. electromechanical VOM
2. TVM or solid-state VOM
3. line
4. (*a*) function; (*b*) range; (*c*) zero; (*d*) zero ohms
5. (*a*) Select the operating function: voltage, current, or resistance; (*b*) select the appropriate range of measurement; (*c*) set the pointer at zero on the left-hand side of the scales; (*d*) set the pointer at infinity on the right-hand side of the ohms scale.
6. (*a*) common or ground; (*b*) dc; (*c*) ac and resistance
7. true
8. 2

2 EXPERIMENT

RESISTOR COLOR CODE AND USE OF OHMMETER

OBJECTIVES

1. To determine the value of resistors from their EIA (Electronic Industries Association) color code
2. To read the resistance value at a specified point on each of the ohmmeter scales on an electronic voltmeter
3. To measure resistors of different values
4. To measure the resistance across each combination of two of the three terminals of a potentiometer and to observe the range of resistance change as the shaft of the potentiometer is varied throughout its entire range

INTRODUCTORY INFORMATION

Color Code

The *ohm* is the unit of resistance. The symbol for ohm is Ω (Greek letter omega). Resistance values are indicated by a standard color code that manufacturers have adopted. This code involves the use of color bands on the body of the resistor. The colors and their numerical values are given in the resistor color chart, Table 2-1. This code is used for ⅛-W, ¼-W, ½-W, 1-W, 2-W, and 3-W resistors.

The basic resistor is shown in Fig. 2-1. Note the color bands. The color of the first band tells the first significant figure of the resistor. The color of the second band tells the second significant figure. The color of the third band tells the multiplier (number of zeros to be added or the placement of the decimal point). A fourth color band is used for tolerance designation. The absence of the fourth color band means 20 percent tolerance.

In Fig. 2-1 the resistor is coded red, red, black, gold. Its value would be 22 Ω at 5 percent tolerance.

In the case of a resistor whose value is less than 1 Ω, the multiplier is silver (band or dot). In the case of a resistor whose value is greater than 1 Ω but less than 10 Ω, the multiplier is gold.

Resistors used in military electronics carry a fifth band to indicate the reliability level, rated in percent per 1000 hours. The fifth band color code is as follows:

Color	*% Rating*
Brown	1%
Red	0.1%
Orange	0.01%
Yellow	0.001%

TABLE 2-1. Resistor Color Chart

Resistors, EIA and MIL		*Significant figures*	*Color*
Multiplier	*Tolerance, %*		
1	—	0	Black
10	—	1	Brown
100	—	2	Red
1,000	—	3	Orange
10,000	—	4	Yellow
100,000	—	5	Green
10^6	—	6	Blue
10^7	—	7	Violet
10^8	—	8	Gray
10^9	—	9	White
0.1	5	—	Gold
0.01	10	—	Silver
—	20	—	No color

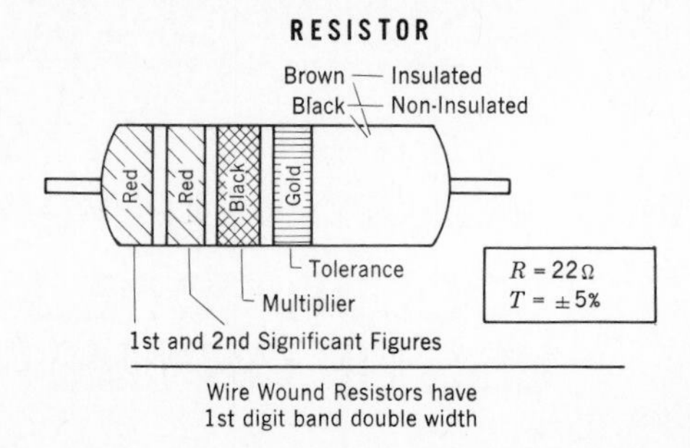

Fig. 2-1. Resistor color code (*Centralab, a Division of Globe-Union, Inc.*).

Wirewound, high-wattage resistors usually are not color-coded, but have the ohmic value and wattage rating printed on the body of the resistor.

Resistors with brown body color are insulated; those with black body color are not insulated.

In writing the values of resistors, the following designations are employed:

- k, a multiplier which stands for 1000
- M, a multiplier which stands for 1,000,000

For example, 33 kilohms (33 kΩ) stands for 33,000 Ω; 1.2 megohms (1.2 MΩ) stands for 1,200,000 Ω.

Variable Resistors

In addition to fixed-value resistors, variable resistors are used extensively in electronics. There are two types of variable resistors, the *rheostat* and the *potentiometer*. Volume controls used in radio and the contrast and brightness controls of television receivers are typical examples of potentiometers.

A rheostat is essentially a two-terminal device. Its circuit symbol is shown in Fig. 2-2. Points *A* and *B* connect into the circuit. A rheostat has a maximum value of resistance, specified by the manufacturer, and a minimum value, usually zero ohms. The arrowhead in Fig. 2-2 indicates a mechanical means of adjusting the rheostat so that the resistance, measured between points *A* and *B*, can be set to any intermediate value within the range of variation.

The circuit symbol for a potentiometer (Fig. 2-3*a*) shows that this is a three-terminal device. The resistance between points *A* and *B* is fixed. Point *C* is the variable arm of the potentiometer. The arm is a metal contactor which moves along the uninsulated surface of the resistance element, selecting different lengths of resistive surface. Thus, the longer the surface between points *A* and *C*, the greater is the ohms resistance between these two points. Similarly, the resistance between points *B* and *C* varies as the length of element included between points *B* and *C*.

The axiom which states that the whole is equal to the sum of its parts applies to a potentiometer as well as it does to

Fig. 2-2. A rheostat is a variable resistor with two terminals.

geometric figures. In this case it is apparent that the resistance R_{AC} from *A* to *C*, plus the resistance R_{CB} from *C* to *B*, make up the fixed resistance R_{AB} of the potentiometer. The action of the arm, then, is to increase the resistance between *C* and one of the end terminals, and at the same time to decrease the resistance between *C* and the other terminal, while the sum of the two resistances R_{AC} and R_{CB} remains constant.

A potentiometer may be used as a rheostat if the center arm and one of the end terminals are connected into the circuit and the other end terminal is left disconnected. Another method of converting a potentiometer into a rheostat is to connect a piece of hookup wire between the arm and one of the end terminals, for example, *C* can be connected to *A*. The points *B* and *C* now serve as the terminals of a rheostat. (When two points in a circuit are connected by hookup wire, these points are said to be *shorted* together.)

Measuring Resistance

This is one of the functions of an EVM. Each manufacturer provides operating instructions for the use of the particular instrument. Hence, it will be necessary to refer to the instruction manual before using any electronic voltmeter. You should be thoroughly familiar with the operation of the ohmmeter function before attempting to use it in this experiment.

To measure resistance, the function switch should be set to Ohms. Next, before you use the ohms function of an electronic meter, you should adjust to their proper settings the Ohms and the Zero controls of the meter. You are then ready to make resistance and continuity checks. Now to measure the resistance between two points, say *A* and *B*, one of the ohmmeter leads is connected to point *A*, the other to point *B*. The meter pointer then indicates, on the ohms scale, the value of resistance between *A* and *B*. If the meter reading is zero ohms, points *A* and *B* are *short-circuited*, or simply "shorted." If, however, the meter pointer does not move (that is, if the indicator points to infinity on the ohms scale), points *A* and *B* are *open-circuited*, that is, there is an infinite resistance between them.

Reading the Ohms Scale

All EVMs contain a basic ohms scale (Fig. 2-4) from which readings are made directly on the $R \times 1$ range of the meter. Figure 2-4 shows that the ohms scale is nonlinear; that is, the

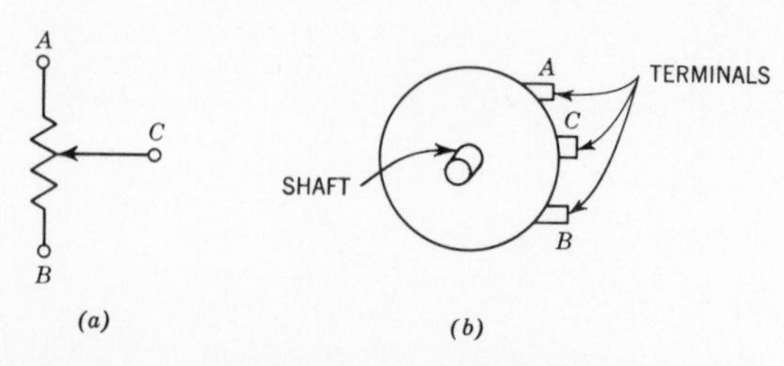

Fig. 2-3. A potentiometer is a three-terminal device. (*a*) Circuit symbol; (*b*) end view showing shaft and terminals.

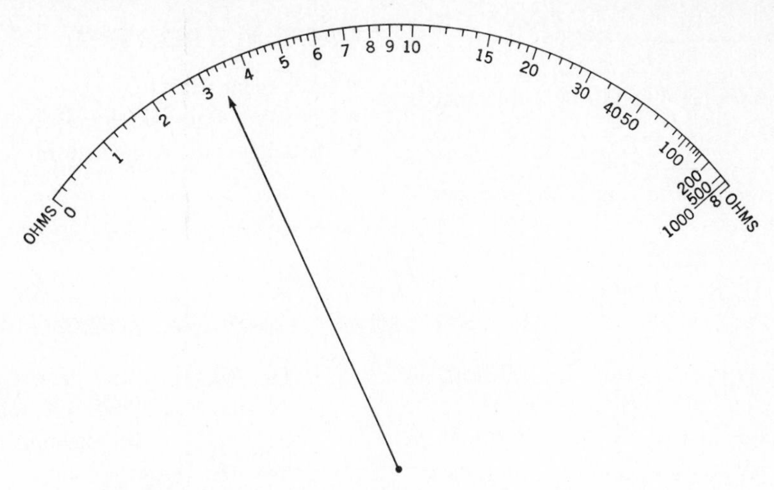

Fig. 2-4. Ohmmeter scale. On $R \times 1$, pointer shows 3.4 Ω.

arc distance between consecutive graduations is not equal. Thus the arc distance between 0 and 1 is much greater than the distance between 9 and 10, though each arc represents, in this case, a change of 1 Ω.

As in the case of the voltmeter scale in Experiment 1, the operator is required mentally to supply numbers for the unnumbered calibrations. If the pointer is on the second graduation to the right of 3, between 3 and 4, as in Fig. 2-4, the corresponding ohms value on the $R \times 1$ range is 3.4 Ω.

Note that the ohms scale becomes fairly crowded to the right of the 100-Ω division. If a resistance greater than 100 Ω is to be measured with some degree of accuracy, the meter range should be switched to $R \times 10$, $R \times 100$, or $R \times 1000$, depending on the actual resistance to be measured. These three ranges, $R \times 10$, $R \times 100$, and $R \times 1000$, will usually be found on the meter. In the $R \times 10$ range, any reading made on the basic scale must be multiplied by 10. In the $R \times 100$ range, any reading must be multiplied by 100, etc.

NOTE: After switching from one ohms range to another, the settings of the Zero and Ohms adj. should be checked and reset if necessary.

SUMMARY

1. The unit of resistance is the ohm.
2. The body of a fixed carbon resistor is color-coded to specify its ohmic value and tolerance.
3. Twelve colors are contained in the color chart. These give the values of the significant figures of resistance and the tolerance. Refer to Table 2-1 for the resistor color chart.
4. The first three of four parallel color bands on the body of a carbon resistor designate its resistance, the fourth band its tolerance (see Fig. 2-1).
5. High-wattage wirewound resistors are not color-coded, but have the resistance and wattage value printed on the body of the resistor.
6. Variable resistors are of two types, the rheostat and the potentiometer.
7. A rheostat is a two-terminal device whose resistance value may be varied between the two terminals.
8. A potentiometer is a three-terminal device. The resistance between the two end terminals is fixed. The resistance between the center terminal and either end terminal is variable.
9. Ohmmeters or the ohms functions of a VOM or EVM are used to measure resistance and continuity.
10. The scale of an ohmmeter is nonlinear.
11. Ohmmeters or the ohms function of a VOM or EVM have several ohms ranges ($R \times 1$, $R \times 10$, $R \times 100$, etc.).

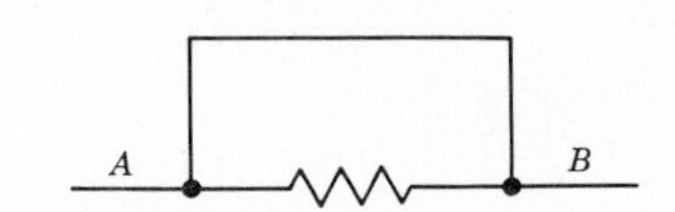

Fig. 2-5. Resistor short-circuited.

SELF-TEST

Check your understanding by answering these questions:

1. A color code is used to indicate the ____________ ____________ of a carbon resistor.
2. If the color red appears in either the first or second color band on a resistor, it stands for the number (significant figure) ____________.
3. If the color yellow appears in the third band on a resistor, it stands for the multiplier ____________.
4. If the color ____________ appears on the fourth band on a resistor, it indicates a tolerance value of 10 percent.
5. A resistor coded brown, black, black, gold has a value of ____________ ohms and a tolerance of ____________ percent.
6. A high-wattage resistor is color-coded in the same way as a low-wattage resistor. ____________ (true/false)

7. A potentiometer has ____________ terminals.
8. A rheostat is a ____________ ____________.
9. A resistor whose value is 120 Ω and whose tolerance is 20 percent is color-coded ____________ ____________ ____________ ____________.
10. If a resistor measures *infinite* ohms, the resistor is ____________-circuited.

MATERIALS REQUIRED

- Equipment: EVM
- Resistors: Ten assorted values; 10,000-Ω potentiometer
- Miscellaneous: A piece of hookup wire and hand tools

PROCEDURE

1. Determine the value of each resistor supplied from its color code. Fill in the information required in Table 2-2.
2. Refer to the instruction manual of the EVM for the procedure in measuring resistance. Zero the ohmmeter. Measure each resistor with the ohmmeter, and fill in the results in the row "Measured value." The coded value and the measured value should agree within the tolerance range of the resistor. Indicate percentage of *accuracy* between coded and measured values.
3. Measure and record the resistance of a small piece of uninsulated wire. ____________ Ω
4. **a.** Measure and record the resistance of one of the resistors supplied. ____________ Ω
   b. Connect the uninsulated wire across this resistor, as in Fig. 2-5. We say that the resistor has been short-circuited.
   c. Measure and record the value of this combination (i.e., the resistance from *A* to *B*). ____________ Ω
5. Remove the wire from the resistor. Cut the wire in half. Place the pieces of wire near each other, but not touching. (This is an open circuit.) Measure and record the resistance between the two pieces. ____________ Ω
6. Examine the potentiometer assigned to you. Orient it so that the rotatable shaft comes out toward you. Call the terminals of the potentiometer *A*, *B*, and *C* as in Fig. 2-3*b*. Measure and record in Table 2-3 the total resistance R_{AB} between *A* and *B*. Vary the arm of the potentiometer, while keeping the ohmmeter connected across *AB*. Does the total resistance vary? Indicate the effect in Table 2-3.
7. Connect the ohmmeter terminals across *AC*. Turn the potentiometer control completely clockwise (CW). Measure and record the resistance R_{AC} (between points *A* and *C*), and also the resistance R_{BC} (between points *B* and *C*). Compute and record the value of $R_{AC} + R_{BC}$.
8. Now observe how the resistance R_{AC} varies as the potentiometer arm is turned from its CW position to complete CCW position; how R_{BC} varies over the same range. Record the CW and CCW values for R_{AC} and R_{BC}. Compute and record $R_{AC} + R_{BC}$ in each case.
9. Set the control one-quarter of the way CW. Measure and record R_{AC} and R_{BC} in Table 2-3. Compute and record $R_{AC} + R_{BC}$.
10. Set the control three-quarters of the way CW. Measure and record R_{AC} and R_{BC} in Table 2-3. Compute and record $R_{AC} + R_{BC}$.
11. Set up an experiment and study the effect on the resistance R_{AC} when point *B* is shorted to *C*; the resistance R_{BC} when *A* is shorted to *C*. Set up a table and list the data for your experiment.

TABLE 2-2. Resistor Measured versus Color-coded Values

	Resistor									
	1	*2*	*3*	*4*	*5*	*6*	*7*	*8*	*9*	*10*
1st color										
2d color										
3d color										
4th color										
Coded value, ohms										
Tolerance, %										
Measured value, ohms										
Accuracy										

TABLE 2-3. Potentiometer Measurements

Step	*Setting of potentiometer control*	R_{AB}, Ω	R_{AC}, Ω	R_{BC}, Ω	$R_{AC} + R_{BC}$ *Computed value*
6	Vary over its range		X	X	X
7	Completely CW	X			
8	CW to CCW	X			
9	¼ CW	X			
10	¾ CW	X			

QUESTIONS

1. What resistance is in the center of your ohms scale, $R \times 1$ range?
2. At which end of the scale are resistance readings more reliable, the crowded or the uncrowded end?
3. To what range should you shift if your readings are at the crowded end of the scale?
4. What is the symbol used for ohms?
5. For carbon resistors, what is the color code for
   (*a*) 0.27 Ω?
   (*b*) 2.2 Ω?
   (*c*) 39 Ω?
   (*d*) 560 Ω?
   (*e*) 33,000 Ω?
6. What is meant by a short circuit? An open circuit?
7. How much resistance would you measure across the terminals of a short circuit? An open circuit?
8. How does a potentiometer differ from a rheostat?
9. How can a potentiometer be used as a rheostat?
10. Explain the significance of your measurements in procedural steps 6 to 10.
11. State your exact procedure in procedural step 11. Include your table and readings.
12. Comment on the results of your findings in procedural step 11.

Answers to Self-Test

1. ohmic value
2. 2
3. 10,000
4. silver
5. 10 Ω; 5 percent
6. false
7. 3
8. variable resistor
9. brown; red; brown; no color
10. open

3

EXPERIMENT

DRY CELLS AND MEASUREMENT OF DC VOLTAGE

OBJECTIVES

1. To measure the voltage of a dry cell using a voltmeter
2. To measure the voltage across the combinations which result from connecting dry cells in series and in parallel

INTRODUCTORY INFORMATION

Measuring DC Voltage

Electronics technology is inconceivable without measurement of electrical quantities. Thus, circuit analysis and understanding of circuit function are facilitated by the basic measurements of voltage, current, and resistance.

Technicians are the "know-how" people on a team which may include scientists, engineers, and skilled craftsmen. (Of course, technicians must also "know why.") They will therefore be the people primarily involved in making the measurements which will test the scientists' or engineers' theories. They will test the operation and characteristics of experimental circuits. They will make measurements in repairing a device which is not working. It is therefore essential that technicians know the instruments of this technology, and that they understand how to use them. They must be aware of the effect an instrument may have on the circuit and the errors of measurement, and they must also know how to interpret the results of their measurements.

Measurement of dc voltage is basic in all electronics work. In this experiment we will be concerned with learning how to use the voltage function of an electronic voltmeter. We will then employ this newly acquired skill in studying the characteristics of dry-cell arrangements.

Voltage is defined as electrical pressure. It is the *difference* in electrical pressure between two points. It is also called potential difference and electromotive force (emf). The voltage *across* two points is measured with a voltmeter. An electronic voltmeter combines the functions of measuring resistance and voltage, and in some instances current. Hence its voltage-measuring functions may be used here. The student should refer to the manufacturer's operating instructions for the particular meter before using it in this experiment.

Generally, the following facts apply, regardless of the make of VTVM or TVM:

1. A VTVM must be turned on and allowed to warm up. A TVM requires no warmup.
2. The meter test leads should be plugged into the proper meter terminal jacks. The black lead is the common or ground lead, and the probe is the voltage or "hot" lead.
3. The + or − polarity switch is set to agree with the polarity of the voltage being tested. For example, when measuring voltages which are positive with respect to common, the function selector switch is set to the + dc position.
4. The dc voltage probes are shorted together and the Zero adjustment is used to set the meter pointer to zero. Zero is usually the left-hand side of the meter scale. (On zero-center meters, zero is the middle of the scale.)
5. The common lead is connected to the return (negative) of the circuit under test. This test lead is usually connected first.
6. The range selector switch should be set to the highest dc voltage range in measuring an unknown dc voltage. This is to avoid damaging the meter. If measurement indicates that the voltage is too small to determine its value accurately, a lower dc voltage range should be selected.
7. The dc voltage probe is then connected to the circuit under test. (The insulated probe and not the metal prod should be held when using the voltmeter.)
8. The voltage value is read from the proper voltage scale. The maximum voltage for a specific voltage range is at the right end of the scale.
9. The dc voltage scale on an electronic voltmeter is separate from the *ohms* scale. The meter pointer moves in an arc from left to right.
10. The dc voltage scale is linear, with equal spacings for equal voltage changes. There may be two or three scale calibrations along the dc scale, making it possible to read voltage on all the dc ranges of the instrument. The greater the deflection toward the right, the higher the voltage that is being measured, and the greater the meter accuracy. A voltage scale should be selected that will provide maximum pointer deflection without exceeding full-scale deflection.

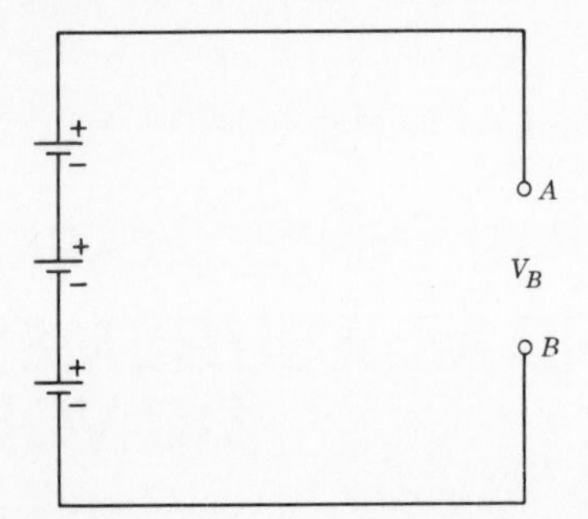

Fig. 3-1. Dry cells connected in series-aiding. The negative terminal of one cell is connected to the positive terminal of the next. The battery voltage V_B is measured across the terminals *AB*.

Dry Batteries

Dry batteries consist of arrangements of primary cells, called "dry cells." The familiar flashlight "battery" is really a dry cell. Individual dry cells produce a low voltage.

By connecting dry cells in a *series-aiding* arrangement (Fig. 3-1), we produce a dry battery whose voltage is the sum of the dry-cell voltages. By connecting two or more dry cells in *parallel* (Fig. 3-2), we produce a battery whose voltage is the same as that of an individual dry cell. However, the permissible current drain of this parallel-cell battery is proportionately greater than that of any dry cell used alone in the same application. Dry batteries usually consist of a series-parallel arrangement of cells (Figs. 3-3 and 3-4). These arrangements make possible a battery whose voltage is higher than that of an individual dry cell, and whose life is longer.

Dry cells and batteries are not always connected in series-aiding. They are sometimes used in *series-opposing*. The student should therefore understand the effects of such a connection.

Consider the arrangement of Fig. 3-5. The two batteries are in series-opposing, not in series-aiding. That is, the resultant voltage V_B delivered across *AB* is equal to the difference between the battery voltages. In the illustration, terminal *A* is 9 volts (V) positive with respect to terminal *C*. Terminal *B* is 3 V positive with respect to *C*. Hence, terminal *A* is 6 V positive relative to terminal *B*. Point *A* is therefore labeled "+," and point *B* "−."

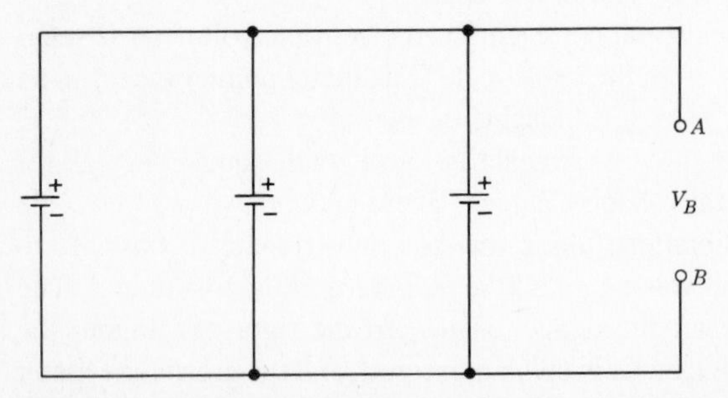

Fig. 3-2. Dry cells connected in parallel. The negative terminals of the cells are connected together, and the positive terminals of the cells are connected. The output voltage V_B is measured across the terminals *AB*.

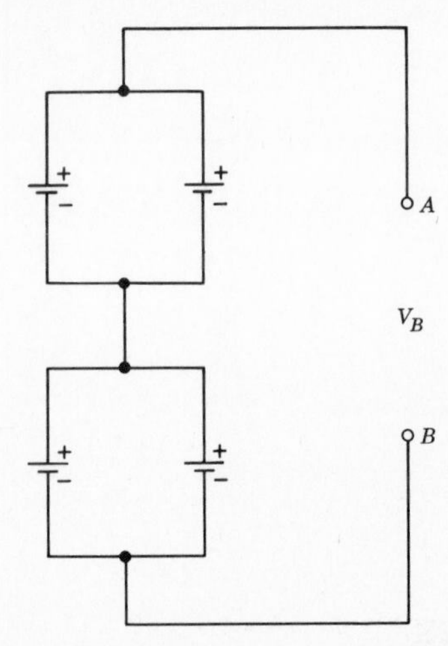

Fig. 3-3. Dry cells connected in series-parallel.

SUMMARY

1. An electronics technician must measure voltage, current, and resistance in a circuit in order to determine whether the circuit is working properly.
2. He or she also makes circuit measurements in experimental circuits to determine their characteristics.
3. The term "electronic voltmeter" is understood to mean a transistorized voltmeter (TVM) or a vacuum-tube voltmeter (VTVM). Either instrument can be used to measure voltage and resistance. In addition, current can be measured with a TVM.
4. Voltage is the difference in electrical pressure between two points. Voltage is also called potential difference, or electromotive force (emf).
5. Voltage is measured by connecting a voltmeter *across* the two points where a potential difference is present.
6. In measuring an unknown voltage, the range-selector switch is first set to the highest voltage range; then, if necessary, a lower range is selected.
7. The *insulated probe* and not the metal tip is held in using a voltmeter.

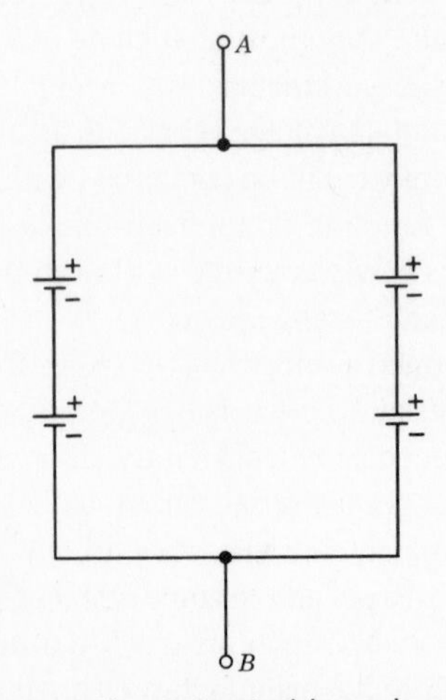

Fig. 3-4. Four dry cells connected in series-parallel (another arrangement).

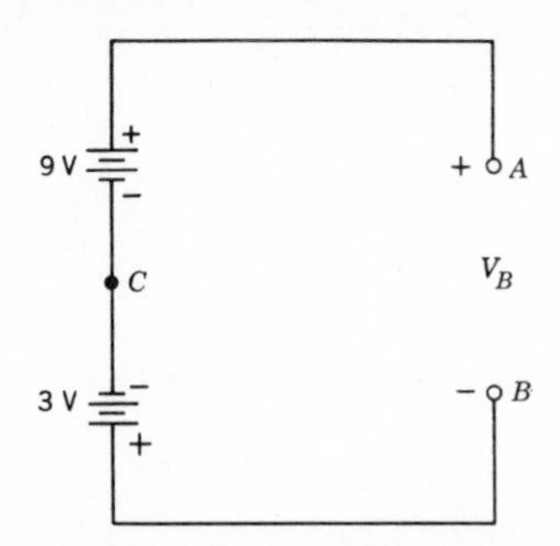

Fig. 3-5. Two batteries connected in series-opposing. The negative terminal of one is connected to the negative terminal of the other. The positive terminals are free. The resultant voltage V_B is measured across *AB*.

8. The basic dry batteries consist of combinations of individual cells called dry cells.
9. The total voltage of a battery produced by adding dry cells in *series-aiding* is the sum of the individual cell voltages.
10. If similar dry cells are connected in parallel, the total voltage of this battery is the same as the voltage of any one cell. The advantage gained in this connection is that the life of the battery is increased (by a factor approximately equal to the *number* of parallel cells).

SELF-TEST

Check your understanding by answering these questions:

1. In an electronic circuit, three measurable quantities are ________, ________, and ________.
2. A TVM can be used for measuring ________, ________, and ________.
3. The person *primarily* concerned with making circuit measurements is the ________. This person must also know how to interpret these measurements.
4. The polarity switch on an electronic voltmeter may be set to ________ or ________.
5. The ________ ________ switch of an electronic voltmeter is used to select the desired voltage range.
6. The dc voltage scale is ________ (linear/nonlinear).
7. The basic unit for forming dry batteries is the ________ ________.
8. The total voltage of dry cells connected in series depends on whether they are connected in series-aiding or series-opposing. ________ (true/false)

MATERIALS REQUIRED

- Equipment: EVM (electronic voltmeter)
- Miscellaneous: Four 1½-V dry cells, connecting wires

PROCEDURE

NOTE: Before using the electonic voltmeter, be certain that you have read and are familiar with the operating instructions. Set the meter switches properly and have them checked by an instructor. You may then proceed.

1. Connect the negative lead of the meter to the negative terminal of the cell, positive lead to positive. Measure and record in Table 3-1 the voltage of each of the dry cells supplied. Use that range of the voltmeter where you get the maximum pointer deflection without going off the scale.

2. Repeat the measurements of step 1 using the next higher range of the voltmeter. Record in Table 3-2.

3. Connect cells 1 and 2 in series-aiding. Measure and record in Table 3-3 their total voltage (*A* to *B*).

TABLE 3-1. Dry-Cell Voltage Measurement—Optimum Range

Dry-cell number	1	2	3	4
Voltage				

TABLE 3-2. Dry-Cell Voltage Measurement—Another Range

Dry-cell number	1	2	3	4
Voltage				

TABLE 3-3. Dry-Cell Arrangements

Step	3	4	5	6	7	8	9	10*a*	10*b*	10*c*
Voltage *A* to *B*										

4. Connect cells 1, 2, and 3 in series-aiding. Measure and record their total voltage (*A* to *B*).
5. Repeat this process, measuring and recording the voltage of four cells in series-aiding.
6. Connect two cells in parallel. Measure and record their voltage (*A* to *B*).
7. Connect three cells in parallel. Measure and record their voltage (*A* to *B*).
8. Connect four cells in series-parallel (Fig. 3-4). Measure and record their voltage (*A* to *B*).
9. Connect the cells in the series-parallel arrangement of Fig. 3-3. Measure and record their voltage (*A* to *B*).
10. Connect dry cells in series-opposing (Fig. 3-6*a*). Measure and record their output voltage (*A* to *B*). Repeat for Fig. 3-6*b* and *c*. Before connecting the meter, determine which is the positive terminal, *A* or *B*.

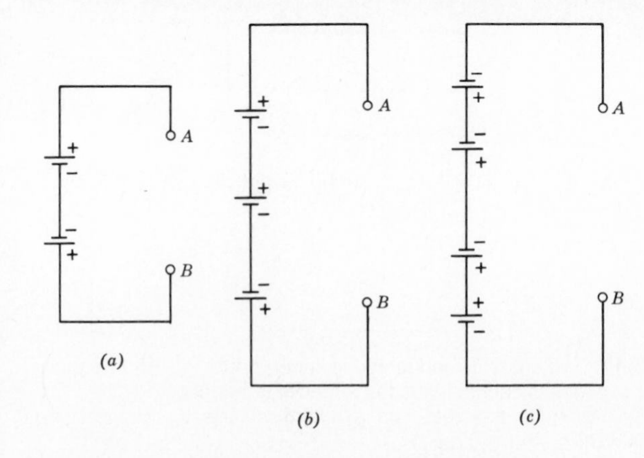

Fig. 3-6. Dry cells connected in series-opposing.

QUESTIONS

1. Name four precautions which must be observed in measuring voltage.
2. List the dc voltage ranges on your EVM.
3. How much voltage is required to give full-scale deflection of the pointer on a 300-V range?
4. In the voltage measurements of dry cells in this experiment, why must the hot lead of the meter be connected to *A* and the common lead to *B*?
5. What would happen to a dry cell or battery if the positive and negative terminals were shorted?
6. What arrangement of six dry cells gives the maximum life, regardless of voltage?
7. What arrangement of six dry cells gives the maximum voltage?
8. Draw a practical arrangement of ten 1.5-V dry cells to give a battery of 7.5 V.
9. Explain the difference in connection between two dry cells connected in series-aiding and in series-opposing.

Answers to Self-Test

1. voltage; current; resistance
2. voltage; current; resistance
3. electronics technician
4. + or −
5. range selector
6. linear
7. dry cell
8. true

4

EXPERIMENT

FAMILIARIZATION WITH DC POWER SUPPLY

OBJECTIVES

1. To identify the controls and switches of a dc power supply
2. To turn the supply on and measure its output voltage
3. To measure the range of voltage change as the output control is varied from minimum to maximum

Fig. 4-1. Front panel of a variable regulated low-voltage, high-current dc power supply (*Hewlett-Packard*).

INTRODUCTORY INFORMATION

DC voltage and current sources are required for experiments in the laboratory. Electronic, regulated, variable power supplies satisfy this requirement and are used extensively in both school and industrial laboratories.

DC Power Supply

A variable voltage-regulated supply is one which can be manually adjusted to deliver any required voltage within its range of operation. The voltage delivered by this supply remains constant despite changes in load current, within specified limits. Thus, one manufacturer states that a given supply will deliver 0 to 400 V at 150 mA. This means that the output voltage can be varied from 0 to 400 V, and that the current drawn must not be greater than 150 milliamperes (mA).

A power supply may have facilities for providing two or more independent dc voltages, in which case the instrument will have separate controls and separate output terminals.

The polarity of the dc terminals on the supply is usually marked either −, +, or gnd and V^+. A red jack is conventionally used for the positive and a black for the negative terminal of the supply.

Low-Voltage, High-Current DC Source

Figure 4-1 shows the front panel of a variable, regulated, low-voltage, high-current dc power source. This and equivalent sources are used for solid-state experiments. This all-solid-state device is line-operated (120 V ac) and requires no warmup time.

The specific model shown has two switch-selected output ranges, the first 0 to 20 V at a maximum current of 1.5 A, the second 0 to 40 V at 0.75 A. The panel meter displays, respectively, either output voltage or current. A dual concentric output control is both a coarse and a fine control of output voltage. One additional control requires mention. This is the "current" control. Maximum clockwise rotation of this control will provide the maximum current which the supply can deliver. Or the experimenter can limit the supply current by setting the current control at the required level. The purpose of the current control is to protect the experimental setup from excessive current. An on-off switch turns the equipment on and off.

There are many other types of low-voltage, high-current sources, including one with two sets of controls which delivers two independent voltages at two independent sets of terminals.

High-Voltage, Low-Current DC Sources

These sources are available for circuits that require higher voltages and lower currents. They normally provide variable, regulated dc voltages and fixed low-level, high-current ac voltages.

Figure 4-2 is the front panel of a supply which provides two dc and two ac outputs. One of the dc outputs is regulated

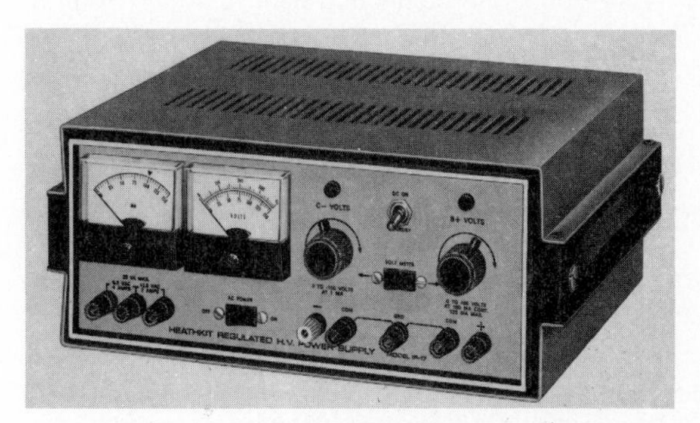

Fig. 4-2. Front panel of a variable regulated high-voltage, low-current dc power supply (*Heath Company*).

and variable from 0 to 400 V at 100 mA maximum. The other dc output is nonregulated and variable from 0 to 100 V at 1 mA. The nonregulated supply *will not maintain* a steady preset output voltage when current is drawn from it. Its output voltage will drop when loaded.

The three ac output jacks at the lower left provide either 6.3 or 12.6 V, depending on which two jacks are used. The left jack and the center jack deliver 6.3 V. The right jack and the center jack provide 12.6 V.

The supply shown is line-operated and has an on-off switch on the front panel and a dc-on switch which also includes a standby position. Other controls include voltage control for the regulated V^+ volts output; voltage control for the nonregulated C^- volts output; a meter switch which can be set to measure either the regulated or nonregulated dc voltage delivered at the respective output jacks; and a current meter to measure the output current.

Using the DC Power Supply

The following general precautions should be observed in using this device.

1. Read the operating instructions carefully and be certain that you understand them before turning a power supply on for the first time.
2. Never short the dc or ac output terminals, or you may damage the supply. To prevent the terminals from shorting, keep the leads issuing from these terminals from making contact.
3. Set the output voltage of the supply at the required level *before* you connect the supply leads to the experimental circuit. After the leads are connected to the circuit, again measure the output voltage to verify that it is still at the preset level.
4. If any component on the experimental circuit appears to be overheating after power is applied, turn the supply off and determine the cause before you proceed with the experiment.
5. Do not grasp the uninsulated output terminals of a supply when it is on. Your body will receive the output voltage, and you may experience a severe electrical shock.
6. To prevent damage to the power supply, do not switch the supply on and off excessively. If an experiment requires power to be interrupted frequently, use an external switch on the breadboard to apply and remove power from the circuit.
7. Do not operate the supply beyond its *rated* current capacity. If the current meter indicates that you are exceeding the current capability of the power supply, turn it off and check the experimental circuit to determine why it is drawing excessive current.

SUMMARY

1. Electronic dc voltage and current sources are used in the laboratory. They are called power supplies.
2. Electronic dc supplies derive their power from the ac line. We say they are line-operated.
3. A voltage-regulated supply is one which maintains a preset voltage level, within its range of operation, despite variations in output current and line voltage.
4. A nonregulated supply is one whose output voltage changes with changes in load current and line voltage.
5. In the school laboratory are generally found two types of supplies:
   (*a*) Variable low-output-voltage, high-output-current source needed for solid-state experiments
   (*b*) Variable high-output-voltage, low-output-current source used in experiments requiring high voltage
6. Power-supply output terminals should *not* be shorted.
7. Power supplies should not be operated beyond their rated current capacity.

SELF-TEST

Check your understanding by answering these questions:

1. The output voltage of dc supplies found in the laboratory is made ____________ to accommodate a variety of voltage needs.
2. If a power-supply voltage remains constant at a preset level despite changes in supply current or line voltage, we say the supply is ____________.
3. Variable low-voltage, high-current supplies are normally used for ____________ ____________ circuits.
4. The output terminals of a power supply must *not* be ____________ ____________ when the supply is on.
5. The ____________ capacity of a supply should not be exceeded.

MATERIALS REQUIRED

- Power supply: Variable dc voltage-regulated power supply
- Equipment: Electronic voltmeter
- Resistors: 250-Ω 4-W
- Miscellaneous: SPST switch; breadboarding device

PROCEDURE

1. Draw the front panel of the power supply assigned to you. Label all switches, controls, and output terminals. Read the operating instuctions and proceed only after you clearly understand them. In Table 4-1 name and describe the function of each switch, control, and terminal.
2. **Power on.** If necessary, permit the supply to warm up. Set the regulated voltage control for maximum voltage as indicated on the panel meter of the power supply. In Table 4-2 indicate the panel meter voltage. Does the panel current meter show current?

CAUTION: Be certain that no output terminals are short-circuited.

TABLE 4-1. Power Supply Controls

Control, Switch or Terminal	*Function*

3. Turn **on** an electronic voltmeter. After it is properly zeroed, set the meter on a voltage range equal to or just greater than the maximum rated voltage of the power supply. Connect the negative lead of the meter to the negative terminal of the regulated supply, the positive meter lead to the positive terminal. Measure the output voltage and record it in Table 4-2.
4. Reduce the supply voltage to its minimum level. Measure, both on the panel meter and on the electronic voltmeter, the lowest voltage the power supply provides. Record these measurements in Table 4-2.

NOTE: Reduce the range of the EVM when necessary to secure a more accurate measurement.

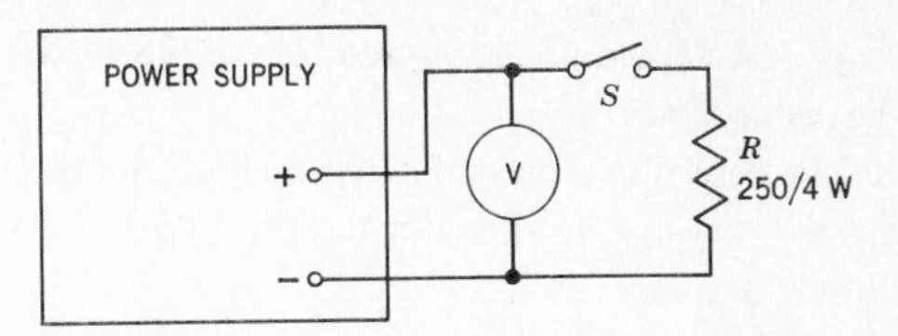

Fig. 4-3. Circuit to determine load–no-load characteristic of a regulated power supply.

5. Set the output of the regulated supply to half its maximum voltage. Record in Table 4-2 the EVM and panel meter readings. Indicate also if the panel current meter shows current.
6. Set the measured output of the regulated supply at 20 V. Connect the circuit of Fig. 4-3. V is an electronic voltmeter used to measure the output voltage of the power supply. S is an on-off switch on the breadboard. R is a 250-Ω 4-W resistor. Switch S is **open**. Does the panel current meter show current?
7. Close switch S. Measure and record in Table 4-2 the output voltage of the supply. Record also if the panel current meter shows current.
8. If there is another dc voltage source on the power supply, check it in a manner similar to steps 2 to 7. Construct a new Table 4-3, similar to Table 4-2, and enter all your measurements.

TABLE 4-2. Voltage and Current Characteristics of a Power Supply

		Output Voltage		*Supply Current*	
Step	*Position of Output Voltage Control*	*Panel Meter Volts*	*EVM Volts*	*Yes*	*No*
2,3	Maximum				
4	Minimum				
5	Half maximum				
6	X	X	20 V		
7	X	X			

QUESTIONS

1. What is the dc voltage rating of the power supply assigned to you? Current rating? If there is more than one dc output, give the ratings of each.
2. What is meant by short-circuiting the output terminals of a power supply?
3. Why must we be careful to keep from short-circuiting the output terminals?
4. In addition to the voltage source (battery or power supply), is it necessary to have a complete circuit for voltage to be present? Relate your answer to the measurements in this experiment.
5. Compare your voltage measurements in Table 4-2 with (*a*) the panel meter and (*b*) the EVM. Are they substantially the same?
6. How can you tell which meter is correct, the panel meter or the electronic voltmeter?

7. When is it necessary to switch the voltmeter ranges in measuring voltage?
8. Refer to your measurements in Table 4-2, steps 6 and 7. Does the voltage of the supply hold at 20 V, with and without load current?
9. Refer to Table 4-2. Under what conditions did the supply register current? What supplied this current?
10. Name at least two advantages an electronic dc power supply has over a dry cell.

Answers to Self-Test

1. variable
2. regulated
3. solid-state
4. short-circuited (or shorted together)
5. current

5

EXPERIMENT

RESISTANCE OF CONDUCTORS AND INSULATORS

OBJECTIVES

1. To identify conductive and nonconductive materials and components
2. To measure the resistance of a variety of conductive and nonconductive electrical and electronic components

INTRODUCTORY INFORMATION

Conductors and Insulators

The useful effects of electricity result from the movement of electric charges in a circuit. This movement of electric charges is called *current*. Electrical charges move easily through paths called *conductors*, but it is very difficult for these charges to move through *insulators*. Another way of saying this is to state that conductors are materials which permit current to flow easily with little electrical pressure (voltage) applied, while insulators are materials which permit very little or no current to flow. Copper is an example of a familiar conductor, while rubber is an insulator.

A complete (closed) circuit or path for direct current consists of a voltage source such as a battery or dc power supply, a conductive load such as an electric bulb, and connecting conductors (copper wire) (Fig. 5-1). In such a closed circuit there is electric current, and if sufficient current flows, the bulb will light. If rubber cords instead of copper wires were used to tie the battery to the bulb, the bulb would not light, because rubber is an insulator and does not permit current to flow through it in the circuit. The first illustration describes a *complete* or *closed* circuit, the second an *incomplete* or *open* circuit.

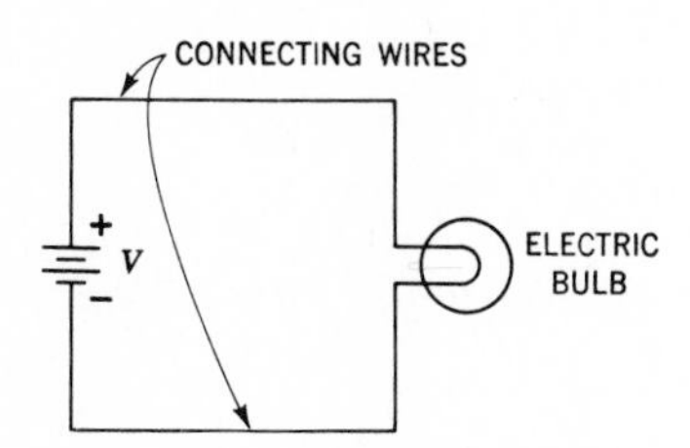

Fig. 5-1. A complete (closed) electric circuit containing a battery V, connecting wires, and an electric bulb as load.

The tungsten filament in the electric bulb is a conductor, but it is not as good a conductor as the connecting copper wires. Tungsten is used for the filament because its properties cause it to glow and give off light when it is heated to incandescence by an electric current.

It is apparent that not all conductive materials are equally good conductors of electricity. They are used in constructing electrical components because of their additional desirable characteristics.

Wire Conductors

The ability of a conductor to permit current flow is designated conductance (G), while opposition to current is called resistance (R). The ohm (Ω) is the unit of resistance; the *mho* is the unit of conductance. Various materials exhibit different resistivity. Thus the resistance of round wire of the same diameter increases in the following order, depending on the material of the wire: silver, copper, gold, aluminum, tungsten, nickel, iron, nichrome, etc. That is, silver has the greatest conductivity, and hence the least resistance. The conductivity of copper is less than that of silver; therefore the resistance of copper wire is greater than that of silver for the same cross-sectional area and length. However, copper wire is widely used as a conductor because of its relatively low cost and low resistance.

The resistance of round wire depends not only on the material from which it is made, but also on its length, cross-sectional area, and temperature. A formula which relates all these factors is:

$$R = \rho \frac{l}{A} \tag{5-1}$$

where R = resistance of the wire in ohms
l = its length in feet
A = its cross-sectional area in circular mils (cmil)
ρ = its resistivity (depends on the material) at a specified temperature

It should be noted that the area A in circular mils of a round wire whose diameter is D mils is:

$$A_{\text{cmil}} = (D_{\text{mils}})^2 \tag{5-2}$$

An example will explain how the formula is applied.

Problem. If the resistivity ρ of copper wire at 20°C equals 10.37, find the resistance R of 100 ft of such wire whose diameter $D = 0.005$ in.

Solution. A mil is defined as 0.001 in. Hence here $D = 5$ mils. Therefore, $A_{cmil} = D^2 = 25$. Now since

$$R = \rho \frac{l}{A}$$

$$R = 10.37 \times \frac{100}{25} = 41.48\ \Omega$$

Round wire is specified by gage. The Bureau of Standards has published a table of American Wire Gages (AWG) to which the technician can refer for calculations. AWG numbers decrease in diameter from gage #0000 to gage #40. The diameter of AWG 0000 is 460 mils. The diameter of AWG 40 is 3.1 mils. The diameter of AWG 36 is 5 mils. The resistance of 1000 ft of AWG 0000 at 20°C is 0.0490 Ω, and the resistance of 1000 ft of AWG 36 at 20°C is 414.8 Ω. A technician who wished to calculate the resistance of 100 ft of AWG 36 copper wire would simply refer to the AWG table. In this case, since the resistance of 1000 ft of AWG #36 is 414.8 Ω, 100 ft would have a resistance of

$$\frac{100}{1000} \times 414.8 = 41.48\ \Omega$$

This is the same result as that obtained by applying Eq. (5-1).

In addition to giving the resistance of 1000 ft of each gage wire, AWG tables also give the weight of 1000 ft of each wire and its current capacity. Current capacity depends on the cross-sectional area A and on the type of insulation used.

Wire Components

Some components are made by winding a length of wire around a form. For example, ceramic high-power resistors consist of a wire winding around a ceramic core. Iron-core inductors (chokes) are wound around an iron core. Transformers consist of several windings around some specified core. If an ohmmeter is placed across each winding of these components, it will measure the resistance of that winding. Therefore, one way of determining if a wire winding is continuous, that is, if it is not broken, is to measure its resistance across the two end terminals. The normal resistance is usually specified by the manufacturer. If a winding measures infinite resistance, it is open.

Carbon Conductors—Resistors

In a previous experiment we learned how to determine, from their color code, the resistance of carbon resistors. In this connection we must note that carbon is also a conductor of electricity. A complete circuit which contains a battery or other dc power source, connecting wires, and a resistor (Fig. 5-2) will permit direct current to flow.

The physical construction of R will determine its resistance. Here as in circular wire the resistance of R will vary inversely with the cross-sectional area of the carbon element in the resistor, and directly with the length of the element. By controlling the diameter and length of the inner carbon element, we can control the resistance of R.

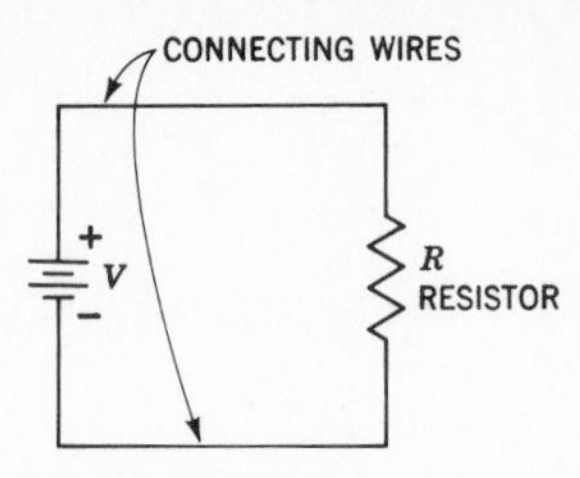

Fig. 5-2. A complete (closed) electric circuit containing a battery V, connecting wires, and a resistor R as load.

Comparing a resistor with a copper wire, we note that the resistance of the wire is *distributed* throughout its entire length. In a small carbon resistor, the resistance is lumped between the two terminals.

In the circuit of Fig. 5-2 the current can be controlled by controlling the value of R, in ohms. With V_B held constant, the higher the resistance of R, in ohms, the lower the current will be. That is, as the resistance of R increases, the opposition to current increases and the current decreases.

NOTE: Any object whose resistance can be measured with an ohmmeter is a conductor of direct current.

Insulators

All metals are conductors of electricity, though it is evident from the preceding discussion that not all metals conduct equally. There are some materials whose resistivity is so high that they permit *very*, *very* little current to flow. These materials are called *insulators* or *nonconductors*. Rubber was previously mentioned as an insulator. Wood, glass, paper, mica, bakelite, and air are other examples of insulators.

If the leads of an ohmmeter are connected across an insulator, say a rubber cord, the meter will register infinite resistance. An infinite resistance will not permit direct current to flow in a circuit. Thus Fig. 5-3 is an open circuit, because direct current will not flow through the insulator.

Insulating materials are used to cover copper wires in electricity and electronics. The purpose of insulating electrical wires is to prevent them from making electrical contact (shorting) with any other conductor or component which they may touch accidentally.

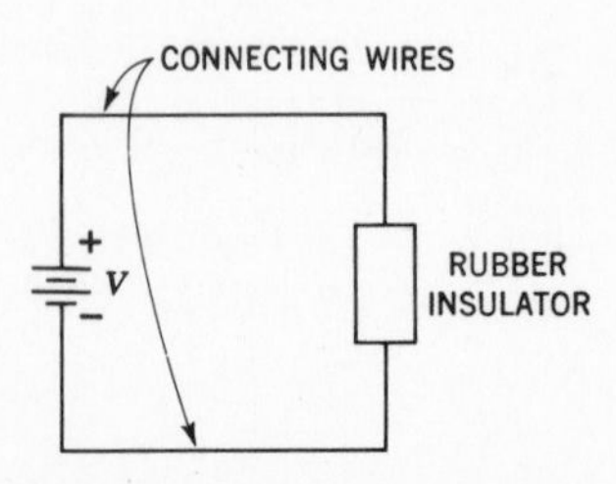

Fig. 5-3. An open circuit because the rubber insulator will not permit direct-current flow.

NOTE: Wires whose insulation has melted or become frayed are safety hazards and should not be used.

Before leaving the subject of insulators, we should note that insulators may break down and become conductors if the voltage across them is high enough. Thus an air gap between two voltage terminals which are 1 cm apart will break down if the voltage across these terminals equals or exceeds 30,000 V. Another descriptive term frequently used for this phenomenon is "arc-over." For example, in the high-voltage section of a television receiver, arc-over may occur under certain abnormal conditions.

Semiconductors

This discussion would not be complete without mention of *semiconductors*. A transistor is an example of a device made of semiconductors. Another familiar semiconductor device is an IC (integrated circuit chip).

Silicon and germanium are the materials from which semiconductors are made. These two elements, in their pure state, are insulators. However, when they have been "doped" with certain impurities, such as gallium or antimony, they can be made somewhat conductive, and they are then called semiconductors. The resistivity of semiconductors depends on the nature and amount of impurity injected. In general, however, the resistivity of semiconductors is said to be between that of insulators and conductors. Semiconductor devices are also referred to as solid-state devices, and that branch of electronics dealing with semiconductors is known as solid-state electronics.

The Human Body as a Conductor

The human body is a conductor of electricity. The most certain proof is the electric shocks that people have experienced when coming in accidental contact with the + and − terminals of a voltage source. Electric current will flow through the body, and the higher the voltage, the higher the current. Electric shock is *dangerous* and can be *fatal*. That is why the student of electricity and electronics should follow the safety precautions taught in order to avoid shock.

It should be mentioned that the resistance of the body at a specific point in time can be measured with an ohmmeter. The method to follow will be given under Procedure.

SUMMARY

1. Conductors are materials which permit direct current to flow through them with ease. Silver, copper, and gold are examples of conductors.
2. Insulators are materials which do not permit direct current to flow through them. Rubber, glass, and mica are examples of insulators.
3. Conductors vary in their conductivity. Thus copper is a better conductor than nichrome.
4. Round copper wire is used to connect electrical components in a handwired circuit.
5. The resistance of wire varies directly with its length and inversely with its cross-sectional area. Resistance also depends on the material from which the conductor is made.
6. The Bureau of Standards publishes a table of American Wire Gages (AWG) which gives the diameter in mils, area in circular mils, resistance in ohms of 1000 ft, and weight of 1000 ft of every copper wire gage made.
7. Insulators can withstand voltages up to a critical voltage. For voltages at or higher than the critical voltage, insulators break down and permit current flow by arc-over or arc-through.
8. The breakdown voltage of an insulator depends on its material.
9. The resistance of a good conductor is very low and that of an insulator very high (infinite).
10. Semiconductors are solid-state devices whose resistivity lies between that of a conductor and insulator. Silicon and germanium diodes and transistors are examples of semiconductors.
11. The human body is a conductor of electricity. Body contact with electrical wires must be avoided to prevent shock.

SELF-TEST

Check your understanding by answering these questions:

1. The following materials are examples of good electrical conductors: ________; ________; ________; ________; ________.
2. The following materials are examples of electrical insulators: ________; ________; ________; ________; ________.
3. For current to flow in a circuit, it must have: (*a*) ________; (*b*) ________; and (*c*) ________.
4. A circuit in which current flows is called a ________ circuit.
5. Current will not flow in an ________ circuit.
6. AWG 22 round copper wire has a ________ (larger/smaller) diameter than #12 wire.
7. The resistance of #10 wire is ________ (higher/lower) than that of #40.
8. A 100-ft piece of #22 wire has ________ the resistance of a 50-ft piece of the same wire.
9. An insulator may ________ ________ if it is subjected to a voltage higher than its breakdown voltage.
10. Semiconductors will, under the proper conditions, permit electric current to flow. ________ (true/false)
11. The diameter of #25 round copper wire is 17.9 mils. The circular mil area of this wire is ________.
12. The diameter of #22 round copper wire is 25.35 mils. Its resistivity is 10.58. The resistance of 1000 ft of this wire is ________.

MATERIALS REQUIRED

- Equipment: EVM
- Resistors: One carbon resistor, any value
- Capacitors: 0.01 μF
- Silicon rectifier: 1N2615 or equivalent
- Miscellaneous: 12-in length of #40 copper wire, single strand; 6-in length of #40 nichrome wire; 12-in length of #40 nichrome wire; piece of rubber ¼ × 2 in; piece of wood ¼ × 2 in; piece of plastic ¼ × 2 in; 8-H choke

PROCEDURE

1. Identify each of the 10 objects listed in Table 5-1.
2. Set your electronic voltmeter on Ohms and zero the meter at both ends of the scale if necessary, as in Experiment 2.
3. Measure the resistance of object 1 (resistor) and record this value in Table 5-1, column *A*. Observe which lead you connected to the right-hand terminal of the resistor and which to the left.
4. Reverse the leads of the ohmmeter and again measure the resistance of object 1. Record this value in Table 5-1, column *B*.
5. Check in the proper column if this is a conductor or insulator of direct current.
6. Repeat steps 3 through 5 for every object shown in Table 5-1. Check your meter zero if you switch to another resistance range.

NOTE: Be certain that you hold the insulated probes on the meter leads when you measure resistance. This becomes *particularly important* when you are using the *higher* ranges of the meter ($R \times 1000$, etc.). The reason is that the ohmmeter will read your body resistance if you hold the metal tips of the meter leads. This measurement of body resistance will affect the resistance reading of the object you are measuring, particularly if the object has high resistance.

Body Resistance

7. Set the ohmmeter on the $R \times 10{,}000$ or $R \times 100{,}000$ range and check that it is properly zeroed.
8. Measure your body resistance by grasping the metal tip of one lead with one hand, and the metal tip of the other lead with your other hand. Record. Body resistance = ________ Ω.

TABLE 5-1. Resistance of Conductors and Insulators

	Resistance, Ω			
Object	*A*	*B*	*Conductor*	*Insulator*
1. Resistor				
2. Capacitor				
3. 12-in length #40 copper wire				
4. 12-in length #40 nichrome wire				
5. 6-in length #40 nichrome wire				
6. Piece of rubber ¼ × 2 in				
7. Piece of wood ¼ × 2 in				
8. Silicon rectifier 1N2615				
9. 8-H choke				
10. Piece of plastic ¼ × 2 in				

QUESTIONS

Refer to your experimental results, where possible, in answering these questions.

1. How can one determine whether an object is a conductor of direct current?
2. Did reversing the meter leads affect the resistance of the objects in Table 5-1? List the object(s), if any, whose resistance was affected.
3. What conclusion can you draw about the conductivity of the silicon rectifier (1N2615)?
4. Which is a better conductor of direct current, the 2-in copper wire or the 2-in nichrome wire? Why?
5. Which of your measurements, if any, tend to confirm the

statement that the resistance of a wire varies directly with its length, all other factors being equal [Eq. (5-1)]?

6. Which of your measurements, if any, tend to confirm the statement that not all metals have the same conductivity?
7. Briefly explain how you could experimentally confirm the accuracy of Eq. (5-1) in predicting the resistance of round copper wire.
8. Explain how you can zero an ohmmeter at both ends of its scale.
9. Is the ohmic resistance of your body higher or lower than that of air? How do you know?
10. Assume the wires used in the circuit of Fig. 5-1 are rubber-covered. Why is it necessary to trim the rubber off the ends of the wires before connecting them in the circuit?

Answers to Self-Test

1. silver; copper; gold; aluminum; tungsten, etc.
2. wood; glass; paper; mica; air, etc.
3. (*a*) voltage source; (*b*) conductive load; (*c*) connecting conductive wiring
4. closed
5. open
6. smaller
7. lower
8. twice
9. arc-over (or breakdown)
10. true
11. 320.41
12. 16.46 Ω

6

EXPERIMENT

DIRECT-CURRENT MEASUREMENT AND CONTROL OF CURRENT

OBJECTIVES

1. To connect a current meter to measure direct current in a circuit
2. To measure the effect of resistance in controlling current in a circuit
3. To measure the effect of voltage in controlling current in a circuit

INTRODUCTORY INFORMATION

Resistance, Voltage, and Current

In Experiments 2 and 3 the uses of the ohmmeter and voltmeter for measuring resistance and voltage were studied. From the nature of the measurements, it was apparent that there were components called resistors whose resistance could be measured directly with an ohmmeter. The quantity of ohms of resistance was not dependent on the connection of that resistor in a circuit. The characteristic of resistance was associated with the component itself.

Similarly, in the measurement of emf (electromotive force; i.e., voltage), it was evident that voltage was a characteristic of some voltage source (dry cell, battery, power supply, etc.), and that it could also exist independently without need for a complete electric circuit.

Electric current differs from voltage and resistance in that it cannot exist by itself. A view of current is that it is the result of a movement of electric charges. A voltage source by itself is insufficient to create current. A voltage source *and* a closed (complete) path are required for the movement of these charges. The movement of electric charges, then, is restricted to the closed path (circuit) within which a voltage source can act.

The quantity of current in a circuit is dependent on the amount of voltage applied by the source of emf and on the nature of the conductive path. If the path offers little opposition, the current is larger than it would be in a circuit where the opposition to current is greater. Opposition to direct current is called resistance. Resistance is measured in ohms. Current, then, can be controlled by the amount of resistance in the circuit.

Measuring Direct Current

Direct current in a circuit may be measured by means of a dc ammeter. When the magnitude of the current to be measured is small, a milliammeter [1 milliampere (mA) = one-thousandth of an ampere] or a microammeter [1 microampere (μA) = one-millionth of an ampere] is used. In measuring current the circuit must be physically broken or opened and the meter inserted in series with the circuit. Suppose, for example, it is required to measure current in the circuit of Fig. 6-1. The circuit is first broken at A (Fig. 6-2). The milliammeter is then inserted in series with the circuit at the two open leads (Fig. 6-3).

Polarity must be observed, that is, the *common* meter lead (negative) must be connected to the point of lower potential, and the "hot" lead (positive) to the point of higher potential. When the meter is connected properly, the pointer will move in an arc from left to right (except, of course, when a *digital* meter is used. In a digital meter there is no pointer. A readout indicates the measured quantity.) If the pointer deflects in the opposite direction, the meter leads must be reversed. In Fig. 6-3 the negative lead of the meter is connected to B, the positive lead to A. After the measurement has been made, the meter is removed from the circuit, and the original circuit connections are restored.

CAUTION: The current meter must *not* be connected across (that is, in parallel with) any component. *It must always be connected in series* with the component to measure the current in the component. Failure to observe this rule may result in serious damage to the meter.

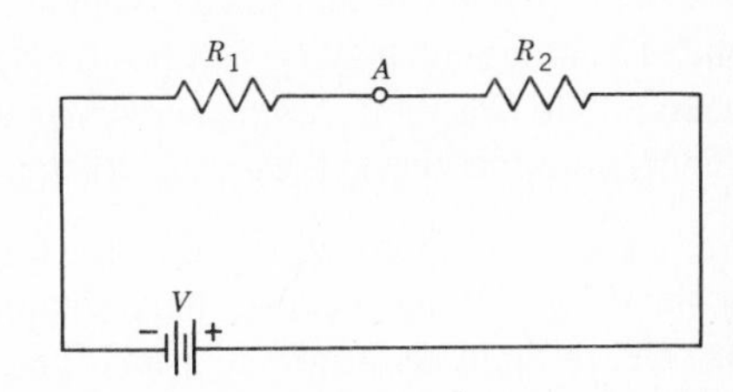

Fig. 6-1. Circuit where current is to be measured.

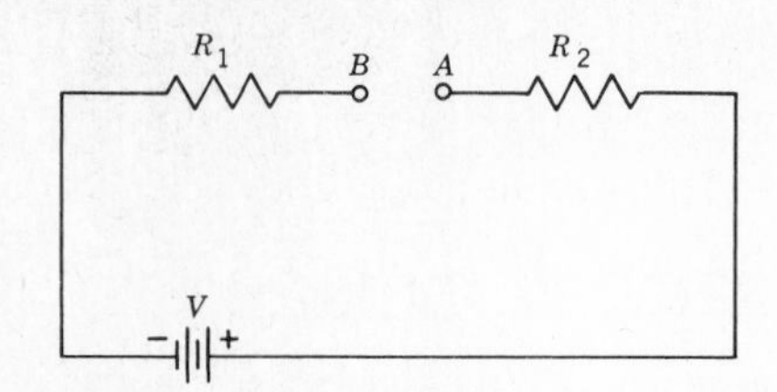

Fig. 6-2. Breaking the circuit at point *A*.

SUMMARY

1. Opposition to current is a characteristic of a resistor. *Resistance can exist by itself.*
2. Voltage is a characteristic of an electromotive force (emf) source. *Voltage can exist by itself.*
3. *Current cannot* exist by itself. For current to exist, a voltage source and a closed path (circuit) are required.
4. In measuring direct current in a circuit, the circuit must be *broken* and a current meter must be connected in *series* with the circuit.
5. In connecting a current meter in a circuit, meter polarity must be observed.
6. Proper polarity is observed if the pointer of a current meter moves upscale, that is, from left to right, when the meter is connected in a circuit.

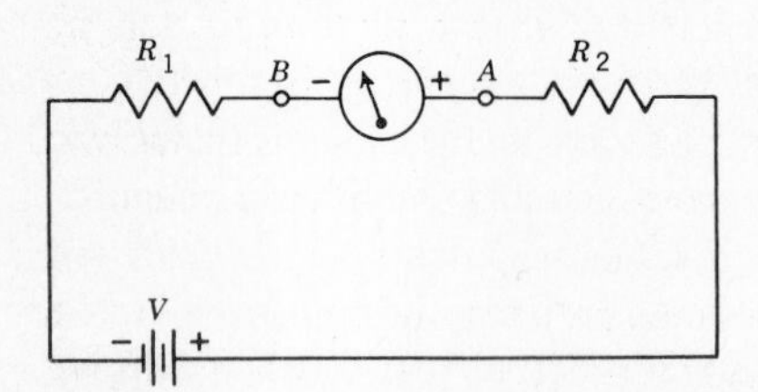

Fig. 6-3. Inserting a milliammeter to measure current.

7. The quantity of current in a circuit can be controlled by the amount of applied voltage and by the amount of resistance in the circuit.

SELF-TEST

Check your understanding by answering these questions:

1. To measure current in a circuit, it is first necessary to ____________ the circuit.
2. The current meter is then connected ____________ ____________ with the circuit.
3. Current can exist only in a ____________ circuit.
4. ____________ must be observed in connecting a current meter in a dc circuit.
5. The ____________ lead of a milliammeter is the common; the ____________ lead of a milliammeter is frequently called the *hot* lead.
6. A current meter must never be connected ____________ a component. It must always be in ____________ with the component.
7. The current in a circuit varies with the (*a*) ____________ and (*b*) ____________.

MATERIALS REQUIRED

- Power supply: Four 1½-V dry cells
- Equipment: 0–10-mA milliammeter, EVM
- Resistors: ½-W, three 1000-Ω
- Miscellaneous: Connecting wires

PROCEDURE

Controlling Current by Resistance

1. Measure and record in Table 6-1 the resistance of one of the three 1000-Ω resistors.
2. Connect this resistor in the circuit (Fig. 6-4). The power source consists of four dry cells connected in series-aiding. The circuit is *incomplete* because it is *open* between points *P* and *Q*. (That is, the positive lead of the battery is not connected.)

CAUTION: Observe polarity in connecting the milliammeter. The negative terminal of the milliammeter is connected to the negative end of the four-cell battery. The positive lead of the milliammeter goes to resistor R_1. *Do not complete* the circuit until the connections are checked by an instructor.

3. After the hookup is approved by an instructor, connect the positive lead of the battery, thus completing the circuit. The milliammeter now registers the current in the circuit. Read and record the current.
4. Open the circuit by removing R_1. Does the milliammeter still show current? Record in Table 6-1 the current, if any. If there is no current, record "none."
5. Now connect two 1000-Ω resistors (R_1 and R_2) in series, outside of the circuit. Measure their resistance and record in Table 6-1.

CAUTION: The resistance of a component must *never* be measured while it is in a circuit in which there is current. The

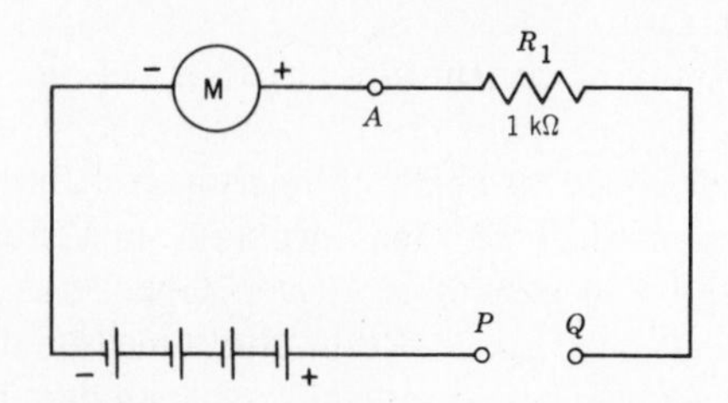

Fig. 6-4. Milliammeter connected in an open circuit.

TABLE 6-1. Controlling Current by Resistance. Voltage Source Constant

Step	*Resistance,* Ω	*Circuit Condition*	*Current,* mA	*Milliammeter*
1,3		Closed		*A*
4	Same as 3	Open		*A*
5		Open	X	X
6	Same as 5	Closed		*A*
7		Open	X	X
8	Same as 7	Closed		*A*
9	Same as 7	Closed		*B*
10	Same as 7	Closed		*C*
11	Same as 7	Closed		*D*

circuit must first be opened. The required resistance is then measured *outside* of the circuit. If this rule is not observed, the ohmmeter can be *seriously* damaged.

6. Connect R_1 and R_2 in the circuit (Fig. 6-5). Measure and record the current now in the circuit.

NOTE: A battery symbol is now used to replace the four dry cells.

7. Remove R_1 and R_2 from the circuit and add a third 1000-Ω resistor, R_3, in series. Measure and record the resistance of this series combination of resistors.
8. Connect R_1, R_2, and R_3 in the circuit (Fig. 6-6). Measure and record the current.

Current in a Series Circuit

9. Shift the position of the milliammeter so that it is now at *B*, as in Fig. 6-7. Observe meter polarity. Measure and record the current.

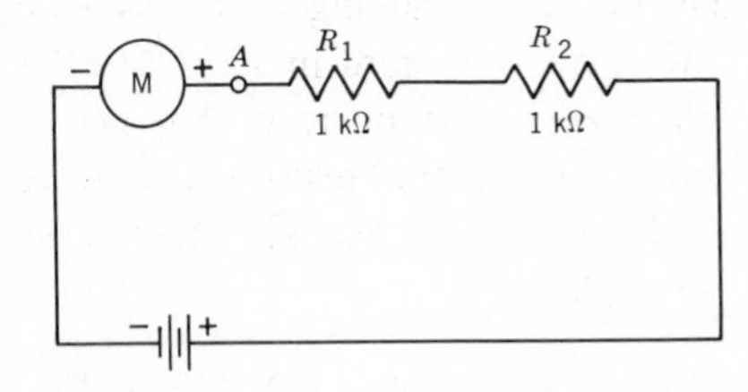

Fig. 6-5. Measuring current in a circuit containing two resistors connected in series.

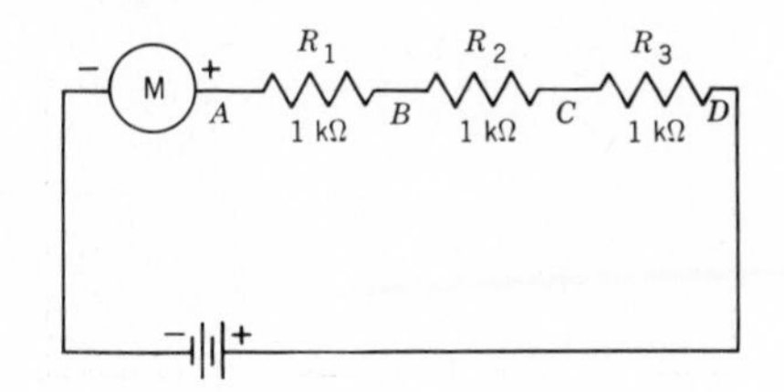

Fig. 6-6. Measuring current in a circuit containing three resistors connected in series.

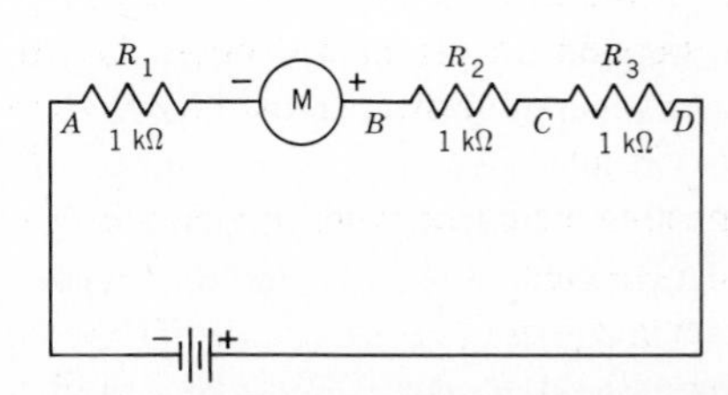

Fig. 6-7. Measuring current in a series circuit.

10. Shift the position of the milliammeter so that it is now at *C*, as in Fig. 6-8. Measure and record the current.
11. Shift the position of the milliammeter so that it is now at *D*, as in Fig. 6-9. Measure and record the current.

Controlling Current by Voltage

12. Remove R_2 and R_3, leaving R_1 in the circuit of Fig. 6-10. The voltage source still consists of four 1½-V cells connected in series-aiding. Measure and record in Table 6-2 the voltage source *V*. Measure and record the current.
13. Remove one of the dry cells, leaving three connected in series. Measure and record the voltage source. Measure and record the current.
14. Remove another dry cell, leaving two connected in series. Measure and record the voltage source. Measure and record the current.
15. Remove another dry cell, leaving one connected in the circuit. Measure and record the voltage source. Measure and record the current.
16. Remove the last dry cell. Complete (close) the circuit. Measure and record current, if any.

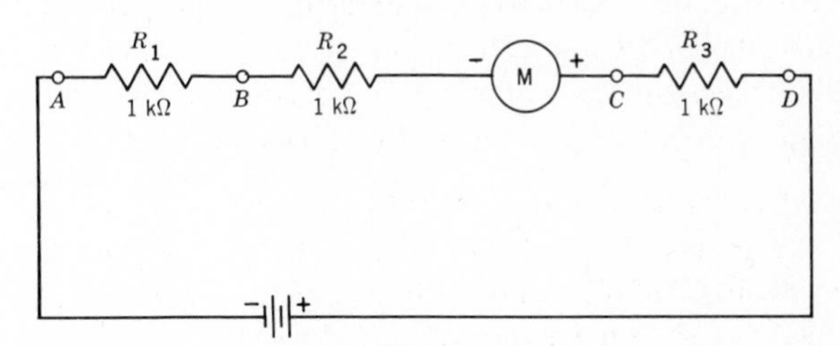

Fig. 6-8. Measuring current in a series circuit.

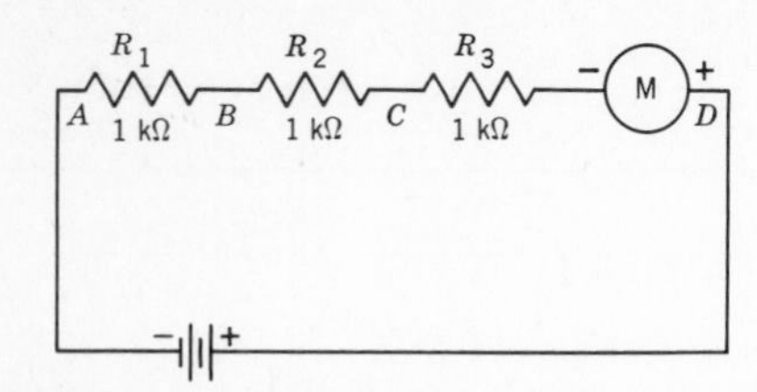

Fig. 6-9. Measuring current in a series circuit.

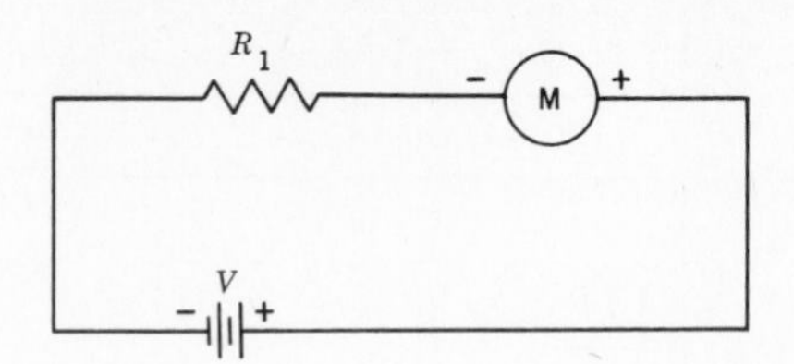

Fig. 6-10. Circuit for determining the effect of voltage on current.

QUESTIONS

1. Under what conditions will there be current in a circuit? Justify your answer by referring to specific data in Tables 6-1 and 6-2.
2. Explain how a milliammeter (or ammeter) is connected to measure current in a series circuit.
3. What precautions must be observed in measuring current to prevent damaging the meter?
4. Why must resistance *never* be measured in an *active* circuit, that is, in a circuit carrying current?
5. (*a*) State a conclusion showing the relationship between current and resistance in a circuit, if the voltage of the source is held constant. Refer to specific measurements in the experiment to substantiate this rule.
   (*b*) Formulate this rule mathematically, using the symbols I for current and R for resistance.
6. (*a*) State a conclusion showing the relationship between current and voltage in a circuit if the resistance is held constant. Refer to specific measurements in the experiment to substantiate this rule.
   (*b*) Formulate this rule mathematically, using the symbols I for current and V for voltage of the source.
7. What do your measurements in steps 9 through 11 prove concerning current in a series circuit?

Answers to Self-Test

1. break (or open)
2. in series
3. closed
4. polarity
5. negative; positive
6. across; in series
7. (*a*) applied voltage; (*b*) resistance

TABLE 6-2. Controlling Current by Voltage. Resistance in the Circuit Remains Constant

Step	*No. of Dry Cells in Series*	*Voltage of Source*	*Current*, mA
12	4		
13	3		
14	2		
15	1		
16	0	0	

7

EXPERIMENT

OHM'S LAW

OBJECTIVE

To determine experimentally the mathematical relationship between the current I in a resistor, the voltage V across the resistor, and the resistance in ohms R of the resistor (Ohm's law).

INTRODUCTORY INFORMATION

Experiment 6 established that there is a definite relationship between current, voltage, and resistance in a circuit. Stated nonmathematically, one conclusion that was reached is: There is direct current I in a closed circuit containing a dc voltage source V and resistance R, and the amount of current decreases as the resistance increases, if the voltage source remains constant. Another conclusion that was reached is: In a closed circuit containing resistance R and dc voltage source V, the amount of direct current I increases as the voltage increases, if the resistance remains constant.

These conclusions are important, but the form in which they are stated is qualitative (descriptive) and not quantitative (mathematical). Modern science seeks quantitative relationships between cause and effect wherever possible. This is because a quantitative rule or formula not only states that a change will occur if one of several related quantities is permitted to vary, but also makes it possible to predict how *much* of a change will occur.

It is possible, following the method in Experiment 6, to formulate a mathematical statement (that is, to write a formula) for the relationships among I, V, and R. To do this, it is necessary to make many *precise* measurements involving the two variables (V and R). Then, applying logic and mathematical reasoning, a formula is written which will fit the measured facts. Additional measurements are then made to verify the formula or to modify it, as the case may be.

An example will illustrate the process. The circuit of Fig. 7-1 was used to study the relationship between I and V, for a constant value of $R = 100\ \Omega$. A voltmeter was employed to measure the voltage applied to the circuit and an ammeter to measure the current. The following results were obtained and tabulated in Table 7-1.

The exact relationship between V and I is quite apparent from Table 7-1, for the ratio of V/I is equal to 100 in each case. The equation which would appear to fit the facts of Table 7-1 is

$$\frac{V}{I} = 100 \quad \text{or} \quad I = \frac{V}{100} \tag{7-1}$$

To test the accuracy of Eq. (7-1), the researcher would compute the value of I_1 for a voltage V_1 other than any of the voltages already used, then measure the current I in the circuit with V_1 applied. If I and I_1 were equal, the researcher would check the accuracy of Eq. (7-1) for other voltage values V_2, V_3, etc., until convinced that the equation was a valid statement of the facts.

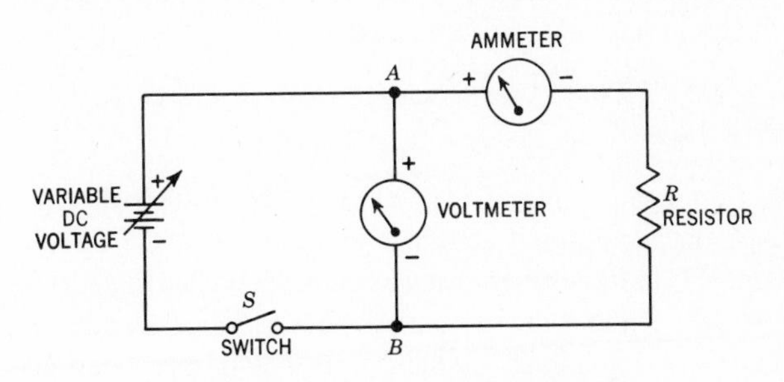

Fig. 7-1. Theoretical circuit for verifying Ohm's law.

TABLE 7-1. Deducing a Formula for I when $R = 100\ \Omega$

R	$100\ \Omega$					Formula Relating V, I, and R *for* $R = 100\ \Omega$	*Formula Test*	
V, volts	100	200	300	400	500		$V = 150$	550
I, amperes	1	2	3	4	5	$\frac{V}{I} = 100$	$I = 1.5$ computed	5.5
V/I	100	100	100	100	100	or $I = \frac{V}{100}$	$I = 1.5$ measured	5.5

Up to this point, it has been established that for the data presumed,

$$I = \frac{V}{100}$$

is valid for a circuit where $R = 100\ \Omega$. How must this equation be modified if $R = 200\ \Omega$, $300\ \Omega$, $400\ \Omega$, etc.?

Substituting $R = 200\ \Omega$, then $R = 300\ \Omega$, etc., in the experimental circuit of Fig. 7-1, and following the method just outlined, new formulas are established, which show that:

for $R = 200\ \Omega$, $$I = \frac{V}{200} \qquad (7\text{-}2)$$

for $R = 300\ \Omega$, $$I = \frac{V}{300} \qquad (7\text{-}3)$$

for $R = 400\ \Omega$, $$I = \frac{V}{400} \qquad (7\text{-}4)$$

etc.

It now is intuitively obvious that formulas (7-1), (7-2), (7-3), etc., may be generalized as follows:

$$I = \frac{V}{R} \qquad (7\text{-}5)$$

Of course this generalization, Eq. (7-5), must still be verified for values of V and R not used in the experiment. Once the verification is complete, Eq. (7-5) can be stated as a law which gives the mathematical relationships among I, V, and R in a closed circuit.

NOTE: In the circuit of Fig. 7-1 it was assumed that the ammeter had no resistance, and that R was the total resistance in the circuit. In fact, the ammeter does have some resistance, but it is so low (in this case a fraction of 1 Ω) that it can be ignored here and the resistance of R *can* be assumed to be the total resistance in the circuit.

Ohm's Law

Now, refer again to Fig. 7-1 and to Eq. (7-5). Assume that the voltage drop measured across R is V, and that the circuit current is I. In Experiment 6 it was determined that in a series circuit the current is the same everywhere. Hence the current in R is also I. Equation (7-5) then states that the direct current I in a resistor is directly proportional to the voltage V across the resistor and inversely proportional to the resistance R, where I is measured in amperes, V in volts, and R in ohms.

Equation (7-5) is *Ohm's law*, first formulated by the scientist Georg Simon Ohm. This law is the foundation on which the science of electricity and electronics is based.

Measurement Errors

In the preceding discussion it was assumed that all the measurements made experimentally were 100 percent accurate. In practice, this is never so. Errors do occur and for several reasons.

One possible error results from reading a meter scale incorrectly. This is trivial and can be corrected by exercising greater care, and by taking the average of a number of the same measurement. Interpolating between calibrated markers may be another source of error. A digital voltmeter eliminates these particular errors.

Parallax is another source of error which can be easily corrected. It occurs when a meter reading is taken from an off-center position; that is, when the line of sight between the viewer and the meter pointer is not perpendicular to the meter scale. Figure 7-2 illustrates the error of parallax. When the viewer is in position P_1, the line between the eye and the meter pointer A is perpendicular to the meter scale. This gives a correct reading of 5. However, if the viewer is in position P_2, the reading will be 7, an error due to parallax. To eliminate errors of parallax, a mirror strip is sometimes placed just below the meter scale. The correct reading position is the one in which the pointer cannot be distinguished from its reflected image in the mirror.

Meter reading errors can be eliminated by using meters with a numerical readout, that is, *digital* meters.

There are other errors which are not so obvious, and which cannot be corrected so easily. For example, there are inherent errors in the instruments used. The instrument manufacturer usually specifies the percentage of instrument error. The general-purpose EVM and other meters used in the school laboratory usually have between 2 and 5 percent error, except for digital multimeters, which are much more precise. For greater accuracy, laboratory precision instruments are required. These are highly precise instruments whose inherent error is held to a fraction of a percent.

Another source of error results from the process of inserting an instrument in a circuit to make a measurement. If the instrument alters circuit conditions in any way, incorrect readings may be obtained. Insertion errors will be discussed in greater detail in later experiments.

The fact that errors do occur is mentioned at this point because in this experiment the student will attempt to develop

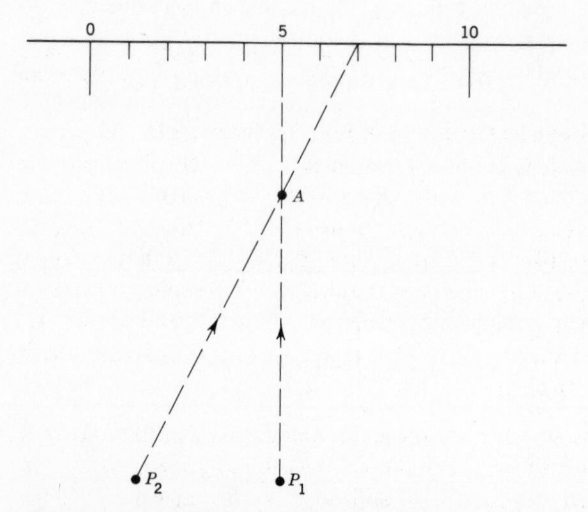

Fig. 7-2. Error of parallax occurs when the line of sight of the viewer and the meter pointer are not perpendicular to the meter scale.

the equation for Ohm's law from experimental data. You can expect that your data may contain some errors of measurement.

SUMMARY

1. The relationship between the voltage V applied to a closed circuit by some electromotive force such as a battery, the total resistance R, and the current I in that circuit is given by the formula $I = V/R$.
2. The relationship between the voltage drop V across a resistor R and the current I in that resistor is given by the formula $I = V/R$.
3. The equation $I = V/R$ is a mathematical statement of Ohm's law.
4. To verify Ohm's law experimentally, many measurements must be made and the results of the measurements must be substituted in Eq. (7-5) to ascertain that the facts actually fit the formula.
5. One set of data is obtained by measuring I while holding the measured value of V constant and varying the measured value of R. The data obtained should fit the formula $I = V/R$.
6. Another set of data is obtained by measuring I while holding the measured value of R constant and varying the measured value of V. The data obtained should also fit the formula $I = V/R$.
7. Measurement errors do occur, and these must be considered in attempting to establish the accuracy of a formula such as Ohm's law.
8. Among the measurement errors that may occur are: (*a*) incorrect reading of the meter scale, (*b*) incorrect meter readings due to parallax, (*c*) errors resulting from the accuracy of the instrument used, and (*d*) errors introduced by inserting the instrument in the circuit (insertion or loading errors).

SELF-TEST

Check your understanding by answering these questions:

1. The current in a fixed resistor is __________ proportional to the voltage across that resistor.
2. If the voltage across a resistor is held constant, the current in that resistor is __________ __________ to its resistance.
3. The formula which gives the mathematical relationship between I, V, and R in a closed circuit is $I =$ __________.
4. The formula in question 3 is called __________ __________.
5. If the voltage across a 10,000-Ω resistor is 125 V, the current in the resistor is __________ A.
6. If the voltage across a resistor is 60 V, and the current in the resistor is 0.05 A, the value of the resistor is __________ Ω.
7. What is the voltage across a 1500-Ω resistor, if the current in the resistor is 0.12 A? __________ V
8. In reading a meter, the line of sight between the viewer and the meter pointer should be __________ to the meter scale.
9. The error of parallax may be eliminated by placing a __________ __________ just below the meter scale. The correct reading position occurs when the pointer and its reflected image in the mirror __________.

MATERIALS REQUIRED

- Power supply: Source of regulated variable dc voltage
- Equipment: EVM, 0–10-mA milliammeter
- Resistors: ½-W 10,000-Ω; 1000-Ω
- Miscellaneous: 10,000-Ω 2-W potentiometer, connecting wires

PROCEDURE

NOTE: In the experimental procedure which follows, use Fig. 7-3*a* if a 0- to 400-V dc voltage supply is available. However, if a 0- to 40-V supply is employed, use Fig. 7-3*b*.

1. Connect the circuit of Fig. 7-3. **Power off.** Switches S_1 and S_2 are **open**, as shown. Zero the ohmmeter and connect it across R, a 10,000-Ω potentiometer connected as a rheostat (points C to D). Adjust R so that the resistance between points C and D is 2000 Ω. *Remove* the ohmmeter from the circuit. Set the voltage control on the variable dc voltage source for 0 V, maximum counterclockwise (CCW) position. Set the voltmeter on the 10-V range.

CAUTION: Be certain that the voltmeter and milliammeter are connected with the proper polarity as shown. Have your circuit checked and approved by an instructor before proceeding.

2. *After* the circuit is approved by an instructor, turn power **on**. Close switches S_1 and S_2. Slowly adjust the voltage control until 6 V is measured on the voltmeter across R (that is, V_{AB}). Measure the current I in amperes and record in Table 7-2. (Convert milliamperes to amperes before entering I in Table 7-2.)
3. Measure and record in turn the current I (in amperes) at each of the voltage settings shown in Table 7-2, for $R =$ 2000 Ω. Reset the voltmeter to the 50-V range for voltages greater than 10 V.
4. Compute the ratio V/I for corresponding values of V and I and record in Table 7-2. (For example, if at 6 V the measured current I is 0.003 A, then $V/I = 6/0.003 =$ 2000. Therefore, record 2000 in the column under 6 V, in the row "V/I.")
5. If your measurements have been correct, you should now be able to deduce a formula relating V and I when $R =$

TABLE 7-2. Measurements to Verify Ohm's Law

R	*2000 Ω*					*Formula Relating V, I, and R*	*Formula Test*		
V, volts	6	8	10	12	14	when $R = 2000$	*V*	9	18
I, amperes						$V/I =$	*I* measured		
V/*I*						$I = \frac{V}{\ }$	*I* computed		
R	*4000 Ω*					*Formula Relating V, I, and R*	*Formula Test*		
V, volts	8	12	16	20	24	when $R = 4000$	*V*	10	30
I, amperes						$V/I =$	*I* measured		
V/*I*						$I = \frac{V}{\ }$	*I* computed		
R	*6000 Ω*					*Formula Relating V, I, and R*	*Formula Test*		
V, volts	12	18	24	30	36	when $R = 6000$	*V*	15	36
I, amperes						$V/I =$	*I* measured		
V/*I*						$I = \frac{V}{\ }$	*I* computed		
R	*8000 Ω*					*Formula Relating V, I, and R*	*Formula Test*		
V, volts	16	20	24	28	32	when $R = 8000$	*V*	22	40
I, amperes						$V/I =$	*I* measured		
V/*I*						$I = \frac{V}{\ }$	*I* computed		

2000. Record this formula in Table 7-2, in the form $V/I = ?$ Also record this formula in the form $I = V/?$

6. Using the formula derived in step 5, calculate the current you would expect in the circuit when $R = 2000\ \Omega$ and $V = 9$ V; $V = 18$ V. Record the predicted (computed) values in Table 7-2 under "Formula test, *I* computed." Show sample calculation.

   Now measure the current in the circuit when $V = 9$ V and $V = 18$ V and record in Table 7-2.

QUESTION: If your formula is correct, how should the computed and measured values compare?

7. **Power off.** Switches S_1 and S_2 **open**. Turn the voltage control on the variable dc source to zero volts (completely counterclockwise). Place the ohmmeter across *R* and readjust *R* so that the resistance between points *C* and *D* is 4000 Ω. Remove the ohmmeter.
8. Follow the experimental procedure set forth above for each value of voltage shown in Table 7-2 for $R = 4000$ Ω. It is not necessary to show your other computations.
9. Follow the experimental procedure above for each value of voltage shown in Table 7-2 at $R = 6000\ \Omega$, $R = 8000$ Ω. **Power off.**
10. Draw a graph of *I* (measured) versus *V* applied at $R = 2000\ \Omega$; $R = 4000\ \Omega$; $R = 6000\ \Omega$; $R = 8000\ \Omega$. Use the same axes, with *I* plotted along the vertical and *V* along the horizontal axis. Identify the value of *R* on each graph.

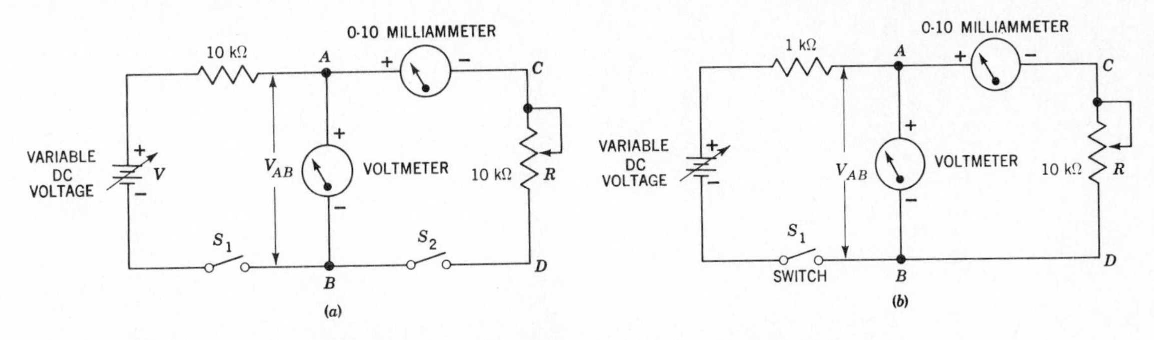

Fig. 7-3. Experimental circuits for verifying Ohm's law: (*a*) when dc voltage source is variable from 0 to 400 V; (*b*) when dc voltage source is variable from 0 to 40 V.

QUESTIONS

1. From the data in Table 7-2, what general conclusion can you draw about the relationship between I and V, when R is any value?
   (*a*) State the conclusion in words.
   (*b*) Write the conclusion as a formula.
2. (*a*) Is it necessary to confirm further the general formula set forth in answer to question 1*b*?
   (*b*) How would you confirm the general formula in question 1*b*?
3. What is the similarity, if any, between the graphs drawn in procedural step 10?
4. (*a*) From the graphs drawn in step 10, what is the current I in the circuit when $V = 9$ V and $R = 2000\ \Omega$?
   (*b*) How does this compare with the measured value?
5. (*a*) From the graphs in step 10, can you find the value of I when $V = 15$ V and $R = 7500\ \Omega$?
   (*b*) If not, how can you predict the value of I for the conditions given?
6. For a constant value of R, what is the effect on I of:
   (*a*) Doubling V?
   (*b*) Tripling V?
   (*c*) Halving V?
7. For a constant value of V, what is the effect on I of:
   (*a*) Doubling R?
   (*b*) Halving R?
   (*c*) Tripling R?
8. Referring to your data, discuss the errors, if any, in your experimental results.
9. What is the accuracy of your
   (*a*) Ohmmeter?
   (*b*) Voltmeter?
   (*c*) Milliammeter?
10. Explain in your own words what is meant by the error of parallax in electrical measurements.

Extra Credit

11. If the ohms function of your EVM were inoperative (and you had no other ohmmeter), how could you determine the value of an unknown resistance, assuming you had the equipment listed under Materials Required? Explain in detail.
12. Plot the graph showing the relationship between I and R for a constant voltage V. What is the curve called?

Answers to Self-Test

1. directly
2. inversely proportional
3. $I = V/R$
4. Ohm's law
5. 0.0125
6. 1200
7. 180
8. perpendicular
9. mirror strip; coincide

8

EXPERIMENT

THE SERIES CIRCUIT

OBJECTIVES

1. To determine experimentally what the total resistance R_T is, in a circuit in which the resistors R_1, R_2, R_3, etc., are connected in series
2. From the experimental data, to formulate a mathematical rule which gives the total resistance R_T of series-connected resistors

INTRODUCTORY INFORMATION

Determining R_T for Series-Connected Resistors

In electronic circuits there may be one or more resistors connected in series, in parallel, in a series-parallel arrangement, or in more complex combinations. The technician must be able to analyze such circuits in order to be able to determine and predict the effect of a resistor or combination of resistors in controlling current. Therefore, the technician must understand the laws which govern these circuits.

With the knowledge that the student has already gained it is possible to determine experimentally the quantitative effect of the series-connected resistors R_1, R_2, R_3, etc., on current.

Figure 8-1 shows a voltage source V applied across a resistor R_1. You have already determined that if the values of V and R_1 are given, you can predict the current in this circuit by substituting the values of V and R_1 in Ohm's law:

$$I = \frac{V}{R} \tag{8-1}$$

Now, knowing the values of V, R_1, R_2, R_3, etc., is it possible to predict the current I in the circuit containing: Two resistors R_1 and R_2 connected in series (Fig. 8-2*a*)? Three resistors R_1, R_2, and R_3 connected in series (Fig. 8-2*b*)? Any number of resistors connected in series?

What are the "facts" on which an educated guess may be made? In Experiment 6 it was established that the current I is everywhere the same in a series circuit. This, on reflection, now appears self-evident—for there is one and only one path for current in a series circuit. Moreover, in Experiment 6, measurement showed that the current I in a circuit decreased as more resistors were added in series, if the applied voltage was kept constant. It would appear, then, that the effect of adding resistors in series is to "increase" the opposition to current in a circuit. Again, this conclusion now seems evident, for since there is only one path for current, each resistor must exercise its effect in controlling current. It is the cumulative effect, the equivalent or total resistance R_T of the series combination, that is controlling current.

R_T, for a specific circuit, may be determined by measuring the current I_T in that circuit and solving for R_T by substituting the measured values I_T and V (applied) in Ohm's law. Thus,

$$R_T = \frac{V}{I_T} \tag{8-2}$$

Equation (8-2) suggests that a *single resistor* R_T may be substituted for two or more resistors connected in series, for the effect of R_T on controlling current is the same as the effect of the series-connected resistors.

An experimental procedure for determining if there is a law by which R_T may be determined readily suggests itself. Assume a group of resistors whose values are known or whose values may be determined by measurement with an ohmmeter. Connect two of these resistors in series. Apply a measured voltage V across the circuit and measure the current I_T. Compute the equivalent resistance R_T of this combination, using Ohm's law. Suppose, for the sake of discussion, that $R_T = R_1 + R_2$ in this case. Now, connect two other resistors in the circuit and experimentally determine their equivalent resistance. Repeat this procedure for two more resistors. Analyze the results. Is there a readily identifiable pattern? Can it be stated analytically by formula, as, for

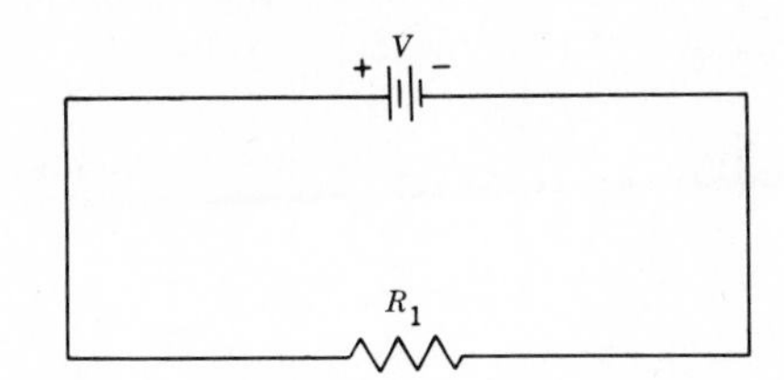

Fig. 8-1. Voltage V applied across resistor R_1.

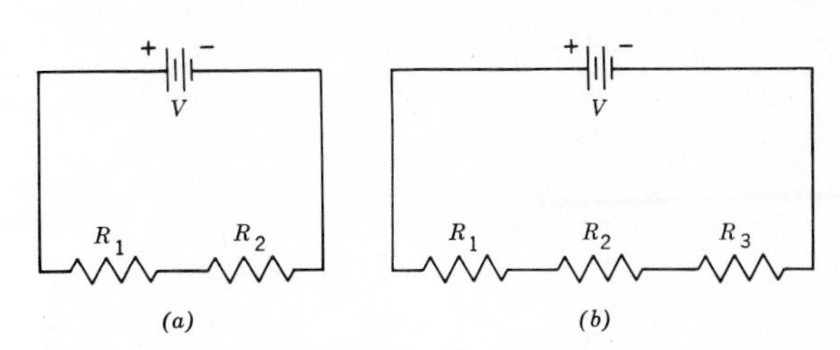

Fig. 8-2. Voltage V applied across (*a*) two resistors; (*b*) three resistors connected in series.

example, $R_T = R_A + R_B$? if it can, experimentally determine if the formula still holds for three resistors in series, four resistors in series, etc.

If the formula holds for any number of resistors, it may be accepted as a general law for resistors connected in series.

NOTE: The value of R_T may also be measured directly by disconnecting the voltage source and measuring across the series-connected resistors with an ohmmeter. This method may be used to confirm the results of the preceding procedure.

SUMMARY

1. The effect of two or more resistors connected in series in a closed circuit containing a voltage source V is to offer more resistance to current than can any one of these resistors acting alone in the circuit.
2. The effective or total resistance R_T of series-connected resistors R_1, R_2, etc., in a circuit containing a voltage source V may be determined experimentally by measuring the applied voltage V and the current I. Then R_T may be found by substituting the measured values of V and I in Ohm's law. Thus

$$R_T = \frac{V}{I}$$

This is method 1.
3. Two or more series-connected resistors may be replaced by a single resistor R_T determined as in method 1, since the effect of R_T in controlling current is the same as the effect of the series-connected resistors.
4. The value R_T of series-connected resistors may also be measured directly with an ohmmeter. However, in using an ohmmeter to measure total resistance in a series circuit, the power source must be disconnected from the resistors. This is method 2.
5. A general formula for determining the value R_T of series-connected resistors may be deduced from experimental data gathered using either method 1 or method 2. But measurements must be made involving many different combinations of resistors. In all cases the resistance of individual resistors must be measured, together with the total resistance of each series combination. The measured values must then be compared with the values predicted by the general formula to see if they are equal.

SELF-TEST

Check your understanding by answering these questions:

1. The current in a circuit containing four series-connected resistors measures 0.025 A. If the voltage applied to the circuit is 10 V, the total resistance R_T of the four series-connected resistors is: R_T = ____________ Ω.
2. One hundred resistors were measured with an ohmmeter, and each was found to be 100 Ω. The total resistance R_T of any two of these resistors, connected in series, measured 200 Ω. R_T of any three of these resistors connected in series measured 300 Ω; R_T of any four of these resistors connected in series measured 400 Ω. What formula is suggested by these measurements, for any number n of these resistors connected in series?
R_T = ____________ Ω.
3. The total resistance R_T of series-connected resistors is ____________ than the resistance of any one of these resistors taken by itself.
4. In a series-connected closed circuit there is just ____________ path for current.
5. Current in a series circuit is the ____________ everywhere.

MATERIALS REQUIRED

- Power supply: Regulated variable direct current
- Equipment: EVM, 0–10-mA milliammeter, ohmmeter (or ohms scale of an EVM)
- Resistors: ½-W 330-, 470-, 1200-, 2200-, 3300-, and 4700-Ω
- Miscellaneous: SPST switch

PROCEDURE

Determining R_T for Series-Connected Resistors

Ohmmeter Method 1

1. Measure the resistance of each of the resistors supplied and record its value in Table 8-1.

2. Connect the series resistors of Fig. 8-3 in the combinations shown in Table 8-2 and measure and record the total resistance R_T (from A to B) of each combination.

3. From Table 8-1, compute the sum of the measured values of R_1, R_2, R_3, etc., of each combination and record in Table 8-2 in the column labeled "Computed value $R_1 + R_2 + \cdots$."

4. In the last column of Table 8-2, write a formula for R_T which fits the measurements for each combination. For example, if measured values of R_1 = 300 Ω, R_2 = 450 Ω, and R_T = 750 Ω, then

$$R_T = R_1 + R_2$$

5. Write a general formula for R_T in Table 8-2, for any number of series-connected resistors.

Determining R_T for Series-Connected Resistors

Milliammeter-Voltmeter Method 2

6. Connect the circuit of Fig. 8-4. R_1 and R_2 are the resistors in combination 1, Table 8-2. S_1 is open. Adjust

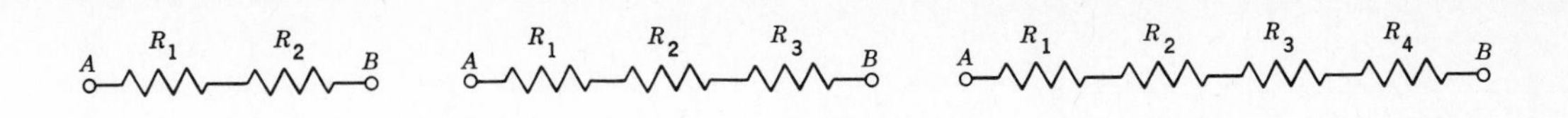

Fig. 8-3. Series-connected resistor combinations.

TABLE 8-1. Measured Values of Experimental Resistors

Rated value, Ω	330	470	1200	2200	3300	4700
Measured value, Ω						

TABLE 8-2. Determining Total Resistance of Series-Connected Resistors

	Rated Value, Ω					R_T, Ω	*Computed Value*	*Formula*
Combination	R_1	R_2	R_3	R_4	R_5	*Measured Value*	$R_1 + R_2 + \cdots$	*for* R_T
1	330	1200	X	X	X			
2	1200	2200	X	X	X			
3	3300	4700	X	X	X			
4	330	470	1200	X	X			
5	1200	2200	3300	X	X			
6	330	470	1200	2200	X			
7	330	470	1200	2200	3300			

General formula: R_T =

TABLE 8-3. Total Resistance of Series-Connected Resistors—Method 2

Combination	*V Applied,* V	*I Measured,* A	$R_T = \frac{V}{I}$	*Computed Value* $R_1 + R_2 + \cdots$ (Ω)	*Formula for* R_T
1					
2					
3					
4					
5					
6					
7					

General formula: R_T =

TABLE 8-4. Design Values for Series-Connected Resistors

	Measured Values					
Combination	R_1	R_2	R_3	R_4	*V Applied, Design Value*	*I Measured*
1						
2						
3						

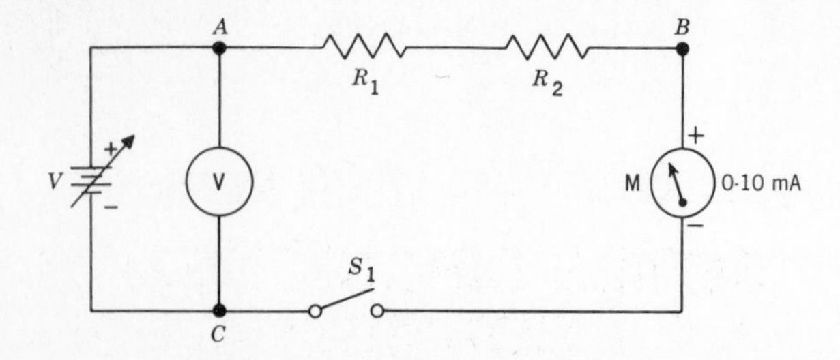

Fig. 8-4. Experimental circuit for determining R_T.

the power supply for $V = 10$ V as measured with an EVM set on the proper voltage range.

7. Close switch S_1, applying power to the circuit. Check the voltmeter. The voltage V must be maintained at 10 V for this load and for each subsequent load. Record this voltage in the column labeled "V applied," Table 8-3.

   Measure the current. Record this value in the column labeled "I measured" in Table 8-3.
8. Open the circuit (S_1 **off**). Between points A and B replace series combination 1 with series combinations 2, 3, 4, 5, 6, and 7 in turn. Measure and record in Table 8-3 the current through each combination, maintaining the measured V applied of 10 V.
9. From the measured values of I and V applied in Table 8-3, calculate R_T and record in the column labeled $R_T = V/I$ for each combination of series resistors.
10. Record from Table 8-2, in the column labeled "Computed value $R_1 + R_2 + \cdots$," the values of each combination, in the corresponding column in Table 8-3.
11. Write a formula for R_T (in the last column of Table 8-3) in terms of R_1, R_2, R_3, etc., which fits the measurements for each combination in Table 8-3.
12. Write a general formula R_T for any number of series-connected resistors, in terms of the resistors.

Design Problem (extra credit)

13. Theoretically design a series circuit with the resistors in this experiment, using first
    a. Two resistors, then
    b. Three resistors, then
    c. Four resistors

    so chosen that each combination *a* through *c* draws 5 mA, at an applied voltage V to be determined. Record in Table 8-4 the measured values of R_1, R_2, etc., selected in each combination. Show your computations. Record also in Table 8-4 the design (computed) value of V, for each combination.
14. Experimentally verify the design values for V and record your current measurements in Table 8-4.

QUESTIONS

1. Explain in your own words two methods for determining, by measurement, the total resistance R_T of resistors connected in series.
2. Why was it necessary to measure individually the resistance of each resistor used in the experiment?
3. Do the results of your measurements in Tables 8-1, 8-2, and 8-3 prove that it is possible to write a general formula in terms of R_1, R_2, etc., for total resistance of series-connected resistors R_1, R_2, etc.? Explain.
4. (*a*) Why must highly precise laboratory-standard instruments be used in attempting to establish or verify a scientific law?
   (*b*) What is the accuracy of your voltmeter? Ohmmeter?
5. (*a*) How do the ohmmeter-measured values of R_T compare with the computed values of R_T in Table 8-2?
   (*b*) How do the values of R_T derived from the voltampere method compare with the computed values in Table 8-3?
6. What effect, if any, would there be on the total resistance R_T of combination 4, Table 8-2, if we interchanged the positions of R_1, R_2, and R_3 in the circuit (Fig. 8-3)?
7. Write a formula R_T for series-connected resistors R_1, R_2, R_3 . . . , in terms of R_1, R_2, R_3, etc. Explain the formula in your own words.

Answers to Self-Test

1. 400
2. 100 *n*
3. greater
4. one
5. same

9

EXPERIMENT

SERIES-CIRCUIT DESIGN CONSIDERATIONS

OBJECTIVES

1. To design a series circuit which will meet specified resistance requirements
2. To design a series circuit which will meet specified voltage and current requirements
3. To design a series circuit which will meet specified current and resistance requirements
4. To construct the circuit and check, by measurement, the design specifications

INTRODUCTORY INFORMATION

Designing a Circuit to Meet Specified Resistance Requirements

The law for total resistance of series-connected resistors can be applied to the solution of simple design problems. An example will indicate the techniques to be used in circuit design at this level.

Problem 1. A technician has a stock of the following resistors: four 56 Ω, five 100 Ω, three 120 Ω, two 180 Ω, two 220 Ω, 330 Ω, 470 Ω, 560 Ω, 680 Ω, and 820 Ω. He needs a resistance value of 1000 Ω for a circuit he is designing. Find at least four combinations of resistors from those in stock, using the least possible number of components, which will satisfy the design requirement.

Solution. Experiment 8 developed the generalization that the total resistance R_T of series-connected resistances R_1, R_2, R_3, etc., is equal to the sum of these resistances. Stated as a formula,

$$R_T = R_1 + R_2 + R_3 + \cdots \tag{9-1}$$

The technician can use Eq. (9-1) to solve problem 1. Thus

$$1000 = R_1 + R_2 + R_3 + \cdots$$

1. By inspection it is apparent that the 820-Ω resistor and the 180-Ω resistor connected in series will add to 1000 Ω. Hence this is one solution. It is also the solution which requires the least number of components, just two.
2. Another solution is to connect the 680-Ω, 220-Ω, and 100-Ω resistors in series. Here three components are used.
3. Another solution is to connect a 560-Ω and two 220-Ω resistors in series. Again three components are used.
4. A fourth solution is to connect a 470-Ω, 330-Ω, and two 100-Ω resistors in series. Here four components are used.

There are other combinations which will add to 1000 Ω, so that the technician has a fairly wide choice. The restriction that he use the least number of components, however, does limit the choice.

Finally, the technician should connect the resistors and measure their total resistance with an ohmmeter, to confirm the solution.

Designing a Circuit to Meet Specified Voltage and Current Requirements

Ohm's law and the law for total resistance of series-connected resistors can be applied to the solution of this type of design requirement. Again a problem will illustrate the procedures to be employed.

Problem 2. A technician has a 15-V battery and the same stock of resistors as in problem 1. She must design a circuit using the least number of components in which current must be 0.01 A. Show the circuit arrangement she can use, including the values of all resistors.

Solution. Assume a closed series circuit is used. Two of the circuit conditions are known, namely, voltage and current. Using Ohm's law, the total resistance in the circuit can be found. Thus

$$I = \frac{V}{R_T}$$

and

$$R_T = \frac{V}{I} \tag{9-2}$$

Substituting in Eq. (9-2) the known values of V and I, we get

$$R_T = \frac{15}{0.01} = 1500\ \Omega$$

This is the circuit resistance which will hold circuit current at 0.01 A. It is now necessary to find a combination of resistances from those in stock which will add to 1500 Ω. Since the sum of 820 and 680 = 1500, the required resistors are R_1 = 820 Ω and R_2 = 680 Ω, connected as in Fig. 9-1.

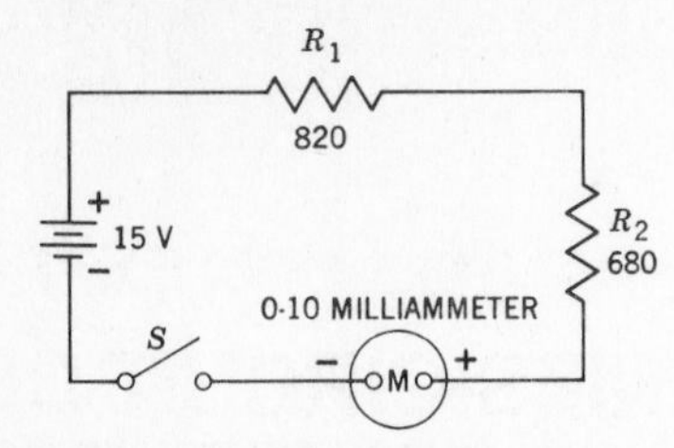

Fig. 9-1. Circuit used to limit current to 0.01 A.

As a final step the technician should connect the circuit of Fig. 9-1 and verify by measurement that there is indeed 0.01 A of current in the circuit.

The procedure to be followed, then, in this type of problem is to:

1. Solve for R_T by substituting the known values of V and I in the formula

$$R_T = \frac{V}{I}$$

2. Find the combination of resistances whose sum will add to the given value of R_T, using the equation

$$R_T = R_1 + R_2 + R_3 + \cdots$$

3. Connect the circuit using the combination of resistors determined in step 2 and verify by measurement that the circuit conditions have been met.

Designing a Circuit to Meet Specified Current and Resistance Requirements

As in the preceding design problem, Ohm's law and the law for R_T of series-connected resistors are applied here. Again, a problem will illustrate the process.

Problem 3. The series circuit in Fig. 9-2 is required to draw 0.05 A. To what voltage should the technician set the variable dc source to achieve this level of current?

Solution.

1. Find the total resistance R_T.

$$R_T = 330 + 470 + 560 = 1360\ \Omega$$

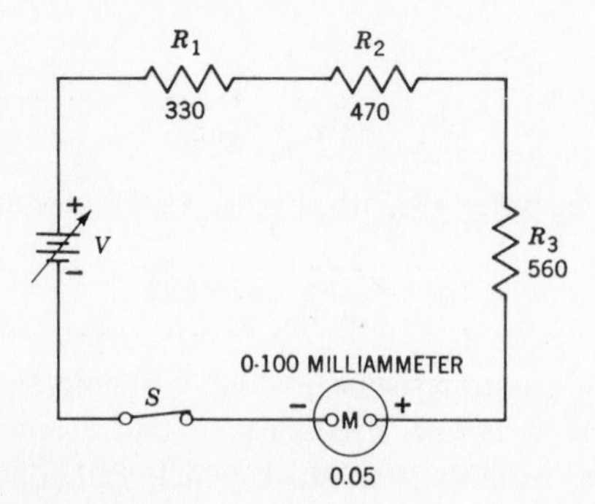

Fig. 9-2. Determining the voltage to which V must be set to permit 0.05 A of current in the circuit.

2. Substitute the values $I = 0.05$ and $R_T = 1360$ in Eq. (9-3), which was derived from Ohm's law:

$$V = I \times R \qquad (9\text{-}3)$$
$$V = (0.05)(1360) = 68.00\ \text{V}$$

3. Connect the circuit of Fig. 9-2 and set the power-supply voltage V to 68 V. The milliammeter should read 0.05 A (50 mA).

SUMMARY

1. If it is required to make up a resistance value R_T, not otherwise available, from series-connected resistors whose resistance values are R_1, R_2, R_3, etc., Ω, find the combination of resistors which will satisfy the formula

$$R_T = R_1 + R_2 + R_3 + \cdots$$

2. If it is required to design a series circuit which will meet specified voltage (V) and current (I) requirements, find the value R_T which will satisfy the given voltage and current, by substituting V and I in the equation

$$R_T = \frac{V}{I}$$

Then select the resistors R_1, R_2, etc., whose resistance sum is R_T.

$$R_T = R_1 + R_2 + R_3 + \cdots$$

3. If it is required to design a series circuit which will meet specified current (I) and resistance (R_T) values, select first those resistors whose resistance sum is R_T. Thus

$$R_T = R_1 + R_2 + R_3 + \cdots$$

Then solve for the unknown voltage V by substituting the given values of I and R_T in the equation

$$V = I \times R_T$$

4. After the circuit has been designed theoretically, connect it and measure the unknown quantity to see that it is in fact the required design value.

SELF-TEST

Check your understanding by answering these questions:

1. The formula which gives the total resistance of series-connected resistors is R_T = __________.
2. The formula which gives the relationship between the applied voltage V, the current I, and the resistance R of a closed circuit is V = __________.
3. To design a circuit powered by a battery of V volts which draws I amperes, it is necessary to find __________. This is done by substituting V and I in the formula __________.
4. To design a circuit which draws I amperes through a

resistance of R ohms, it is necessary to find __________. This is done by substituting I and R in the formula __________ .

5. The final step in the design of a circuit, after the parameters have been determined, is to __________ the circuit and __________ the quantities involved.

MATERIALS REQUIRED

- Power supply: Regulated variable direct current
- Equipment: EVM, 0–10-mA milliammeter; 0–100-mA milliammeter, ohmmeter (or ohms scale of an EVM)
- Resistors: ½-W 330-, 470-, 1200-, 2200-, 3300-, 4700-Ω
- Miscellaneous: SPST switch

PROCEDURE

1. From the resistors listed under Materials Required in this experiment, select three resistors whose total color-coded resistance value R_T, when they are connected in series, is 2000 ohms. List their color-coded resistance values in Table 9-1.
2. Connect these resistors in series and measure their combined resistance with an ohmmeter. Record this measured value in Table 9-1, in the column headed R_T Measured.
3. Repeat steps 1 and 2 for every value of R_T Required listed in Table 9-1. Use as many resistors as may be required.
4. Design a series circuit (like Fig. 9-1) with any combination of resistors in this experiment which will permit 0.02 A of current in the circuit when the dc supply is set at 40 V. Show your computations. List in Table 9-2 the design values of the resistors selected.
5. Connect the circuit in step 4. Set the dc supply at 40 V. Measure the circuit current and record in Table 9-2.
6. Repeat steps 4 and 5 for every value of voltage and current listed in Table 9-2.
7. Design a series circuit with the resistors in this experiment, using first (*a*) two resistors, then (*b*) three resistors, then (*c*) four resistors, so chosen that each combination *a* through *c* draws 4 mA at an applied voltage V to be determined analytically. Record in Table 9-3 the *measured* values of R_1, R_2, etc., selected for each combination. Show your computations. Also record in Table 9-3 the design value of V, determined analytically, for each combination.
8. Experimentally verify the design values for V and record your current measurements in Table 9-3.

TABLE 9-1. Measured versus Color-Coded Values of Series-Connected Resistors

R_T Required, Ω	Color-Coded Value of Resistors Whose Sum Will Satisfy R_T						R_T Measured, Ω
	R_1	R_2	R_3	R_4	R_5	R_6	
2,000							
5,300							
7,500							
10,000							
11,000							

TABLE 9-2. Circuit Design for Specified Values of V and I

V Applied, V	Circuit Current I, A		Resistor Design Values, Ω					
	Required	Measured	R_1	R_2	R_3	R_4	R_5	R_6
40	0.02							
30	0.01							
5.5	0.001							
16	0.02							
11.4	0.001							

TABLE 9-3. Circuit Designed to Draw 4 mA

Combination	Measured Values, Ω				V Applied, Design Value, V	I Measured, mA
	R_1	R_2	R_3	R_4		
1						
2						
3						

QUESTIONS

1. Refer to Table 9-1. Compare the R_T Required values with the R_T Measured values. If they are not the same, explain why.
2. Three ½-W resistors 1000 Ω, 5000 Ω, and 10,000 Ω with tolerance band color-coded gold are connected in series. What is the range of their total resistance, as measured with a meter whose error is 0%, that is, whose accuracy is 100 percent?
3. Refer to Table 9-2. Compare the required and measured values of I. Explain any difference between each set of values.
4. Consider the measured values of current I, in Table 9-2, for *each* value of applied voltage. Indicate if the R_T of the selected resistors is equal to, less than or greater than the design value, and explain why you think so.
5. Refer to Table 9-3. Compare the current *I measured* with the design requirement of 4 mA for every combination listed in the table. Explain any differences between the two current values for each combination.

Answers to Self-Test

1. $R_T = R_1 + R_2 + R_3 + \cdots$
2. $V = I \times R$
3. resistance; $R = V/I$
4. voltage V across R; $V = I \times R$
5. connect; measure

10

EXPERIMENT

VOLTAGE-DIVIDER CIRCUITS (UNLOADED)

OBJECTIVES

1. To develop, by analysis, a general law for computing the voltage across each resistor in an unloaded fixed resistive voltage divider
2. To confirm, experimentally, the law set forth in objective 1
3. To determine, by analysis, the voltage with respect to common, at each point in a variable resistive voltage divider
4. To confirm, experimentally, the results of objective 3

INTRODUCTORY INFORMATION

Series-Connected Voltage-Divider Circuits

Ohm's law finds immediate application in the analysis and design of voltage-divider circuits. Resistive voltage dividers can be very simple circuits, or complex arrangements of resistors serving one or more "loads." In this experiment we will be concerned with "unloaded" dividers, that is, with *circuits which are not required to deliver current to an external load.*

The simplest dc voltage divider consists of two resistors R_1 and R_2 connected in series, across which a dc voltage V is applied (Fig. 10-1). Assume that V is 100 V and that the resistors R_1 and R_2 are 75,000 Ω and 25,000 Ω, respectively. The voltages V_1 across R_1 and V_2 across R_2, measured with an electronic voltmeter, are 75 V and 25 V, respectively. The 100-V source has thus been divided by the circuit of Fig. 10-1 to produce two lower voltages.

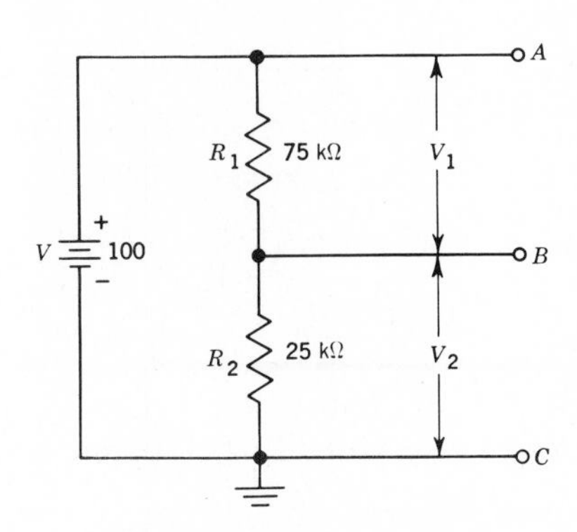

Fig. 10-1. DC voltage divider.

Figure 10-1 can be modified by the addition of one or more resistors to produce any number of lower voltages, measured across individual resistors, or measured with respect to some common point such as C. An arrangement of resistors to produce specific voltages can be achieved by the "cut-and-try" method or by analysis of the circuit. The cut-and-try method is too time-consuming and inefficient. Circuit analysis is a fast and effective means of finding the values of the resistors that will produce a required result.

In circuit analysis, the laws of electricity are applied to secure a "mathematical" solution of the problem. For example, in Fig. 10-1, the current I may be determined by substituting the values of V and R_T in Eq. (10-1).

$$I = \frac{V}{R_T} \tag{10-1}$$

where V is the applied voltage

and $$R_T = R_1 + R_2$$

Since $V = 100$ V, and $R_1 + R_2 = 100{,}000\ \Omega$,

$$I = \frac{100}{100{,}000} = 0.001 \text{ A}$$

Now $$V_1 = I \times R_1$$

and $$V_2 = I \times R_2 \tag{10-2}$$

Therefore $V_1 = (0.001)(75{,}000) = 75$ V

and $V_2 = (0.001)(25{,}000) = 25$ V

A general solution (formula) may be found to simplify our work. Consider Fig. 10-2. It is required to find V_1, V_2, V_3, and V_4. Assume that I is the current in this circuit. Then

$$V = I \times R_T \tag{10-3}$$

where $$R_T = R_1 + R_2 + R_3 + R_4$$

Since
$$\begin{aligned} V_1 &= I \times R_1 \\ V_2 &= I \times R_2 \\ V_3 &= I \times R_3 \\ V_4 &= I \times R_4 \end{aligned} \tag{10-4}$$

we can find the ratio of V_1, V_2, etc., to V. Thus

$$\frac{V_1}{V} = \frac{I \times R_1}{I \times R_T} = \frac{R_1}{R_T} \tag{10-5}$$

and $$V_1 = V \times \frac{R_1}{R_T} \tag{10-6}$$

Similarly $$V_2 = V \times \frac{R_2}{R_T}$$

$$V_3 = V \times \frac{R_3}{R_T} \tag{10-7}$$

$$V_4 = V \times \frac{R_4}{R_T}$$

We have thus derived a simple formula [Eqs. (10-6) and (10-7)] for finding the voltage across any resistor in a series circuit. Stated in words, it is: The voltage V_1 across the resistor R_1 is equal to the product of the applied voltage V and the ratio of the resistance of R_1 to the total resistance R_T of the series circuit. This formula applies to a series circuit containing any number of resistors.

As an example, let us apply the formula to Fig. 10-1.

$$V_1 = V \times \frac{R_1}{R_T} = 100 \times \frac{75{,}000}{100{,}000} = 75\ V$$

Similarly, $$V_2 = 100 \times \frac{25{,}000}{100{,}000} = 25\ \text{V}$$

Another example is given to illustate how the formula may be used to design a voltage divider. Using a 250-V source, we wish to find the values of resistors R_1 through R_4 connected in a simple series circuit to provide 25, 50, 75, and 100 V across R_1, R_2, R_3, and R_4, respectively. Assume that the current I in this circuit must be limited to 0.001 A.

Solution.

1. First, we find R_T:

$$R_T = \frac{V}{I} = \frac{250}{0.001} = 250{,}000\ \Omega$$

2. Next, we can rewrite Eqs. (10-6) and (10-7) as follows:

$$R_1 = \frac{V_1}{V} \times R_T$$

$$R_2 = \frac{V_2}{V} \times R_T$$

$$R_3 = \frac{V_3}{V} \times R_T \tag{10-8}$$

$$R_4 = \frac{V_4}{V} \times R_T$$

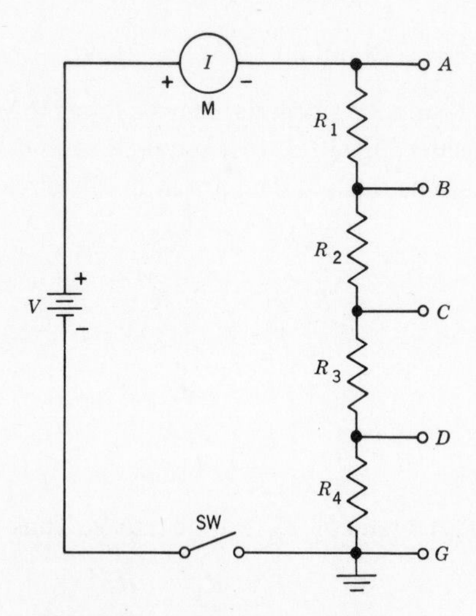

Fig. 10-2. Another dc voltage divider.

3. Substituting $V = 250$ V, $V_1 = 25$ V, $R_T = 250{,}000\ \Omega$, etc., in Eqs. (10-8) gives

$$R_1 = \frac{25}{250} \times 250{,}000 = 25{,}000\ \Omega$$

$$R_2 = \frac{50}{250} \times 250{,}000 = 50{,}000\ \Omega$$

$$R_3 = \frac{75}{250} \times 250{,}000 = 75{,}000\ \Omega$$

$$R_4 = \frac{100}{250} \times 250{,}000 = 100{,}000\ \Omega$$

These are the required values. The circuit of Fig. 10-2 may be connected using these values of R_1, R_2, etc., and V, and the required voltages may be verified by measurement with an EVM.

In analyzing the voltage-divider circuits of Figs. 10-1 and 10-2, we have considered the voltages developed across individual resistors of the divider. Another view of the divider is to consider the voltages at the junction points of the network, relative to a common point. In Fig. 10-1, point C is the common or ground return of the circuit. In Fig. 10-2, point G is ground. Now consider Fig. 10-2. What is the voltage at A relative to ground (G)? B to ground? C to ground? D to ground?

These voltages may be determined by a modification of Eqs. (10-6) and (10-7). First, the voltage from A to G is obviously the applied voltage V. Now, the voltage V_{BG} from B to ground is

$$V_{BG} = V \times \frac{R_2 + R_3 + R_4}{R_T}$$

and $$V_{CG} = V \times \frac{R_3 + R_4}{R_T} \tag{10-9}$$

and $$V_{DG} = V \times \frac{R_4}{R_T}$$

Another method of determining V_{BG} and V_{CG} in Fig. 10-2 is to solve for the voltages V_1, V_2, V_3, and V_4 by the formulas (10-6) and (10-7). Then

$$V_{BG} = V_2 + V_3 + V_4$$
$$V_{CG} = V_3 + V_4 \tag{10-10}$$
and $$V_{DG} = V_4$$

Variable Voltage-Divider Circuit (Unloaded)

Suppose it is required to set up a divider, as in Fig. 10-1, with a total resistance of 10,000 Ω, whose divider ratio is such that

$V_1 = 69$ V and $V_2 = 31$ V. Solution of this circuit, by the method just described, yields the results

$$R_1 = 6900\ \Omega$$
$$R_2 = 3100\ \Omega$$

It would normally be expensive to secure resistors of these exact values. To overcome this difficulty, we make use of a potentiometer. You will recall that a potentiometer is a three-terminal variable resistor. The resistance between the two outer terminals is fixed at the rated value of the potentiometer. The center terminal or arm is connected to a slider which makes contact with the resistive material of the potentiometer. The arm can be varied manually to select different values of total resistance. Thus, if R_1 and R_2 of Fig. 10-1 are replaced with a 10,000-Ω potentiometer, the corresponding circuit is that of Fig. 10-3.

As the arm B moves toward A, the resistance R_1 (AB) decreases, while the resistance R_2 (BC) increases. As the arm B moves toward C, R_1 increases and R_2 decreases. When B is at A, $R_1 = 0$ and $R_2 = 10{,}000\ \Omega$; when B is at C, $R_2 = 0$ and $R_1 = 10{,}000\ \Omega$.

Thus, by manually adjusting the position of the slider, we can set the ratio R_1/R_2 as we choose, and therefore we have a means for setting the voltage V_1 at any value between zero and the total voltage V across the potentiometer. In this process we have not changed the total resistance of the potentiometer (resistance from A to C).

In actual practice, if a potentiometer is used to develop a voltage which cannot be achieved by a combination of ordinary resistors, an EVM is connected across the arm and one of the end terminals. The arm of the potentiometer is then varied until the desired voltage is measured. This practical arrangement eliminates computation while achieving the desired voltages.

It is possible to limit the range of voltage variation by placing a potentiometer in series with one or more resistors. Thus, in Fig. 10-4 the voltage variation (range) from B to C is from 25 to 75 V.

It should be noted that these results are true only if no current is drawn from any tap of the divider.

Variable voltage dividers are used as volume controls in radios, contrast controls in television receivers, speed controls in electronic motor-control circuits, voltage regulator controls, etc.

SUMMARY

1. The voltage across each resistor in a series-connected resistive voltage divider may be determined by formula. Thus in Fig. 10-2, the voltage across any resistor in the divider, say V_1 across R_1, may be found by substituting R_1, R_T (the sum of R_1, R_2, R_3, etc.), and the applied battery voltage V in

$$V_1 = V \times \frac{R_1}{R_T}$$

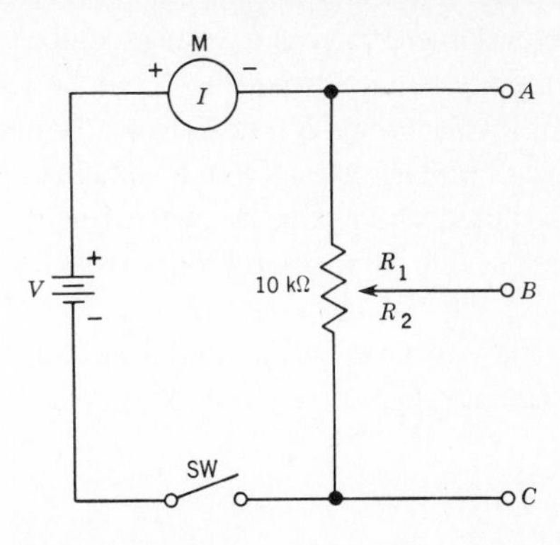

Fig. 10-3. Potentiometer as a variable voltage divider.

2. A somewhat longer method to determine the voltage across any resistor in a series-connected divider, say V_1 across R_1 in Fig. 10-2, is to find the total resistance. Thus here

$$R_T = R_1 + R_2 + R_3 + R_4$$

Next, solve for the current I in the circuit. Here

$$I = \frac{V}{R_T}$$

Now, knowing I, to find the voltage drop across R_1, simply multiply R_1 by I. Thus,

$$V_1 = I \times R_1$$

NOTE: Methods 1 and 2 are essentially the same, for if we substitute for I in the last formula, we get

$$V_1 = (I) \times R_1 = \left(\frac{V}{R_T}\right) \times R_1 = V \times \frac{R_1}{R_T}$$

3. If it is required to find the voltage to common, or to any reference, from any junction in a series-connected voltage divider, the same methods 1 or 2 may be used. For example, to find the voltage V_{CG} from C to G in Fig. 10-2, we actually find the voltage across the series-connected resistors R_3 and R_4. Applying the formula,

$$V_{CG} = V \times \frac{R_3 + R_4}{R_T}$$

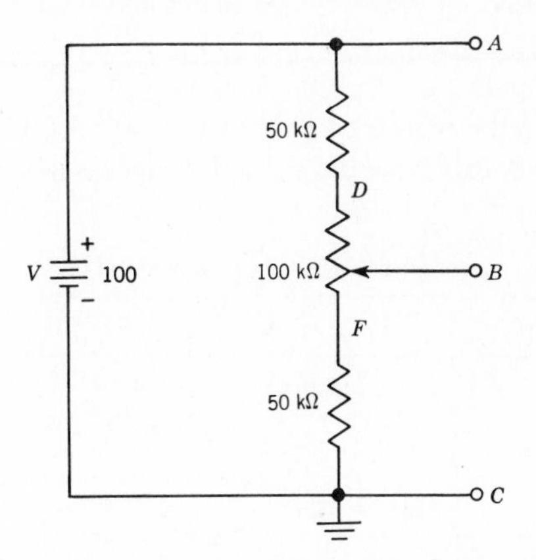

Fig. 10-4. Limiting the range of variation.

4. Variable voltage-divider circuits can be formed by utilizing a potentiometer across a voltage source, as in Fig. 10-3. Here the voltage from *B* to *C* can be varied all the way from *V* volts (when *B* is at *A*) to 0 V (when *B* is at *C*).
5. The range of voltage variation of a voltage divider can be limited to any desired value by connecting a potentiometer in series with voltage-dropping resistors, as in Fig. 10-4. Here the voltage range of V_{BC} is the difference between the voltage drop V_{DC} and the voltage drop V_{FC}, and the actual range of variation of V_{BC} is

$$V_{FC} \leqq V_{BC} \leqq V_{DC}$$

SELF-TEST

Check your understanding by answering these questions:

1. In Fig. 10-1, if the positions of resistors R_1 and R_2 were reversed in the circuit, the voltage across the 75-kΩ resistor would be ____________ V.
2. In Fig. 10-2, R_T = 150 kΩ and R_3 = 30 kΩ. If the applied voltage *V* is 225 V, the voltage across R_3 = ____________ V.
3. In Fig. 10-2, R_T = 150 kΩ, R_1 = 35 kΩ, and *V* = 300 V. The voltage V_{BG} = ____________ V.
4. In Fig. 10-2, R_T = 100,000 Ω, V_{BC} = 20 V, and *V* = 80 V. The value of R_2 = ____________ Ω.
5. In the variable voltage divider (Fig. 10-3), *V* = 35 V. The range of variation of V_{BC} is from ____________ V (maximum) to ____________ V (minimum).
6. In the variable voltage divider (Fig. 10-4), the battery voltage *V* = 60 V. The values of the resistors are the same as those shown. The range of V_{BC} is from ____________ V maximum to ____________ V minimum.
7. In Fig. 10-2, R_1 = 10,000 Ω, R_2 = 22,000 Ω, R_3 = 6800 Ω, R_4 = 2200 Ω, and *V* = 164 V. The voltage V_1 = ____________ V, V_2 = ____________ V, V_3 = ____________ V, and V_4 = ____________ V.
8. For the same conditions as in question 7, V_{CG} = ____________ V, V_{BG} = ____________ V, and V_{BD} = ____________ V.

MATERIALS REQUIRED

- Power supply: Regulated variable direct current
- Equipment: EVM, 0–10-mA milliammeter, ohmmeter
- Resistors: ½-W 2200-, 3300-, 4700-, and 5600-Ω
- Miscellaneous: 10,000-Ω 2-W potentiometer, SPST switch

PROCEDURE

1. Connect the circuit of Fig. 10-2, using the rated values R_1 = 2200 Ω, R_2 = 3300 Ω, R_3 = 4700 Ω, and R_4 = 5600 Ω. **Power off.** *M* is a 0–10-mA milliammeter.
2. Set the power supply so that *V* = 40 V and maintain it at this level. Close switch *S*, applying power to the circuit. Measure, and record in Table 10-1, the applied voltage *V*, the current *I*, and the voltages V_1, V_2, V_3, and V_4 across R_1, R_2, R_3, and R_4, respectively. Measure also and record the voltages V_{BG}, V_{CG}, and V_{DG}.
3. Compute, and record in Table 10-1, *I*, V_1, V_2, etc. (as in step 2). Use Eqs. (10-6), (10-7), and (10-10). Show your computations.
4. Adjust the level of the power supply so that the measured current *I* = 0.001 A. Measure and record the applied voltages *V*, V_1, V_2, V_3, V_4, V_{BG}, V_{CG}, and V_{DG}.
5. Compute and record the applied voltage *V* which would cause the measured current *I* (0.001 A). Compute and record also the voltages V_1, V_2, etc. (as in step 4), which would result from application of the voltage *V* to the circuit. Show all your computations.
6. **Power off.** Connect the circuit (Fig. 10-3). Set the power-supply level so that the applied voltage *V* = 20 V and maintain it at this level. *M* is a 0–10-mA milliammeter.
7. Close *S*. Measure, and record in Table 10-2, the voltage *V*, the current *I*, and the voltages from *B* to *C* (V_{BC}) and from *A* to *B* (V_{AB}) as the arm of the potentiometer is varied from (*a*) its maximum position (*B* at *A*), to (*b*) three-fourths of maximum, (*c*) center position (approximately), (*d*) one-fourth of maximum, and (*e*) its minimum position (*B* at *C*).

   Add and record the voltages V_{AB} and V_{BC} at each position of the arm.
8. With *V* still at 20 V, set the arm of the potentiometer so that the voltage, V_{BC} = 9 V. Measure and record in Table 10-3, *V*, *I*, V_{BC}, and V_{AB}.
9. Open switch *S*. Do not change the setting of the arm.

TABLE 10-1. Voltage-Divider Measurements

Step		*V*	*I*	V_1	V_2	V_3	V_4	V_{BG}	V_{CG}	V_{DG}
2	Measured	40								
3	Computed	—								
4	Measured		0.001							
5	Computed		—							

TABLE 10-2. Variable Voltage-Divider Measurements

	Position of Arm	Measured Values				Computed Values
		V	I	V_{BC}	V_{AB}	$V_{AB} + V_{BC}$
(a)	Max B at A					
(b)	¾ of max					
(c)	Center					
(d)	¼ of max					
(e)	Min B at C					

TABLE 10-3. Variable Voltage-Divider Values

Measured							Computed	
V	I	V_{BC}	V_{AB}	R_{BC}	R_{AB}	R_{AC}	R_{BC}	R_{AB}
20		9						

Measure and record the resistance from the arm to C (R_{BC}), from the arm to A (R_{AB}), and across the potentiometer (R_{AC}).

10. In Table 10-3, record the computed R_{BC} and R_{AB}, knowing that $V = 20$ V, the resistance of the potentiometer $= 10{,}000\ \Omega$, and the measured voltage $V_{BC} = 9$ V. Show your computations.

Design Problems (extra credit)

11. **a.** Design a circuit which will deliver a voltage which may be varied in the range 15 to 30 V (approximately). Select components from those used in this experiment. Draw the diagram showing the computed values of V and of the required resistors. Show your computations.
b. Connect the circuit. Record the minimum and maximum output voltages measured from the arm of the potentiometer to ground.
Min volts __________. Max volts __________.

12. **a.** Design a circuit which will deliver a voltage which may be varied in the range 7.2 to 27.2 V (approximately). Select components from those used in this experiment. Draw the diagram showing the computed values of V and of the required resistors. Show your computations.

HINT: It will be necessary to parallel the potentiometer with a resistor. To solve this problem it will be necessary to know the total resistance R_T of two resistors connected in parallel. This is given by the formula

$$R_T = \frac{R_1 \times R_2}{R_1 + R_2}$$

where R_1 and R_2 are the two parallel resistors.

b. Connect the circuit. Record the minimum and maximum output voltages measured from the arm of the potentiometer to ground.
Min volts __________. Max volts __________.

QUESTIONS

1. Refer to Table 10-1, steps 2 and 3. How do the measured values of V_1, V_2, V_3, and V_4 relate to the computed values? Explain any discrepancies.
2. Refer to Table 10-1, steps 4 and 5. How do the measured and computed values of V compare? Explain any differences.
3. Explain two methods by which the voltages V_1, V_2, etc., in the circuit (Fig. 10-2) may be computed.
4. Indicate, by reference to specific measurements and computations, if the results of your experiment confirm Eq. (10-10). Explain any discrepancies.
5. Refer to Table 10-2. What is true about V_{AB} and V_{BC}, regardless of the setting of the arm of the potentiometer?
6. *(a)* Refer to your measurements in Table 10-3. Find the ratios V_{BC}/V_{AB} and R_{BC}/R_{AB}.
(b) Are these ratios equal? Should they be? Why?
7. Refer to Table 10-3. What is the relationship between the measured values R_{AB}, R_{BC}, and R_{AC}?
8. How do the computed and measured values of resistance in Table 10-3 compare? Explain any discrepancies.
9. What is meant by an unloaded voltage divider?
10. Refer to Table 10-3. *Explain* the effect on I (measured) as the arm of the potentiometer is varied.

Answers to Self-Test

1. 75
2. 45
3. 230
4. 25 kΩ
5. 35 to 0
6. 45 to 15
7. 40; 88; 27.2; 8.8
8. 36; 124; 115.2

11

EXPERIMENT

CURRENT IN A PARALLEL CIRCUIT

OBJECTIVES

To determine experimentally that the total current I_T in a circuit containing resistors connected in parallel is:

1. Greater than the current in any branch
2. Equal to the sum of the currents in each of the parallel branches

INTRODUCTORY INFORMATION

Branch Currents

In considering the series circuit in Experiment 6, it was established that a closed circuit is required for current, that current ceases when the circuit is open, and that current in a series circuit is the same everywhere. What are the characteristics of a parallel circuit?

Figure 11-1 shows three resistors connected in parallel and a voltage V applied across them. If the line connecting the battery to the parallel network is broken at X or at Y, and an ammeter is inserted in the circuit at X or Y (Fig. 11-2), the ammeter will measure current. This "line" current is drawn by the three resistors from the battery.

A simple experiment suggests an important characteristic of a parallel circuit. If, in Fig. 11-2, resistor R_1 is removed from the circuit, the line current measured by the ammeter decreases. If R_2 is then removed from the circuit, the line current decreases further. What remains is a simple series circuit consisting of V, R_3, and the ammeter. The "line" current is now the current drawn by R_3 from V; it may be computed directly by Ohm's law.

The results of this experiment indicate that in Figs. 11-1 and 11-2 there are indeed three conductive paths for current, namely, R_1, R_2, and R_3. When all three paths are closed, there is maximum line current. When path R_1 is broken, there is less line current because only two paths remain, namely, R_2 and R_3, and so on. The individual paths are called *branches* or *legs* of the parallel circuit.

One characteristic of a parallel resistive circuit, then, is that the total current I_T in the circuit is greater than the current in any branch. The conclusion that follows is that each branch current in a parallel resistive circuit is less than the total or line current I_T.

A numerical example will illustrate this characteristic. Assume in Fig. 11-3 that the voltage source V is 6.6 V. Note that 6.6 V appears across each branch resistor in the circuit, that is, the voltage across R_1 is 6.6 V, that across R_2 is 6.6 V, and that across R_3 is 6.6 V also. The individual currents in each branch, I_1 in R_1, I_2 in R_2, and I_3 in R_3, may be calculated by Ohm's law. Thus $I_1 = 3.3$ mA, $I_2 = 2.2$ mA, and $I_3 = 0.66$ mA. The total current I_T must be greater than the largest branch current, so in this case I_T must be greater than 3.3 mA. It will be shown in the next section that I_T is in fact equal to the sum of the branch currents, that is,

$$I_T = 3.3 + 2.2 + 0.66 = 6.16 \text{ mA}$$

Total Current in a Parallel Circuit

It is apparent that the current I_1 in R_1 combines with the current I_2 in R_2 and I_3 in R_3 to form the "line" current, and that this is the total current I_T drawn from the battery.

We say that the current "combines." Does this mean that I_T is the sum of the individual currents I_1, I_2, and I_3 drawn respectively by R_1, R_2, and R_3? This question may be answered experimentally by measuring I_1, I_2, and I_3 and the line current I_T in Fig. 11-3 and determining the relationship

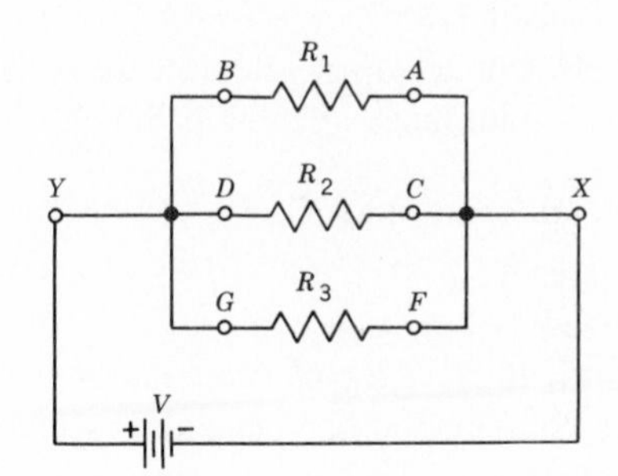

Fig. 11-1. Voltage applied across three resistors connected in parallel.

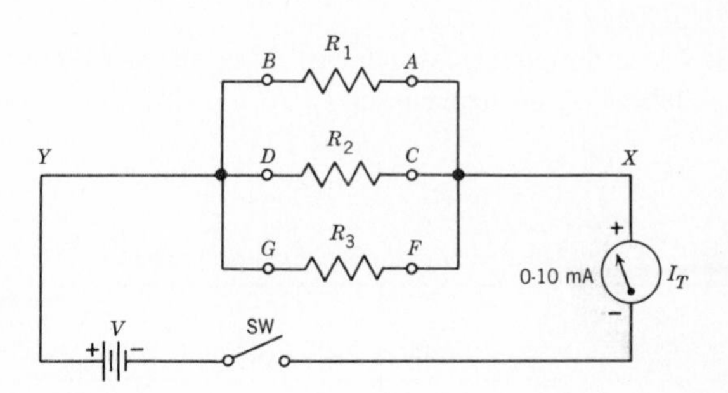

Fig. 11-2. Measuring total current in a circuit containing parallel resistors.

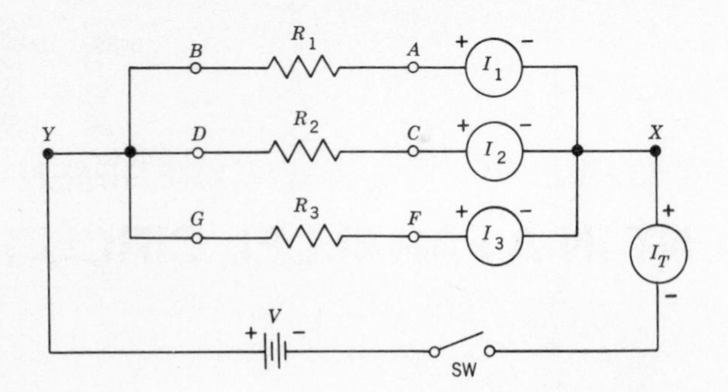

Fig. 11-3. Circuit to determine the relationship between branch currents and total current in a parallel circuit.

between them. I_T is the current measured by breaking the circuit at point X or Y and inserting an ammeter there. The branch current I_1 is measured by breaking the circuit at A or B, and inserting an ammeter there; I_2 is measured by breaking the circuit at C or D and inserting an ammeter there; and I_3 is measured by breaking the circuit at G or F and inserting an ammeter there.

If the experiment just suggested is properly performed, it should establish the validity of the formula

$$I_T = I_1 + I_2 + I_3 \cdots \tag{11-1}$$

which states that the sum of the branch currents I_1, I_2, etc., in a parallel circuit equals the total current I_T.

SUMMARY

1. In a parallel resistive circuit the individual current in each branch is less than the total line current.
2. The total line current is greater than each individual branch current.
3. The total line current is in fact equal to the sum of all the branch currents.
4. The voltage across each branch is the same.

SELF-TEST

Check your understanding by answering these questions:

1. In Fig. 11-3, $I_1 = 1$ A, $I_2 = 2$ A, $I_3 = 3$ A. The total current I_T must therefore be greater than __________ A.
2. In Fig. 11-3, for the branch currents listed in question 1, $I_T =$ __________ A.
3. In Fig. 11-3 the voltage V_1 across R_1, the voltage V_2 across R_2, and the voltage V_3 across R_3 must all be __________ __________.
4. The voltage across each branch of a parallel network must be __________.

MATERIALS REQUIRED

- Power supply: Regulated variable direct current
- Equipment: EVM, 0–10-mA milliammeter, ohmmeter (or ohms scale of an EVM)
- Resistors: ½-W 2200-, 3300-, 4700-, 5600-, and 10,000-Ω
- Miscellaneous: SPST switch

PROCEDURE

1. Measure the resistance of each of the resistors supplied and record its value in Table 11-1.
2. Connect the circuit of Fig. 11-3, using combination 1 of resistors shown in Table 11-2. Set the output of the voltage source V at 10 V and maintain it at this level under load until you have completed the measurements in Table 11-2.
3. Close the circuit (switch **on**). Measure and record in Table 11-2 the total current I_T. Also measure and record the current I_1 in R_1, I_2 in R_2, and I_3 in R_3. Compute I_T by adding I_1, I_2, and I_3 and record.
4. Remove R_1 from the circuit. Measure and record I_T, I_2, and I_3. Compute and record I_T as in step 3.
5. Remove R_2 from the circuit and also leave R_1 *out* as in step 4. Measure and record I_T and I_3. Compute and record I_T as in step 3.
6. **Power off.** (Switch open.) Replace the resistors in combination 1 with those in combination 2. Voltage is still 10 V.
7. Repeat step 3.
8. Repeat step 4.
9. Repeat step 5.
10. **Power off.** Replace the resistors in combination 2 with those in combination 3. Voltage V is still 10 V.
11. Repeat step 3.
12. Repeat step 4.
13. Repeat step 5. **Power off.**

TABLE 11-1. Measured Values, Experimental Resistors

Rated value, Ω	2200	3300	4700	5600	10,000
Measured value, Ω					

TABLE 11-2. Measured and Computed Values in Parallel Circuit

Combination	Step	Rated Values			Measured Values					Computed Values
		Ohms			Volts	Amperes				Amperes
		R_1	R_2	R_3	V	I_T	I_1	I_2	I_3	$I_T = I_1 + I_2 + I_3$
1	3	2200	3300	10,000						
	4		3300	10,000			—			
	5			10,000			—	—		
2	7	3300	4700	5,600						
	8		4700	5,600			—			
	9			5,600			—	—		
3	11	4700	5600	10,000						
	12		5600	10,000			—			
	13			10,000			—	—		

QUESTIONS

1. How do the individual branch currents of parallel-connected resistors compare with the total current drawn by the circuit (Fig. 11-3)?
2. Support your answer to question 1 by referring specifically to the measurements you recorded in Table 11-2.
3. What is the effect on total current of parallel-connected resistors of:
   (*a*) Increasing the number of resistors in parallel?
   (*b*) Decreasing the number of resistors in parallel?
4. Support your answers to question 3 by referring specifically to your measurements in Table 11-2.
5. Do the results of your measurements in Tables 11-1 and 11-2 prove that it is possible to write a general formula relating current in the individual branches of parallel-connected resistors with total current? If yes, write a general formula for I_T and explain it in your own words.
6. Support your answer to question 5 by comparing your computed and measured values. Comment on any discrepancies.

Answers to Self-Test

1. 3
2. 6
3. equal to *V*
4. equal

12

EXPERIMENT

TOTAL RESISTANCE OF A PARALLEL CIRCUIT

OBJECTIVE

To verify experimentally that the total resistance R_T of resistors connected in parallel is given by the formula

$$\frac{1}{R_T} = \frac{1}{R_1} + \frac{1}{R_2} + \frac{1}{R_3} + \cdots$$

INTRODUCTORY INFORMATION

Total Resistance in a Parallel Circuit

The resistance R_T which battery V "sees" in Fig. 12-1 limits the current in the circuit to the value I_T. A single resistor with value R_T could be used to replace R_1, R_2, and R_3, for R_T would also limit the current to I_T for the same value of applied voltage V. R_T, then, is the total or equivalent resistance of the parallel network.

The value of R_T for any parallel network (Fig. 12-2) may be measured directly by placing an ohmmeter across the network at points X and Y.

NOTE: If the resistors are connected in a network together with a voltage source V (Fig. 12-1), the circuit must first be broken at X or Y to remove the voltage before measuring R_T.

R_T may also be calculated by measuring I_T, as in Fig. 12-3, measuring V, the applied voltage, and substituting I_T and V in Eq. (12-1).

$$R_T = \frac{V}{I_T} \tag{12-1}$$

One characteristic of a parallel circuit may easily be deduced from our measurements. Since the total current I_T is greater than the currents in any of the branch circuits, the total resistance R_T must be smaller than the smallest branch resistance, for resistance varies inversely as the current according to Ohm's law.

It seems logical that there must be a more definite relationship betwen R_1, R_2, etc., and R_T in a parallel circuit. This relationship, however, is not immediately apparent from the measurement process. And so, reasoning from the facts which we have determined by measurement, we must call on analysis to find that relationship. Specifically, we wish to find the formula, if any, from which R_T may be calculated if the values of R_1, R_2, . . . are known.

Assume that the resistance of the wires connecting V to R_1, R_2, and R_3 in Fig. 12-1 is negligible, that is, zero. This is a valid assumption where short lengths of interconnecting wire are used. Therefore, it is apparent that electrically points B, D, and G are all the same as point Y, and that points A, C, and F are all the same as point X. Hence, if we wish to measure the voltage across R_1 or R_2 or R_3, we place our voltmeter across XY in all three cases. It follows, therefore, that the voltage across R_1, R_2, and R_3 is the same. In the case of Fig. 12-1, it is the applied voltage V.

This line of reasoning can be extended to other parallel circuits. It suggests this fundamental principle which can be readily verified by measurement:

The voltage across each of the branches of a parallel circuit is the same.

Refer to Figs. 12-1 and 12-3. We can now solve for the branch currents analytically. For

$$I_1 = \frac{V}{R_1} \qquad I_2 = \frac{V}{R_2} \qquad I_3 = \frac{V}{R_3} \tag{12-2}$$

Equation (11-1) states that

$$I_T = I_1 + I_2 + I_3 + \cdots$$

Now, substituting the values $I = V/R$, etc., from Eqs. (12-2), we get

$$I_T = \frac{V}{R_1} + \frac{V}{R_2} + \frac{V}{R_3} + \cdots$$

$$= V\left(\frac{1}{R_1} + \frac{1}{R_2} + \frac{1}{R_3} + \cdots\right) \tag{12-3}$$

Equation (12-1) states that

$$R_T = \frac{V}{I_T}$$

and from Eq. (12-1) it follows that

$$I_T = \frac{V}{R_T}$$

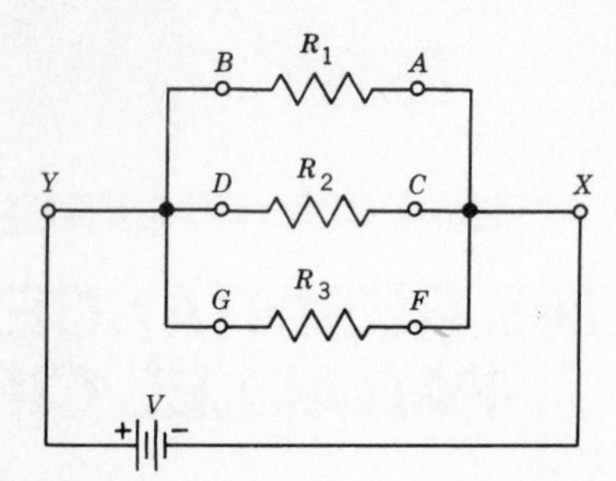

Fig. 12-1. Voltage applied across three resistors connected in parallel.

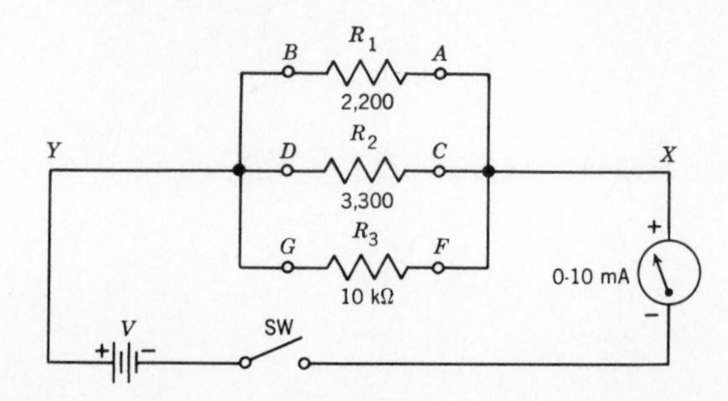

Fig. 12-3. Measuring total current in a circuit containing parallel resistors.

Substituting V/R_T for I_T in Eq. (12-3), we find that

$$\frac{V}{R_T} = V\left(\frac{1}{R_1} + \frac{1}{R_2} + \frac{1}{R_3} + \cdots\right) \tag{12-4}$$

Dividing by V to simplify Eq. (12-4), we get

$$\frac{1}{R_T} = \frac{1}{R_1} + \frac{1}{R_2} + \frac{1}{R_3} + \cdots \tag{12-5}$$

Taking the inverse of both sides of Eq. (12-5), we have

$$R_T = \frac{1}{\dfrac{1}{R_1} + \dfrac{1}{R_2} + \dfrac{1}{R_3} + \cdots} \tag{12-6}$$

Equations (12-5) and (12-6) express the analytical relationship between the total or equivalent resistance R_T, and the individual branch resistances R_1, R_2, R_3, etc., connected in parallel. Though this relationship was derived analytically, our analysis was based on experimentally observed facts. Are the results valid, and can they be verified by measurement?

A method for verifying Eq. (12-5) or (12-6) is to measure the total resistance R_T of various combinations of resistances connected in parallel by either of the two methods which have been discussed. We then substitute the measured values of the individual resistances in Eq. (12-5) or (12-6) and calculate the value of R_T. If the measured and calculated values of R_T agree, we have experimental evidence confirming the validity of the derived formulas.

Measuring Individual Resistances in a Parallel Circuit

Suppose we need to measure the resistance R_1 in the parallel network of Fig. 12-2. How may this be achieved? Obviously it *cannot* be done by placing an ohmmeter across R_1 in the network, because this procedure would give the measurement of R_T, not of R_1. We can measure R_1 by disconnecting it from the parallel network, and measuring it outside of the circuit. Or we can disconnect one lead of R_1, say at point A, thus removing the effect of the network. We can then measure R_1 by placing an ohmmeter across it.

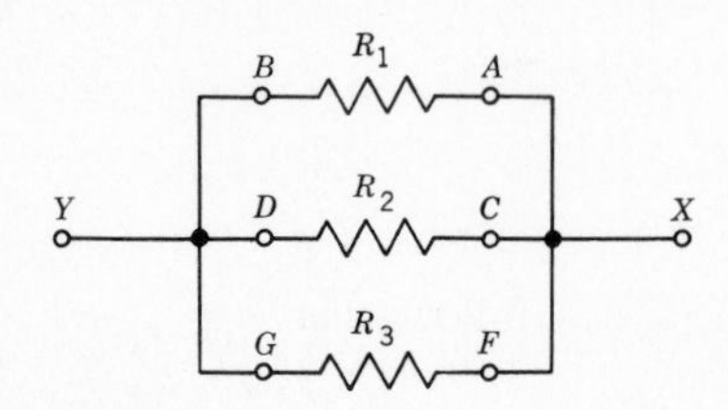

Fig. 12-2. Three parallel resistors.

SUMMARY

1. The voltage V across each branch (i.e., each resistor in Figs. 12-1 and 12-3), of a parallel circuit is the same.
2. The total or equivalent resistance R_T of two or more resistors connected in parallel, as in Fig. 12-3, may be determined experimentally by measuring the total current I_T, measuring the voltage V across the parallel network, and substituting the measured values in the formula

$$R_T = \frac{V}{I_T}$$

3. Another method of determining experimentally the total resistance R_T of two or more parallel-connected resistors, as in Fig. 12-2, is to place an ohmmeter across the parallel circuit. The meter measures R_T.
4. Resistance should never be measured when there is power applied to the circuit. So if the parallel resistance of R_1, R_2, and R_3 in Fig. 12-3 is required, power must first be removed (by opening the switch S in Fig. 12-3).
5. A formula which expresses the relationship between R_T and R_1, R_2, R_3, etc., of parallel-connected resistors is:

$$R_T = \frac{1}{\dfrac{1}{R_1} + \dfrac{1}{R_2} + \dfrac{1}{R_3} + \cdots}$$

6. Another way of writing the formula for R_T is:

$$\frac{1}{R_T} = \frac{1}{R_1} + \frac{1}{R_2} + \frac{1}{R_3} + \cdots$$

7. To measure one of two or more resistors connected in parallel, say R_1 in Fig. 12-2, disconnect one lead of R_1 from the circuit. Then measure the resistance of R_1.

SELF-TEST

Check your understanding by answering these questions:

1. In the circuit of Fig. 12-1, $I_T = 0.02$ A. $V = 50$ V. The total resistance R_T equals __________ Ω.
2. For the conditions in question 1, it is possible for R_1 to equal 1000 Ω. __________ (true/false)
3. For the conditions in question 1, the voltage V across R_2 = __________ V.
4. In order to measure the resistance of R_3 in Fig. 12-2, simply place the ohmmeter leads across GF and read the resistance. __________ (true/false)
5. In Fig. 12-2 the resistance of $R_1 = 100\ \Omega$, $R_2 = 200\ \Omega$, $R_3 = 300\ \Omega$. R_T = __________ Ω.

MATERIALS REQUIRED

- Power supply: Regulated variable direct current
- Equipment: EVM, 0–10-mA milliammeter, ohmmeter (or ohms scale of an EVM)
- Resistors: ½-W 2200-, 3300-, 4700-, 5600-, and 10,000-Ω
- Miscellaneous: SPST switch

PROCEDURE

1. Measure the resistance of each of the resistors supplied and record its value in Table 12-1.

2. Connect resistors R_1, R_2, etc., in parallel as in Fig. 12-2, using combination 1 of resistors shown in Table 12-2.

3. With an ohmmeter measure the total resistance of this combination and record its value in Table 12-2.

4. Repeat the procedures in steps 2 and 3 for each of the combinations shown in Table 12-2.

5. Compute the value of R_T for each combination in Table 12-2 by substituting the measured values of R_1, R_2, etc., from Table 12-1 in the formula

$$R_T = \frac{1}{\dfrac{1}{R_1} + \dfrac{1}{R_2} + \dfrac{1}{R_3} + \cdots}$$

and record the computed values in Table 12-2. Show your computations.

6. Connect the circuit of Fig. 12-3 using the resistor values of combination 1 from Table 12-2. Just two resistors are required in combination 1. Therefore leave out R_3. Close the switch and set the output of the voltage source V at 10 V, and maintain it at this level, under load, until you have completed the measurements in Table 12-3.

7. Measure and record, in Table 12-3, the applied voltage V and the total current I_T.

8. Repeat the procedures in steps 6 and 7 in turn for each of the combinations in Table 12-2.

9. For each combination compute the total resistance of the parallel-connected resistors by substituting the measured values of V and I_T in the formula

$$R_T = \frac{V}{I_T}$$

Record your results in Table 12-3. Show all your computations.

TABLE 12-1. Measured Values, Experimental Resistors

Rated value, Ω	2200	3300	4700	5600	10,000
Measured value, Ω					

TABLE 12-2. R_T of Parallel-Connected Resistors

Combination	Rated Value, Ω					Measured Value of R_T, Ω	Computed Value of R_T, Ω
	R_1	R_2	R_3	R_4	R_5		
1	2200	3300	X	X	X		
2	2200	3300	4700	X	X		
3	2200	3300	4700	5600	X		
4	2200	3300	4700	5600	10,000		

TABLE 12-3. R_T Determined by Voltage-Current Method

Combination	Measured Values		Computed Values
	V, volts	I_T, A	R_T, Ω
1			
2			
3			
4			

QUESTIONS

1. What is the effect on the total resistance of parallel-connected resistors of:
   (*a*) Increasing the number of resistors in parallel?
   (*b*) Decreasing the number of resistors in parallel?
2. Support your answer to question 1 by referring specifically to the measurements you recorded in Tables 12-2 and 12-3.
3. Do the measured values of R_T in Table 12-2 agree with the computed values of R_T in the same table using Eq. (12-5)? Refer specifically to your measurements.
4. Do the results of your measurements in Tables 12-1 and 12-2 prove that it is possible to write a general formula for total resistance of parallel-connected resistors? If yes, write a general formula for R_T and explain it in your own words.
5. Support your answer to question 4 by comparing your computed and measured values in Table 12-2. Comment on any discrepancies.
6. Compare the computed values of R_T in Table 12-3 with the computed values of R_T in Table 12-2 for the same combinations.
7. What was the purpose of steps 6 to 9 in this experiment?
8. What are the *three* methods you used in this experiment to determine the total resistance R_T of parallel-connected resistors?

Answers to Self-Test

1. 2500 Ω
2. false
3. 50
4. false
5. $54\frac{6}{11}$ Ω

13 EXPERIMENT

PARALLEL-CIRCUIT DESIGN CONSIDERATIONS

OBJECTIVES

1. To design a parallel circuit which will meet specified voltage, current, and resistance requirements
2. To construct the circuit and check, by measurement, the design parameters

INTRODUCTORY INFORMATION

Designing a Parallel Circuit to Meet Specified Resistance Requirements

The formulas for total resistance of parallel-connected resistors can be applied to the solution of simple design problems of this type. An example will indicate the techniques to be used.

Problem 1. A technician has a stock of the following color-coded resistors: four 68-Ω, five 82-Ω, two 120-Ω, three 180-Ω, two 330-Ω, 470-Ω, 560-Ω, 680-Ω, and 820-Ω. He needs a resistance value of 37 Ω for a circuit he is designing. Find a combination of resistors, using the least possible number of components, which will satisfy the design requirement. Assume the measured values of the resistors is the same as the color-coded.

Solution. Since the value of 37 Ω is less than the resistance of the smallest resistor in stock, a parallel arrangement will be required (recall from a previous experiment that the total resistance of a parallel circuit is less than the resistance value of the smallest resistor). The formula for total resistance R_T of a parallel circuit is

$$R_T = \frac{1}{\frac{1}{R_1} + \frac{1}{R_2} + \frac{1}{R_3} + \cdots} \qquad (13\text{-}1)$$

Suppose that two resistors satisfy the design requirement. Equation (13-1) can then be written as follows:

$$R_T = \frac{R_1 \times R_2}{R_1 + R_2} \qquad (13\text{-}2)$$

That is, the total resistance of two parallel resistors is equal to the product of their resistances divided by the sum of their resistances. The use of this formula can save computation time.

Let it be assumed that two resistors will meet the design requirements and that the resistors are 68 Ω and 82 Ω. Substituting these assumed values in Eq. (13-2) gives

$$R_T = \frac{68 \times 82}{68 + 82} = \frac{68 \times 82}{150} = 37.2\ \Omega$$

It is evident then that the two values selected meet the problem requirements, for when connected in parallel their total resistance is 37 Ω, approximately. Trial and error will show that no other combination of two resistors in stock will yield 37 Ω.

It is not always possible to select the two required resistors so easily. Another method is suggested. Suppose we assume that 68 Ω is one of the resistors. We wish to find another resistor R_X which, when paralleled with 68 ohms, will yield an equivalent resistance of 37 Ω. Substitute the known values of R_T and R_1 in Eq. (13-2):

$$37 = \frac{68 \times R_X}{68 + R_X}$$

Solving this first-degree equation yields the value $R_X = 81$ Ω. It is evident then that the 82-Ω resistor in stock is the required R_X.

Problem 2. Assume the same stock as in problem 1. The technician must design a circuit requiring a 60-Ω resistor.

Solution. By inspection it is evident that 60 Ω is one-half the resistance value of 120 Ω. By manipulating Eq. (13-1), it can be shown that if n equal-valued resistors, say R_1, are connected in parallel, their total resistance R_T is equal to $(1/n) \times R_1$. That is, a general formula for the total resistance R_T of n parallel-connected resistors each of which measures R_1 ohms is

$$R_T = \frac{R_1}{n} \qquad (13\text{-}3)$$

Using Eq. (13-3) yields the solution to this problem, namely two 120-Ω resistors connected in parallel.

The technician could also use three 180-Ω resistors connected in parallel, for their total resistance is also 60 Ω. However, here three components are used rather than the two 120-Ω resistors.

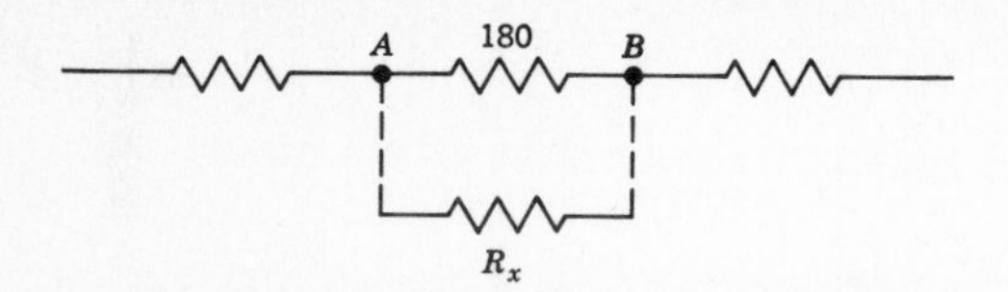

Fig. 13-1. What must be the value of R_X, in parallel with 180 Ω, if the resultant resistance R_{AB} = 45 Ω?

Problem 3. In the circuit of Fig. 13-1, a technician measures the resistance between points *A* and *B* and finds it is 180 Ω. Her circuit requires that the resistance between *A* and *B* must be 45 Ω. How can she modify the circuit to meet the required value, assuming she has the same stock of resistors as in problem 1?

Solution. Assume that there is a resistor R_X which, when connected in parallel with 180 Ω, will bring the resistance R_{AB} down to 45 ohms. Now substitute the known values in Eq. (13-2).

$$45 = \frac{180 \times R_X}{180 + R_X}$$

The solution of this first-degree equation for R_X is

$$R_X = 60\ \Omega$$

What is required, then, is a 60-Ω resistor. From problem 2 it is evident that two 120-Ω resistors, connected in parallel with the 180-Ω resistor, will yield the required result, namely

$$R_{AB} = 45\ \Omega$$

The procedure, then, for designing a parallel circuit having a specified resistance value R_T from a group of known resistors is to apply Eqs. (13-1), (13-2), and (13-3) to the resistors at hand and find a combination which, analytically, equals or is closest to the required value. After this is done the technician should connect the resistors in parallel and measure their total resistance with an ohmmeter to confirm the solution.

Designing a Parallel Circuit to Meet Specified Resistance and Current Requirements

Ohm's law, and the formulas for total resistance in a parallel circuit, are applied in the solution of this type of problem. Again, an example will illustrate the procedure.

Problem 4. In the circuit of Fig. 13-2*a*, determine analytically the voltage value *V* to which the variable dc power supply must be set to permit a total current of 0.20 A.

Solution. In order to find the applied voltage *V*, it is first necessary to determine the total resistance R_T in the circuit. When that is known, *V* can be found by using the formula $V = I \times R_T$ derived from Ohm's law.

1. *Procedure to find R_T*: The 680-Ω resistor between *A* and *B* is connected in parallel with the two series-connected resistors, 1200 and 820 Ω. The first step is to replace the 1200- and 820-Ω resistors with a single resistor R_1 whose resistance value is equal to the sum of the two resistors. Thus

$$R_1 = 1200 + 820 = 2020\ \Omega$$

The resulting equivalent circuit is then Fig. 13-2*b*, whose resistance and current characteristics are the same as those of Fig. 13-2*a*.

R_T may now be calculated by use of Eq. (13-2).

$$R_T = \frac{680 \times 2020}{680 + 2020} = 509\ \Omega$$

2. *Procedure to find V*: Now

$$V = I_T \times R_T$$
$$= 0.2 \times 509 = 101.8\ \text{V} = 102\ \text{V}$$

The required voltage is 102 V, approximately.

The technician must now connect the circuit of Fig. 13-2*a* and apply a measured 102 V from the dc supply. The measured current should be 0.2 A.

NOTE: Practically, the procedure the technician could have followed without attempting to find the value of *V* analytically, was to connect the circuit of Fig. 13-2 and adjust the output of the dc supply until the ammeter measured 0.2 A. He or she then could measure the dc voltage, which should be 102 V, approximately.

Designing a Parallel Circuit to Meet Specified Voltage and Current Requirements

Ohm's law and the formulas for total current and total resistance are applied in the solution of this type of problem. The following examples will illustrate the procedures to use.

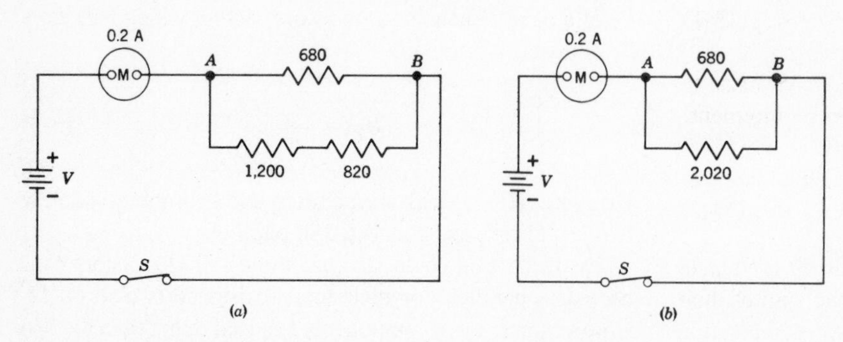

Fig. 13-2. What value of *V* will permit 0.2 A of current in the circuit?

Problem 5. In the circuit of Fig. 13-3, a current-divider network is required which will permit 0.02 A in resistor R_1 and 0.03 A in resistor R_2. A dc voltage source of 75 V powers the circuit. What values of R_1 and R_2 must be used to meet the circuit requirements?

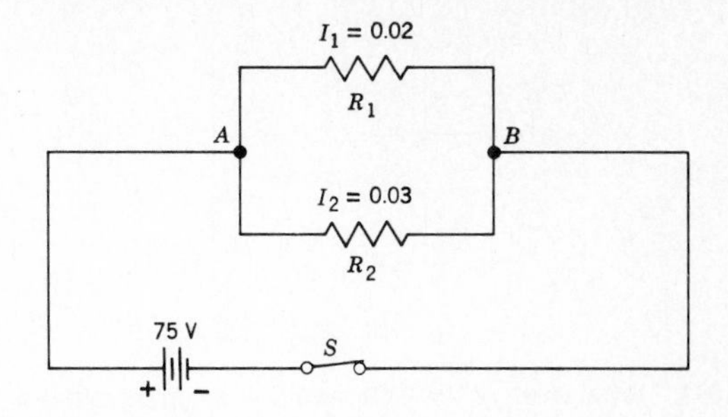

Fig. 13-3. What are the values of the resistors R_1 and R_2 which meet the voltage and current specifications of the circuit?

Solution. The voltage across each branch of a parallel network is the same. Therefore the voltage V_1 across R_1 equals the voltage V_2 across R_2. In this case,

$$V_1 = V_2 = 75 \text{ V}$$

By Ohm's law,

$$R_1 = \frac{V_1}{I_1} = \frac{75}{0.02} = 3750\ \Omega$$

$$R_2 = \frac{V_2}{I_2} = \frac{75}{0.03} = 2500\ \Omega$$

The required resistors, therefore, are 3750 and 2500 Ω. The technician should therefore connect 3750- and 2500-Ω resistors in the circuit of Fig. 13-3 and measure the branch currents I_1 and I_2 to verify the solution.

Problem 6. A technician must design a circuit which will deliver 0.25 A of current from a 15-V supply. Determine analytically what combination of resistors, using the least number of components, will satisfy the design requirements. The stock of resistors which the technician has on hand includes two 120-, three 150-, three 180-, 330-, 470- and three 680-Ω resistors.

Solution. The total resistance which will satisfy the requirements of the problem is

$$R_T = \frac{15}{0.25} = 60\ \Omega$$

A series combination will not give the required solution, but a parallel arrangement may. By inspection, it is apparent from the stock of resistors that two 120-Ω resistors in parallel will give the required value of 60 Ω. Another solution is a parallel combination of three 180-Ω resistors, but of course this requires one more component than the first combination. Hence the required solution is the combination of two parallel-connected 120-Ω resistors.

SUMMARY

1. If a circuit requires a resistance value R_T which is less than the smallest resistance value of a stock of resistors, it may be possible to approximate the required value by paralleling two or more of the available resistors. The procedure is to choose a resistor R_1 whose resistance is slightly higher than the required R_T, then find from the formula

$$R_T = \frac{R_1 \times R_2}{R_1 + R_2}$$

the value of R_2 which will yield R_T. If R_2 is in stock, the two resistors have been found. Otherwise it may be necessary to repeat the process. Finally, if two resistors whose parallel combination is R_T cannot be found, it may be necessary to try three resistors, or four.

2. The equivalent resistance R_T of n equal-valued resistors R_1, connected in parallel, is

$$R_T = \frac{R_1}{n}$$

3. If it is necessary to determine analytically the voltage V to which a dc voltage source must be set to permit I amperes of current in a circuit containing R ohms of resistance, V may be found by the formula

$$V = I \times R$$

4. It is possible to determine experimentally the voltage V to which a dc voltage source must be set to permit I amperes of current in a circuit containing R ohms of resistance. A circuit is connected containing the variable dc source in series with an ammeter and the resistance R. The voltage is varied unitl I current is measured. Then the measured voltage V, at the terminals of the dc source, is the required voltage.

5. If it is required to determine analytically what resistance R to connect in a circuit which will permit I amperes from a voltage source V, find R from the formula

$$R = \frac{V}{I}$$

If the value of R is available in stock, that solves the problem. Otherwise determine, using series or parallel resistors, a combination that will satisfy the required R.

SELF-TEST

Check your understanding by answering these questions:

1. What two resistors in parallel will yield a total resistance R_T of 30 Ω? Give at least two solutions.
(*a*) __________ Ω in parallel with __________ Ω;
(*b*) __________ Ω in parallel with __________ Ω.

2. 22-, 33-, and 47-Ω resistors are connected in parallel. Their total resistance will be __________ than 22 Ω.

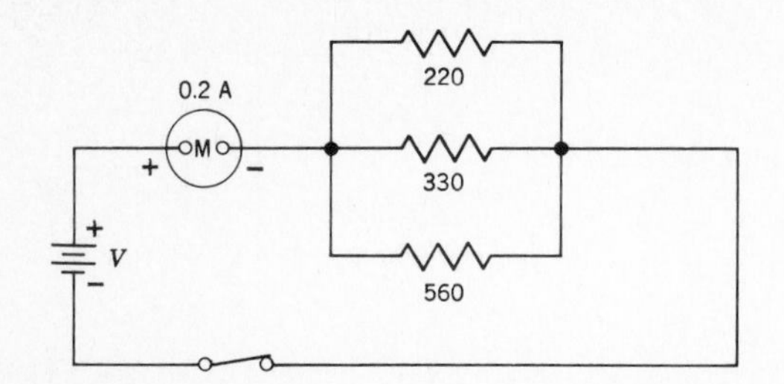

Fig. 13-4. What value of V will permit 0.2 A of current in this circuit?

3. A 10-Ω resistance may be constructed from __________ 50-Ω resistors by paralleling these resistors.

4. In the circuit of Fig. 13-4, what is the voltage V which will permit 0.2 A? __________ V
5. In the circuit of Fig. 13-3, there is 0.003 A of current. The total resistance R_{AB} between points A and B is __________ Ω.

MATERIALS REQUIRED

- Power supply: Regulated variable dc voltage source
- Equipment: EVM, 0–100-mA milliammeter
- Resistors: ½-W 560-, 1200-, 1800-, 3300-, 5600-, 10,000-Ω
- Miscellaneous: SPST switch

PROCEDURE

1. Measure the resistance of each of the resistors supplied and record its value in Table 13-1.
2. Determine analytically what parallel combination of the resistors supplied will yield the value 457 Ω (approximately). Use the color-coded values of the resistors for this computation and show your computations. Record the color-coded values of R in Table 13-2.
3. Connect the parallel combination and measure the total resistance R_T with an ohmmeter. Record the measured value in Table 13-2.
4. Repeat steps 2 and 3 for each value of R_T listed in Table 13-2.
5. Using the resistors supplied, design a parallel circuit combination, like Fig. 13-1, which will yield 0.02 A (approximately) when 40 V is applied. Show your computations. Record the color-coded values of R in Table 13-3.
6. Connect the circuit and measure the current I_T. Record this value in Table 13-3.

TABLE 13-1. Measured Value of Experimental Resistors

Rated value, Ω	560	1200	1800	3300	5600	10,000
Measured value, Ω						

TABLE 13-2. Designing a Parallel Resistive Network for a Required R_T

R_T Required, Ω	*Combination of Parallel Resistors*			*Measured Value R_T, Ω*
	R_1	R_2	R_3	
457				
720				
315				
3590				

TABLE 13-3. Designing a Circuit Which Will Yield I_T with V Known

	V Applied, V			*I_T, A*		*Parallel Resistors, Ω*	
Step	*Given Value*	*Computed Value*	*Measured Value*	*Required Value*	*Measured Value*	R_1	R_2
5, 6	40	X		0.02			
7, 8	X			0.05		1200	1800

7. Determine, analytically, what value of voltage V will be required for a circuit similar to Fig. 13-1 if $R_1 = 1200\ \Omega$, $R_2 = 1800\ \Omega$, and $I_T = 0.05$ A. Show your computations. Record the computed value of V in Table 13-3.
8. Connect the circuit and adjust the supply until there is 0.05 A in the circuit. Measure the supply voltage V and record this measured value in Table 13-3.
9. Design a two-branch parallel circuit from the resistors supplied so that 0.03 A of current is in one branch, 0.02 A in the other branch. Draw the circuit, and show your computations, including the voltage V Computed. Record the color-coded values of the resistors in Table 13-4.
10. Connect the circuit. Set the dc supply at the computed value of voltage V. Measure and record the current I_1 in R_1 and I_2 in R_2.
11. Theoretically design a parallel circuit with the resistors in this experiment, using
    a. Two resistors
    b. Three resistors
    c. Four resistors
    so chosen that each combination **a** through **c** draws 10 mA total current I_T at an applied voltage V to be determined. Show your computations, and record in Table 13-5 the measured values of R_1, R_2, R_3, and R_4 chosen and the design (computed) value for V for each combination.
12. Experimentally verify the design values for V and record your current measurements in Table 13-5.

TABLE 13-4. Designing a Current-Divider Circuit

Branch Current, A				*Resistance, Ω*		*Voltage, V*	
Required		*Measured*		R_1	R_2	*Computed*	*Measured*
I_1	I_2	I_1	I_2				
0.03	0.02						

TABLE 13-5. Design Circuit: Given I_T and R, Find V

Combination	*Measured Values*				*V Applied, Design Value*	*I Measured*
	R_1	R_2	R_3	R_4		
1						
2						
3						

QUESTIONS

1. How did the measured values of the resistors in this experiment compare with their color-coded values? Refer to Table 13-1 for your answer.
2. Why was it necessary to measure the resistors supplied in this experiment?
3. In Table 13-2, how did the measured value of R_T compare with the required value? Explain any discrepancies.
4. Did the measurement in step 6 confirm the design values in step 5? If not, why not? Refer specifically to the data in Table 13-3.
5. Did the measured values in step 8 confirm the design values in step 7? If not, why not? Refer specifically to the data in Table 13-3.
6. Did the measured values in step 10 confirm the design values in step 9? If not, why not? Refer specifically to the data in Table 13-4.
7. Design a three-branch parallel circuit which will divide a total current of 0.22 A in the ratio 2:3:6. If the total resistance of the three parallel resistors is 109 Ω, find the value of each resistor and find also the applied voltage. Draw the circuit and show your computations.

Answers to Self-Test

1. (*a*) 60, 60; (*b*) 50, 75
2. less
3. 5
4. 21.4
5. 25,000 Ω

14 EXPERIMENT

TOTAL RESISTANCE OF SERIES-PARALLEL CIRCUITS

OBJECTIVES

1. To determine and verify experimentally the law for total resistance R_T of a series-parallel combination of resistors
2. To verify that the voltage across each leg in a parallel circuit is the same as the voltage across the parallel circuit
3. To design a series-parallel network which will meet specified requirements

INTRODUCTORY INFORMATION

Total Resistance of a Series-Parallel Circuit

Figure 14-1 shows a series-parallel arrangement of resistors. In this circuit, R_1 is in series with the parallel circuit between points B and C, which in turn is in series with R_3. What is the total resistance between points A and D? Obviously we can measure R_T with an ohmmeter, or R_T can be found by the voltage-current method described in Experiment 8. Is it possible, however, to write a formula by which R_T of a series-parallel circuit may be computed without measurement? It is possible to determine a formula for R_T. The method we will use to develop a formula involves *analysis* and *measurement* to verify the results of our analysis.

Refer to Fig. 14-1. If R_1 and R_3 were removed and if the parallel circuit between points B and C were left standing alone, we could compute the total resistance R_{T2} between B and C by the formula for parallel resistors. We can therefore replace the parallel network in Fig. 14-1 by its equivalent resistance R_{T2}, and the circuit will now take the form of Fig. 14-2. It is apparent that the total resistance R_T of the series circuit (Fig. 14-2) is the same as the total resistance R_T between points A and D (Fig. 14-1).

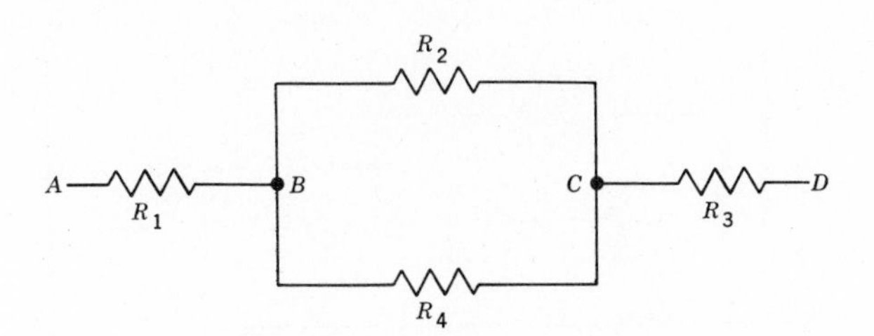

Fig. 14-1. Series-parallel combination of resistors.

Our analysis suggests that to find the total resistance of a series-parallel network, replace the parallel circuit by its equivalent resistance R and treat the resultant network like a simple series circuit. In the case of Fig. 14-1, the total resistance R_T between points A and D would be

$$R_T = R_1 + R_{T2} + R_3 \qquad (14\text{-}1)$$

Can this method be extended to include a series-parallel circuit involving more than one parallel network, as for example in Fig. 14-3? Observation and comparison show that the circuit between points A and D is identical with the circuit of Fig. 14-1. Hence, we can compute the resistance R_{T1} of Fig. 14-3 between points A and D using Eq. (14-1). We can now replace the section of Fig. 14-3 between A and D with R_{T1}. The resulting Fig. 14-4 is the equivalent of Fig. 14-3.

Figure 14-4 can further be simplified by replacing R_6 and R_7 with their series resistance $R_{6\text{-}7}$, where

$$R_{6\text{-}7} = R_6 + R_7$$

The resultant Fig. 14-5 is the equivalent of Figs. 14-4 and 14-3. Comparison with Fig. 14-1 shows that the total resistance for Fig. 14-5 may now be computed by Eq. (14-1).

It is apparent, therefore, that series-parallel networks containing more than one parallel circuit may be simplified by replacing each parallel circuit with its equivalent resistance. Then the total resistance R_T of the network (see Fig. 14-3) is

$$R_T = R_1 + R_{T2} + R_3 + R_{T4} + R_8 + \cdots \qquad (14\text{-}2)$$

where R_{T2} is the equivalent (total) resistance of the first parallel circuit, R_{T4} is the equivalent resistance of the second parallel circuit, etc., and R_1, R_3, R_8, etc., are resistors in series with the parallel circuits.

Verifying Eqs. (14-1) and (14-2) Experimentally

A final step in the process of establishing Eqs. (14-1) and (14-2) as a valid "law" requires experimental verification. A simple method of verification suggests itself. Measure the value of each resistance before it is connected in the circuit.

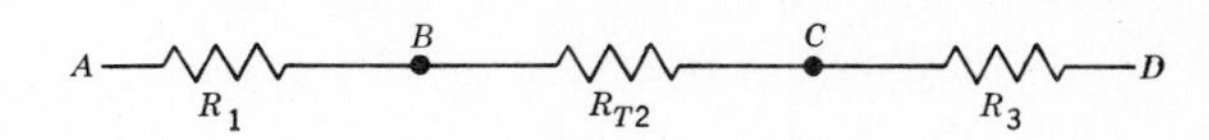

Fig. 14-2. Equivalent circuit of Fig. 14-1.

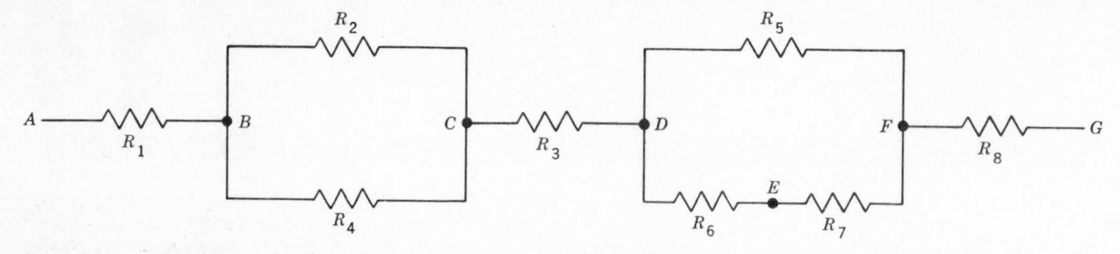

Fig. 14-3. Another series-parallel combination of resistors.

Measure the resistance R_T of the total combination. Compute R_{T2}, R_{T4}, etc., for each of the parallel circuits. (Substitute the measured values of the appropriate resistors in the formula for parallel resistance and solve for R_{T2}, R_{T4}, etc.) Now add the measured values R_1, R_3, R_8, etc., to the computed values R_{T2}, R_{T4}, etc. If the measured value of

$$R_T = R_1 + R_{T2} + R_3 + R_{T4} + R_8 + \cdots$$

we have a confirmation of Eq. (14-2). Additional measurements, using other combinations, must be made to establish the validity of Eqs. (14-1) and (14-2).

Voltage across Each Leg of a Parallel Circuit

Consider the circuit of Fig. 14-6. The circuit between points B and D consists of R_2 in parallel with the series-connected combination R_4 and R_5. To determine the voltage across resistor R_2, we measure the potential difference between points B and D. To determine the voltage across the series combination of R_4 and R_5, we also measure the emf across points B and D. Moreover, the voltage across the parallel circuit between B and D is measured in the same way. This is so because a common connection joins one side of all the legs of a parallel circuit and another common connection joins the opposite sides of all the legs.

Since a voltage check across any or all legs must be made between the same two common points, the voltage across any or all legs must be the same. We reached this conclusion in Experiment 11. We will verify it in this experiment.

SUMMARY

1. In a series-parallel network like that in Fig. 14-1, the total resistance R_T of the network measured across the end terminals (A and D) can be found by replacing each parallel combination by its equivalent resistance R_{T2}, leaving an equivalent series circuit (Fig. 14-2). R_T may then be computed by using the series-resistance formula:

$$R_T = R_1 + R_{T2} + R_3 + \cdots$$

2. In determining the equivalent resistance R_{T2} of one of the parallel combinations such as R_2, R_4 in Fig. 14-1, the formula for parallel resistors is used.
3. A parallel branch may sometimes have two or more series resistors, such as branch R_6–R_7 in Fig. 14-4. In that case R_6 and R_7 are combined like series resistors, and their equivalent series resistance $R_{6\text{-}7} = R_6 + R_7$ is substituted in the circuit, leaving a network like that in Fig. 14-5. Then Fig. 14-5 is transformed into its series equivalent form, Fig. 14-2.
4. In a parallel network the voltage across each branch is the same. Thus in Fig. 14-6, the voltage V_2 across R_2 equals V_{BD}, and this voltage is equal to the voltage $V_{4\text{-}5}$ across the series combination of R_4 and R_5. That is

$$V_2 = V_{BD} = V_{4\text{-}5}$$

SELF-TEST

Check your understanding by answering these questions:

1. In Fig. 14-1, $R_1 = 120\ \Omega$, $R_2 = 280\ \Omega$, $R_3 = 330\ \Omega$, and $R_4 = 470\ \Omega$. What is the value of R_{T2}, measured from B to C? __________ Ω

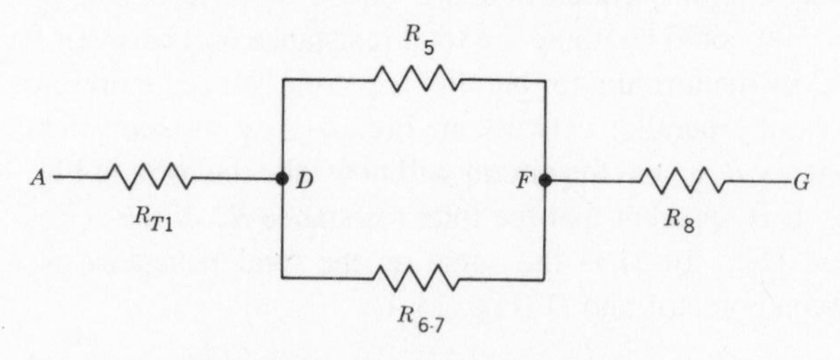

Fig. 14-5. Equivalent circuit of Figs. 14-4 and 14-3.

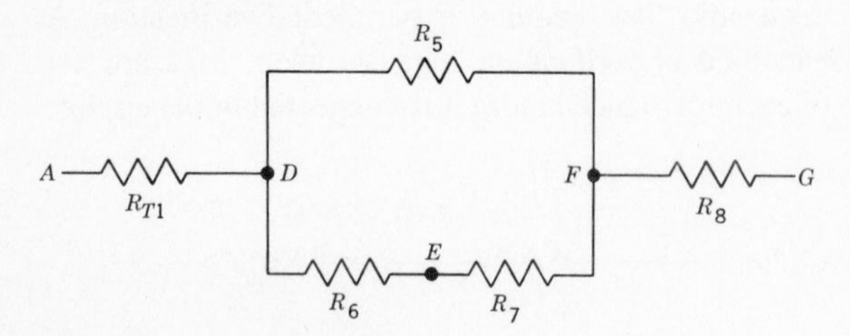

Fig. 14-4. Equivalent circuit of Fig. 14-3.

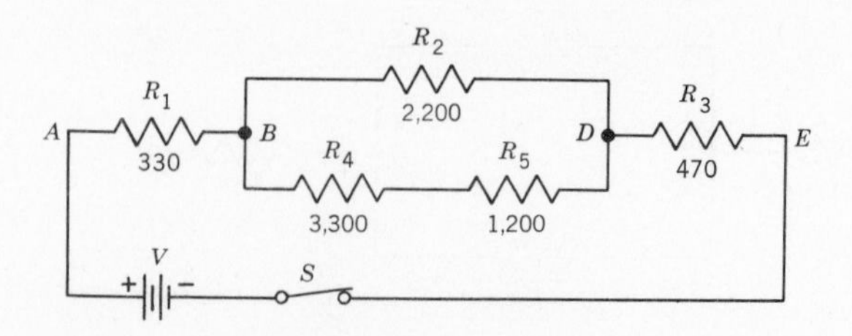

Fig. 14-6. What is the voltage across R_2?

2. For the same conditions as in question 1, what is the value of R_T, measured from A to D? __________ Ω
3. In Fig. 14-3, $R_1 = 33\ \Omega$, $R_2 = 56\ \Omega$, $R_4 = 68\ \Omega$, $R_3 = 100\ \Omega$, $R_5 = 120\ \Omega$, $R_6 = 47\ \Omega$, $R_7 = 73\ \Omega$, and $R_8 = 470\ \Omega$. What is the equivalent resistance R_{T2} between points B and C? __________ Ω
4. For the same conditions as in question 3, what is the equivalent resistance $R_{6\text{-}7}$ in the lower branch of the parallel network between points D and F? __________ Ω
5. For the same conditions as in question 3, what is the equivalent resistance R_{T3} of the parallel network between points D and F? __________ Ω
6. For the same conditions as in question 3, what is the total resistance R_T between points A and G? __________ Ω
7. In Fig. 14-6, what is the total resistance R_T in the circuit? __________ Ω
8. In Fig. 14-6, if $V = 228$ V, what is the total current in the circuit? __________ A
9. If in Fig. 14-6 the current in R_2 is 0.0675 A, what is the voltage V_{BD} across the series combination of R_4 and R_5? __________ V

MATERIALS REQUIRED

- Power supply: Regulated variable direct current
- Equipment: EVM, 0–10-mA milliammeter, ohmmeter (or ohms scale of an EVM)
- Resistors: ½-W 330-, 470-, 560-, 1200-, 2200-, 3300-, 4700-, and 10,000-Ω
- Miscellaneous: SPST switch

PROCEDURE

Total Resistance of a Series-Parallel Network

1. Measure the resistance of each of the resistors supplied and record its value in Table 14-1.
2. Connect the circuit of Fig. 14-1 using the values in Table 14-1. This is combination 1. With an ohmmeter measure and record in Table 14-1 the total resistance R_T (between points A and D) and the resistance R_{T2} (between points B and C).
3. Add the measured values R_1, R_{T2}, and R_3 and record their sum under column "Computed value, R_T:(*a*)."

   Sample calculation. If the measured values of $R_1 = 300\ \Omega$, $R_{T2} = 300\ \Omega$, $R_3 = 500\ \Omega$, then $R_T{:}(a) = R_1 + R_{T2} + R_3 = 300 + 300 + 500 = 1100$.

4. Calculate the value of R_{T2} by substituting the measured values R_2 and R_4 in the formula $1/R_{T2} = 1/R_2 + 1/R_4$ and record this value under the column "Computed value, R_{T2}."

   Sample calculation. If the measured values of $R_2 = 400$ and $R_4 = 1200$, then

$$\frac{1}{R_{T2}} = \frac{1}{R_2} + \frac{1}{R_4} = \frac{1}{400} + \frac{1}{1200} = \frac{4}{1200}$$

   and "Computed value, R_{T2}" = 300.

5. Calculate the value of R_T by adding the measured values of R_1 and R_3 and the computed value of R_{T2} (from step 4) and record this value under the column "Computed value, R_T:(*b*)."

   Sample calculation. Since the measured values of $R_1 = 300$, $R_3 = 500$, and the *computed value* of $R_{T2} = 300$, $R_T{:}(b) = R_1 + R_{T2} + R_3 = 300 + 300 + 500 = 1100$.

NOTE: In the sample calculations, the computed values R_T:(*a*) and R_T:(*b*) are exactly equal. Of course, we assumed perfect accuracy in our meter and in our measurements. This

TABLE 14-1. Ohmmeter Method for Determining R_T in Series-Parallel Network

	R_1	R_2	R_3	R_4	R_5	R_6	R_7	R_8
Rated Value, Ω	330	470	560	1200	2200	3300	4700	10,000
Measured Value, Ω								

	Measured Value			*Computed Value*			
Combination	R_T	R_{T2}	R_{T4}	R_T:(*a*)	R_{T2}	R_{T4}	R_T:(*b*)
1 (Fig. 14-1)			X			X	
2 (Fig. 14-1)			X			X	
3 (Fig. 14-3)							
4 (Fig. 14-3)							

is the ideal situation which the research technician strives for but never quite attains.

6. Repeat steps 2 through 5 for combination 2, using the rated values $R_1 = 470\ \Omega$, $R_2 = 2200\ \Omega$, $R_3 = 1200\ \Omega$, and $R_4 = 3300\ \Omega$.
7. Connect the circuit of Fig. 14-3, using the values in Table 14-1. This is combination 3. Measure and record the total resistance R_T between points *A* and *G*, the resistances R_{T2} between points *B* and *C*, and the resistance R_{T4} between points *D* and *F*.
8. Add the measured values R_1, R_{T2}, R_3, R_{T4}, and R_8, and record their sum under column "Computed value, R_T:(*a*)."
9. Calculate the value of R_{T2} using the measured values of R_2 and R_4. Calculate also the value of R_{T4} by substituting the measured values of R_5, R_6, and R_7 in

$$\frac{1}{R_{T4}} = \frac{1}{R_5} + \frac{1}{R_6 + R_7}$$

Record these values in Table 14-1.
10. Calculate the value of R_T by adding the measured values R_1, R_3, and R_8 and the computed values R_{T2} and R_{T4}, and record under the column "Computed value, R_T:(*b*)."
11. Repeat steps 7 through 10 for combination 4, using the rated values $R_1 = 470\ \Omega$, $R_2 = 2200\ \Omega$, $R_3 = 1200\ \Omega$, $R_4 = 3300\ \Omega$, $R_5 = 10{,}000\ \Omega$, $R_6 = 560\ \Omega$, $R_7 = 4700\ \Omega$, and $R_8 = 330\ \Omega$.

Voltage across Each Leg of a Parallel Circuit

12. Connect the circuit of Fig. 14-6. *S* is open. Set the output of the regulated power supply to 20 V and maintain it at this level.
13. Close *S*. With an EVM set on the proper voltage range, measure the voltage across R_1, R_2, and R_3 and the series combination of R_4 and R_5. Measure also the voltage across the parallel network V_{BD}. Record in Table 14-2.

Design Problem (extra credit)

14. Theoretically design a series-parallel circuit with the resistors in this experiment, using first
    a. Four resistors, then
    b. Six resistors, then
    c. Eight resistors,
    so chosen that each combination **a** through **c** draws 5 mA total current I_T. Draw each of the three circuits, showing on the diagram the value of each resistor chosen.

NOTE: Use the measured values from Table 14-1. Show your computations and record in Table 14-3 the design (computed) values of R_T, *V*, and I_T in each case.

15. Connect each circuit, in turn. Determine by measurement and record in Table 14-3 (under "Measured values") *V*, R_T, and I_T.

TABLE 14-2. Branch Voltage in Series-Parallel Network

V Applied	V_1 (*across* R_1)	V_2 (*across* R_2)	$V_{4\text{-}5}$ (*across* R_4-R_5)	V_3 (*across* R_3)	V_{BD}

TABLE 14-3. Design Problem, Extra Credit

	Design Values			*Measured Values*		
Circuit	*V*	R_T	I_T	*V*	R_T	I_T
a						
b						
c						

QUESTIONS

1. In measuring the resistance R_{T2} in step 2, was it necessary to open the circuit at *B* or *C*? Why?
2. Why is it necessary to disconnect power before measuring resistance in a circuit?
3. Refer to Table 14-1. Are the measured value R_T and the computed values R_T:(*a*) and R_T:(*b*) the same? Should they be? Explain.
4. What conclusion can you draw from your measurements in Table 14-1 concerning the total resistance of a series-parallel circuit? Refer to specific measurements to substantiate your conclusion.

5. What is the rule for determining the total resistance R_T of a series-parallel circuit?
6. What can you say about the voltage across each leg of a parallel circuit? Refer to specific measurements to substantiate your statement.
7. What information do you need to compute the current in each leg of a parallel network?
8. Refer to Fig. 14-6. If I_2 is the current in R_2, $I_{4\text{-}5}$ the current in R_4 and R_5, and I_T the total current, what is the relationship between $I_2 \times R_2$, $I_{4\text{-}5} \times (R_4 + R_5)$, and $I_T \times (R_{T2})$?
9. Explain two methods you may use for measuring the total resistance R_T of a series-parallel network, such as that in Fig. 14-1.
10. Refer to Table 14-2. Does there appear to be any relationship between V applied and V_1, V_2, and V_3? If yes, what is the relationship?

Extra Credit

11. Comment on the results of your design values and measured values in the design problem. Explain any discrepancies between measured and predicted results.

Answers to Self-Test

1. 176 Ω
2. 626 Ω
3. 30.8 Ω
4. 120 Ω
5. 60 Ω
6. 694 Ω
7. 2278 Ω
8. 0.1 A
9. 148 + V

15

EXPERIMENT

KIRCHHOFF'S VOLTAGE LAW (FOR ONE GENERATOR)

OBJECTIVES

1. To determine by analysis the relationship between the sum of the voltage drops across series-connected resistors, and the applied voltage
2. To verify experimentally the relationship determined in objective 1

INTRODUCTORY INFORMATION

The solution of complex electric circuits is facilitated by the application of Kirchhoff's laws. These laws were formulated and published by the physicist Gustav Robert Kirchhoff (1824–1887), and they established the basis for modern network analysis. They are applicable to circuits with one or more voltage sources. In this experiment we will be concerned with a single voltage source.

Voltage Law

Consider the simple series circuit, Fig. 15-1. We have previously established that the series-connected resistors R_1, R_2, R_3, and R_4 may be replaced by their total or equivalent resistance R_T without affecting the current I_T in the circuit. The current I_T may then be computed by Ohm's law, for

$$I_T = \frac{V}{R_T} \qquad (15\text{-}1)$$

From Eq. (15-1), it also follows that

$$V = I_T \times R_T \qquad (15\text{-}2)$$

In the series-connected circuit of Fig. 15-1

$$R_T = R_1 + R_2 + R_3 + R_4 \qquad (15\text{-}3)$$

Substituting the equivalent value for R_T from Eq. (15-3) in Eq. (15-2) gives

$$V = I_T(R_1 + R_2 + R_3 + R_4)$$

or

$$V = I_T \times R_1 + I_T \times R_2 + I_T \times R_3 + I_T \times R_4 \qquad (15\text{-}4)$$

One of the characteristics of a series circuit is that current is the same everywhere in the circuit. I_T is, therefore, the total current in R_1, R_2, R_3, and R_4. It follows, therefore, that

$I_T \times R_1 = V_1$, the voltage drop across R_1
$I_T \times R_2 = V_2$, the voltage drop across R_2
$I_T \times R_3 = V_3$, the voltage drop across R_3

and $I_T \times R_4 = V_4$, the voltage drop across R_4

Equation (15-4) may now be rewritten as follows:

$$V = V_1 + V_2 + V_3 + V_4 \qquad (15\text{-}5)$$

Equation (15-5) is a mathematical expression of Kirchhoff's voltage law for resistors connected in series in a closed circuit.

It is apparent that Eq. (15-5) may be generalized for circuits containing one or more series-connected resistors in a closed circuit. The law also applies in the case of series-parallel circuits (Fig. 15-2). Here, $V = V_1 + V_2 + V_3 + V_4 + V_5$, where V_1, V_3, and V_5 are the voltage drops across R_1, R_4, and R_8, respectively. V_2 and V_4 are the voltages across the parallel circuits between A and B and between C and D, respectively.

Expressed in words, Eq. (15-5) states that in a closed circuit or loop, the applied voltage equals the sum of the voltage drops in the circuit.

The concept of algebraic signs or polarity is helpful in solving electrical network problems. An example will illustrate the convention employed in assigning a + or − sign to a voltage in a circuit.

Consider the circuit of Fig. 15-3. Assume that current in this circuit is synonymous with electron flow, and that electrons move from a negative to positive potential. The arrow in Fig. 15-3 shows the direction of current, and the − and + signs indicate the following: point A is negative with respect to B; point B is negative with respect to C, etc. This is consistent with our assumption of current in this circuit.

Now to establish the algebraic sign for the voltages in the closed circuit, move in the direction of assumed current.

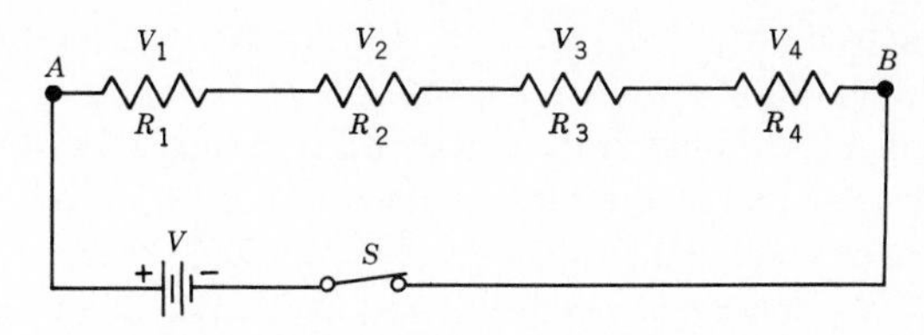

Fig. 15-1. $V = V_1 + V_2 + V_3 + V_4$

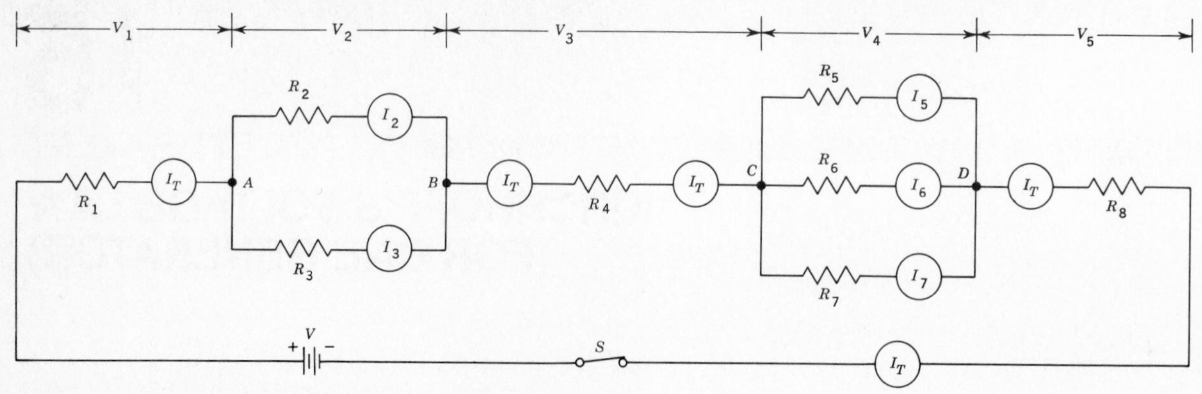

Fig. 15-2. Application of Kirchhoff's voltage law to a series-parallel circuit. $V = V_1 + V_2 + V_3 + V_4 + V_5$.

Consider as positive any voltage source or voltage drop whose + (positive) terminal is reached first, and negative any voltage source or drop whose − (negative) terminal is reached first. Starting at point A in Fig. 15-3 and moving in the direction of current, we have $-V_1$, $-V_2$, $-V_3$, $-V_4$, and $+V$. With this convention in mind, Kirchhoff's voltage law may now be generalized as follows.

The algebraic sum of the voltages in a closed circuit equals zero.

Applying the convention on signs and Kirchhoff's law to the closed circuit of Fig. 15-3 and starting at point A, we may write as follows:

$$-V_1 - V_2 - V_3 - V_4 + V = 0 \qquad (15\text{-}6)$$

Is this equation consistent with Eq. (15-5)? Yes, for by transposing the terms on the right side of Eq. (15-5) to the left side, we get $V - V_1 - V_2 - V_3 - V_4 = 0$, a result which is identical with that in Eq. (15-6).

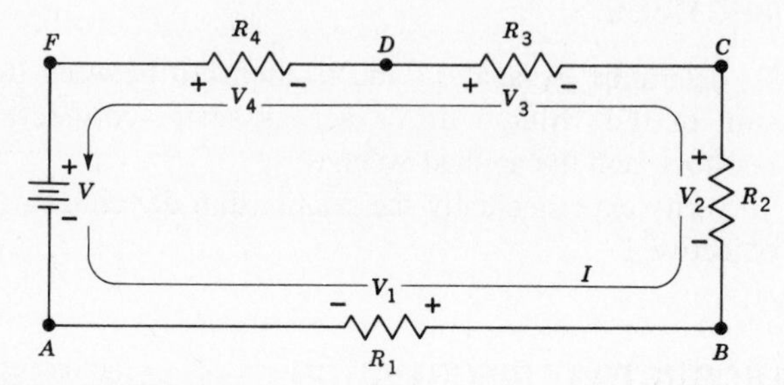

Fig. 15-3. Convention for assigning polarity to voltages in a closed circuit.

SELF-TEST

Check your understanding by answering these questions:

1. In Fig. 15-1, $V_1 = 10$ V, $V_2 = 12$ V, $V_3 = 20$ V, and $V_4 = 15$ V. The applied voltage V must then equal __________ V.
2. In Fig. 15-2, $V_1 = 15$ V, $V_2 = 20$ V, $V_4 = 27$ V, $V_5 = 9$ V, and $V = 100$ V. The voltage $V_3 =$ __________ V.

SUMMARY

Kirchhoff's voltage law may be expressed in two ways.

1. The sum of the voltage drops in a closed circuit equals the applied voltage.
2. The algebraic sum of the voltages in a closed circuit equals zero.

MATERIALS REQUIRED

- Power supply: Regulated variable direct current
- Equipment: EVM, 0–10-mA milliammeter
- Resistors: ½-W 330-, 470-, 1200-, 2200-, 3300-, 4700-, 5600-, and 10,000-Ω
- Miscellaneous: SPST switch

PROCEDURE

Voltage Law

1. Connect the circuit of Fig. 15-1 using the values R_1, R_2, R_3, and R_4 shown in Table 15-1.
2. Adjust the output of the power supply so that $V = 40$ V. Measure and record this voltage in Table 15-2. Measure also and record the voltages V_1, V_2, V_3, and V_4 across R_1, R_2, R_3, and R_4, respectively. Add V_1, V_2, V_3, and V_4 and enter the sum in the proper column in Table 15-2.
3. Connect the circuit of Fig. 15-2 using the values R_1, R_2, etc., shown in Table 15-1.
4. Adjust the output of the power supply so that $V = 40$ V. Measure this voltage and record in Table 15-2. Measure also and record the voltages V_1, V_2, V_3, V_4, and V_5 as designated in Fig. 15-2. Add V_1, V_2, V_3, etc., and enter the sum in the proper column.

Design Problem (Extra Credit)

5. Design a series-parallel circuit like that in Fig. 15-2 using all the resistors listed in this experiment, with an applied voltage of 35 V, drawing a total current of 5 mA, approximately. Draw a diagram of the circuit, including the values of all resistors. Show all your computations.

6. Connect the circuit and set the power-supply voltage to 35 V. Measure the total current I_T and indicate this value on the circuit diagram.

TABLE 15-1. Color-Coded Values, Experimental Resistors

	R_1	R_2	R_3	R_4	R_5	R_6	R_7	R_8
Rated value, Ω	330	470	1200	2200	3300	4700	10,000	5600

TABLE 15-2. Verifying Kirchhoff's Voltage Law

Step	V	V_1	V_2	V_3	V_4	V_5	$V_1 + V_2 + V_3 + V_4 + V_5$
2						X	
4							

QUESTIONS

1. How does the sum of V_1, V_2, V_3, and V_4 (step 2) in Table 15-2 compare with V? Explain the difference, if any.
2. How does the sum of V_1, V_2, . . . , V_5 (step 4) in Table 15-2 compare with V? Explain the difference, if any.
3. In your own words state Kirchhoff's voltage law. Also write the equation for it.
4. Discuss the results of the design problem, step 6.
5. What made it impossible for you to design a circuit, as required, which would give *exactly* 5 mA of current?

Answers to Self-Test

1. 57
2. 29

16

EXPERIMENT

KIRCHHOFF'S CURRENT LAW

OBJECTIVES

1. To determine by analysis the relationship between the sum of the currents entering any junction of an electric circuit and the current leaving that junction
2. To verify experimentally the relationship determined in objective 1

INTRODUCTORY INFORMATION

Current Law

In Experiment 11 you verified that the total current I_T in a circuit containing resistors connected in parallel is equal to the sum of the currents in each of the parallel branches. This was one demonstration of Kirchhoff's current law, limited to a parallel network. The law is perfectly general, however, and is applicable to any circuit. It states that the *current entering any junction of an electric circuit is equal to the current leaving that junction*.

Consider the series-parallel circuit (Fig. 16-1). Designate the total current as I_T. I_T enters the junction at A in the direction indicated by the arrow. The currents leaving the junction at A are I_1, I_2, and I_3 as shown. The currents I_1, I_2, and I_3 then enter the junction at B, and I_T leaves the junction at B. What is the relationship between I_T, I_1, I_2, and I_3? Analysis will help establish this relationship.

The voltage across the parallel circuit $V_{AB} = I_1 \times R_1 = I_2 \times R_2 = I_3 \times R_3$. The parallel network may be replaced by its equivalent resistance R_T, in which case Fig. 16-1 is transformed into a simple series circuit and $V_{AB} = I_T \times R_T$. It follows, therefore, that

$$I_T \times R_T = I_1 \times R_1 = I_2 \times R_2 = I_3 \times R_3 \tag{16-1}$$

Equation (16-1) may be rewritten as

$$\begin{aligned} I_1 &= I_T \times \frac{R_T}{R_1} \\ I_2 &= I_T \times \frac{R_T}{R_2} \\ I_3 &= I_T \times \frac{R_T}{R_3} \end{aligned} \tag{16-2}$$

Adding I_1, I_2, and I_3 gives

$$I_1 + I_2 + I_3 = I_T \times R_T \left(\frac{1}{R_1} + \frac{1}{R_2} + \frac{1}{R_3}\right)$$

But

$$\frac{1}{R_1} + \frac{1}{R_2} + \frac{1}{R_3} = \frac{1}{R_T}$$

Therefore

$$I_1 + I_2 + I_3 = I_T \times R_T \times \frac{1}{R_T} = I_T$$

and

$$I_T = I_1 + I_2 + I_3 \tag{16-3}$$

Equation (16-3) is a mathematical statement of Kirchhoff's law, applied to the circuit of Fig. 16-1. In general, if I_T is the current entering a junction of an electric circuit, and $I_1, I_2, I_3,$

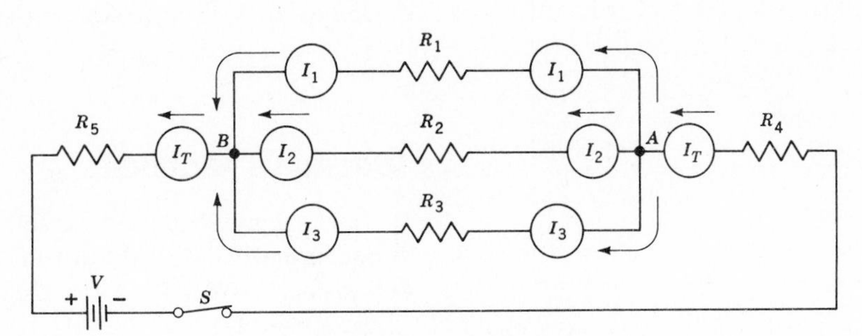

Fig. 16-1. $I_T = I_1 + I_2 + I_3$.

. . . , I_n are the currents leaving that junction (or vice versa), then

$$I_T = I_1 + I_2 + I_3 + \cdots + I_n \qquad (16\text{-}4)$$

Kirchhoff's current law is usually stated in another way, namely, *the algebraic sum of the currents entering and leaving a junction is zero*. It will be recalled that this is similar to the formulation of Kirchhoff's voltage law, that *the algebraic sum of the voltages in a closed path or loop is zero*. Just as it was necessary to agree on a polarity convention for voltages in a loop, so it is necessary to agree on a current convention at a junction.

If the current entering a junction is considered positive (+) and the current leaving a junction is considered negative (−), then the statement that the algebraic sum of the currents entering and leaving a junction is zero can be shown to be identical with Eq. (16-4). Consider the circuit of Fig. 16-2. The total current I_T enters the junction at A and is considered +. The currents I_1 and I_2 leave the junction at A and are designated −. Then,

$$+I_T - I_1 - I_2 = 0 \qquad (16\text{-}5)$$

and

$$I_T = I_1 + I_2 \qquad (16\text{-}6)$$

It is evident that the two statements of Kirchhoff's current law lead to the same algebraic formula.

An example will show how Kirchhoff's current law may be applied in network analysis. Suppose in Fig. 16-3 that I_1 and I_2 are currents entering the junction at A and are respectively +5 and +3 A. I_3, I_4, and I_5 are currents leaving A. I_3 and I_4 are respectively 2 and 1 A. What is the value of I_5? Applying Kirchhoff's current law,

$$I_1 + I_2 - I_3 - I_4 - I_5 = 0$$

and substituting the known values of current, we get

$$5 + 3 - 2 - 1 - I_5 = 0$$

$$8 - 3 = I_5$$

and

$$I_5 = 5 \text{ A}$$

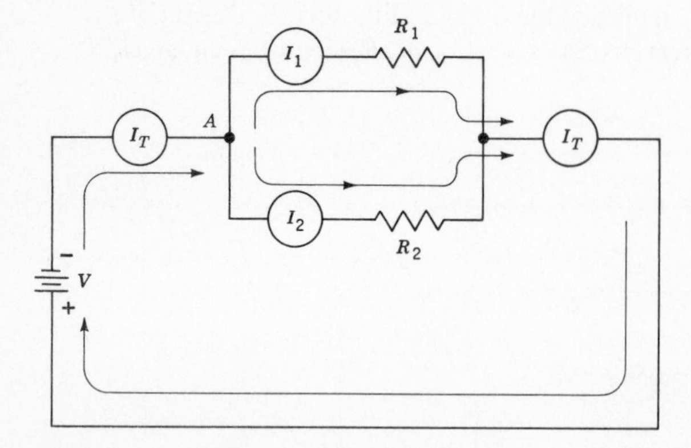

Fig. 16-2. The algebraic sum of the currents entering and leaving a junction is equal to zero: $I_T - I_1 - I_2 = 0$.

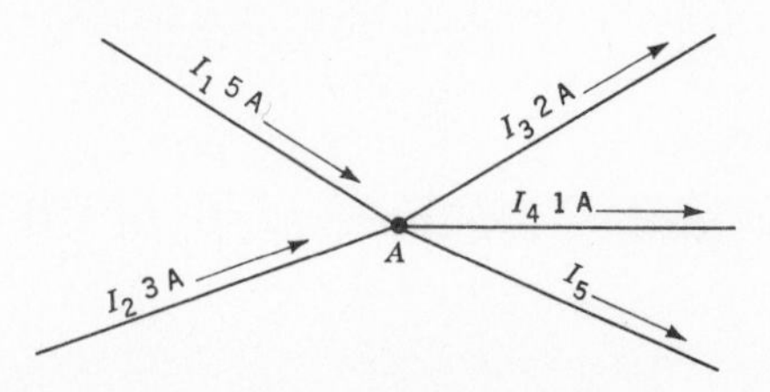

Fig. 16-3. Currents entering and leaving a junction: $I_1 + I_2 - I_3 - I_4 - I_5 = 0$.

SUMMARY

1. Kirchhoff's current law states that the current entering any junction of an electric circuit is equal to the current leaving that junction.
2. An analytical statement of Kirchhoff's current law requires the assignment of polarity to current entering a junction (assume it is +) and to current leaving a junction (assume it is −).
3. With the convention of 2, Kirchhoff's current law may be stated as follows: The algebraic sum of the currents entering and leaving a junction is zero. Thus, for Fig. 16-1, at junction A

$$I_T - I_1 - I_2 - I_3 = 0$$

SELF-TEST

Check your understanding by answering the following questions:

1. In Fig. 16-1 the current entering the junction at A is 0.55 A. The current I_1 = 0.25 A, I_2 = 0.1 A. The current I_3 must therefore equal ________ A.
2. In Fig. 16-1 the current leaving the junction at B = 1.25 A. The sum of currents I_1, I_2, and I_3 must be ________ A.
3. At junction B (Fig. 16-1) the polarities which must be assigned to the currents in applying Kirchhoff's current law are: I_1 is ________; I_2 is ________; I_3 is ________; and I_T is ________.
4. The equation which describes the relationship among the currents at junction A in Fig. 16-3 is ________ ________.
5. In Fig. 16-3, I_2 = 10 A, I_3 = 4 A, I_4 = 4 A, I_5 = 4 A. I_1 = ________ A.

MATERIALS REQUIRED

- Power supply: Regulated variable dc source
- Equipment: EVM, 0–10-mA milliammeter
- Resistors: ½-W 330-, 470-, 2200-, 3300-, 4700-, 5600-, and 10,000-Ω
- Miscellaneous: One SPST switch

PROCEDURE

Current Law

NOTE: Before connecting the milliammeter to measure current, disconnect the power supply by opening switch S. Follow this procedure each time the meter is moved. Observe meter polarity.

1. Connect the circuit of Fig. 16-4 with $V = 40$ V. Use the values R_1, R_2, etc., in Table 16-1. Measure and record in Table 16-2 the currents I_T at A; I_2, I_3, and I_T at B; I_T at C; and I_5, I_6, I_7, and I_T at D. Add the currents I_2 and I_3 and record the sum in the proper column. Add also the currents I_5, I_6, and I_7, and record the sum in the proper column.
2. Connect the circuit of Fig. 16-1, using the values of R in Table 16-1. Adjust the output of the power supply so that $V = 40$ V. Measure and record in Table 16-3 the currents I_T at A; and I_1, I_2, I_3, and I_T at B. Add the currents I_1, I_2, and I_3 and record their sum in Table 16-3.

Design Problem (Extra Credit)

3. Design a series-parallel circuit with three legs in the parallel circuit, such that the currents in the legs I_1, I_2, and I_3 are in the ratio 1:2:3 (approximately), and the total current in the circuit is 6 mA. Use resistors listed in this experiment which will give the closest approximation to the desired ratio. Draw a diagram of the circuit showing the values of resistance chosen and the design voltage V. Show all your computations. Measure the required currents and record them in Table 16-4. Record also the ratio of the currents I_1:I_2:I_3.

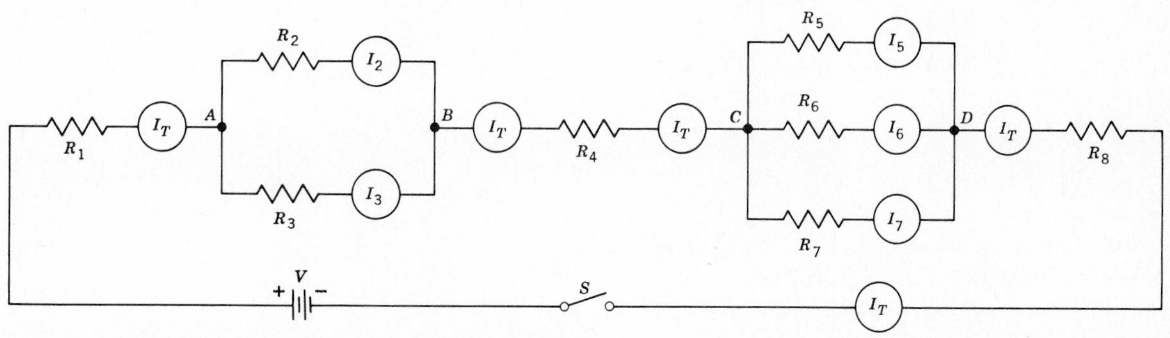

Fig. 16-4. Experimental circuit to verify Kirchhoff's current law.

TABLE 16-1. Color-Coded Values of Experimental Resistors

	R_1	R_2	R_3	R_4	R_5	R_6	R_7	R_8
Rated value, Ω	330	470	1200	2200	3300	4700	10,000	5600

TABLE 16-2. Verifying Kirchhoff's Current Law

	I_T at A	I_2	I_3	I_T at B	I_T at C	I_5	I_6	I_7	I_T at D	$I_2 + I_3$	$I_5 + I_6 + I_7$
mA											

TABLE 16-3. More Data Verifying Kirchhoff's Current Law

	I_T at A	I_1	I_2	I_3	I_T at B	$I_1 + I_2 + I_3$
mA						

TABLE 16-4. Design Problem Data

	I_1	I_2	I_3	I_T	I_1:I_2:I_3
mA					

QUESTIONS

1. How does the sum of I_2 and I_3 in step 1 compare with I_T at A? I_T at B? Explain any discrepancies.
2. How does the sum of I_5, I_6, and I_7 in step 1 compare with I_T at C? I_T at D? Explain any differences.
3. Comment on the results of your measurements in step 2. Explain any unexpected result.
4. In your own words state Kirchhoff's current law and write its equation.
5. What information do you need to compute I_2 and I_3 in Fig. 16-4?
6. Discuss the results of the design problem (step 3).

Answers to Self-Test

1. 0.2
2. 1.25
3. I_1 is $+$; I_2 is $+$; I_3 is $+$; I_T is $-$
4. $I_1 + I_2 - I_3 - I_4 - I_5 = 0$
5. 2

17 EXPERIMENT

VOLTAGE-DIVIDER CIRCUITS (LOADED)

OBJECTIVES

1. To determine analytically the effects of a load on the voltage relationships in a resistive voltage-divider circuit
2. To confirm experimentally the results of objective 1

INTRODUCTORY INFORMATION

In the simple dc voltage-divider circuits studied in Experiment 10, no load current was drawn. The only current was the "bleeder" current in the divider network itself. In electronics, divider networks are frequently used as a source to supply voltage to a load, which draws current. When this is the case, the divider-voltage relationships which were obtained for no-load conditions no longer hold. The actual changes depend on the amount of current drawn and on the circuit connections. Consider the circuit of Fig. 17-1. Suppose V is a constant-voltage source which maintains 90 V across the divider network without load and under load. Without load, points A, B, and C are 30, 60, and 90 V, respectively, with respect to G. Moreover, a 3-mA bleeder current I_1 is drawn. Now if we add a load resistor R_L (at point B) which draws 2 mA of load current I_L, the voltages at A and B will be different from the condition of no load. See Fig. 17-2 for the load circuit. We can compute the voltages from the circuit information.

Assume that there is a bleeder current I_1 when a load current of I_L is drawn by R_L. By Kirchhoff's law we can set up the equation

$$I_1(R_2 + R_3) + (I_1 + I_L)(R_1) = 90 \qquad (17\text{-}1)$$

Solving for I_1 (since this is the only unknown), we get

$$I_1(R_1 + R_2 + R_3) = 90 - (R_1)I_L \qquad (17\text{-}2)$$

$$I_1 = \frac{90 - (I_L \times R_1)}{R_1 + R_2 + R_3} \qquad (17\text{-}3)$$

Substituting 10,000 for R_1, for R_2, and for R_3, and $I_L = 2/1000$ in Eq. (17-3), we get

$$I_1 = \frac{7}{3}\text{ mA} \qquad (17\text{-}4)$$

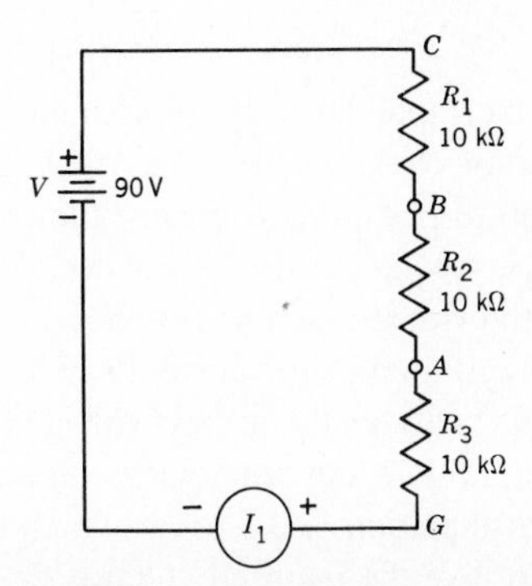

Fig. 17-1. Voltage divider without load.

It now follows that the voltage from A to G is

$$V_{AG} = I_1(10{,}000) = \frac{70}{3} = 23\tfrac{1}{3}\text{ V} \qquad (17\text{-}5)$$

and the voltage from B to G is

$$V_{BG} = 46\tfrac{2}{3}\text{ V} \qquad (17\text{-}6)$$

It becomes apparent that both the voltages and the bleeder current are affected when a load is added to a voltage-divider circuit.

To verify this effect experimentally, it is necessary to have a voltage source across the divider which will remain constant with and without load. In Fig. 17-2 the divider network consists of R_1, R_2, and R_3 connected to a voltage source.

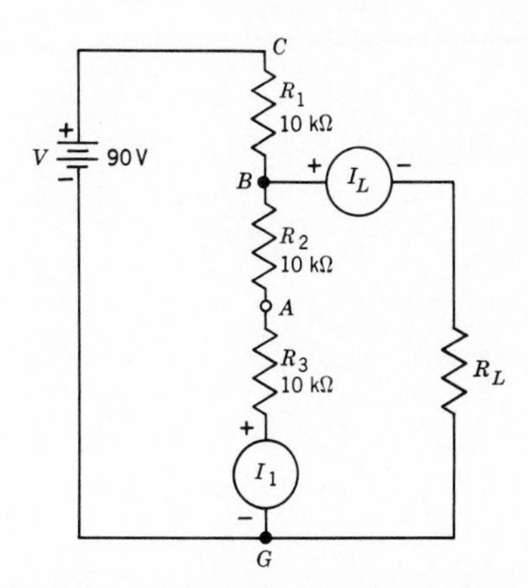

Fig. 17-2. Voltage divider with fixed load.

When a load resistor (R_L) is added as shown, the voltage source is readjusted if necessary to maintain the same voltage across the network as under no-load conditions. The required measurements are made under no-load and load conditions.

SUMMARY

1. A series-resistive circuit, such as that in Fig. 17-1, is a simple voltage divider without load.
2. The voltage, say V_1 across R_1, in such a divider may be computed by use of the formula

$$V_1 = V \times \frac{R_1}{R_T}$$

   Where V is the applied battery voltage and R_T is the sum of the series-connected resistors. The voltages at A, B, and C with relation to G can be determined in this manner.
3. If a load R_L drawing current I_L is added to such a circuit, Fig. 17-2, the voltages at A and B with relation to G, will change, as will the bleeder current I_1.
4. To determine analytically the new voltages in this voltage divider, Kirchhoff's current and voltage laws must be applied. The equation or equations which result can then be solved to give the required voltages.

SELF-TEST

Check your understanding by answering these questions:

1. In the circuit of Fig. 17-1, assume R_1 = 10 kΩ, R_2 = 18 kΩ, and R_3 = 22 kΩ. V = 150 V. Then the (*a*) voltage V_1 across R_1 = ________ V; (*b*) V_{BG} (across $R_2 + R_3$) = ________ V; (*c*) V_{AG} (across R_3) = ________ V.
2. In question 1, bleeder current I_1 = ________ mA.
3. In the circuit of Fig. 17-2, R_1 = 10 kΩ, R_2 = 18 kΩ, R_3 = 22 kΩ, and V = 150 V. I_L = 5 mA. The bleeder current I_1 = ________ mA.
4. For the conditions in question 3, (*a*) V_1 = ________ V; (*b*) V_{BG} = ________ V; (*c*) V_{AG} = ________ V.

MATERIALS REQUIRED

- Power supply: Regulated, variable direct current
- Equipment: EVM, 0–10-mA milliammeter, ohmmeter
- Resistors: ½-W, three 1200-Ω
- Miscellaneous: 10,000-Ω 2-W potentiometer, SPST switch

PROCEDURE

1. Connect the circuit of Fig. 17-3. Maintain a constant voltage V of 10 V in the output of the power supply.
2. With zero load current (i.e., with the lead to the 10,000-Ω rheostat open), measure the bleeder current I_1 (mA) and record in Table 17-1. Measure also and record the voltages V_{BG} and V_{AG}.
3. Connect the rheostat in the circuit and adjust it to draw 2 mA of load current while maintaining V = 10 V, measured. Measure and record the bleeder current and the voltages V_{BG} and V_{AG}. Open the load resistor R_L, but do not vary the setting of the arm. Measure and record the resistance to which it was set to draw 2 mA of load current. Reconnect R_L after this measurement.
4. Repeat step 3 for conditions of 4 mA and 6 mA of load current.
5. Compute, and record in Table 17-1, the bleeder current I_1, the voltages V_{BG} and V_{AG}, and the load resistance R_L for each of the load conditions in the experiment. Show your computations.

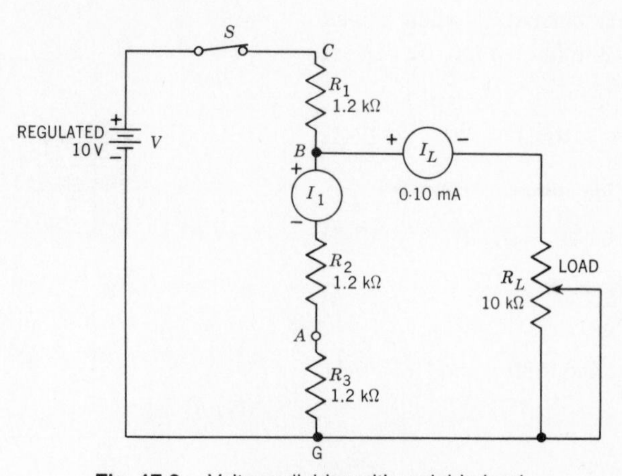

Fig. 17-3. Voltage divider with variable load.

TABLE 17-1. Effect of Load on *V* and *I* Relationships in Divider

		Measured Values				Computed Values			
V	I_L (load current), mA	I_1, mA	V_{BG}, V	V_{AG}, V	R_L, Ω	I_1, mA	V_{BG}, V	V_{AG}, V	R_L, Ω
10	0								
10	2								
10	4								
10	6								

QUESTIONS

1. Refer to the data in Table 17-1. How does load current vary with load resistance R_L? Explain why.
2. Refer to Table 17-1. What is the effect on bleeder current I_1 as the load current increases? Explain why.
3. What is the effect on the voltages V_{AG} and V_{BG} at the divider taps as the load current increases (Table 17-1)? Explain why.
4. Compare the computed values in Table 17-1 with the measured values. Explain any differences.

Answers to Self-Test

1. (*a*) 30; (*b*) 120; (*c*) 66
2. 3
3. 2
4. (*a*) 70; (*b*) 80; (*c*) 44

18

EXPERIMENT

VOLTAGE- AND CURRENT-DIVIDER CIRCUIT DESIGN CONSIDERATIONS

OBJECTIVES

1. To design a voltage divider which will meet specified voltage and current requirements
2. To design a current divider which will meet specified current and voltage requirements
3. To construct the circuits, and confirm, by measurement, the design parameters

INTRODUCTORY INFORMATION

Designing a Voltage Divider for Specific Loads

The electronics technician may be called upon to design a voltage divider which will meet specified requirements of voltage, load current, and bleeder current. The process to follow will be illustrated by an example.

Example. Design a voltage-divider circuit for a 300-V power supply. The loads which must be served are 0.05 A at 300 V, and 0.04 A at 250 V. The bleeder current under these conditions is 0.01 A.

NOTE: A "rule of thumb" value of bleeder current is 10 percent (approximately) of load current.

Solution. Draw a diagram, as in Fig. 18-1. Label the known load currents and voltages. Designate the *unknown resistors* as R_1 and R_2. (The loads are shown as resistors drawing the required currents.)

Apply Kirchhoff's current law and show the currents entering and leaving the junctions A, B, and C. The total current I_T which must be supplied is $I_T = 0.05 + 0.04 + 0.01 = 0.1$ A. I_T is shown entering the junction at C and leaving the junction at A.

The current I_1 in R_1 is the bleeder current. Now

$$I_1 \times R_1 = 250 \text{ V} \tag{18-1}$$

or

$$R_1 = \frac{250}{I_1} \Omega \tag{18-2}$$

Substituting the value of bleeder current (0.01) for I_1 gives

$$R_1 = \frac{250}{0.01} \Omega = 25{,}000 \Omega \tag{18-3}$$

The current in R_2 is 0.05 A, the sum of the bleeder current and load 1 current. Moreover, as is evident from Fig. 18-1, the voltage from A to B is 50 V. Therefore

$$(0.05)(R_2) = 50 \text{ V} \tag{18-4}$$

and

$$R_2 = \frac{50}{0.05} \Omega = 1000 \Omega \tag{18-5}$$

We have thus determined the proper values of R_1 and R_2 which will sustain a load of 0.04 A at 250 V, a load of 0.05 A at 300 V, a bleeder current of 0.01 A, from a 300-V power supply.

Designing Current-Divider Circuits

It is sometimes necessary for the technician to design current-divider circuits. To avoid trial and error methods, analysis is again suggested. The actual steps in the analytical method will depend on the nature of the design problems. But in general certain steps must be taken to facilitate the solution. These are:

1. Draw a circuit diagram labeling all the known values and the unknowns.
2. Apply Ohm's and Kirchhoff's laws and set up one or more equations which will reflect the electrical relationships involved.
3. Solve these equations for the unknown component values.

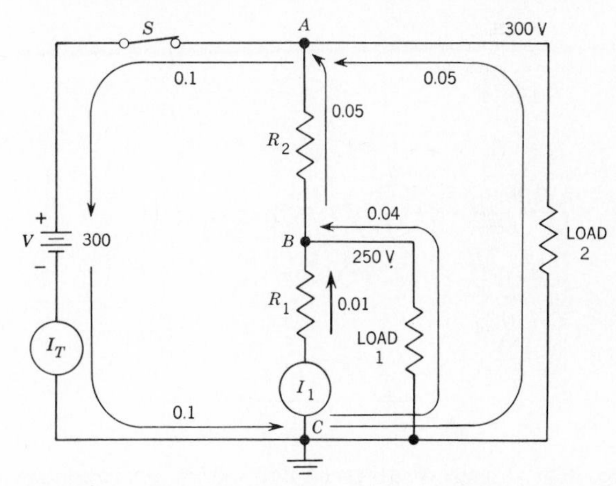

Fig. 18-1. Voltage-divider circuit for illustrative example.

4. Construct the circuit from the computed design values and by measurement determine if the design requirements have been met.

Several examples will illustrate this process.

Example 1. It is required to design a three-branch parallel network, like that in Fig. 18-2, such that a 100-V source V will deliver a total current I_T of 0.9 A. A further requirement is that the current I_T be distributed among R_1, R_2, and R_3, such that $I_1:I_2:I_3 = 2:3:4$. What values of R_1, R_2, and R_3 will satisfy these conditions?

Analysis. We know V, I_T, and the ratios $I_1:I_2:I_3$. If we could find the actual values of I_1, I_2, and I_3, we would then know two of the Ohm's law values for each of the branches, namely V and I. From V and I, R can be found, since

$$R = \frac{V}{I} \tag{18-6}$$

Solution. We can set up two equations for I from the current ratios which will help us.

$$\frac{I_1}{I_2} = \frac{2}{3} \qquad \frac{I_1}{I_3} = \frac{2}{4} \tag{18-7}$$

Solving for I_2 and I_3 in terms of I_1 yields

$$I_2 = \frac{3}{2} I_1 \qquad I_3 = \frac{4}{2} I_1 \tag{18-8}$$

Now applying Kirchhoff's current law,

$$I_T = I_1 + I_2 + I_3 = 0.9 \tag{18-9}$$

and substituting in Eq. (18-9) the values of I_2 and I_3 from Eqs. (18-8) gives

$$I_1 + \frac{3}{2} I_1 + \frac{4}{2} I_1 = 0.9 \tag{18-10}$$

Solving,

$$\frac{9I_1}{2} = 0.9$$

$$I_1 = 0.2 \text{ A} \tag{18-11}$$

Now substituting 0.2 for I_1 in Eqs. (18-8) gives

$$I_2 = 0.3 \text{ A} \qquad I_3 = 0.4 \text{ A} \tag{18-12}$$

Now we can find the values of R_1, R_2, and R_3. Thus

$$R_1 = \frac{V}{I_1} = \frac{100}{0.2} = 500\ \Omega$$

$$R_2 = \frac{V}{I_2} = \frac{100}{0.3} = 333.3\ \Omega \tag{18-13}$$

$$R_3 = \frac{V}{I_3} = \frac{100}{0.4} = 250\ \Omega$$

Example 2. Design a divider network which will supply three resistive loads from a 125-V source such that loads 1, 2, and 3 receive 0.02 A, 0.05 A, and 0.06 A, respectively. All three loads are in parallel, and the resistance of load 1 is 3000 Ω.

Solution. Draw the three parallel loads R_1, R_2, and R_3. Since the current and resistance of load 1 are given, we can find the voltage across load 1, and so across each of the other two parallel loads. Thus

$$V_1 = I_1 \times R_1 = 0.02(3000) = 60 \text{ V} \tag{18-14}$$

We have a 125-V source. Therefore we will have to drop 65 V before we reach the three parallel loads. A series-parallel circuit, such as that in Fig. 18-3, will serve our requirements. It is now a simple matter to find the values of R_4 and loads R_2 and R_3.

First, solving for R_2 and R_3,

$$R_2 = \frac{V_{AB}}{I_2} = \frac{60}{0.05} = 1200\ \Omega$$

$$R_3 = \frac{V_{AB}}{I_3} = \frac{60}{0.06} = 1000\ \Omega \tag{18-15}$$

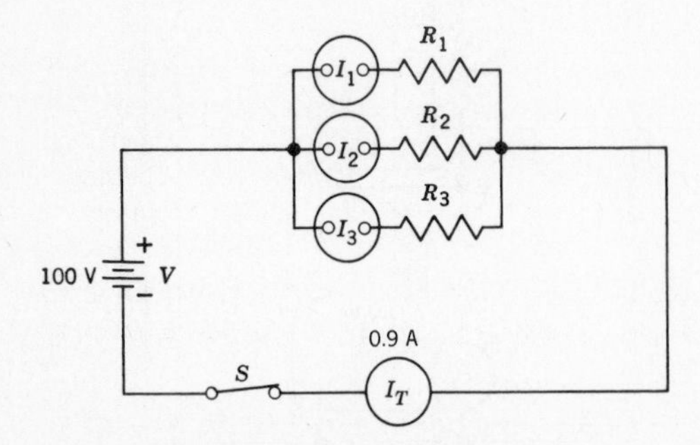

Fig. 18-2. Current-divider circuit. If $I_T = 0.9$ A, what are the values of I_1, I_2, and I_3?

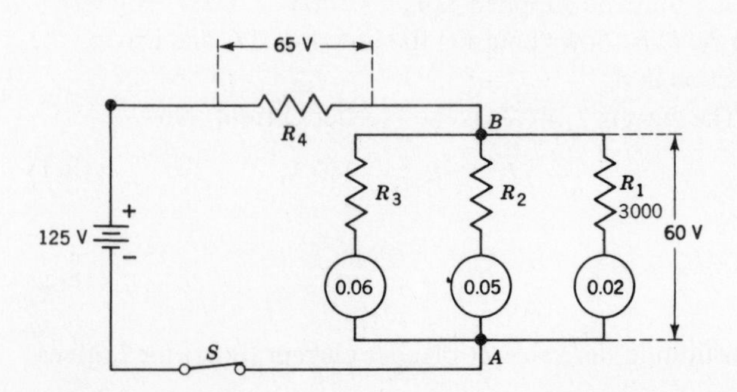

Fig. 18-3. A voltage- and current-divider network designed to supply three loads from a 125-V source.

Finally, the total current I_T in R_4 may be found by Kirchhoff's current law:

$$I_T = I_1 + I_2 + I_3 = 0.02 + 0.05 + 0.06$$
$$= 0.13 \text{ A}$$

Therefore

$$R_4 = \frac{65}{0.13} = 500\ \Omega \tag{18-16}$$

SUMMARY

1. In designing either voltage- or current-divider circuits, an analytical solution is found first. Ohm's and Kirchhoff's laws are applied in the analysis.
2. The general steps in analyzing a design problem are:
   (*a*) Draw a diagram of the proposed circuit, labeling all the knowns and unknowns.
   (*b*) Set up one or more equations which reflect the electrical relationships involved.
   (*c*) Solve these equations for the values of the unknown components.
   (*d*) Construct the circuit from the computed design values and determine by measurement if the design requirements have been fulfilled.
3. In designing voltage-divider circuits for specified load currents, a rule-of-thumb value for bleeder current is 10 percent (approximately) of load current.

SELF-TEST

Check your understanding by answering these questions:

1. It is required to design a voltage-divider circuit like that in Fig. 18-1, so that a 200-V source V can supply two loads, a 0.1-A load 1 at 50 V and a 0.4-A load 2 at 200 V. Assume a bleeder current I_1 of 0.05 A. The values of R_1 and R_2 which will accomplish this result are:
   (*a*) R_1 = ________ Ω
   (*b*) R_2 = ________ Ω
2. For the same conditions as in question 1, the resistances of the two loads are:
   (*a*) Resistance of load 1 is ________ Ω
   (*b*) Resistance of load 2 is ________ Ω
3. In a divider circuit similar to that in Fig. 18-3, it is required to supply three equal-load resistors, drawing a total current of 0.15 A, at a load voltage of 20 V. The supply source V is 50 V. The values of R_1, R_2, R_3, and R_4 which will achieve this are:
   (*a*) R_1 = ________ Ω
   (*b*) R_2 = ________ Ω
   (*c*) R_3 = ________ Ω
   (*d*) R_4 = ________ Ω

MATERIALS REQUIRED

- Power supply: Regulated, variable dc source
- Equipment: EVM, 0–10-mA milliammeter
- Resistors: ½-W, as required for procedural steps 1 to 6
- Miscellaneous: SPST switch, and potentiometer(s) as required for procedural steps 1 to 6

PROCEDURE

1. Design a voltage-divider circuit for a 35-V regulated power supply which must feed a 9-mA load at 20 V. The bleeder current should be 1 mA (approximately). Draw the circuit diagram, showing all values of voltage, current, and resistance. Show your computations, and have your instructor check the solution before you connect the circuit.

2. Select the required resistors from your kit. If your kit does not contain a design-value resistor, adjust a potentiometer connected as a rheostat to the desired value, or make up the resistor from a combination of other resistors.

3. Connect the circuit. Measure the required voltages and currents and record them in Table 18-1, which you will construct for the required data.

4. Design a circuit which will feed three parallel loads from a 40-V source. The total current drawn by the three loads must equal 7 mA, and the load currents must be in the ratio 1:2:4. The load voltage is 20 V. Draw the circuit diagram, showing all values of voltage, current, and resistance. Show your computations, and have an instructor check your solution before you connect the circuit.

5. Select the required resistors from your kit. If your kit does not contain a design-value resistor, adjust a potentiometer connected as a rheostat to the desired value, or make up the resistor from a combination of other resistors.

6. Connect the circuit. Measure the required voltages and currents and record them in Table 18-2, which you will construct for the required data.

QUESTIONS

1. In the circuit of Fig. 18-1 what would be the effect on the voltage V_{BC} of:
   (*a*) Increasing load 1 current? Why?
   (*b*) Increasing load 2 current? Why?
2. In the circuit of Fig. 18-1, for the design requirements shown, will the value of current in load 2 affect the values of R_1 and R_2? Explain.
3. How did the measured values of current in step 3 compare with the design values? Refer specifically to your measurements in Table 18-1. Explain any discrepancies.

4. How did the measured values of current in step 6 compare with the design values? Refer specifically to your measurements in Table 18-2. Explain any discrepancies.
5. Write a list of Materials Required for procedural steps 1 to 6.

Extra Credit

6. Write a general formula for R_1, and another for R_2 (Fig. 18-1). Designate the supply voltage as V, the voltage for load 1 as V_1, the bleeder current as I_1, load 1 current as I_{L1}, and load 2 current as I_{L2}.

Answers to Self-Test

1. (*a*) 1000
   (*b*) 1000
2. (*a*) 500
   (*b*) 500
3. (*a*) 400
   (*b*) 400
   (*c*) 400
   (*d*) 200

19

EXPERIMENT

DEFECT ANALYSIS BY VOLTAGE, CURRENT, AND RESISTANCE MEASUREMENT

OBJECTIVES

1. To investigate an analytical system of troubleshooting electric circuits by voltage, current, and resistance measurements
2. To apply these troubleshooting techniques in locating a defective part in a:
   (*a*) Series circuit
   (*b*) Parallel circuit
   (*c*) Series-parallel circuit

INTRODUCTORY INFORMATION

The job of electronics technicians has many aspects. They must be competent in many areas. One of the interesting and challenging tasks they face is the repair of electronic equipment which has become defective.

Modern equipment is too complex to follow hit-or-miss troubleshooting procedures. What is required is a logical, analytical, scientific troubleshooting system which will assure locating a defective component efficiently and quickly. Some of the elements of such a system will be considered in this experiment.

Electronic equipment is made up of direct-current and alternating-current circuits. Direct-current circuits consist of one or more sources of power (voltage and current), a network of resistors, and other elements with which you will become familiar in due time. If trouble has been isolated to a dc network, how is the defective component found? One way is to replace the entire network of components with known good parts. This, however, may be too time-consuming and costly. A better method is to find the defective part and replace it. Our attention, then, will be directed here to an analytical system of locating a defect in a dc circuit containing a power source, resistors, and interconnecting wiring.

Defects in a DC Circuit

Consider the circuit (Fig. 19-1). A power source V supplies 100 V through a switch S to three resistors, $R_1 = 2000\ \Omega$, $R_2 = 3000\ \Omega$, and $R_3 = 5000\ \Omega$, connected in series. When S is open, as shown, there is no complete path for current. There is *no* current I. Since there is no current, there is no voltage drop ($I \times R$) across any of the resistors. A current meter M connected in the circuit under these conditions would measure zero current. A voltmeter connected across R_1, R_2, or R_3 would measure 0 V in each case.

When S is closed, there is a complete circuit. Under normal conditions, the measured current is 10 mA. The measured voltage drop across R_1 is 20 V. Across R_2 it is 30 V, and across R_3 it is 50 V. These values of voltage and current can be computed before measurement by applying the laws of electricity. Thus

$$I = \frac{V}{R_T} = \frac{100}{10{,}000} = \frac{10}{1000} = 10 \text{ mA} \qquad (19\text{-}1)$$

The voltage V_1 across R_1 is

$$R_1 = I \times R_1 = \frac{10}{1000} \times 2000 = 20 \text{ V} \qquad (19\text{-}2)$$

Similarly, the other voltages, V_2 and V_3, can be evaluated.

In Fig. 19-1, if any one of the components changes value or becomes defective, circuit conditions are altered. Measurement would show values of current and voltage different from the expected (computed) values. This condition would affect the operation of any circuit or device dependent on normal operation of Fig. 19-1. For example, if M were a small motor which required 10 mA to make it turn, a change in the circuit which appreciably reduced the current below 10 mA would make the motor unable to run.

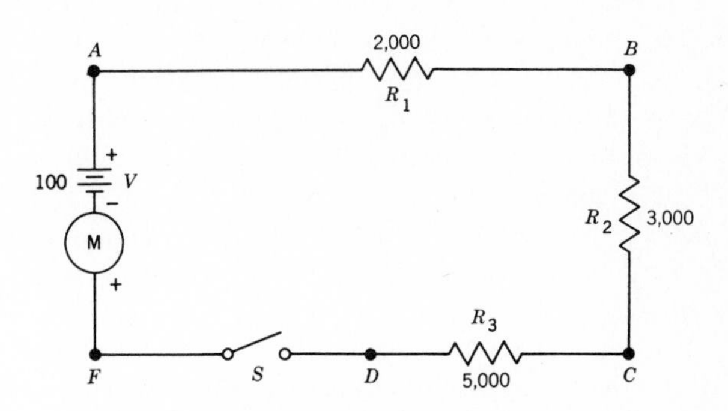

Fig. 19-1. Troubleshooting a series dc circuit.

Changes in circuit operation can be caused by changes in the characteristics of individual components.

Resistors. The resistance of a resistor can increase or decrease. A resistor may become *open* or short-circuited. An open resistor acts like an open switch. It results in an *incomplete* or open circuit. A short-circuited resistor acts like an interconnecting wire which has *zero* resistance.

Switch. Mechanical troubles in a switch may cause the switch to remain *open* at all times, *shorted* at all times (that is, acting as though it were always closed), or operate intermittently (work properly sometimes, but not always).

Interconnecting Wires. The inner conductive wire, covered by insulation, may break, opening the circuit as an open switch would. An interconnecting wire may also become intermittent in use.

Power Source. A battery develops high internal resistance, with time. The terminal voltage under load (that is, when it is supplying current to a circuit) will therefore drop to a lower value than that for which the battery is rated.

The output of an electronic power supply which is not regulated will also drop under load. That is, the no-load voltage will be higher than the voltage under load. Though this is a normal characteristic, it must be taken into consideration.

Finally, a power source may become totally inoperative and produce no voltage.

Checking Components

Resistors, switches, and connecting wires, out of the circuit, may be checked with an ohmmeter. The resistance of a resistor is measured to determine if it equals its rated value. Tolerance must be considered. A 1200-Ω ± 10 percent resistor is considered within tolerance if it measures within the range 1080 to 1320 Ω.

The "continuity" of a switch may be checked by measuring resistance across its terminals. When the switch is open, the ohmmeter should show *infinite* resistance. When the switch is closed, the reading should be *zero* ohms.

A continuity check of an interconnecting wire will show whether it is open (infinite resistance) or closed (zero ohms). Flexing the wire while the meter is connected across it may reveal an intermittent condition.

The resistors and interconnecting wires in the series circuit of Fig. 19-1 may be checked individually or all together, without removing them from the circuit. Switch S must be open, however, because resistance must *never* be measured in a circuit to which power is applied. The resistance between points A and B must measure 2000 Ω; between B and C, 3000 Ω; and between C and D, 5000 Ω. Measuring the circuit in this manner checks the resistors and interconnecting wires.

If a resistance check between points A and D shows 10,000 Ω, we may assume that the components and interconnecting wires between A and D are good.

The power source may be checked with a voltmeter. The terminal voltage of the supply (or battery) must be measured under load; that is, S must be closed, permitting the load (R_1, R_2, and R_3 in this case) to draw current.

Dynamic Troubleshooting Measurements

The resistance of resistors and the continuity of switches and interconnecting wires may measure "good" out of the circuit. Yet these parts may break down or change under load. A *dynamic* test of their operation under load is therefore necessary.

Dynamic measurements may be required for another reason. It may be inconvenient to remove parts from a circuit or even to unsolder one end of a component to free it for resistance measurement. In this case, also, circuit measurements under load are indicated.

In a dynamic test, voltage and current checks are made to determine if the measured values equal the computed values. If they are different, analysis of the measurements should provide the clue to the defective part.

Troubleshooting Axioms and Troubleshooting Assumption

The conclusions which may be drawn from dynamic tests are predicated on two basic troubleshooting axioms. They are:

1. In a resistive circuit which is operating normally, the voltage across individual linear resistors and the current in every part of the circuit are distributed according to Ohm's and Kirchhoff's laws.
2. Conversely, if the voltage across individual linear resistors or the current in any part of the circuit is not distributed according to Ohm's and Kirchhoff's laws and the indicated values of resistance, then the circuit is not operating normally, that is, there is a defect in the circuit.

In addition to the two axioms, the following basic troubleshooting assumption is helpful in an evaluation of circuit measurements: The trouble in a resistive circuit which is not operating normally is due to *one* and *only one* defective component.

This assumption greatly simplifies troubleshooting analysis and, from a statistical standpoint, is relatively accurate. Of course, if the circuit still operates abnormally after *one* defective component is found and is replaced with a known good component, we proceed to look for a second *single* possible defective component, etc. *In all cases the technician should determine the cause for the initial failure and correct it.*

Troubleshooting a Series Circuit

A circuit illustration will serve to demonstrate the troubleshooting process. Consider Fig. 19-1. Voltage V measures 100 V. Current in the circuit measures 4 mA, rather than 10 mA. Voltage V_1 across R_1 is 8 V, V_2 cross R_2 is 12 V, V_3 across R_3 is 80 V. Since current is everywhere the same in a series circuit, by applying Ohm's law we can compute the value of each resistor from the measured values. The computed values are: $R_1 = 2000\ \Omega$, $R_2 = 3000\ \Omega$, and $R_3 = 20{,}000\ \Omega$. It is apparent, therefore, that R_3 has increased in value to 20,000 Ω. Replacing R_3 with a 5000-Ω resistor will restore the circuit to normal operation.

The measured values can be interpreted in another way. Since current has decreased from 10 mA to 4 mA, it follows that the total resistance of the circuit has increased. The troubleshooting assumption states that there is just one defect in the circuit. Hence, just *one* resistor has increased in value, namely, R_3, because R_3 is the only resistor across which the voltage has increased despite a decrease in circuit current.

The following conclusion can be applied to any series circuit: In a series circuit a *decrease* in current results from an increase in total resistance. The defective resistor is the one across which an *increased* voltage is measured.

It can be readily demonstrated that if the current in a series circuit *increases*, the change results from a *decrease* in total resistance. The resistor whose value has *decreased* is the only one across which a *decreased* voltage is measured. Across each of the other resistors, the voltage has increased.

These conclusions may be drawn about a series circuit only if the applied voltage V measures the same as the indicated circuit value. If the measured value of V is not the same, it must be determined why, before any other step is taken. Hence, *one* measurement which must be made is a voltage check of the source V.

If the current in Fig. 19-1 measures zero, it is because the circuit is open and there is no complete path for current. A logical procedure to follow is to check the voltage of the source V first. If that is normal, power is removed and an ohmmeter may be used to check the continuity of the switch, connecting wires, and the resistance of the resistors.

There is another method to find the discontinuity (open) in the circuit. The negative lead of the voltmeter is connected to the (−) terminal of the battery. Switch S is closed. The positive lead is then used to probe the circuit. First place the positive lead at A. If the battery and interconnecting leads are good, the full voltage V, 100 V in this case, will be measured. Next, shift the positive lead to B. If R_1 and the interconnecting wires are good, the meter will read 100 V. However, if either R_1 or the interconnecting wires is open, the voltage will be zero.

We continue probing the circuit from points A to B to C to D until we find a point where the voltage is zero. The trouble lies between that point and the point immediately preceding it, where the voltage read the full value V.

NOTE: If the voltage from D to F reads 100 V, then switch S is defective (open).

Troubleshooting a Parallel Circuit

Consider the parallel circuit, Fig. 19-2. By Kirchhoff's laws, $I_T = I_1 + I_2 + I_3$ and $I_1 = V/R_1$, $I_2 = V/R_2$, $I_3 = V/R_3$.

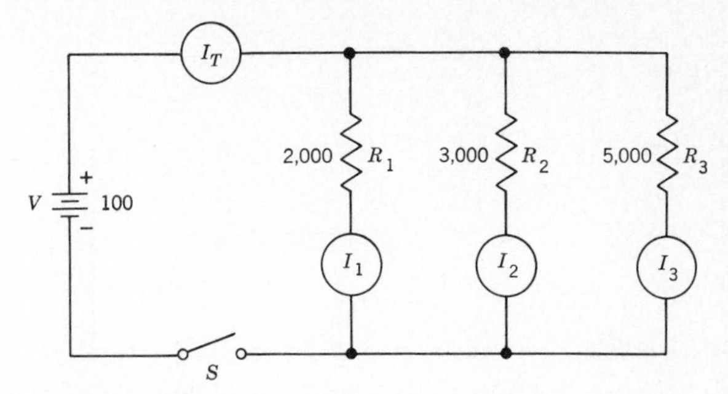

Fig. 19-2. Troubleshooting a parallel dc circuit.

It is therefore, a simple matter to compute the values I_1, I_2, I_3, and I_T from the circuit voltage V and the rated values of R_1, R_2, and R_3. In Fig. 19-2, the computed values are $I_1 = 50$ mA, $I_2 = 33\frac{1}{3}$ mA, $I_3 = 20$ mA, and $I_T = 103\frac{1}{3}$ mA.

Suppose V measures 100 V as required and the total current I_T measures $83\frac{1}{3}$ mA. It is apparent that one of the branch currents has decreased, and that this decrease results from an increase in the resistance of that branch. Measurement shows that I_2 and I_3 are normal, but $I_1 = 30$ mA. R_1 is therefore defective, and its resistance has increased from 2000 to 3333 Ω. Replacing R_1 with a 2000-Ω resistor will restore the circuit to normal operation.

If V measures 100 V, and I_T has increased to 170 mA, we conclude that one of the branch currents has increased, and that this increase results from a decrease in resistance in that branch.

Measurement shows that I_1 and I_3 are normal, but that $I_2 = 100$ mA. Therefore, R_2 is defective because its resistance has decreased from 3000 to 1000 Ω.

The following conclusions may be drawn about a parallel circuit: If the source voltage measures the required voltage V, and the measured value of I_T is:

1. *Smaller* than the computed value, the decrease in I_T results from an *increase* in resistance of a resistor in one of the parallel branches. That resistor is defective whose branch current is *smaller* than the computed value.
2. *Greater* than the computed value, the increase in I_T results from a *decrease* in resistance of a resistor in one of the parallel branches. That resistor is defective whose branch current is *greater* than the computed value.

As in the case of the series circuit, these conclusions are valid only if the applied voltage V measures the same as the indicated circuit value. Hence, one measurement which must be made is a voltage check of V. The other measurements in the troubleshooting process include I_T, and measurement of the branch currents until the defective resistor is found.

Troubleshooting a Series-Parallel Circuit

In the series-parallel circuit (Fig. 19-3), computation reveals that the normal resistance between points B and C, $R_{BC} = 2000\ \Omega$. R_T of the circuit $= 9000\ \Omega$, $I_T = 10$ mA, V_{AB} (across R_1) $= 20$ V, $V_{BC} = 20$ V, $V_{CD} = 50$ V, $I_2 = 6\frac{2}{3}$ mA, and $I_3 = 3\frac{1}{3}$ mA.

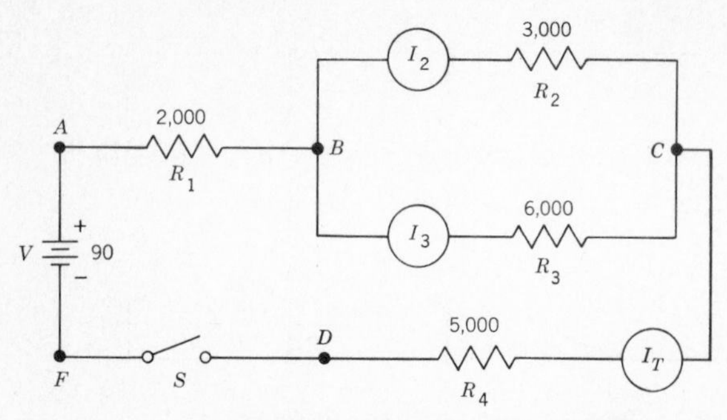

Fig. 19-3. Troubleshooting a series-parallel dc circuit.

Suppose V measures 90 V as required and I_T measures 8 mA. It is apparent that R_T has *increased* and that this increase is due to an increase in the resistance of one of the resistors. The procedure for finding the defective resistor is the same as in the case of a series circuit. The voltage is measured across each of the resistors and across each parallel circuit. The defective resistor is the one, or is one of the branch resistors of the parallel circuit, across which an *increased* voltage is measured.

For example, in Fig. 19-3, if the measured values are $I_T = 8$ mA, $V = 90$ V, $V_{AB} = 16$ V, $V_{BC} = 34$ V, and $V_{CD} = 40$ V, the trouble must be in R_2 or R_3. Now a measurement of branch currents I_2 and I_3 will reveal the defective resistor. Thus, if $I_2 = 2\frac{1}{3}$ mA and $I_3 = 5\frac{2}{3}$ mA, the trouble is in R_2, which has changed value from 3000 to 14,570 Ω.

If the current in a series-parallel circuit *increases*, the change results from a *decrease* in total circuit resistance R_T. The resistor whose value has decreased, or the parallel circuit whose voltage has *decreased*, is the only one across which a *decreased* voltage is measured. Trouble in a branch of a parallel circuit is found in the same manner as in the parallel circuit in Fig. 19-2.

The method of troubleshooting a series-parallel circuit, then, involves:

1. Computation of the normal circuit values of current and voltage
2. Measurement of the actual values of current and voltage
3. Analysis of the measurements, applying the principles of both series- and parallel-circuit troubleshooting techniques

SUMMARY

1. In a dc circuit, troubles in operation may result from any defective component in the circuit. Thus resistors may open or change value; switches may open, short, or become intermittent; interconnecting wires may open or become intermittent; power sources may develop high internal resistance or become totally inoperative.
2. Components can be checked with an ohmmeter for resistance and continuity. Accordingly, a resistor should measure within the tolerance of its color-coded value. An open switch will measure infinite resistance across its terminals; a shorted or closed switch will measure zero resistance. Interconnecting wires, if open, will measure infinite resistance.
3. The output voltage of a power source should be checked under load. If it is lower than the rated value, it either has developed a high internal resistance, is defective for other internal reasons, or is being excessively loaded (that is, too much current is being drawn).
4. Ohmmeter checks of components are considered static checks. Dynamic tests of all components made when they are operating under load, that is, with power applied, are more reliable than ohmmeter tests.
5. In a dynamic test, voltage and current checks are made in a circuit to determine if the measured values are the same as the computed values. If they are different, analysis of the measurements should help identify the defective part.
6. One troubleshooting assumption is that if the voltage across linear resistors or the current in such a circuit is not distributed according to Ohm's and Kirchhoff's laws, then there is a defect in the circuit.
7. In the troubleshooting process it also simplifies matters to assume that there is just *one* defect in the circuit.
8. In a *series* circuit a *decrease* in current results from an *increase* in total resistance. The defective resistor is the one across which an *increased* voltage is measured.
9. In a *series* circuit an *increase* in current results from a *decrease* in total resistance. The resistor whose value has *decreased* is the one across which a *decreased* voltage is measured.
10. In troubleshooting any circuit, check the output voltage of the power source to determine if it is the same as its rated value.
11. If the current in a *series* circuit is *zero*, some component is open, or the output of the power source is zero volts.
12. In a *parallel* circuit if the source voltage is normal and the measured value of I_T is:
    (*a*) *Smaller* than the computed value, the decrease in I_T results from an *increase in resistance* of a resistor in one of the parallel branches. *That resistor is defective whose branch current is smaller than the computed value.*
    (*b*) *Greater* than the computed value, the increase in I_T is the result of a *decrease in resistance* of one of the resistors in the parallel branches. *That resistor is defective whose branch current is greater than the computed value.*
13. If the current in a *series-parallel* circuit has *decreased*, there has been an *increase* in the *total resistance* of the circuit.
14. If the current in a *series-parallel* circuit has *increased*, there has been a *decrease* in the *total resistance* of the circuit.
15. In troubleshooting a series-parallel circuit, the same procedures are followed as in troubleshooting series circuits and parallel circuits, including voltage and current measurements and analysis of these measurements, applying the principles of both series- and parallel-circuit troubleshooting techniques.

SELF-TEST

Check your understanding by answering these questions:

1. In the circuit of Fig. 19-1, the color-coded value of R_1 is 2200 Ω, R_2 is 4700 Ω, and R_3 is 8200 Ω. V = 150 V. When the switch is closed, milliammeter M reads 4.5 mA. The voltage V_1 (measured across R_1) = 10.0 V, V_2 = 103 V, V_3 = 37.0 V.
   (*a*) If the circuit were normal, the values of current and voltage would be:
   (1) I = ______ mA
   (2) V_1 = ______ V
   (3) V_2 = ______ V
   (4) V_3 = ______ V
   (*b*) What has happened to the total resistance in the circuit? ______ (increased/decreased)
   (*c*) The defective component is ______.
2. In the circuit of Fig. 19-1, the resistors are color-coded the same as in question 1 and V = 150 V. When the switch is closed, milliammeter M reads *0* current. The measured voltages V_{AF} = 150 V, V_{BF} = 150 V, V_{CF} = 150 V. The components which are possibly defective (open) are: ______ or ______ or ______.
3. In the circuit of Fig. 19-2, the color-coded values of the resistors are: R_1 = 1200 Ω, R_2 = 1800 Ω, R_3 = 3000 Ω, and V = 300 V. When the switch S is closed, the measured currents are as follows: I_T = 0.4 A, I_1 = 0.133 A, I_2 = 0.167 A, and I_3 = 0.1 A.
   (*a*) If the circuit were normal, the total current I_T would measure ______ A.
   (*b*) The total resistance in the circuit has therefore ______ (increased/decreased).
   (*c*) The normal currents should be:
   (1) I_1 = ______ A
   (2) I_2 = ______ A
   (3) I_3 = ______ A
   (*d*) The defective resistor is ______ whose resistance has ______ (increased/decreased).
4. In the circuit of Fig. 19-2, when switch S is closed, I_T = 0, I_1 = 0, I_2 = 0, I_3 = 0, V = 100 V. With switch S open, an ohmmeter check from the positive terminal of V to the right-hand terminal of S shows continuity. Assuming all the wiring is good, the defective component must be (*a*) ______ which is (*b*) ______ (open/shorted).
5. In the circuit of Fig. 19-3 the measured voltages are as follows: V_{AB} = 18 V, V_{BC} = 27 V, V_{CD} = 45 V. The defective component is (*a*) ______ which is (*b*) ______ (open/decreased resistance).

MATERIALS REQUIRED

- Power supply: Regulated variable direct current
- Equipment: EVM, 0–10-mA milliammeter, 0–100-mA milliammeter
- Resistors: ½-W 2200-, 2700-, 3300-, 4700-, 5100-, 5600-, 6800-, 8200-, 10,000-, and 12,000-Ω

NOTE: Your instructor will assign resistors as required from the coded values shown above. One or more of these resistors will be defective.

- Miscellaneous: SPST switch, interconnecting wires

NOTE: These parts may also be either good or defective.

PROCEDURE

1. Connect a series circuit similar to that in Fig. 19-1, using the components assigned to you by your instructor.

NOTE TO INSTRUCTOR: Resistors R_1, R_2, and R_3 should be selected so that their normal total resistance is no greater than 10,000 Ω. The smallest normal-valued resistor should be no less than 2200 Ω.

Apply 40 V across the circuit and maintain the voltage V at this level. One of the components will be defective, but it will not be possible to determine which simply by observation.

Draw a circuit diagram and on it show the color-coded value of each resistor. Also record the rated values in Table 19-1. Compute the total resistance R_T, the value of voltage which should appear across each resistor under normal operation, and the circuit current I, and record in Table 19-1. Show your computations.

2. Troubleshoot the circuit, entering in Table 19-1 each measurement you actually make. Analyze your measurements, and under "Analysis" in Table 19-1 record the defect which you believe exists. State the reasoning which led to your conclusion.

Verify your conclusion by replacing the defective component with a known good part and measure the current I. Indicate under "Verification and measurements" the measured current I and whether the defect was found. If you were incorrect in your analysis, reevaluate your conclusion and find the defective part. List the defective part in the column headed "Defect."

3. Connect a parallel circuit similar to that in Fig. 19-2, using the components assigned by your instructor.

NOTE TO INSTRUCTOR: The normal values of R_1, R_2, and R_3 should be so selected that the total current I_T is no greater than 30 mA. The smallest normal-valued resistor should be no smaller than 2200 Ω.

Apply 30 V across the circuit and maintain the voltage V at this level. Draw a circuit diagram and on it show the color-coded value of each resistor. Also record these rated values in Table 19-2. Compute the currents I_1, I_2, I_3, and I_T which should be found under normal operation, and record in Table 19-2. Show your computations.

4. Troubleshoot the circuit as in step 2. Enter the measurements you actually make, your analysis, suspected defect, verification, etc., in Table 19-2.
5. Connect a series-parallel circuit similar to that in Fig. 19-3, using the components assigned by your instructor. Apply 40 V across the circuit and maintain the voltage V at this level. Draw a circuit diagram, and on it show the color-coded values of each resistor. Also record the rated values in Table 19-3. Compute the total resistance R_T, total current I_T, the voltages V_{AB}, V_{BC}, and V_{CD}, and the currents I_2 and I_3, and record in Table 19-3. Show your computations.
6. Troubleshoot the circuit, as in steps 2 and 4. Enter the measurements you actually make, your analysis, suspected defect, verification, etc., in Table 19-3.

TABLE 19-1. Measurements in Troubleshooting a Defective Series Circuit

Resistance			*Voltage*			*Current*			
	Rated	*Computed*		*Computed*	*Measured*		*Computed*	*Measured*	*Defect*
R_T	—		V	40		I_T			
R_1		—	V_1			I_1	—	—	
R_2		—	V_2			I_2	—	—	
R_3		—	V_3			I_3	—	—	

Analysis:

Verification and measurements:

TABLE 19-2. Troubleshooting a Defective Parallel Circuit

Resistance			*Voltage*			*Current*			
	Rated	*Computed*		*Computed*	*Measured*		*Computed*	*Measured*	*Defect*
R_T	—		V	30		I_T			
R_1		—	V_1			I_1			
R_2		—	V_2			I_2			
R_3		—	V_3			I_3			

Analysis:

Verification and measurements:

TABLE 19-3. Troubleshooting a Defective Series-Parallel Circuit

Resistance			*Voltage*			*Current*			
	Rated	*Computed*		*Computed*	*Measured*		*Computed*	*Measured*	*Defect*
R_T	X		V	40		I_T			
R_1		X	V_{AB}			X	X	X	
R_2		X	V_{BC}			I_2			
R_3		X	X	X	X	I_3			
R_4		X	V_{CD}			X	X	X	

Analysis:

Verification and measurements:

QUESTIONS

1. State in your own words the two troubleshooting axioms.
2. Is the troubleshooting assumption valid? Why is it used?
3. In a series circuit what is the effect of one increased resistance on:
   (*a*) Current in the circuit?
   (*b*) Voltage across each of the resistors?

NOTE: Assume that the voltage source remains constant.

4. In a parallel circuit, assuming that the voltage source remains constant, what is the effect of one increased resistance on:
   (*a*) Total current in the circuit?
   (*b*) Each branch current?
5. Same as question 4, except discuss the effect of a *decreased* resistance.
6. Same as question 3, except discuss the effect of a *decreased* resistance.
7. In a series-parallel circuit, assuming that the voltage source remains constant, what is the effect of an *increased* series resistance on:
   (*a*) Total current?
   (*b*) Individual branch currents?
   (*c*) Voltage across each resistor and across each parallel circuit?
8. Same as question 7, except discuss the effect of a *decreased* parallel resistance.
9. How can you check whether the following parts are good:
   (*a*) Resistor
   (*b*) Switch
   (*c*) Connecting wire
   (*d*) Battery
10. Explain a dynamic check for finding an open resistor in a:
    (*a*) Series circuit
    (*b*) Parallel circuit

Answers to Self-Test

1. (*a*) (1) 9.93; (2) 21.8; (3) 46.7; (4) 81.5
   (*b*) increased
   (*c*) R_2
2. R_3; *S*; M
3. (*a*) 0.517
   (*b*) increased
   (*c*) (1) 0.25; (2) 0.167; (3) 0.1
   (*d*) R_1; increased
4. (*a*) *S*; (*b*) open
5. (*a*) R_3; (*b*) open

20

EXPERIMENT

INTERNAL RESISTANCE OF A BATTERY

OBJECTIVE

To measure the internal resistance of a dry cell

INTRODUCTORY INFORMATION

Power Loss

Certain losses are associated with a power source which is delivering current to a load. These losses prevent the power source from operating at 100 percent efficiency. We say that this power loss is due to the *internal resistance* of the source. In this experiment we will be concerned with the internal resistance of a dry cell.

Internal Resistance of a Battery

Consider the power source of Fig. 20-1. This is a battery whose output voltage without load is V volts. When current is drawn from the battery, that is, when a load R_L is connected across the battery, its output voltage may drop to some lower value V_O, and the value of V_O will vary, depending on the load current drawn. The higher the current, the lower V_O will become. How can this characteristic of the battery be explained?

To explain this phenomenon it is necessary to assume that there is associated with the battery an internal resistance r_{in} which acts as a lumped resistor in series with V, as in Fig. 20-2. Now when a load R_L is connected across the battery, the voltage across R_L (that is, the V_O of the battery under load) will depend on the *ideal* voltage V (that is, the voltage of the battery without load) and on the values of r_{in} and R_L, because r_{in} and R_L act as a voltage divider across the battery. The higher the load current drawn (that is, the lower the load resistance R_L) the higher will be the voltage drop across r_{in} and the lower will be the output voltage V_O across the battery.

Now, what determines the value of this internal resistance? The internal resistance of a battery (or dry cell) depends on its composition and construction, on its power-handling capacity, on the length of time it has been in use, and on the extent of use it has had. This suggests that r_{in} will vary with the age and use of the battery, and that is true.

Determining the Internal Resistance of a Battery

The internal resistance of a battery can be calculated if certain facts about the battery's operation are known. One set of facts which makes it possible to calculate r_{in} is: (1) the ideal voltage V of the battery (without load); (2) the output voltage V_O under load, and (3) the value of the load R_L. How are V, V_O, R_L, and r_{in} related? Let us see.

Assume in the circuit of Fig. 20-2 that V is the battery's no-load voltage and that V_O is the voltage the battery delivers when a load R_L is connected across it. Assume further that I_L is the load current and that r_{in} is the battery's internal resistance. By Kirchhoff's voltage law, V is equal to the sum of the voltage drops across r_{in} and R_L. That is,

$$V = I_L \times r_{in} + I_L \times R_L \qquad (20\text{-}1)$$

but

$$I_L \times R_L = V_O \qquad (20\text{-}2)$$

therefore

$$V - V_O = I_L \times r_{in}$$

and

$$r_{in} = \frac{V - V_O}{I_L} \qquad (20\text{-}3)$$

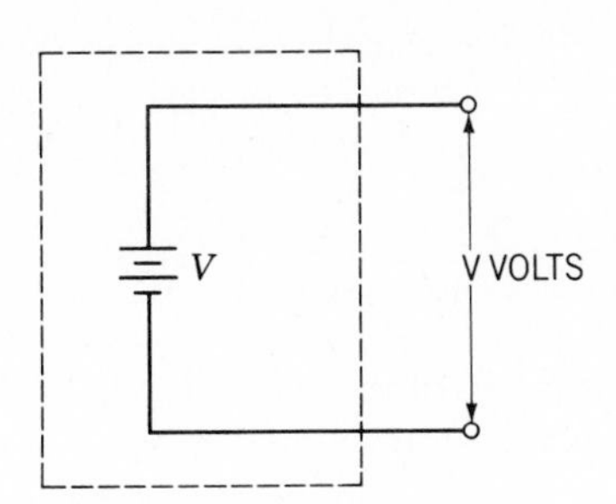

Fig. 20-1. V is a battery whose output voltage without load is V volts.

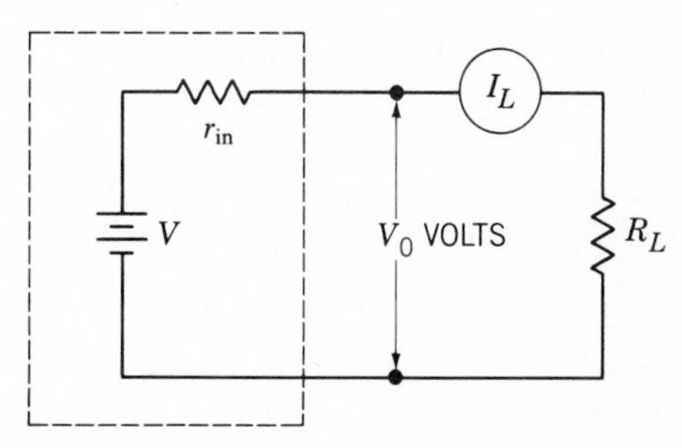

Fig. 20-2. Internal resistance r_{in} of battery V acts in series with the battery. When a current I_L is drawn by a load R_L, the output voltage of V drops to V_O volts.

From Eq. (20-2) we know that

$$I_L = \frac{V_O}{R_L}$$

Therefore, $$r_{in} = \frac{V - V_O}{V_O/R_L} = \frac{V - V_O}{V_O} \times R_L \qquad (20\text{-}4)$$

where $V - V_O$ is the voltage lost across the battery's internal resistance. Equation (20-4) shows the relationship between V, V_O, and R_L. It tells us that if we measure V (without load), V_O (with load), and R_L, and substitute the measured values in Eq. (20-4), we will calculate the value of r_{in}. Of course V and V_O are in volts and R_L and r_{in} in ohms.

Equation (20-3) suggests another means of determining r_{in}. This involves measurement of V and V_O as before, and measurement of I_L in amperes. Then, substituting V, V_O, and I_L in Eq. (20-3) will yield r_{in} in ohms.

It should be noted that our discussion about the internal resistance of a battery applies equally to the internal resistance of a dry cell.

Variation of r_{in}

The internal resistance of a battery or dry cell is not a fixed value but one which will vary with age and extent of use. Thus the difference between a fresh dry cell and one which has been in use for some time is the value of internal resistance. The value of r_{in} for a used dry cell is higher than that of an equivalent fresh cell. This is another way of saying that used cells are not as efficient as fresh cells, for some of the power which was stored in the used cell has been dissipated in use.

How do dry cells of the same type, age, and use compare with respect to r_{in}? By and large the value of r_{in} for such cells tends to be approximately the same, although there are some minor variations. However, not all dry cells are the same. Thus "heavy-duty" cells, alkaline cells, etc., exhibit a lower r_{in} and a longer life than ordinary dry cells.

Wet cells, for example, lead-acid cells, exhibit a lower internal resistance than dry cells. Moreover, their capacity to store and deliver energy is higher than that of dry cells.

Measuring Internal Resistance

The value of the internal resistance of a battery or cell is not measured directly, but indirectly. It is computed from the measured values of V, V_O, R_L, and I_L. And so it is at once apparent that the resolution of the measuring instruments and their accuracy will determine the accuracy of the calculated results. An accurate digital VOM yields results which are close enough for the purposes of this experiment. The student may wish to compare the results obtained by using a digital VOM with those resulting from a nondigital VOM.

SUMMARY

1. The loss which a battery exhibits in delivering current to a load is associated with its *internal resistance*.
2. Internal resistance depends on the composition, construction, and power-handling capacity of a battery, and on its age and use.
3. The older a battery is and the longer it has been in use, the higher its internal resistance.
4. The output voltage V_O of a dry cell or battery depends on its internal resistance and on the amount of current it is delivering to a load. V_O varies inversely with r_{in} and I_L.
5. The measurement of internal resistance is indirect, and is best accomplished in the school lab by use of accurate digital measuring instruments.

SELF-TEST

Check your understanding by answering these questions:

1. Without a load, the voltage measured across the terminals of a battery is ________ (higher/lower) than the voltge V_O delivered by the battery to a load.
2. The higher the current I_L drawn by a load, the ________ (higher/lower) the voltage measured across the battery.
3. The value of internal resistance of a battery ________ (increases/decreases) with use.
4. The voltage loss across the internal resistance of a battery is equal to the voltage measured across the battery ________ load, minus the voltage measured across the battery ________ load.
5. A heavy-duty dry cell has a ________ (higher/lower) internal resistance than an ordinary dry cell, all other conditions being the same.

MATERIALS REQUIRED

- Equipment: Resistance decade box variable in 1-Ω steps from 0 to 99,999 Ω; digital VOM
- Miscellaneous: SPST switch; two 1½-V size D dry cells, never used; one old (weak) 1½-V size D dry cell; dry cell holder

PROCEDURE

NOTE: In the procedure that follows, *keep switch S open* until the instruction to close it is given. Then make the necessary measurements *quickly and open S immediately* to prevent the dry cell from draining.

1. Adjust the controls of the decade box until the resistance R_L measured across the output terminals of the box measures 100 Ω. Note the control settings for this value of R_L. Repeat for R_L = 50 Ω and 25 Ω, respectively.
2. Label the two new dry cells 1 and 2. Label the weak cell 3.

Dry Cell 1

3. Connect the circuit of Fig. 20-3. *S is open*. DVM is a digital voltmeter connected across the terminals *AB* of new dry cell 1. The meter is set on a low range, 2 V or somewhat higher. R_L is the resistance at the output terminals of the resistance decade box. Set the controls of the decade box to $R_L = 100\ \Omega$.
4. With *S* still open, measure and record in Table 20-1 the no-load voltage *V* of the dry cell.
5. Close *S*. Measure the voltage V_O across the dry cell under load. *Open S as soon as the measurement is completed.* Record V_O in Table 20-1.
6. Repeat steps 4 and 5 for $R_L = 50\ \Omega$ and $R_L = 25\ \Omega$, respectively. Open *S*.
7. Calculate and record in Table 20-1 $V - V_O$ and r_{in} for each value of R_L.

Dry Cell 2

8. Repeat steps 3 through 7 for *new* dry cell 2. Record your measurements and calculations in Table 20-2.

Dry Cell 3

9. Repeat steps 3 through 7 for *weak* dry cell 3. Record your measurements and calculations in Table 20-3.

TABLE 20-1. Internal Resistance, Dry Cell 1

	Measured		*Calculated*	
R_L, Ω	V	V_O	$V - V_O$	r_{in}
100				
50				
25				

TABLE 20-2. Internal Resistance, Dry Cell 2

	Measured		*Calculated*	
R_L, Ω	V	V_O	$V - V_O$	r_{in}
100				
50				
25				

TABLE 20-3. Internal Resistance, Dry Cell 3

	Measured		*Calculated*	
R_L, Ω	V	V_O	$V - V_O$	r_{in}
100				
50				
25				

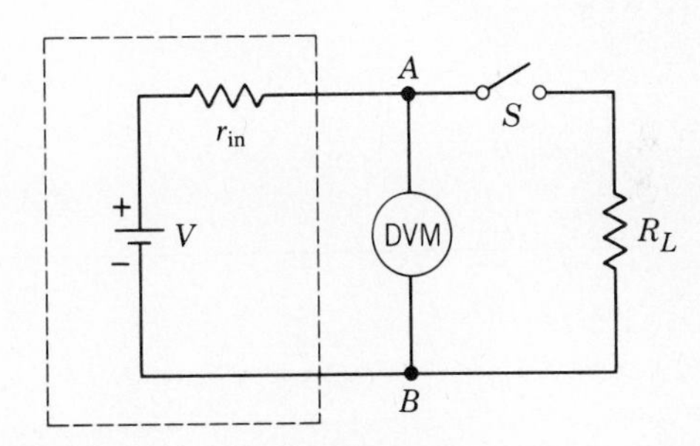

Fig. 20-3. Experimental circuit to determine r_{in} of a dry cell.

QUESTIONS

Refer specifically to your measurements and calculations in Tables 20-1 through 20-3 in answering these questions.

1. How do the values of r_{in} for the three dry cells compare? Comment on any unexpected results.
2. For any one dry cell do the values of r_{in} remain relatively constant for the different values of load resistance? Comment on any unexpected results.
3. How do the values of load current vary with R_L for cell 1? Why? Show your calculations.
4. How do the values of V_O vary with R_L? Why?
5. How do the values of voltage loss across r_{in} vary with a change in load current for cell 2? Show your calculations.
6. What can you say about the internal resistance of a weak dry cell as compared with that of a new, unused cell?

Answers to Self-Test

1. higher
2. lower
3. increases
4. without; with
5. lower

21

EXPERIMENT

MAXIMUM POWER TRANSFER

OBJECTIVES

1. To measure power in a dc load
2. To verify experimentally that the maximum power transferred by a dc source to a load occurs when the resistance of the load equals the internal resistance of the source

INTRODUCTORY INFORMATION

Measuring Power in a DC Load

Power has been defined as the rate of doing work. Electrically, the unit of power is the *watt* (W). The relationships among the power W dissipated by a resistive load R, the voltage V across R, and the current I in R are given by the equations

$$W = V \times I = I^2R = \frac{V^2}{R} \tag{21-1}$$

Here W is given in watts, V in volts, I in amperes, and R in ohms.

Since the power consumed by a resistive load R equals $V \times I$, where V is the voltage across the load and I is the current in the load, the measurement of dc power may be accomplished by measuring V with a voltmeter and I with an ammeter and calculating the product $V \times I$. For example, if 10 V appears across R and 0.25 A flows in R, the power in watts consumed by R is $W = V \times I = 10\,(0.25) = 2.5$ W.

Another means of measuring power is by use of a wattmeter. This is a four-lead instrument connected so that two of the leads carry the current flowing in the load R (just as an ammeter would), while the other two leads are connected across R (as a voltmeter would be connected). The instrument, in effect, measures the current in and the voltage across a load. This type of wattmeter is constructed so that the deflection of the pointer is proportional to the product of V and I. The scale of the wattmeter is calibrated in watts.

NOTE: Technicians rarely need to measure power in a dc circuit. However, if they must, they normally would measure the voltage and current separately and calculate their product.

DC power in a load may also be determined by measuring the current I in the load and the resistance R of the load and substituting the measured values in the formula

$$W = I^2R$$

DC power consumed by a load R may also be calculated by measuring V, the voltage across R, and by measuring the resistance of R, and substituting the measured values in the formula

$$W = \frac{V^2}{R}$$

In this experiment we will determine W dissipated in a load R by measuring V, I, and R and substituting the measured values in the three formulas.

Maximum Transfer of Power

Delivery of dc power to a load resistance R_L involves a dc source V and a circuit network between V and R_L, Fig. 21-1. Note in Fig. 21-1 that the power source V is shown in series with a resistor r_{in}, in series with the load R_L. Of course r_{in} is the internal resistance of the source, just as it was in the preceding experiment. We must take r_{in} into account in considering the amount of power that V can deliver to R_L.

What are the power relationships in the circuit of Fig. 21-1? To answer that question, we must know what the current I is in R_L, and what the values of V and R_L are. The current I is given by

$$I = \frac{V}{r_{in} + R_L} \tag{21-2}$$

Applying the I^2R formula for power, we can see that the power W dissipated by R_L is

$$W = I^2R_L = \left(\frac{V}{r_{in} + R_L}\right)^2 \times R_L$$

$$= \frac{V^2}{(r_{in} + R_L)^2} \times R_L \tag{21-3}$$

A question of considerable importance in electrical technology arises in connection with the circuit of Fig. 21-1. If V is a constant-voltage source with internal resistance r_{in}, is there some value of R_L for which V will deliver maximum power to R_L? To answer this question analytically requires the use of the calculus. However, an experimental method will be helpful in determining that there *is* some value of R_L which

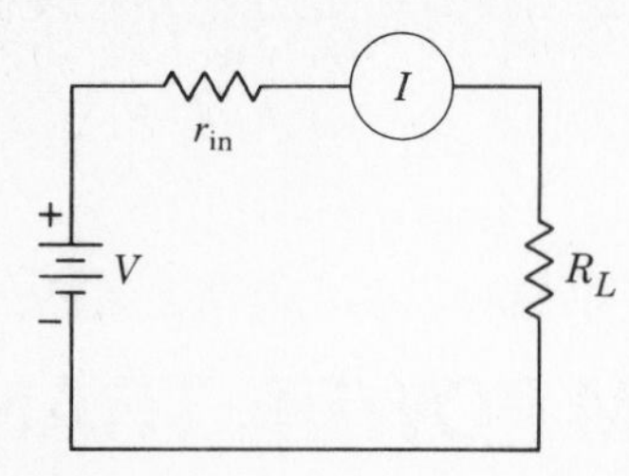

Fig. 21-1. A power supply V with internal resistance r_{in} delivers power to load resistor R_L.

will receive a maximum amount of power in the circuit. The value of both R_L and the power it dissipates will be found in this experiment.

The value of the internal resistance of a constant-voltage source over the range of loads for which it was designed is very low, an ohm or a fraction of an ohm. Attempting to demonstrate experimentally the truth of the maximum power transfer theorem using the internal resistance of a voltage-regulated power supply would require much more power than is normally available to the school laboratory voltage supply. And so in the lab we will add a resistor in series with the voltage supply. This resistor will act as the internal resistance of the supply, and the low r_{in} of the supply will be ignored. The percentage of error that this will introduce will be relatively small, and will not affect our conclusions.

Suppose the voltage V is 100 V, the internal resistance $R = 100\ \Omega$. Let R_L take a series of values as shown in Table 21-1. The power W, calculated by using Eq. (21-3), is also shown in Table 21-1.

The table indicates that as R_L increases from 0 to 100 Ω, the number of watts dissipated by R_L increases from 0 to a maximum of 25. As R_L increases from 100 to 100,000 Ω, the number of watts transferred to R_L decreases from 25 to 0.099.

It would appear that in this circuit maximum transfer of power occurs when the load resistance equals the internal resistance of the generator. Does this conclusion hold for other generators V', with internal resistance R', transferring power to a load R_L? The answer to this question may be determined by taking at random any voltage V', an internal resistance R', varying the load R_L, and then calculating the power in R_L. Again, the result will be that maximum power will be delivered to the load when $R_L = R'$, the internal resistance of the generator.

We may therefore state the law governing maximum transfer of power to a load in a dc circuit:

A generator transfers maximum power to a load when the resistance of the load equals the internal resistance of the generator.

TABLE 21-1. Calculating Maximum Power Dissipated

R_L, Ω	$R + R_L$, Ω	$W = \frac{V^2 R_L}{(R + R_L)^2}$ W
0	100	0
10	110	8.26
20	120	13.9
30	130	17.7
40	140	20.4
50	150	22.2
60	160	23.4
70	170	24.2
80	180	24.7
90	190	24.9
100	200	25
110	210	24.9
120	220	24.8
130	230	24.6
140	240	24.2+
150	250	23.9+
200	300	22.2
400	500	16
600	700	12.25
800	900	9.87
1,000	1,100	8.26
10,000	10,100	0.98
100,000	100,100	0.099

SUMMARY

1. The power W, in watts, dissipated by a resistor R (ohms), across which there is a dc voltage V (volts), is given by the equation $W = V^2/R$.
2. The power W, in watts, dissipated by a resistor R (ohms) in which there is direct current I (amperes) is given by the equation $W = I^2R$.
3. If the dc voltage is V (volts) across a resistor and current is I (amperes), the wattage dissipated by the resistor is given by the equation $W = V \times I$.
4. When power is delivered by a power source V, whose internal resistance is r_{in}, to a load R_L, maximum transfer of power occurs when the load resistance equals the internal resistance of the source, that is, when $r_{in} = R_L$.
5. DC power may be measured directly with a wattmeter, or indirectly by measuring V and R, or V and I, or I and R, and substituting the measured values in the proper formula for power.

SELF-TEST

Check your understanding by answering these questions:

1. The current in a 120-Ω resistor is 0.1 A. The power in watts dissipated by the resistor is W = __________ W.
2. The voltage across a resistor is 12 V, and the current in the resistor is 0.05 A. The power dissipated by the resistor is W = __________ W.
3. The voltage across a 220-Ω resistor is 16.0 V. The power dissipated in the resistor is W = __________ W.
4. A power supply with an internal resistance of 25 Ω delivers power to a 50-Ω load connected across its terminals. If the voltage delivered by the supply without load is 15 V, the power dissipated by the load is W = __________ W.
5. A power supply with an internal resistance of 25 Ω delivers power to a resistive load. If the no-load voltage at the output of the supply is 50 V, the maximum power would be delivered to a load whose resistance R_L = __________ Ω.
6. The power delivered by the supply in question 5 is W = __________ W.

MATERIALS REQUIRED

- Equipment: Regulated variable dc supply; EVM; 0–100-mA milliammeter
- Resistors: ½-W two 100-Ω, 1000-Ω (330-, 470-, 1000-, and 2200-Ω for extra-credit question)
- Miscellaneous: Resistance decade box or 10,000-Ω potentiometer (10,000-Ω potentiometer for extra-credit question); one DPST switch; one SPST switch

PROCEDURE

Measuring Power in a DC Circuit

1. Connect the circuit of Fig. 21-2. I is a milliammeter used to measure load current I_L, and V is the 25-V or equivalent range of an EVM. Set the output of the supply at 10 V.
2. Close S. Measure and record in Table 21-2 the voltage V_{AB} of the supply, the load current I_L in amperes, and the voltage V_L across R_L. Open S. **Power off.**
3. Measure and record the resistance r_{in} and R_L.
4. Calculate and record in Table 21-2 the power W_L consumed by R_L using the measured values:
   a. V_L and I_L
   b. V_L and R_L
   c. I_L and R_L
   In all cases show the formula used.
5. Calculate also and record the power W_T developed by the voltage supply. Show the formula used.

TABLE 21-2. Measuring Power

Step	V_{AB}, V	V_L, V	I_L, A	r_{in}, Ω	R_L, Ω
2, 3					
	Formula			Value, W	
4	(a) W_L =				
	(b) W_L =				
	(c) W_L =				
5	W_T =				

Maximum Power Transfer

6. Connect the circuit of Fig. 21-3. V is a voltage-regulated power supply, R a 1000-Ω resistor, R_L a resistance decade box or a 10,000-Ω potentiometer connected as a rheostat. A double-pole, single-throw switch (two on-off switches ganged together) S_{1A}, S_{1B} is used to open the load so that the resistance of R_L, completely isolated from V, may be adjusted and measured as required. Set the voltage V = 10 V and maintain it at this level during the remainder of the experiment.
7. Open switch S_1. Adjust the rheostat so that R_L is 0 Ω, as measured with an ohmmeter. Close S_1. With an EVM, measure the voltage V_L across R_L, and record in Table 21-3.
8. Open switch S_1. Adjust the rheostat so that R_L measures 100 Ω. Close S_1. With an EVM, measure the voltage V_L across R_L, and record in Table 21-3.
9. Repeat step 8 for every value of R_L shown in the table.

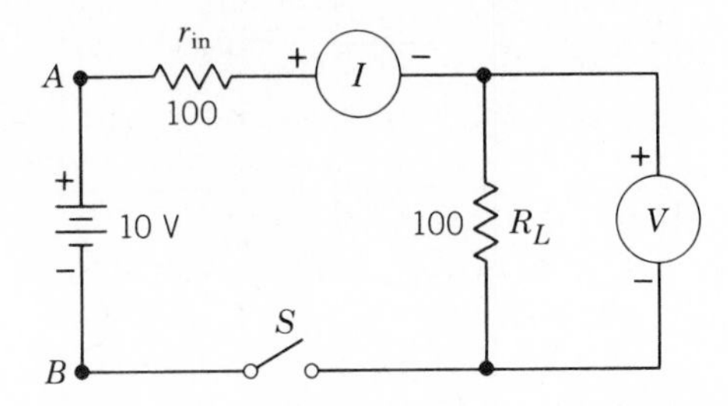

Fig. 21-2. Experimental circuit to measure power dissipated by a load resistor R_L.

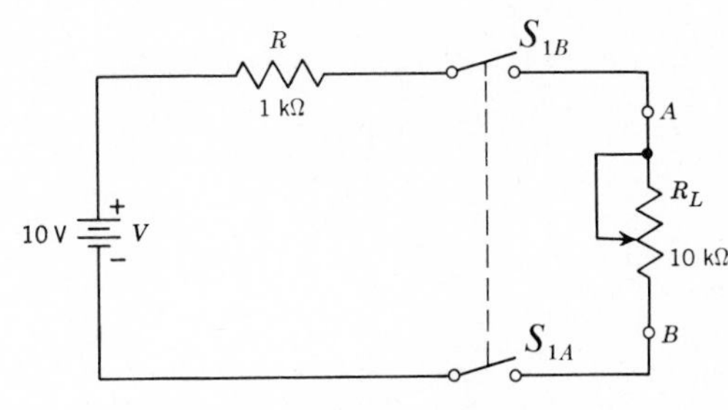

Fig. 21-3. Experimental circuit to determine maximum power transfer to a load.

TABLE 21-3. Experimental Data to Determine Maximum Power in R_L

R_L, Ω	$R + R_L$, Ω	V_L, V	$W = \frac{V_L^2}{R_L}$, mW	$W_T = \frac{V^2}{R + R_L}$, mW
0				
100				
200				
400				
600				
800				
850				
900				
950				
1,000				
1,100				
1,200				
1,500				
1,700				
2,000				
4,000				
6,000				
8,000				
10,000				

10. From the corresponding measured values of V_L and R_L, calculate the power in R_L using the formula $W = V_L^2/R_L$. Convert W into milliwatts and record in Table 21-3. Show a sample calculation.
11. Now calculate W_T using the formula $W_T = V^2/(R + R_L)$. Convert W_T into milliwatts and record in Table 21-3. Show a sample calculation.
12. Draw a graph of W versus R_L. Let R_L be the horizontal, W the vertical axis.
13. Draw a graph of W_T versus R_L.

Extra Credit

14. Describe in detail the experimental procedure you would use to determine the value of R_L for which there is a maximum transfer of power from the circuit of Fig. 21-4 to R_L. Draw a diagram of the experimental circuit. Tabulate your measurements, including a column for R_L

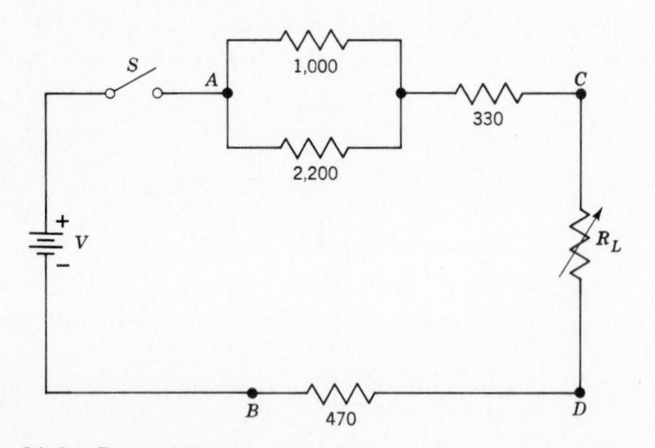

Fig. 21-4. Determining the value of R_L for which there is maximum transfer of power to R_L.

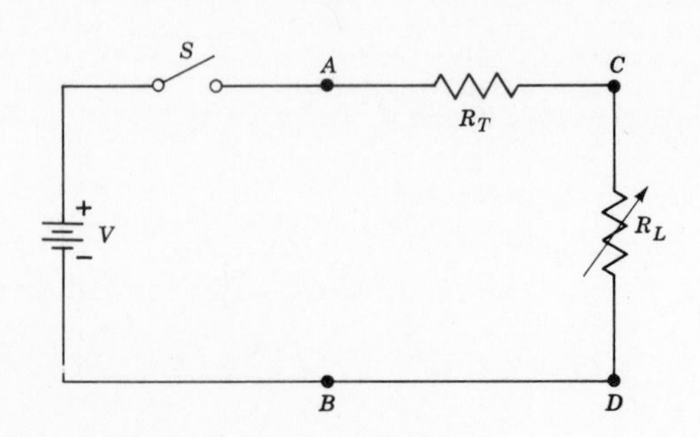

Fig. 21-5. Equivalent circuit for Fig. 21-4.

and for W. When you have determined the value of R_L experimentally, compare it with the computed value of R_L, determined by transforming the circuit of Fig. 21-4 into an equivalent circuit like that in Fig. 21-5.

HINT: Connect the circuit of Fig. 21-4, with switch S open and the power supply disconnected. Short points C and D. Experimentally measure the resistance between points A and B. If the measured value is R_T, then the maximum power transfer should occur when $R_L = R_T$. Select a value for V when you are conducting the experiment that will give readings in the center region of the voltmeter scale when measuring V_L, the voltage across R_L.

QUESTIONS

1. In your experiment, for what value of R_L is there maximum power transfer?
2. Do the measurements and computations in Table 21-3 confirm the maximum power transfer law? Discuss any unexpected results.
3. In Fig. 21-3, how does the voltage across R_L vary with R_L? The current in R_L?
4. In Fig. 21-3, how does W_T vary with R_L?
5. In Fig. 21-3, how does W (the power transferred to R_L) vary with R_L?

Answers to Self-Test

1. 1.2
2. 0.6
3. 1.16
4. 2.0
5. 25
6. 25

22

EXPERIMENT

BALANCED-BRIDGE CIRCUIT

OBJECTIVES

1. To find by analysis the mathematical relationship among the resistors in a balanced-bridge circuit
2. To apply experimentally the principle of the balanced bridge in measuring an unknown resistance

INTRODUCTORY INFORMATION

The volt-ohm-milliammeter and the electronic voltmeter (except the DVOM) are direct-reading instruments utilizing a current-actuated meter movement. The measured quantity, volts, milliamperes, or ohms, is read directly from a calibrated scale along which the pointer travels.

For measuring resistance with greater accuracy than that possible with an ohmmeter, a bridge is used. A resistance bridge employs a highly sensitive galvanometer as an indicating device, together with a calibrated variable-resistance standard and a voltage source in a suitable circuit arrangment. The galvanometer serves as a null indicator which signals a balance condition. Resistance is read from a calibrated scale associated with the standard variable resistance and range multiplier.

Figure 22-1 illustrates the most common type, the Wheatstone bridge. The form shown is the diamond arrangement, which the four resistors in the circuit diagram resemble. The unknown resistance to be measured, R_x, is connected between terminals C and D. R_1 and R_2 are fixed-value precision resistors called *ratio arms*, and R_3 is a variable resistance known as the *standard arm*. The indicator is a highly sensitive zero-centered galvanometer G. Current will flow through G when there is a difference of potential between points A and C. When there is no potential difference between A and C, that is, when $V_{AC} = 0$, the pointer of the galvanometer will return to zero. When $V_{AC} = 0$, the bridge is said to be balanced.

Consider the relationships in the circuit of Fig. 22-1. Let I_1 be the current in R_1, I_2 in R_2, I_3 in R_3, and I_x in R_x. I_1 times R_1 is the voltage from A to B, that is,

$$V_{AB} = I_1R_1$$

Similarly

$$V_{AD} = I_2R_2 \qquad (22\text{-}1)$$

$$V_{CB} = I_3R_3$$

and

$$V_{CD} = I_xR_x$$

To obtain balance ($V_{AC} = 0$), the voltage from A to B must equal the voltage from C to B. That is,

$$V_{AB} = V_{CB} \qquad (22\text{-}2)$$

Therefore, $I_1R_1 = I_3R_3$ and

$$\frac{I_1}{I_3} = \frac{R_3}{R_1} \qquad (22\text{-}3)$$

Similarly

$$V_{AD} = V_{CD}$$

and

$$\frac{I_2}{I_x} = \frac{R_x}{R_2} \qquad (22\text{-}4)$$

At balance, there is no current in G. Therefore

$$I_1 = I_2$$

and

$$I_3 = I_x \qquad (22\text{-}5)$$

Substituting the results of Eqs. (22-5) in Eqs. (22-3) and (22-4), we find that

$$\frac{I_1}{I_x} = \frac{R_3}{R_1}$$

$$(22\text{-}6)$$

and

$$\frac{I_1}{I_x} = \frac{R_x}{R_2}$$

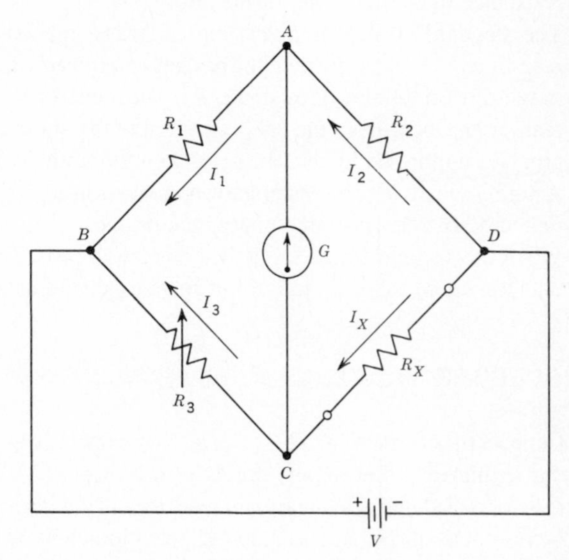

Fig. 22-1. Wheatstone bridge for measuring unknown resistance R_X.

Therefore $$\frac{R_x}{R_2} = \frac{R_3}{R_1} \qquad (22\text{-}7)$$

and $$R_x = \left(\frac{R_2}{R_1}\right) \times R_3 \qquad (22\text{-}8)$$

The fundamental equation (22-8) states that at balance the unknown resistance R_x equals the product of the ratio of the ratio arms (R_2/R_1) and the standard arm R_3.

Maximum accuracy and sensitivity occur when $R_1 = R_2$, that is, when the ratio $R_2/R_1 = 1$. For this condition $R_x = R_3$. If R_3 is a highly accurate decade-type rheostat, the value R_x can be read directly from the calibrated scale of the rheostat when R_3 is adjusted for balance.

The condition $R_2/R_1 = 1$ would limit the range of measurement of R_x to the range of variation of the rheostat, so that if the maximum resistance of $R_3 = 10{,}000\ \Omega$, a resistor whose value is larger than 10,000 Ω could not be measured. To overcome this limitation, a range switch permits selection of different ratio arms. Thus, if $R_2/R_1 = 3$, the maximum value of R_x which may be measured is $3R3$, etc. It should be apparent, though, that any small error of galvanometer imbalance which the eye may not detect is multiplied by the ratio R_2/R_1, 3 in this case, thus increasing the error.

There are other bridge circuits, such as the slide-wire bridge, for measuring resistance. In this experiment, however, we will be concerned with the diamond-type Wheatstone bridge. There will be no attempt to build a bridge of great accuracy. The experiment will simply demonstrate the principle of a balanced bridge.

SUMMARY

1. A Wheatstone bridge is used for measuring resistance values with greater accuracy than that possible with the volt-ohm-milliammeter or electronic voltmeter.
2. The ratio arms R_1 and R_2 in Fig. 22-1 are precision resistors which are switched in to give the desired resistance-range multiplier of the bridge.
3. The standard arm R_3 is a variable calibrated precision resistance by means of which the bridge is *balanced* when measuring an unknown resistance R_x. The value of R_x is read, at balance, from the calibrated dial of the standard arm and multiplied by the setting of the ratio arms.
4. A highly sensitive zero-center galvanometer is used as the indicator. A voltage source powers the bridge.
5. When the voltage across the galvanometer (V_{AC} in Fig. 22-1) is zero, there is no current in the galvanometer, which then reads zero. This is the balance setting at which the resistance of the unknown resistor is read.
6. At balance in a Wheatstone bridge, the product of the resistors of opposite arms is equal, that is, in Fig. 22-1,

$$R_x \times R_1 = R_2 \times R_3$$

and from this condition the value of R_x is

$$R_x = \frac{R_2}{R_1} \times R_3$$

7. R_x, therefore, is equal to the product of the ratio of the ratio arms and the resistance of the standard arm.

SELF-TEST

Check your understanding by answering these questions:

1. A Wheatstone bridge has the following adjustable controls: ____________ and ____________.
2. Power for a portable Wheatstone bridge is provided by a ____________.
3. At balance, the current in the galvanometer is ____________.
4. A galvanometer is a ____________.
5. In Fig. 22-2, the circuit of the experimental Wheatstone bridge, balance occurs when R_3 is set for 3750 Ω. The value of R_x is ____________ Ω.
6. In Fig. 22-1, at balance $R_2/R_1 = 10$ and $R_3 = 1950\ \Omega$. The resistance value of R_x is ____________ Ω.
7. A Wheatstone bridge is known to be in good operating order and can measure resistors in the range 0.1 to 100,000 Ω. In measuring a resistor whose value is about 50,000 Ω, the bridge cannot be balanced for a specific setting of the multiplier (ratio arms). In order to balance the bridge it is necessary to change the setting of the ____________.

MATERIALS REQUIRED

- Power supply: Voltage-regulated variable dc source
- Equipment: Zero-center galvanometer or a 20,000-Ω/V VOM; electronic voltmeter
- Resistors: ½-W, two 5100-Ω (5 percent); assortment of six different values of resistors (minimum value 500, maximum value less than 10,000 Ω)
- Optional: Other resistors as required in step 5
- Miscellaneous: Decade resistance box, variable from 1 to 10,000 Ω, or 10,000-Ω potentiometer; two SPST switches

PROCEDURE

1. Connect the circuit (Fig. 22-2). S_1 and S_2 are open. Adjust the regulated power supply for $V = 6$ V. Set R_3, the resistor decade box, to maximum resistance. (If a decade box is not available, use a 10,000-Ω potentiometer). M is a zero-center galvanometer. (If one is not available, use the most sensitive current range on a 20,000-Ω/V VOM, such as the 50-μA or 60-μA range.) The purpose of R_4 is to reduce the sensitivity of M during the initial stages of adjustment.

2. Connect R_x, the resistor to be measured, between C and D. (R_x must be no smaller than 560 Ω, or greater than 10,000.) Close switch S_1. S_2 is still open. The meter

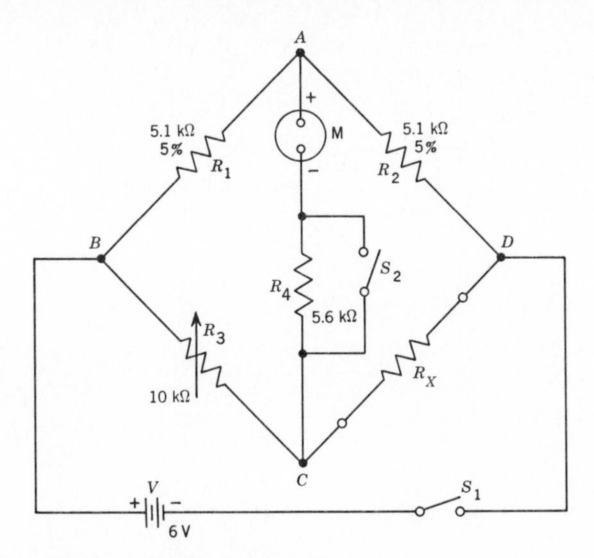

Fig. 22-2. Experimental Wheatstone bridge.

pointer should deflect upscale. If it does not, check to see that the standard arm R_3 is set at maximum.

3. Reduce the resistance of R_3 gradually until M approaches zero. Now close S_2 and readjust R_3 if necessary until M reads zero. Read the value of R_x from the calibrated resistance decade box and record in Table 22-1. Record also the color-coded value of R_x and its rated tolerance.

NOTE: If R_3 is a 10,000-Ω potentiometer connected as a rheostat instead of the resistance decade, remove R_3 from the circuit without varying its setting and measure it with an ohmmeter. Record this as the value R_x in Table 22-1.

4. Open S_1 and S_2. Reset R_3 at maximum resistance. Remove the first R_x from the circuit. Using the bridge-circuit procedure in steps 2 and 3, measure in turn each of six resistors supplied to you. Record the measured and coded values in Table 22-1.

Extra Credit—Design Problem

5. Using the same R_3 as the standard arm, design a bridge circuit which will measure resistances up to
   a. 50,000 Ω (approximately)
   b. 250,000 Ω (approximately)
   Draw the circuit diagram for **a** and **b**, showing required values of all components in the bridge. Secure the necessary parts. If 5 percent resistors are not available, measure the ratio-arm resistors with an ohmmeter and compute the multiplier ratio from the measured values. With bridge **a** measure three resistors greater than 10,000 Ω, but less than 50,000 Ω. With bridge **b** measure three resistors greater than 50,000 Ω, but less than 250,000 Ω.
   Record the results of your measurements in separate tables similar to Table 22-1. Identify the multiplier ratio in each case.

TABLE 22-1. Bridge Measurement of R_X: $R_2/R_1 = 1$

Resistor No.	*1*	*2*	*3*	*4*	*5*	*6*
Rated (code) value						
Percent tolerance						
Measured value						

QUESTIONS

1. What determines the accuracy of measurements made with the experimental resistance bridge?
2. Why is it necessary to use a very sensitive galvanometer as an indicator in this experiment?
3. What is the purpose of the voltage source V? Is the voltage level ($V = 6$ V) of the voltage source critical?
4. Were the results of your measurements within the rated values of the resistors measured? If not, explain why not.

Extra Credit

5. In Fig. 22-1 at balance, $R_1 = R_2 = 5000\ \Omega$; $R_3 = R_X = 6000\ \Omega$; $R_m = 3000\ \Omega$. Find the total resistance which the power supply (V) sees. Show your computations.

Answers to Self-Test

1. standard arm; ratio arms or multiplier
2. battery
3. zero
4. very sensitive zero-center dc meter movement
5. 3750
6. 19,500
7. multiplier

23 EXPERIMENT
SUPERPOSITION THEOREM

OBJECTIVES

1. To learn what is meant by a *linear* circuit
2. To verify experimentally the superposition theorem

INTRODUCTORY INFORMATION

We have applied Ohm's and Kirchhoff's laws to a study of simple resistive circuits, that is, circuits containing series, parallel, or series-parallel combinations of resistors. Moreover, the circuits treated have contained a single voltage source. However, there are more complex circuits in electricity, and these may contain two or more sources. Analysis and solution of such circuits may become difficult, using the methods we have learned. For such complex circuits more powerful analytical tools are helpful.

There is a group of network theorems, including the superposition theorem, Thevenin's theorem, and Norton's theorem, which provides the means for simplifying analysis of complex circuits. In this experiment we will be concerned with the superposition theorem.

Linear Circuit Elements

The resistors we have used in previous experiments are known as linear circuit elements, and the circuits in which they are contained, to the exclusion of other elements, are called linear circuits. What is a linear element? A linear element is one in which the ratio of voltage across to current in the element is constant.

You will recall verifying in Experiment 7 that in a resistor across which a voltage V is applied, the ratio of voltage to current is constant and is given by the formula $R = V/I$ (Ohm's law); see Fig. 23-1. In this formula, R is the resistance in ohms, V is the applied voltage in volts, and I is the resulting current in amperes. In Fig. 23-1, if the applied voltage is doubled, the current I is doubled, and the ratio V/I remains the same. If the applied voltage is increased by a factor of 3, the current I increases by a factor of 3, etc.

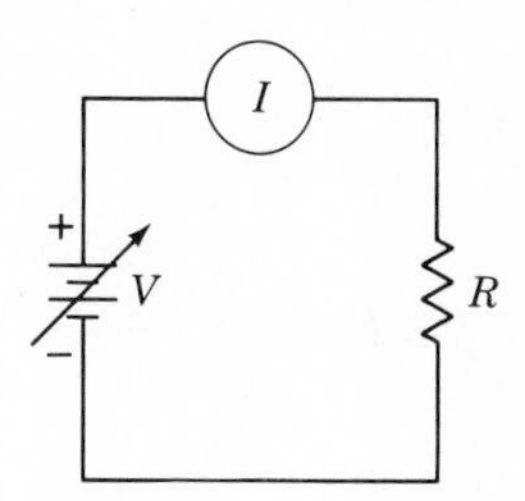

Fig. 23-1. Circuit to demonstrate the relationship between V and I in a resistor.

To show graphically the relationship which exists between V and I in Fig. 23-1, assume that V is a variable dc source and that R is a resistor whose value is fixed, say 1000 Ω. I is a milliammeter used to measure current in the circuit. As the supply voltage is set, in turn, to each of the voltages shown in Table 23-1, the measured current I corresponding to each of these voltages increases from 5 mA to 25 mA, as shown. If each set of corresponding values of V and I is plotted as a point on a graph, and if these points are joined, Fig. 23-2, it is seen that these points lie along a straight line, that is, they are (co)linear. It is this characteristic of voltage *across* and current *in* a resistor which gives rise to the term *linear* element. We will find in a later experiment that there are *special* resistors which are *not* linear. However, for the present, all the resistors we will be using will be linear elements.

TABLE 23-1. Current versus Voltage in a 1000-Ω Resistor

Voltage, V	*Current*, mA
5	5
10	10
15	15
20	20
25	25

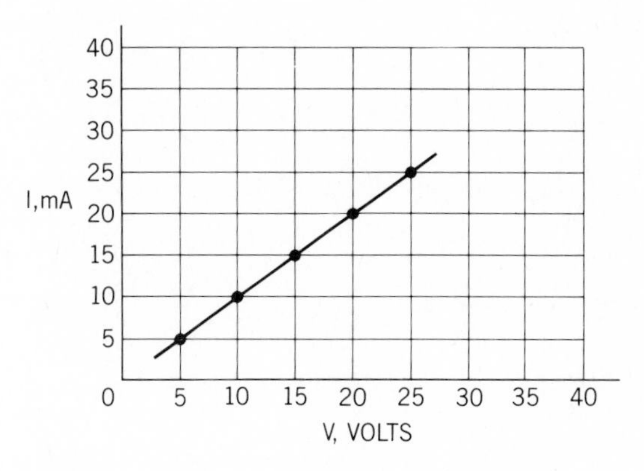

Fig. 23-2. Graph of V versus I in a 1000-Ω resistor.

Superposition Theorem

The superposition theorem states that *in a linear circuit containing more than one voltage source, the voltage across or current in any one element is the algebraic sum of the voltages or currents produced by each source acting alone.*

In order to apply this theorem to the solution of a problem, we must understand what is meant by "each source acting alone." Suppose a network, say Fig. 23-3, has two voltage sources V_1 and V_2, and we wish to find the effect on the circuit of each of these sources acting alone. To determine the effect of V_1 we must replace V_2 by its internal resistance and solve the modified circuit, Fig. 23-4. Note that in Fig. 23-4 we have replaced V_2 by a *short circuit*. This is because the internal resistance of the voltage source, V_2, is so low that it may be neglected entirely. Now to determine the effect of V_2, we replace V_1 by its internal resistance, again a short circuit, as in Fig. 23-5 and then we solve the circuit of Fig. 23-5. Finally we combine algebraically the results of the solution of the modified circuits to give us a complete solution of the original network.

Example

A concrete problem will make the process clear. In Fig. 23-3 we wish to find the current I in R_3. Applying the superposition theorem, we first solve for I_1 in Fig. 23-4. The total current I_{T1} which V_1 supplies this circuit is:

$$I_{T1} = \frac{V_1}{R_1 + R_2 + \dfrac{R_3(R_4 + R_5)}{R_3 + R_4 + R_5}}$$

$$= \frac{20}{100 + 50} = 133.3 \text{ mA}$$

Now since $R_3 = R_4 + R_5$, I_{T1} divides equally in R_3 and the series combination $(R_4 + R_5)$ with which it is in parallel. Therefore $I_1 = 66.67$ mA. Moreover, I_1 flows from B to A in Fig. 23-4. Let us call this direction positive (+).

The total current I_{T2} which V_2 (Fig. 23-5) supplies the circuit is:

$$I_{T2} = \frac{V_2}{R_4 + R_5 + \dfrac{R_3(R_1 + R_2)}{R_3 + R_1 + R_2}}$$

$$= \frac{10}{100 + 50} = 66.7 \text{ mA}$$

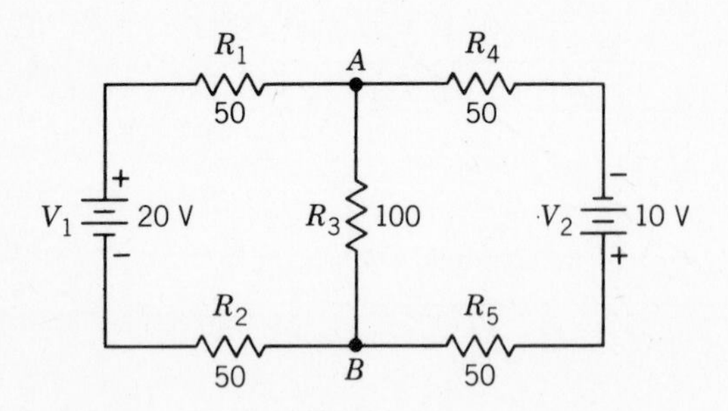

Fig. 23-3. Resistor network with two voltage sources.

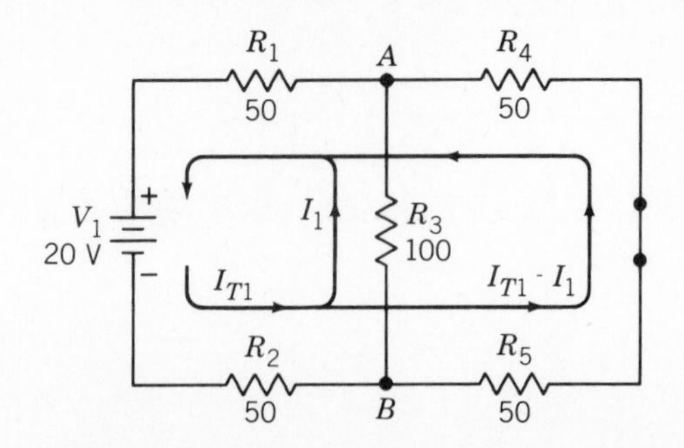

Fig. 23-4. First V_2 is replaced by its internal resistance (a short circuit here) and the effect of V_1 is determined.

Since $R_3 = R_1 + R_2$, I_{T2} divides equally in R_3 and the series combination $(R_1 + R_2)$ with which it is in parallel. Therefore I_2 in $R_3 = 33.33$ mA. Moreover, I_2 flows from A to B in Fig. 23-5, opposite in direction to I_1. We will therefore designate this direction as negative (−). The total current I in R_3 (Fig. 23-3) is the algebraic sum of I_1 and I_2. Therefore $I = I_1 - I_2$, which equals $66.67 - 33.33 = 33.34$ mA. Moreover, the direction that I takes is the same as that of I_1, from B to A.

NOTE: If I_1 and I_2 were found to be flowing in the same direction, the algebraic sum of I_1 and I_2 would have been the actual sum of these two currents.

By a similar process the current in any resistor in the circuit may be found. And similarly, the voltage across any resistor in the circuit may be found by determining the current in that resistor and applying Ohm's law. We will not develop an analytical proof of this theorem, but we will verify it experimentally as it applies to a specific circuit.

SUMMARY

1. A linear circuit element is one in which the ratio of voltage to current in the element is constant.
2. Fixed-value resistors are linear circuit elements.
3. Circuits containing only linear elements are linear circuits.
4. The superposition theorem applies to linear circuits which have two or more voltage sources.

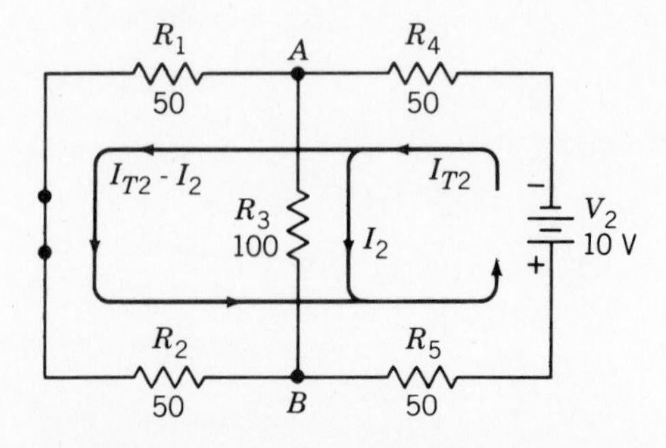

Fig. 23-5. Then V_1 is replaced by its internal resistance (again a short circuit) and the effect of V_2 is determined.

5. To find the current in or voltage across an element in a linear circuit, all the voltage sources except one are removed and replaced by their internal resistances (short circuits in the case of regulated voltage sources), and the effect of the one remaining source on the element in question is determined. This process is repeated for each source. Then the algebraic sum of the effects is determined, giving the required voltage or current.

SELF-TEST

Check your understanding by answering these questions:

1. The superposition theorem may be applied to the circuit of Fig. 23-1. ____________ (true/false)
2. In Fig. 23-3, the graph of current **in** versus voltage **across** R_3 is a ____________ ____________.
3. In Fig. 23-4, the ratio of voltage across R_4 to current in R_4 equals ____________.
4. The 10-V source in Fig. 23-3 causes ____________ (more/less) current in R_3 than would be the case if V_1 were the only voltage source, and V_2 were replaced by a short circuit.
5. In applying the superposition theorem to the solution of the circuit in Fig. 23-3, it is necessary to ____________ V_2 before replacing it by a short ____________.
6. In the circuit of Fig. 23-3, the voltage across R_5 is ____________ V.

MATERIALS REQUIRED

- Power supply: Two regulated voltage-variable dc sources
- Equipment: Digital VOM preferably, or an EVM
- Resistors: ½-W 1200-, 2200-, and 3300-Ω
- Miscellaneous: Three SPST switches

PROCEDURE

Effect of V_1 and V_2 on the Circuit

1. Connect the circuit of Fig. 23-6. Set the voltage supply V_1 at 20 V and V_2 at 15 V. Close S_1, S_2, and S_3. **Power on.**
2. Measure and record in Table 23-2 the voltages across R_1, R_2, and R_3 (respectively, V_{AB}, V_{CD}, and V_{EF}). Show by + or − whether A is positive or negative with respect to B; whether C is positive or negative with respect to D; whether E is positive or negative with respect to F. For example, if A is 3 V negative relative to B, then $V_{AB} = -3$ V, etc.
3. Measure also and record in Table 23-2 the current I_1 in R_1, I_2 in R_2, and I_3 in R_3. A simple way to open the circuit to insert a milliammeter to measure, say I_1, is to open switch S_1 and connect the milliammeter across the terminals of the switch. The meter is now in series with R_1 and will measure I_1. After the measurement is made, close S_1 and measure I_2 by opening S_2 and repeating the process, etc. Identify the direction of current flow in R_1, etc., by giving current the sign opposite to that of the voltage across the resistor. Thus if V_{AB} is −, then I_1 is +.

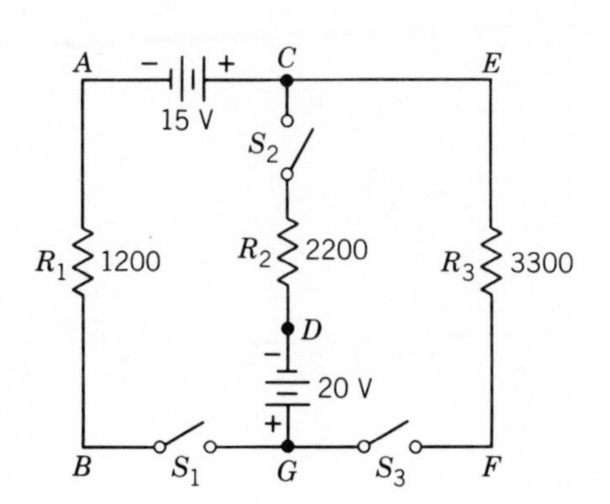

Fig. 23-6. Experimental circuit to verify the superposition theorem.

Effect of V_1 Acting Alone

4. Remove power supply V_2 from the circuit, and short circuit points A to C. The measurements you will now make will be the effect of V_1 acting alone.
5. Repeat steps 2 and 3 and record your measurements in Table 23-3.

TABLE 23-2. Effects of V_1 and V_2

Current, mA	*Voltage,* V
I_1:	V_{AB}:
I_2:	V_{CD}:
I_3:	V_{EF}:

TABLE 23-3. Effect of V_1 Alone

Current, mA	*Voltage,* V
I_1:	V_{AB}:
I_2:	V_{CD}:
I_3:	V_{EF}:

TABLE 23-4. Effect of V_2 Alone

Current, mA	*Voltage,* V
I_1:	V_{AB}:
I_2:	V_{CD}:
I_3:	V_{EF}:

Effect of V_2 Acting Alone

6. Remove the short circuit from points *A* to *C* and reconnect power supply V_2 as in Fig. 23-6.
7. Now remove power supply V_1 from the circuit and short circuit points *D* to *G*. The measurements you will now make will measure the effect on the circuit of V_2 acting alone.
8. Repeat steps 2 and 3 and record your measurements in Table 23-4. **Power off.**
9. Algebraically add I_1 in Table 23-3 to I_1 in Table 23-4 and enter the result in Table 23-5. Similarly record the results of adding, respectively, the values of I_2, I_3, V_{AB}, V_{CD}, and V_{EF}.

TABLE 23-5. Combining Tables 23-3 and 23-4

Current, mA	*Voltage*, V
I_1:	V_{AB}:
I_2:	V_{CD}:
I_3:	V_{EF}:

QUESTIONS

1. How do the respective values of current in Table 23-5 compare with those in Table 23-2? Of voltage?
2. What is the purpose of step 9 in the procedure?
3. Have you verified the superposition theorem for the circuit in Fig. 23-6 in this experiment? Explain.
4. Using the superposition theorem, calculate the value of current in R_2, Fig. 23-6. Show all work.
5. How does the calculated value in answer to question 4 compare with the measured value in Table 23-2?

Answers to Self-Test

1. false
2. straight line
3. 50
4. less
5. remove; circuit
6. 6.67

24

EXPERIMENT

THEVENIN'S THEOREM

OBJECTIVES

1. To determine by analysis the values V_{TH} (Thevenin generator voltage) and R_{TH} (Thevenin generator resistance) in a dc circuit containing a single voltage source
2. To confirm experimentally the values V_{TH} and R_{TH} proposed by Thevenin's theorem in the solution of unbalanced-bridge circuits

INTRODUCTORY INFORMATION

Thevenin's theorem is another analytical tool which is very helpful in the solution of complex linear network problems. By using it, it is possible to determine the voltage or current in a portion of a network, say a resistor. The technique employed involves reducing the network to a single equivalent circuit which acts like the original network on the resistor in question.

Thevenin's Theorem

Thevenin's theorem states that any linear *two*-terminal network may be replaced by a simple equivalent circuit consisting of a Thevenin generator, whose voltage V_{TH}, acting in series with an internal resistance R_{TH}, causes current to flow through the load. Thus, Fig. 24-1*d* is the Thevenin equivalent for the circuit of Fig. 24-1*a* considering R_L as the load. If we knew the values V_{TH} and R_{TH}, the process of finding the current I in R_L would be a simple application of Ohm's law.

The rules for determining V_{TH} and R_{TH} are as follows:

1. The voltage V_{TH} is the voltage "seen" across the load terminals in the original network, with the load resistance removed (open circuit voltage).
2. The resistance R_{TH} is the resistance seen from the terminals of the open load, looking into the original network when the voltage sources in the circuit are replaced by their internal resistance.

The development of the Thevenin equivalent circuit for Fig. 24-1*a* may be readily followed from Fig. 24-1*b*, *c*, and *d*. In Fig. 24-1*b*, R_L has been open circuited and V_{TH} is the voltage which appears across the terminals *AB*. In this case it is obviously 50 V. In Fig. 24-1*c*, we see that the original voltage source V has been replaced by its internal resistance R_1, which equals 5 Ω. The Thevenin resistance R_{TH} is the resistance measured across *AB* in Fig. 24-1*c*. R_{TH} here is 100 Ω. Finally, Fig. 24-1*d*, shows the Thevenin

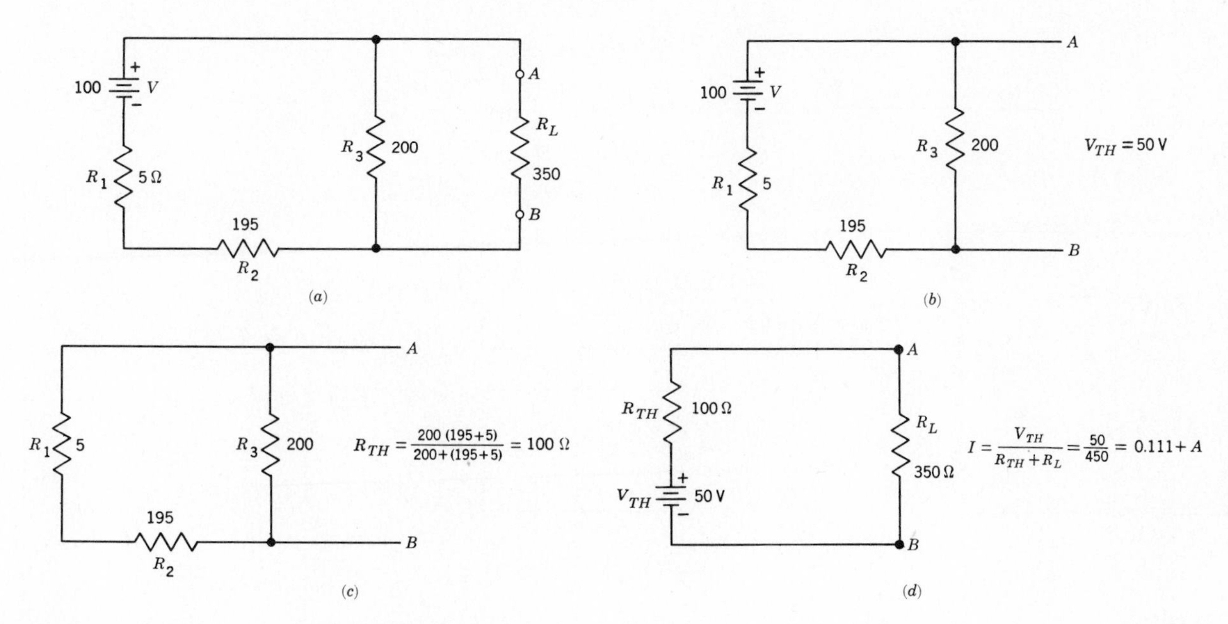

Fig. 24-1. Solution of a network by Thevenin's theorem.

equivalent generator V_{TH} in series with R_{TH}, which replaces the original network. Applying Ohm's law to the circuit of Fig. 24-1*d*, we see that

$$I = \frac{V_{TH}}{R_{TH} + R_L} = \frac{50}{450} = 0.111 \text{ A}$$

This is the required current in R_L.

It might appear, at first, as though this method unnecessarily complicates the problem, for the circuit of Fig. 24-1*a* can be easily solved by Ohm's and Kirchhoff's law. Although this is true in this single instance, the value of Thevenin's theorem becomes apparent if the problem in Fig. 24-1*a* is slightly modified. Suppose it is required to find the current in R_L, in Fig. 24-1*a*, for different values of R_L, as for example, $R_L = 20\ \Omega$, $50\ \Omega$, $100\ \Omega$, $1200\ \Omega$, etc., while the rest of the circuit remains the same. It would be quite laborious to apply Ohm's and Kirchhoff's laws for each individual value of R_L in solving for I.

On the other hand, a single determination of the Thevenin equivalent circuit (Fig. 24-1*d*) would suffice for each value of R_L, because V_{TH} and R_{TH} are independent of the value of R_L. Now it would be much simpler to solve I for the succession of values of R_L, using the Thevenin equivalent generator of Fig. 24-1*d*.

Solving Unbalanced-Bridge Circuit by Thevenin's Theorem

Figure 24-2*a* is an unbalanced-bridge circuit. It is required to find the current I in R_5. Thevenin's theorem lends itself readily to a solution of this problem.

Consider R_5 the load. The problem is to transform the circuit into a Thevenin equivalent generator supplying current to R_5, as in Fig. 24-2*b*.

We open-circuit R_5 to determine V_{TH}, the voltage V_{BC} across BC, Fig. 24-2*c*. To determine the voltage V_{BC}, we can find the voltage V_{BD} from B to D, and V_{CD} from C to D. Then

$$V_{BD} - V_{CD} = V_{BC} = V_{TH} \tag{24-1}$$

The voltage V_{BD} in Fig. 24-2*c* is

$$V_{BD} = \frac{160}{200} \times 60 = 48 \text{ V} \tag{24-2}$$

The voltage V_{CD} is

$$V_{CD} = \frac{120}{180} \times 60 = 40 \text{ V} \tag{24-3}$$

Therefore

$$V_{BC} = 48 - 40 = 8 \text{ V} = V_{TH} \tag{24-4}$$

The resistance R_{TH} of the Thevenin generator V_{TH} is found by replacing V by its internal resistance (assumed to be zero in this case), and solving for R_{BC}. Figure 24-2*d* is the result of shorting A to D and redrawing the circuit in a form which lends itself readily to computation. Hence

$$R_{BC} = R_{TH} = \frac{R_1R_4}{R_1 + R_4} + \frac{R_2R_3}{R_2 + R_3}$$

$$R_{TH} = \frac{40 \times 160}{40 + 160} + \frac{60 \times 120}{60 + 120} \tag{24-5}$$

$$= 32 + 40 = 72\ \Omega$$

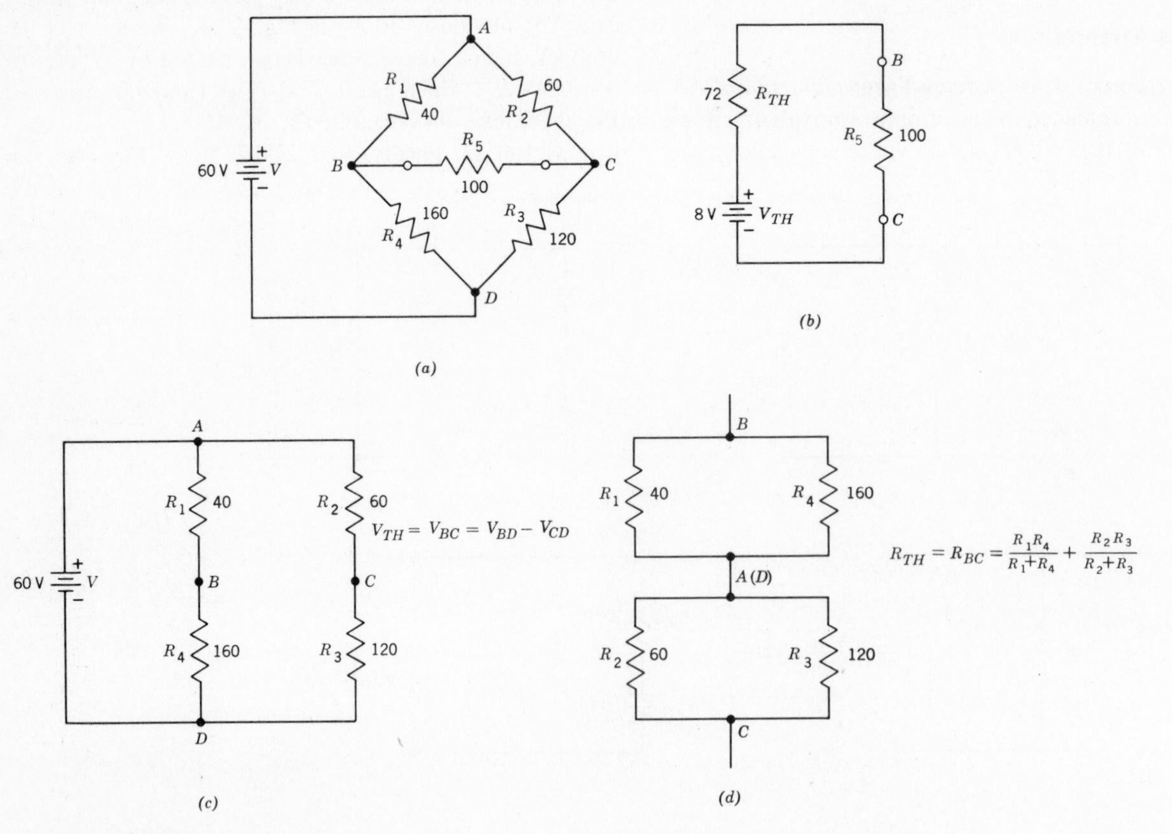

Fig. 24-2. Unbalanced-bridge circuit.

Substituting these values in the Thevenin equivalent circuit (Fig. 24-2*b*) and solving for I, we find

$$I = \frac{V_{TH}}{R_{TH} + R_5} = \frac{8}{172} = 0.0465 \text{ A}$$

Experimental Verification of Thevenin's Theorem

It is possible to determine the values of V_{TH} and R_{TH} for a load R_L in a specific network by measurement. Then we can experimentally set the output of a voltage-regulated power supply to V_{TH} by connecting a resistor whose value is R_{TH} in series with V_{TH} and with R_L. We can measure I in this equivalent circuit. If the measured value of I_L in R_L in the original network is the same as the I measured in the Thevenin equivalent, we have one verification of Thevenin's theorem. For a more complete verification, this process would have to be repeated many times with random circuits.

SUMMARY

1. Thevenin's theorem proposes that any linear two-terminal network may be replaced by a simple equivalent circuit which acts like the original circuit at the load connected to the two terminals.
2. The equivalent circuit consists of a generator or power source called the Thevenin generator (V_{TH}) in series with the Thevenin internal resistance (R_{TH}) in series with the load across the two terminals in question. Thus for the complex circuit Fig. 24-2*a*, the Thevenin equivalent at the terminals *BC* is Fig. 24-2*b*.
3. To determine the Thevenin voltage V_{TH}, open the load at the two terminals in the original network and calculate the voltage at these two terminals. This *open-load* voltage is V_{TH}.
4. To determine the Thevenin resistance R_{TH}, keep the load open at the two terminals in the original network and replace the original power source with its own internal resistance. Then calculate the resistance at the *open-load* terminals, looking back into the original circuit.
5. Thevenin's theorem is applicable to a circuit with one or more power sources.
6. Once the complex network has been replaced by the Thevenin equivalent circuit, solution of the current in the load may be accomplished by Ohm's law. Thus in Fig. 24-1*d* current in the 350-Ω load equals V_{TH} divided by the sum of R_{TH} and R_L.

$$I_{\text{load}} = \frac{V_{TH}}{R_{TH} + R_L}$$

7. In this experiment we will verify Thevenin's theorem experimentally for one set of values of an unbalanced-bridge circuit.

SELF-TEST

Check your understanding by answering these questions:

1. Consider the circuit of Fig. 24-1*a*. $V = 160$ V, $R_1 = 30\ \Omega$, $R_2 = 270\ \Omega$, $R_3 = 500\ \Omega$, and $R_L = 560\ \Omega$. In the equivalent Thevenin circuit, assuming the internal resistance of the supply is zero,
   (*a*) V_{TH} = ________ V
   (*b*) R_{TH} = ________ Ω
   (*c*) I_{load} = ________ A
2. Consider the circuit of Fig. 24-2*a*. $V = 120$ V, $R_1 = 200\ \Omega$, $R_2 = 500\ \Omega$, $R_3 = 300\ \Omega$, $R_4 = 600\ \Omega$, and $R_5 = 112.5\ \Omega$. In the equivalent Thevenin circuit, assuming the internal resistance of the supply is zero,
   (*a*) V_{TH} = ________ V
   (*b*) R_{TH} = ________ Ω
   (*c*) I_5 = ________ A, where I_5 is the current in R_5.

MATERIALS REQUIRED

- Equipment: Regulated variable dc supply; EVM (preferably a digital EVM), milliammeter (or VOM) with a 5- or 10-mA range
- Resistors: ½-W 2700-, 3300-, 5600-, 10,000-, and 15,000-Ω; and others as required in step 7 (optional)
- Miscellaneous: Resistance decade box or, if not available, a 10,000-Ω 2-W potentiometer and a 5600-Ω resistor; two SPST switches

PROCEDURE

1. Measure the resistance of each of the resistors you will use in the experimental circuit, Fig. 24-3, and record the measured values in Table 24-1. Measure also and record the color-coded and measured values of each of three other resistors (R_L) which you will use in step 8.

2. Connect the circuit of Fig. 24-3. S_1 and S_2 are both open. *M* is a milliammeter set on the 1-mA range. Close S_1. Adjust *V* so that the voltage measured with an EVM is 40 V. Close S_2. Measure the current in R_L and record in Table 24-2 under the column "I_L Measured Original Circuit."

3. Open S_2. Switch S_1 remains closed. With an EVM, measure the voltage V_{BC} across *BC*, and record this in Table 24-2 under the heading "V_{TH} Measured."

4. Remove the voltage supply *V* from the circuit. S_2 is still open. Short points *A* to *D*.

NOTE: It is assumed that the internal resistance of the voltage-regulated supply is very low, and hence can be neglected.

S_2 is still open. Measure the resistance between points *B* and *C*, and record it under "R_{TH} Measured."

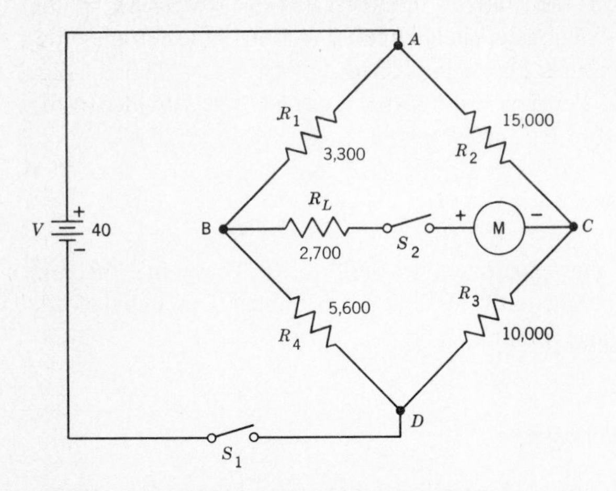

Fig. 24-3. Experimental circuit for verifying Thevenin's theorem, applied to an unbalanced-bridge circuit.

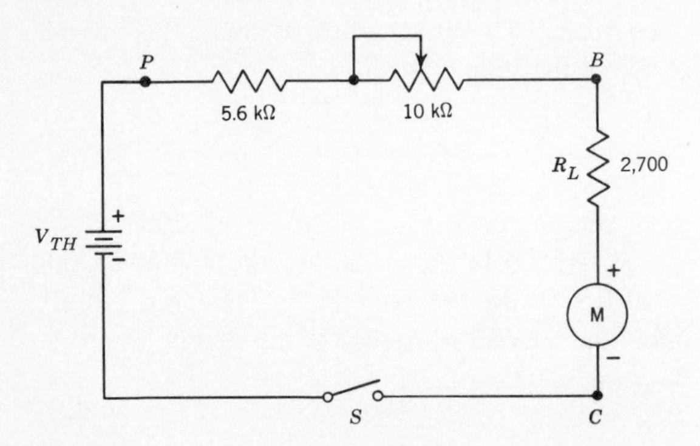

Fig. 24-4. Experimental Thevenin equivalent for Fig. 24-3.

5. Adjust the output of the power supply to V_{TH} measured. Set the resistance of the decade box to the value R_{TH} and connect between points P and B, Fig. 24-4. If a decade box is not available, connect a 5600-Ω resistor in series with a 10,000-Ω potentiometer serving as a rheostat, and adjust the rheostat so that the resistance between points PB (Fig. 24-4) is equal to the measured value of R_{TH}. Now connect R_L, the 2700-Ω load resistor, in series with the milliammeter M between points C and B as in Fig. 24-4.

TABLE 24-1. Measured Resistor Values

Resistor	*Color-coded value, Ω*	*Measured value, Ω*
R_1	3300	
R_2	15,000	
R_3	10,000	
R_4	5600	
R_L	2700	
R_L		
R_L		
R_L		

6. Close switch S. Monitor V_{TH} to see that it has not changed in value. Measure the current in R_L and record under column "I_L Measured Thev." **Power off.**
7. Compute the value of V_{TH} in Fig. 24-3, using the measured values of R from Table 24-1, for the Thevenin equivalent generator for R_L the 2700-Ω resistor, and record in Table 24-2 under column "V_{TH}, computed." Show your computation. Compute also the value of R_{TH} and record under column "R_{TH}, computed." Show your computation.

   Now compute the value of I_L in R_L, using the Thevenin equivalent circuit with the computed values. Record in Table 24-2 under the column "I_L, computed." Show your calculations.

Extra Credit

8. Verify that Thevenin's theorem is true for the circuit of Fig. 24-3 for other values of R_L, substituting for R_L, in turn, each of three resistors whose values are no less than 1200 Ω nor greater than 10,000 Ω. Measure the current I_L in R_L in the original circuit and also in the Thevenin equivalent circuit. Record in Table 24-2 the values of R_L used, and the other pertinent data. Explain in detail the method you used.

TABLE 24-2. Measurements to Verify Thevenin's Theorem

	V_{TH}, V		R_{TH}, Ω		I_L, mA		
					Measured		
R_L	*Measured*	*Computed*	*Measured*	*Computed*	*Thev.*	*Original Circuit*	*Computed*
2700							

QUESTIONS

1. How does the value of I_L measured in the circuit of Fig. 24-3 compare with that in the Thevenin equivalent circuit of Fig. 24-4? Should they be the same? Explain.
2. Discuss the data in Table 24-2 and explain any unexpected results.
3. Does the experiment confirm the validity of Thevenin's theorem for the circuit of Fig. 24-3? Explain.
4. Does the experiment "prove" Thevenin's theorem for any circuit? Explain.
5. What specific advantages, if any, do you see in using Thevenin's theorem in the solution of dc circuits?

Answers to Self-Test

1. (*a*) 100; (*b*) 188; (*c*) 0.134
2. (*a*) 45; (*b*) 338; (*c*) 0.1

25

EXPERIMENT

NORTON'S THEOREM

OBJECTIVES

1. To determine by analysis the values I_N (Norton's generator current) and R_N (Norton's generator resistance) in a dc circuit containing one or two voltage sources
2. To verify experimentally the values I_N and R_N proposed by Norton's threreom in the solution of complex dc networks containing two voltage sources

INTRODUCTORY INFORMATION

Norton's Theorem

Thevenin's theorem simplifies the analysis of complex networks by reducing the original circuit to a simple equivalent circuit involving a constant voltage source, the Thevenin generator (V_{TH}), in series with an internal resistance R_{TH}. This generator delivers current to the circuit load resistance we wish to study. It is then a simple matter to determine the current in the load and the voltage across the load. Norton's theorem utilizes a similar technique of simplification. The Norton generator, however, is a constant current source.

Norton's theorem states that any two-terminal linear network may be replaced by a simple equivalent circuit consisting of a constant-current generator I_N shunted by an internal resistance R_N. Figure 25-1*a* shows the original network as a *block* terminated by a load resistance R_L. Figure 25-1*b* shows the Norton equivalent circuit. The Norton current I_N is distributed between the shunt resistance R_N and the load R_L. The current I_L in R_L may be found from the equation

$$I_L = \frac{I_N \times R_N}{R_N + R_L} \qquad (25\text{-}1)$$

The rules for determining the constants in the Norton equivalent circuit are as follows:

1. The constant current I_N is the current that would flow in the short circuit between the load resistance terminals if the load resistance were replaced by a short circuit.
2. The Norton resistance R_N is the resistance seen from the terminals of the *open* load, looking into the original network, when the voltage sources in the circuit are replaced by their internal resistance. Thus R_N is defined in exactly the same manner as is R_{TH} in Thevenin's theorem.

Applications

Consider the circuit, Fig. 25-2*a*. We wish to find the current I_L in R_L. We can solve for I_L by simple application of Ohm's and Kirchhoff's laws or by Thevenin's theorem. Let us, however, find I_L by Norton's theorem.

The development of the Norton equivalent circuit for Fig. 25-2*a* may be readily followed from Fig. 25-2*b*, *c*, and *d*. In Fig. 25-2*b*, R_L has been short circuited. As a result, R_3 is also "shorted." The Norton current I_N in the short circuit is readily found to be

$$I_N = \frac{V}{R_T} = \frac{100}{195 + 5} = \frac{100}{200} = 0.5 \text{ A} \qquad (25\text{-}2)$$

In Fig. 25-2*c* we see that the original voltage source V has been replaced by its internal resistance R_1, which equals 5 Ω. The Norton resistance R_N is the resistance measured across *AB* in Fig. 25-2*c*. R_N is 100 Ω. Finally, Fig. 25-2*d* shows the Norton equivalent generator I_N in shunt with R_N across the load resistance R_L. Applying Eq. (25-1) we have

$$I_L = \frac{I_N \times R_N}{R_N + R_L} = \frac{0.5(100)}{100 + 350}$$

$$I_L = \frac{50}{450} \qquad (25\text{-}3)$$

and $$I_L = 0.111 + \text{A} \qquad (25\text{-}4)$$

This is the same value we obtained in Experiment 24 when we used Thevenin's theorem to find I_L.

It would now be possible to find the current in any value of R_L in the circuit of Fig. 25-2*a* by using the Norton equivalent generator with the values in Fig. 25-2*d*.

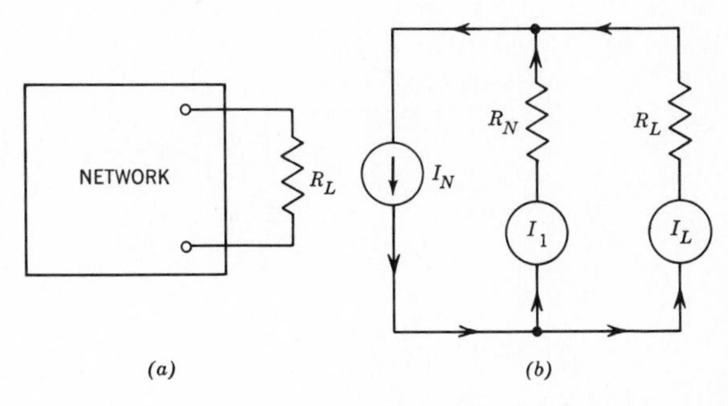

Fig. 25-1. (*a*) Network terminated by a load resistance R_L; (*b*) Norton equivalent generator.

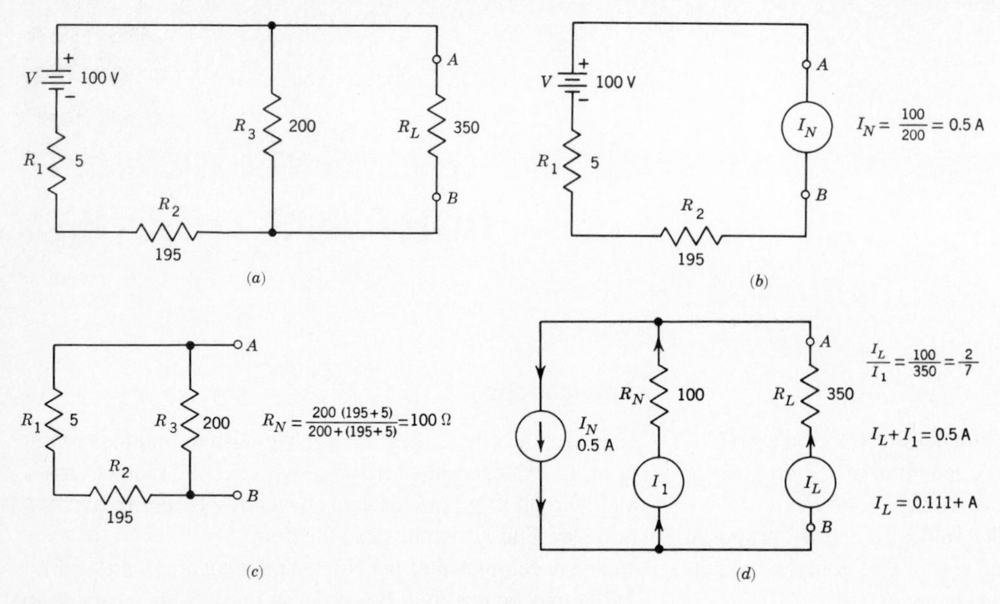

Fig. 25-2. Applying Norton's theorem to the solution of a dc network.

Solution of DC Networks with Two or More Generators

The solution of complex dc networks containing two or more voltage sources is possible using Kirchhoff's laws, Thevenin's theorem, and Norton's theorem. Application of the two latter theorems frequently simplifies analysis and computations. So that the students will have all methods at their disposal, we will solve the circuit of Fig. 25-3 by all three.

By Kirchhoff's Laws

Let us recall Kirchhoff's two laws. These are:

1. The sum of the voltage drops across series-connected resistors in a closed circuit is equal to the applied voltage. This law may be stated in another form, namely, *the algebraic sum of the voltages around any closed circuit loop is zero.*
2. The current entering any junction of an electric circuit is equal to the current leaving that junction. Again, this law may be restated in another form, namely, *the algebraic sum of the currents at any point in a circuit is zero.*

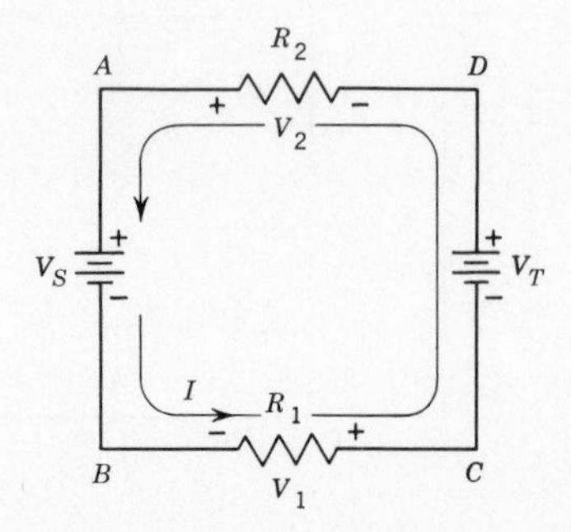

Fig. 25-3. Applying Kirchhoff's voltage law to a closed circuit.

In applying the second form of Kirchhoff's voltage law, certain conventions must be followed. Consider the circuit of Fig. 25-3. The formulation "algebraic sum," etc., requires careful attention to the *signs* of voltage sources and voltage drops. Thus, starting at *A* in Fig. 25-3 and moving in the loop *ABCDA*, we have first the voltage V_S. This is positive at *A*. Hence, the first term in our equation is $+V_S$. Next is the voltage drop V_1 across R_1. Assuming electron flow in the direction of the arrow, the polarity of voltage V_1 is as shown in Fig. 25-3, namely, $-$ to $+$. Hence the next term in our equation is $-V_1$. Similarly, proceeding in the direction *CDA* the polarity of voltages are, respectively, $-V_T$ and $-V_2$. The equation satisfying Kirchhoff's voltage law may now be written as follows:

$$+V_S - V_1 - V_T - V_2 = 0 \qquad (25\text{-}5)$$

NOTE: The opposite *signs* of V_S and V_T show that these batteries are bucking, not aiding, each other, a fact which is quite evident from the circuit diagram.

Now let us apply Kirchhoff's laws to the circuit of Fig. 25-4. It is necessary to set up as many equations as there are independent paths for current (loops). Loop 1 is *ABCDA*. We assume electron current I_1 in this loop in the direction of the arrow as shown. Another complete and independent loop, 2, is *DFGCD*, and electron current I_2 is assumed in the direction shown. I_1 and I_2 constitute the unknowns in this system. Let it be required to find I_2 in R_3.

Start at *A* in loop 1 and write the equation for voltage:

$$+V_1 - (I_1 + I_2)R_2 + V_2 - I_1 \times R_1 = 0 \qquad (25\text{-}6)$$

Substituting the circuit values in Eq. (25-6), we have

$$100 - 2200(I_1 + I_2) + 150 - 1000\,I_1 = 0 \qquad (25\text{-}7)$$

Collecting terms and simplifying, we have

$$3200\,I_1 + 2200\,I_2 = 250 \qquad (25\text{-}8)$$

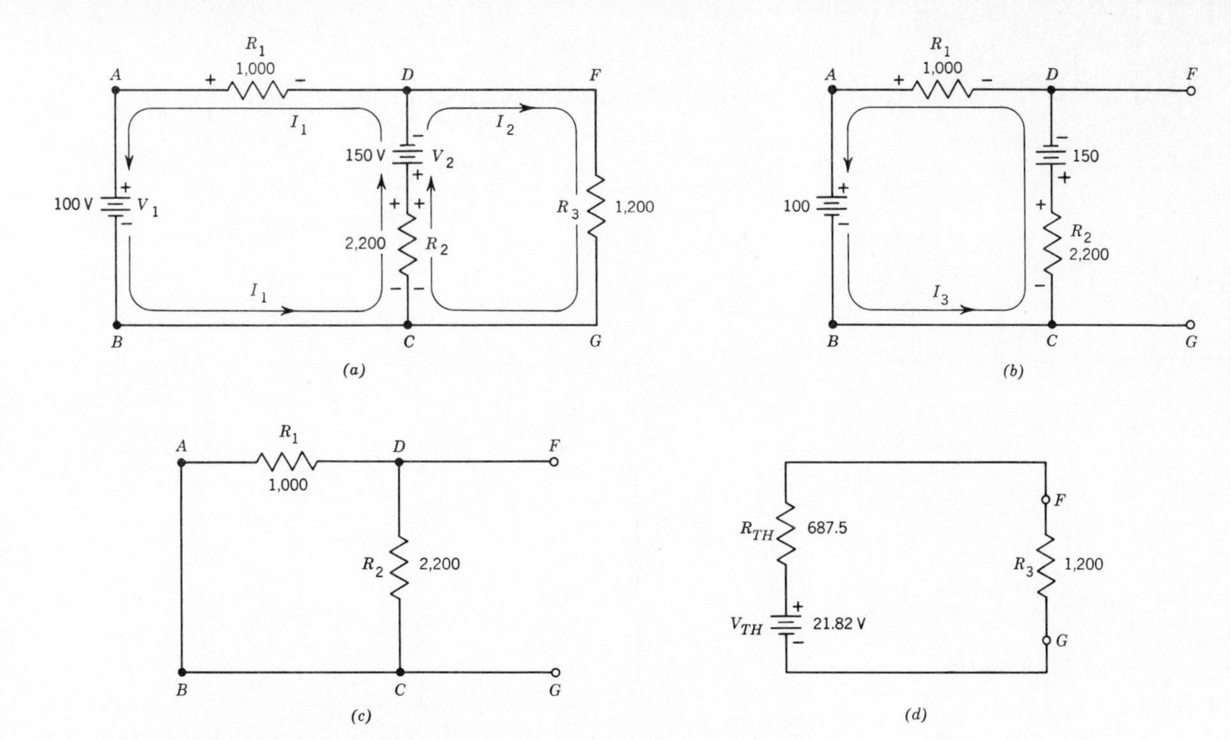

Fig. 25-4. Solution of a complex network by Kirchhoff's laws and Thevenin's theorem.

Now start at D in loop 2. The equation for voltage is

$$-I_2R_3 - (I_1 + I_2)R_2 + 150 = 0 \qquad (25\text{-}9)$$

Substituting the circuit values in Eq. (25-9),

$$-1200\,I_2 - 2200(I_1 + I_2) + 150 = 0 \qquad (25\text{-}10)$$

Collecting terms and simplifying, we get

$$2200\,I_1 + 3400\,I_2 = 150 \qquad (25\text{-}11)$$

Equations (25-8) and (25-11) may now be solved simultaneously for I_1 and I_2, yielding the results

$$I_1 = 0.0861 \text{ A}$$
$$I_2 = -0.0116 \text{ A}$$

The fact that the sign before I_2 is minus simply means that I_2 flows in a direction opposite to that assumed in Fig. 25-4. However, the numerical value is correct. I_2 is the required current in R_3.

Solution by Thevenin's Theorem

Let us solve for I_2 in R_3 (Fig. 25-4*a*) by applying Thevenin's theorem. First we must find V_{TH}, the open-circuit voltage across FG, Fig. 25-4*b*. The open-circuit voltage across FG is $V_{FG} = V_{TH} = -150 + I_3R_2$. We must therefore solve for I_3 before we can determine V_{TH}. Kirchhoff's voltage law applied to Fig. 25-4*b* in the loop $ABCDA$ gives

$$100 - I_3(2200) + 150 - I_3(1000) = 0 \quad (25\text{-}12)$$

and $\quad 3200I_3 = 250$

$$I_3 = \frac{250}{3200} = 0.0781 \text{ A}$$

Now $\quad I_3R_2 = (0.0781)(2200) = 171.82 \text{ V}$

The voltage

$$V_{TH} = -150 + I_3R_2 = 21.82 \text{ V} \qquad (25\text{-}13)$$

Hence, V_{TH} in Fig. 25-4*d* is 21.82 volts.

To find R_{TH}, replace V_1 and V_2 by their internal resistances, which we will assume to be zero, and solve for the resistance across FG (Fig. 25-4*c*).

$$R_{TH} = \frac{2200 \times 1000}{2200 + 1000} = 687.5\ \Omega \qquad (25\text{-}14)$$

Now to solve for I_2 in R_3 of the original network, find I in Thevenin's equivalent circuit (Fig. 25-4*d*). Thus

$$I = I_2 = \frac{V_{TH}}{R_{TH} + R_3} = \frac{21.82}{1887.5} = 0.0116 + \text{A} \qquad (25\text{-}15)$$

We see that the current I_2 in R_3 is 0.0116 A, the same as in the solution of the original circuit using Kirchhoff's laws.

Solution by Norton's Theorem

Let us now solve for I_2 in R_3 (Fig. 25-5*a*, which is the same as Fig. 25-4*a*) by Norton's theorem. The development of Norton's equivalent circuit (Fig. 25-5*d*) may be seen from Fig. 25-5*b* and *c*. In Fig. 25-5*b*, R_3 is replaced by a short circuit. I_N, the current in the short circuit between terminals F and G, is the current in the Norton equivalent generator. To solve for I_N, apply Kirchhoff's laws to loops 1 and 2 of Fig. 25-5*b*. Loop 1 is the path $ABCGFDA$, while loop 2 is the path $DFGCD$.

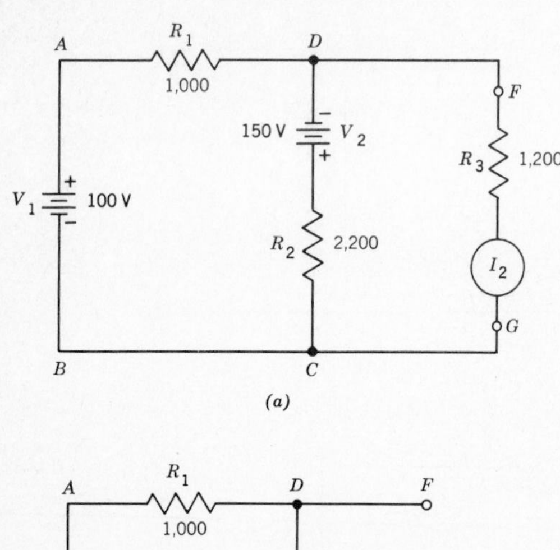

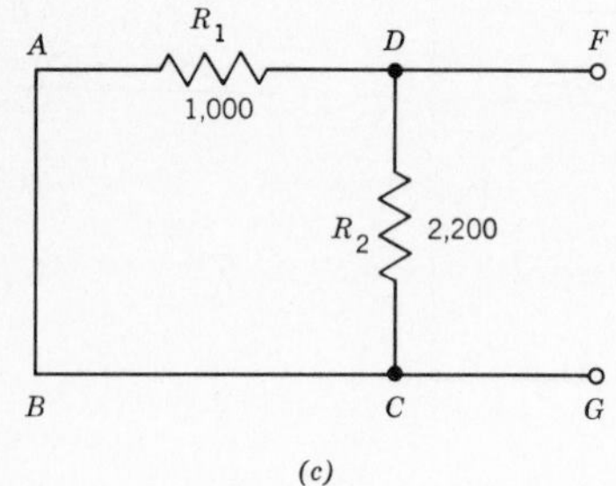

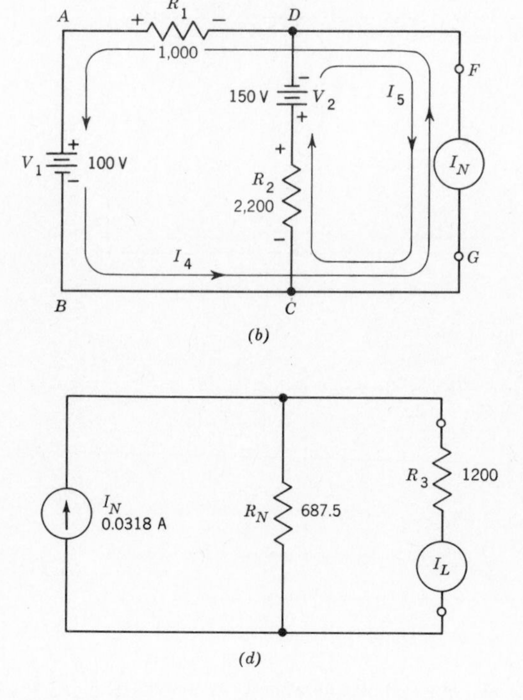

Fig. 25-5. Solution of a complex network by Norton's theorem.

In loop 1,

$$+100 - I_4(1000) = 0$$

$$I_4 = \frac{100}{1000} = 0.1 \text{ A} \tag{25-16}$$

In loop 2,

$$-2200\, I_5 + 150 = 0$$

$$I_5 = \frac{150}{2200} = 0.0682 \text{ A} \tag{25-17}$$

Figure 25-5*b* shows that $I_N = I_4 - I_5$, since the directions of I_4 and I_5 are opposite. Therefore

$$I_N = 0.1000 - 0.0682 = 0.0318 \text{ A} \tag{25-18}$$

The Norton generator resistance R_N is the resistance measured across terminals *FG* in Fig. 25-5*c*. This is the same as the Thevenin resistance R_{TH}, previously computed for Fig. 25-4*c*, and is

$$R_N = 687.5\ \Omega \tag{25-19}$$

Now, to find I_2 in Fig. 25-5*d*, use the formula

$$I_2 = \frac{I_N \times R_N}{R_N + R_L} = \frac{0.0318 \times 687.5}{687.5 + 1200}$$

$$= \frac{0.0318 \times 687.5}{1887.5} = 0.0116 + \text{A} \tag{25-20}$$

Again we find that the value of current in R_3 is the same as that obtained using Kirchhoff's laws and Thevenin's theorem.

SUMMARY

1. Norton's theorem offers another method of solving complex linear circuits. By Norton's theorem a complex two-terminal network may be replaced by a simple equivalent circuit which acts like the original circuit at the load connected to the two terminals.
2. The equivalent circuit is a circuit consisting of a current generator, called the Norton generator (I_N), in parallel with the generator's internal resistance (R_N), across which the load (R_L) is connected. The current I_N of the generator is then divided between R_N and R_L. Figure 25-1*b* shows the Norton generator and its load. This is the Norton equivalent of some complex circuit such as that shown in block form in Fig. 25-1*a*.
3. To determine the Norton current I_N, short-circuit the load and calculate the current through the short in the original circuit. This short-circuited current is I_N. To calculate I_N it may be necessary to use Ohm's and Kirchhoff's laws.
4. To determine the Norton resistance R_N which acts in parallel with the Norton generator, the same technique is applied as in finding the Thevenin resistance in the preceding experiment. Proceed as follows: Open the load at the two terminals in question in the original network. Replace all the voltage sources with their internal resistances. Then calculate R_N, the resistance at the open-load terminals, looking back into the circuit.
5. Norton's theorem applies to a linear circuit with one or more power sources.
6. Once the complex network has been replaced by the Norton equivalent circuit, the current I_L through the load

may be found by applying this formula (see Fig. 25-1*b*):

$$I_L = \frac{I_N \times R_N}{R_N + R_L}$$

7. In this experiment we will verify Norton's theorem experimentally for a dc network containing two voltage sources.

SELF-TEST

Check your understanding by answering these questions:

1. Consider the circuit of Fig. 25-2*a*. $V = 120$ V, $R_1 = 10\ \Omega$, $R_2 = 390\ \Omega$, $R_3 = 600\ \Omega$, and $R_L = 270\ \Omega$. Assume the internal resistance of the battery *V* is zero. In the equivalent Norton's circuit,
   (*a*) I_N = ____________ A
   (*b*) R_N = ____________ Ω
   (*c*) I_L = ____________ A
2. Consider the circuit of Fig. 25-5*a*. $V_1 = 45$ V, $V_2 = 75$ V. Assume the internal resistance of these batteries is zero. $R_1 = 450\ \Omega$, $R_2 = 1500\ \Omega$, and R_3 (the load) $= 470\ \Omega$. In the Norton equivalent circuit,
   (*a*) I_N = ____________ A
   (*b*) R_N = ____________ Ω
   (*c*) I_L = ____________ A

MATERIALS REQUIRED

- Equipment: Two voltage-regulated dc supplies; two multirange milliammeters or VOMs capable of measuring up to 50 mA; EVM
- Resistors: ½-W 3300-, 5100-, 10,000-Ω, and others as required in step 8 (optional)
- Miscellaneous: 10,000-Ω 2-W potentiometer

PROCEDURE

1. Connect the circuit of Fig. 25-6*a*. S_1 is *open*. Turn regulated power supplies V_1 and V_2 **on** and set $V_1 = 40$ V and $V_2 = 20$ V.

2. S_2 is *open*. Close S_1. Be certain that V_1 and V_2 are maintained at 40 V and 20 V, respectively. With milliammeter *M*, measure the current I_L in R_3. Record in Table 25-1, in column "I_L Measured Original Circuit."

3. Close S_2, shorting R_3. Measure the current I_N and record under column labeled "I_N Measured."

NOTE: This is the short-circuit current of the Norton equivalent generator.

4. Remove the voltage supplies V_1 and V_2 from the circuit. Remove R_3 from the circuit. Short points *D* to *F* and *A* to *B*.

NOTE: It is assumed that the internal resistance of the voltage-regulated power supplies is zero.

S_1 is still closed, S_2 open. Measure the resistance between points *F* and *G* and record it under "R_N Measured." This is the resistance in shunt with the Norton equivalent generator.

5. Connect the circuit of Fig. 25-6*b*. S_1 is open.

NOTE: R_N is a 10,000-Ω 2-W adjustable resistor. Set R_N to the resistance measured in step 4.

R_3 is the 5100-Ω load resistance from the previous circuit. M_1 and M_2 are milliammeters to measure the Norton current and the load current, respectively. Turn the voltage output of *V* down to *zero*.

6. Close S_1. Slowly turn up the output of the power supply until the *current* measured by M_1 equals I_N, the Norton generator current measured in step 3. Now, M_2 measures the current I_L, in R_3. Record the current measured by M_2 under "I_L Measured Norton."

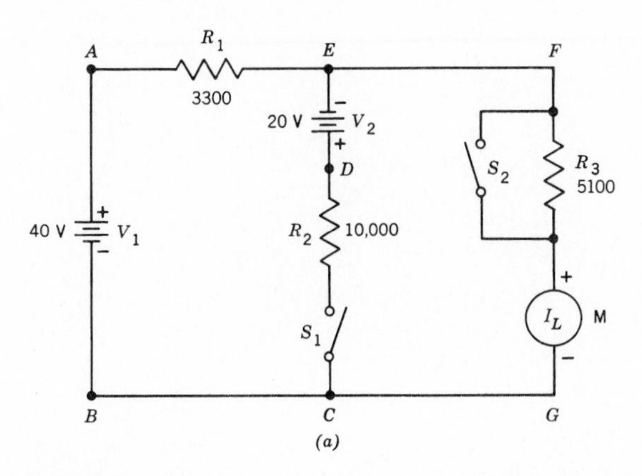

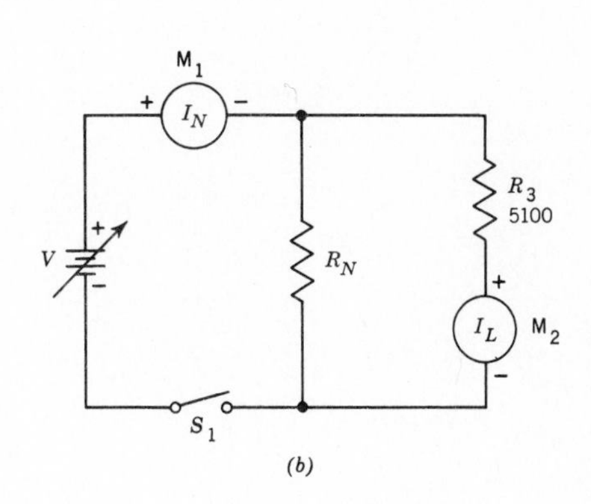

Fig. 25-6. Experimental circuit for verifying Norton's theorem.

7. **Power off.** In the circuit (Fig. 25-6*a*), compute the value I_N for the Norton equivalent generator and record it in Table 25-1, under "I_N Computed." Show your computations.

   Compute also the value R_N, the Norton equivalent generator shunt resistance, and record it under "R_N Computed." Show your computations.

   Now find the value I_L in R_3 using the Norton equivalent generator with the values computed in this step. Record this value of I_L under "I_L Computed." Show your computations.

Extra Credit

8. Verify that Norton's theorem is true for the circuit of Fig. 25-6*a* for other values of R_L by substituting for R_L, in turn, each of three different-valued resistors whose resistance lies between 560 and 12,000 Ω. Measure the current I_L in R_L in the circuit of Fig. 25-6*a* and also in the Norton equivalent circuit of Fig. 25-6*b*. Record in Table 25-1 the values of R_L used and all the other pertinent data. Explain in detail the method you used.

TABLE 25-1. Measurements to Confirm Norton's Theorem

R_L	I_N, mA		R_N, Ω		I_L, mA		
					Measured		
	Measured	*Computed*	*Measured*	*Computed*	*Norton*	*Original Circuit*	*Computed*
5100							

QUESTIONS

1. How does the value of I_L measured in the circuit of Fig. 25-6*a* compare with that measured in the Norton equivalent circuit of Fig. 25-6*b*? With that computed? Should they be the same? Explain.
2. Are the computed and measured values of I_N the same? Should they be? Explain.
3. Are the computed and measured values of R_N the same? Should they be? Explain.
4. Discuss the data in Table 25-1 and explain any unexpected results.
5. Does the experiment confirm the validity of Norton's theorem for the circuit of Fig. 25-6*a*? Explain.
6. Does the experiment "prove" Norton's theorem for any circuit? Explain.
7. If values of R_3 (Fig. 25-6*a*) other than 5100 Ω are used in the experiment, what must be the value of I_N in Fig. 25-6*b*? Explain.

Answers to Self-Test

1. (*a*) 0.3; (*b*) 240; (*c*) 0.141
2. (*a*) 0.05; (*b*) 346; (*c*) 0.021

26

EXPERIMENT

MAGNETIC FIELDS ABOUT BAR AND HORSESHOE MAGNETS

OBJECTIVES

1. To determine experimentally the pattern of magnetic lines of force about a bar magnet; about two bar magnets with like poles near each other; with unlike poles near each other
2. To determine experimentally the pattern of magnetic lines of force about a horseshoe magnet

INTRODUCTORY INFORMATION

In previous experiments we were concerned with a qualitative (descriptive) and quantitative (mathematical) understanding of electricity, and with the measurement of electrical quantities. This concern was dictated by the consideration that an electronic technician deals primarily with the effects of electricity and must have a thorough understanding of electricity and electric circuits. There is another area, namely magnetism, which we must now stop to consider and comprehend, for electricity and magnetism are closely associated. Electricity is generated by a coil of wire rotating in a magnetic field. Moreover, electronics makes extensive use of magnetically operated devices such as relays, meters, transformers, door openers, and alarm systems.

In the experiments on magnetism which follow, no attempt will be made to make measurements of magnetic quantities, for this is generally outside the domain of the electronics technician. Rather, our experiments will be descriptive, dealing with magnetic effects, in an attempt to gain some understanding of the nature of magnetism. For a quantitative discussion of magnetism the student is referred to a standard text on electricity and magnetism.

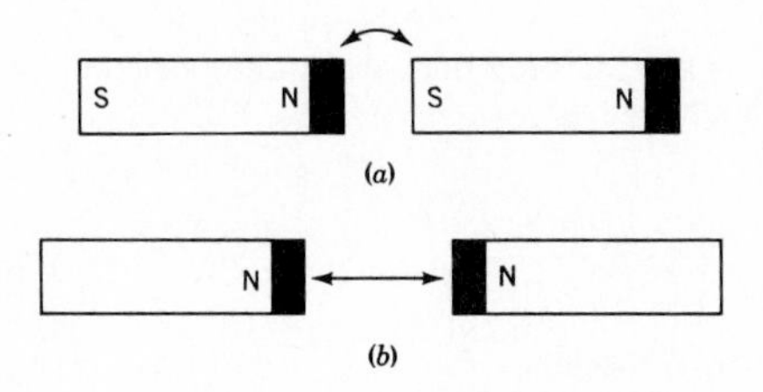

Fig. 26-1. (*a*) Two bar magnets properly oriented are attracted toward each other; (*b*) if one bar magnet is turned 180°, the magnets repel each other.

Some Physical Properties of Magnets

Certain materials, called *magnets*, exhibit unique properties. A magnet attracts iron and other ferrous materials. Between two bar magnets, properly oriented (Fig. 26-1*a*), a force of attraction exists, and one magnet is pulled to the other. If one of the bar magnets is rotated through 180° (Fig. 26-1*b*), the magnets repel each other.

If a small bar magnet, such as that in Fig. 26-2, is freely suspended at its center of gravity, one end of the magnet turns towards the north. If the magnet is spun and permitted to come to rest again, the same end points to the north. We call the north-seeking end of the magnet the north pole, designated N, and the other end, the south-seeking pole, is designated S. A magnetic compass, used for navigation, utilized this characteristic of a magnet as far back as the second century A.D. The earth itself is known to be a huge magnet with magnetic poles close to, but not coincident with, the geographic poles. The needle of a magnetic compass aligns itself with the magnetic poles of the earth. The south magnetic pole of the earth is near the north geographic pole; the north magnetic pole is near the south geographic pole.

If a bar magnet is cut in two (Fig. 26-3), each of the pieces exhibits the properties of a bar magnet. If we then cut each piece into many more pieces, each smaller piece continues to act like a small bar magnet with a north and a south pole. We cannot isolate just a single pole to any tiny piece.

A magnet does not attract some materials. Air, wood, paper, and glass, to mention but a few, will *not* be attracted by a magnet. These materials are *nonmagnetic*. Other materials such as iron and its compounds exhibit marked magnetic properties, and are designated *ferromagnetic*. These substances may be magnetized by placing them in contact with or in the magnetic field of a *natural magnet*. Moreover, some of these continue to act as magnets when the inducing magnet has been removed. Certain mixtures of aluminum, nickel,

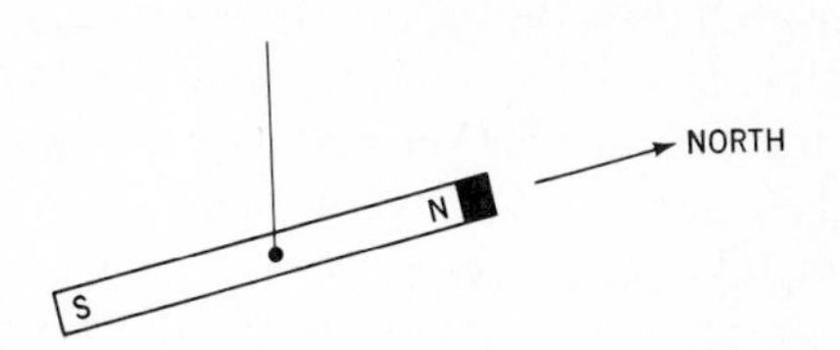

Fig. 26-2. A bar magnet which is free to turn about its center of gravity will point toward the north.

Fig. 26-3. If a bar magnet is cut into two pieces, each piece exhibits the properties of a bar magnet.

and cobalt are used to form very strong permanent magnets, called *alnico* magnets. These are used in radio loudspeakers and other magnetic devices. Hard steel also retains its magnetism. Soft iron is a temporary magnet, for when the inducing magnet is removed, the soft iron loses its magnetism.

Magnetic Field about a Bar Magnet

The force associated with a magnet is said to exist in the magnetic field about the magnet. This magnetic field is presumed to consist of lines of force whose concentration and configuration may be demonstrated experimentally.

Place a cardboard or a piece of glass over a bar magnet and sprinkle iron filings on the cardboard. Tap the cardboard or glass gently so that the iron filings are free to move. The iron filings arrange themselves in a unique pattern (Fig. 26-4). There is a noticeable concentration of the iron filings along curved lines, which we call *lines of force*. The greatest concentration of these lines of force appears at the ends or poles of the magnet. As the lines of force move out from the magnet, the separation between them increases. That is, the magnetic field gets weaker the further we are from the magnet. We would therefore expect that the force of attraction of a magnet is greatest in the immediate vicinity of the magnetic poles where the lines of magnetic force are most heavily concentrated, a fact which is readily verified experimentally.

All the magnetic lines in the field about a magnet constitute the magnetic flux. The stronger the magnetic field, the more lines of force there are in the flux.

A diagrammatic representation of the lines of force about a bar magnet is shown in Fig. 26-5. Note that each field line is a continuous closed loop, leaving the N pole, traveling around the magnet to the S pole and through the magnet back from the S pole to the N pole. The N to S direction of the field lines is purely arbitrary. Note that field lines do *not* cross each other, but push apart. Field lines are most heavily concentrated at the poles.

Figure 26-5 is a two-dimensional view of the magnetic field about a bar magnet. A three-dimensional view would show that the magnetic field is indeed three-dimensional.

Unlike Poles Attract

If the N pole of a magnet is brought near the S pole of a second magnet, there is a force of attraction between the two magnets; the lines of force tend to shorten and pull the two magnets together. The field pattern about the two separated magnets now appears as in Fig. 26-6; it resembles the pattern about a single bar magnet, except for the visibility of the lines between adjacent poles of the two magnets, which are indeed extensions of the lines which complete their loops through the magnets.

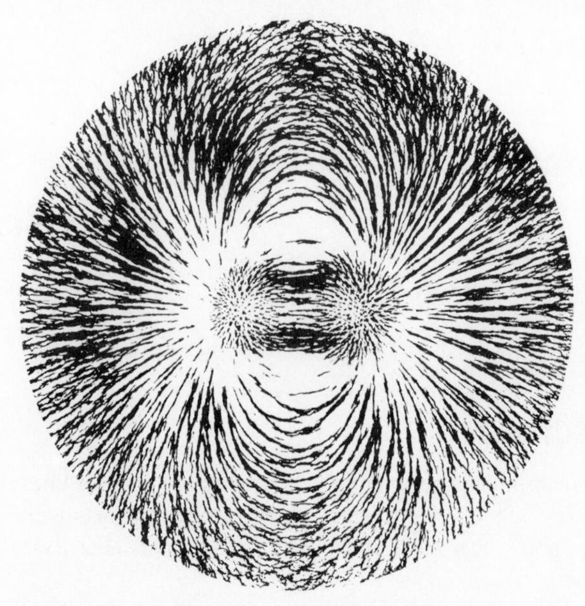

Fig. 26-4. Distribution of iron filings in lines of force about a bar magnet.

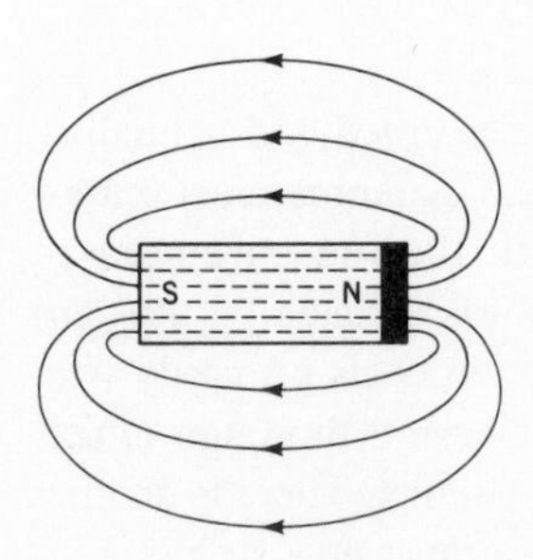

Fig. 26-5. Field of force indicated by magnetic field lines.

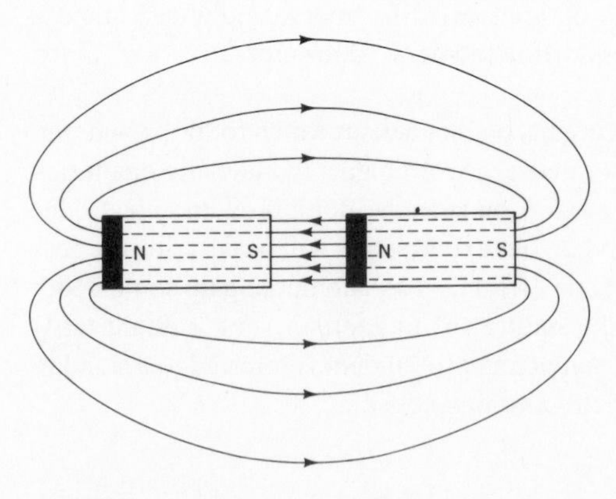

Fig. 26-6. Lines of force between unlike poles reinforce, and the magnets are attracted.

Like Poles Repel

If like poles of two magnets are brought near each other, the two magnets are repelled and push apart. The field pattern

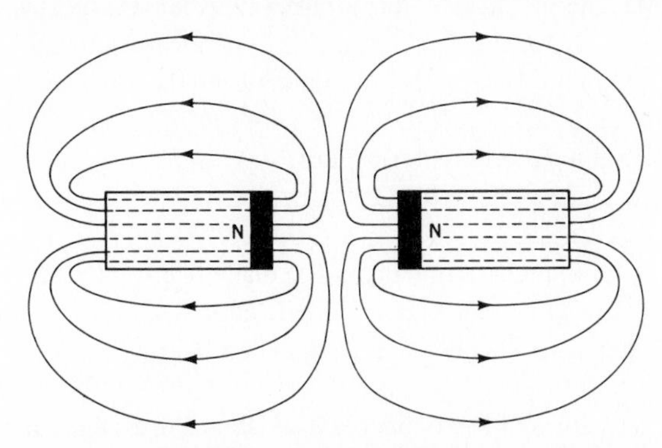

Fig. 26-7. Like poles repel. Field pattern about two magnets when like poles are brought near each other.

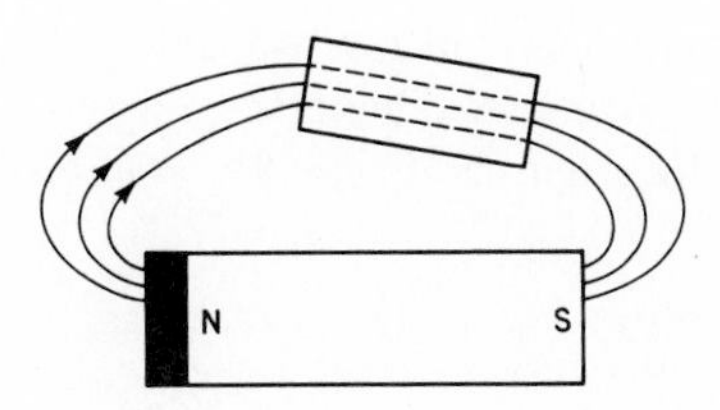

Fig. 26-8. Action of soft iron on magnetic field about a permanent magnet.

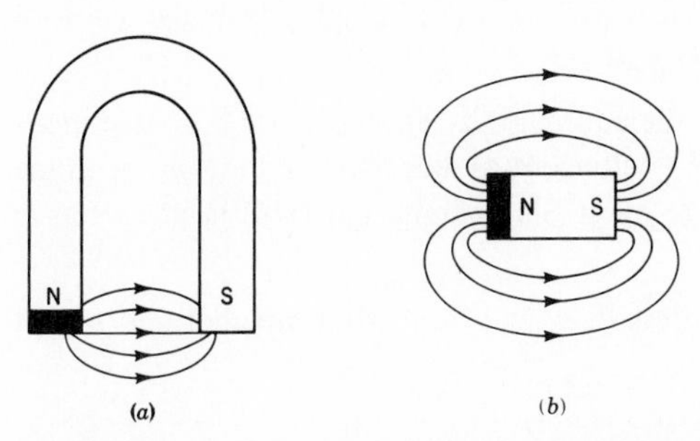

Fig. 26-9. (*a*) The magnetic field in the air gap of a horseshoe magnet is more concentrated than (*b*) the magnetic field in the air gap of a bar magnet when poles are equally spaced.

about these two magnets would appear as in Fig. 26-7. The lines of force close to the adjacent like poles of the two magnets are distorted and push apart.

Magnetic Induction

Magnetic materials more readily admit magnetic lines of force than nonmagnetic materials. When a magnetic material, such as soft iron, is placed in the magnetic field of a permanent magnet, this effect may be noted. The lines of force near the soft iron are attracted to the soft iron as though it were a magnet, and they pass through the soft iron rather than through the space around it (Fig. 26-8). In fact, the soft iron, while in the field of the permanent magnet, acts like a magnet and will attract iron filings to itself. We say that the soft iron has been magnetized by *induction*. It will lose its magnetism when it is removed from the field of the permanent magnet. Soft iron, then, is a temporary magnet. Moreover, since soft iron provides an easier path for magnetic lines of force than does air, soft iron is said to be more permeable than air.

Magnetic Field about a Horseshoe Magnet

A horseshoe magnet may be considered a bar magnet bent in the form of a horseshoe (Fig. 26-9*a*). The magnetic poles remain at the ends of the magnet, but the geometry of the magnet concentrates the lines of force in the air gap between the poles. Compare the lines of force in the air gap of a horseshoe magnet (Fig. 26-9*a*) with those of a bar magnet (Fig. 26-9*b*) whose poles are as far apart as those in the horseshoe magnet. It is apparent that the field in the air gap of the horseshoe magnet is more highly concentrated. Horseshoe magnets with semicircular soft-iron pole pieces are used in meter movements. The magnetic flux in their air gap is uniform and highly concentrated. These are two requirements for a sensitive, linear meter movement.

SUMMARY

1. The force associated with a magnet exists in the three-dimensional magnetic field surrounding the magnet.
2. Though magnetic lines of force cannot be seen, their effects can be observed by the concentration of iron filings along well-defined paths about the magnet.
3. A line of force is the path which an isolated magnetic pole (if such a thing were possible) would travel in moving from one of the poles of the magnet to the other.
4. A magnet has two poles, north and south. If the magnet were suspended at its center of gravity and were free to turn, the *north-seeking* pole of the magnet would point in a northerly direction.
5. The field lines about a magnet are closed loops moving in the direction N to S (by conventional agreement).
6. Lines of magnetic force repel each other and tend to spread apart.
7. The lines of force about a magnet do not cross.
8. Like poles of magnets repel, unlike poles attract.
9. Lines of magnetic force meet less resistance in magnetic than in nonmagnetic materials. Therefore if a piece of soft iron (magnetic) is placed in the field of a magnet, it will distort the field by channeling in itself the lines of force in the vicinity.
10. Magnetic materials may be magnetized by induction, that is, in the presence of a magnetic field. Some will retain their magnetism after the inducing magnet is removed. Others, such as soft iron, will lose their magnetism when the inducing force is removed. Soft iron is therefore in the class of temporary magnets.
11. A knowledge of magnetism is essential to the electronics technician because electronics makes extensive use of magnetically operated devices.

SELF-TEST

Check your understanding by answering these questions:

1. Magnetic lines of force are visible to the naked eye, as is evident from the pattern of iron filings about a magnet. __________ (true/false)
2. The force of a magnet exists in its __________ __________.
3. Magnetic lines will cross each other close to the poles of the magnet. __________ (true/false)
4. The north-seeking pole of a bar magnet which is freely suspended at its center of gravity and is free to turn points to the __________ (north/south) magnetic pole of the earth.
5. Some applications of magnetism in electronics are:
   (*a*) __________
   (*b*) __________
   (*c*) __________
   (*d*) __________
6. Electricity may be generated by a coil of wire __________ in a __________ __________.
7. The magnetic field in the air gap between the poles of a __________ __________ is more concentrated than the field in the air gap between the poles of a __________ __________.
8. Some nonmagnetic materials are:
   (*a*) __________; (*b*) __________;
   (*c*) __________; (*d*) __________.
9. Strong permanent magnets may be made from a mixture of __________, __________, and __________. These are then called __________ magnets.
10. Magnetic lines of force meet less resistance in air than in soft iron. __________ (true/false)

MATERIALS REQUIRED

- Equipment: Polaroid camera and holder, if available
- Miscellaneous: Two bar magnets, magnetic compass, horseshoe magnet, iron filings and cardboard (8½ × 11 in)

PROCEDURE

1. Examine the compass assigned to you. How can you tell which is the N pole?
   Identify the N pole of the compass.
2. The poles of the two bar magnets and the horseshoe magnet assigned to you are unmarked. How can you determine which is the N and which the S pole of each of these magnets?
   Identify the N and S poles of each of these magnets.
3. Place a cardboard over one of the bar magnets assigned to you. Sprinkle iron filings on the cardboard, around the magnet. Gently tap the cardboard until you see a recognizable pattern. This is the pattern of a cross section of the lines of force about the bar magnet. Photograph or draw the pattern for future use. Identify the position of the poles of the magnet and indicate the direction of the lines of force.
4. Now place the two bar magnets, in line, so that the N pole of one is adjacent to and about 1 in away from the S pole of the other. Repeat step 3.
5. Turn one of the bar magnets (in step 4) 180°, so that the N pole of one is adjacent to and about 1 in away from the N pole of the other. The magnets are still in line. Repeat step 3.
6. Place a cardboard over a horseshoe magnet and repeat step 3.

QUESTIONS

NOTE: Submit drawings or photographs of all field patterns you obtained in this experiment.

1. Compare the magnetic field pattern about a bar magnet which you obtained in this experiment with that in Fig. 26-4.
2. From the distribution of iron filings about the bar magnet, was it evident that lines of magnetic force are continuous loops? If not, in which of the experimental patterns was it more evident?
3. How did the field pattern you obtained when you faced like poles of the two bar magnets compare with that in Fig. 26-7? Comment on the differences.
4. How did the field pattern you obtained when you faced unlike poles of two bar magnets compare with that in Fig. 26-6? Comment on the differences.
5. What evidence have you, if any, that magnetic lines of force repel each other?
6. How can you verify experimentally that soft iron is more permeable than air?

Answers to Self-Test

1. false
2. magnetic field
3. false
4. south
5. (*a*) relays; (*b*) transformers; (*c*) electrical generators; (*d*) speakers
6. moving; magnetic field
7. horseshoe magnet; bar magnet
8. (*a*) wood; (*b*) paper; (*c*) glass; (*d*) air
9. aluminum; nickel; cobalt; alnico
10. false

27

EXPERIMENT

MAGNETIC FIELD ASSOCIATED WITH CURRENT IN A WIRE

OBJECTIVES

1. To verify experimentally the existence and direction of a magnetic field about a wire carrying current
2. To determine experimentally the pattern of magnetic lines of force about a solenoid

INTRODUCTORY INFORMATION

Current Producing a Magnetic Field

The physicist Hans Christian Oersted observed that a compass needle was deflected when placed in the vicinity of a wire carrying an electric current. Moreover, he observed that the direction in which the compass needle pointed depended on its position relative to the wire and on the direction of the current.

These observations may be illustrated with the help of the diagram in Fig. 27-1. *W*, the wire carrying current, is perpendicular to the plane of the table through which the wire is drawn. A compass placed flat on the table in position 1 deflects as shown. The direction of the compass needle changes when the compass is moved to positions 2, 3, and 4.

When current in the wire is shut off, the compass needle always points to the north, regardless of its position on the table relative to the wire. Moreover, when current in the wire is turned on again but reversed, the compass needle is again deflected by the force about the wire. However, this time the needle points in the opposite direction from its original orientation in positions 1 through 4 in Fig. 27-1.

These findings point to the conclusions that

1. A magnetic field is developed about a wire carrying current.
2. The direction of the magnetic field depends on the direction of current in the wire.
3. The magnetic field appears to be circular about the wire.

Subsequent researchers showed that the magnetic field lies in a plane perpendicular to the current-carrying wire, that the magnetic field developed by the current *is* circular with the current-carrying wire at the center, and that the magnetic field intensity is greatest close to the wire.

One other fact may be noted. A magnetic field is developed about any moving electric charge, whether that moving charge is a negative electron or is positive. Moreover, the motion of the electric charge need not be restricted to a wire; an electron beam moving in the vacuum of a cathode ray tube has associated with it a magnetic field which has the same characteristics as the field about a wire carrying current.

Experiments with the effects of the magnetic field about a wire lead to a rule which makes it possible to predict the direction of the lines of force. That rule, based on conventional current flow (from the positive terminal of a battery, through the wire, to the negative terminal) is known as the *right-hand rule*. In this text, electron current rather than conventional current is used, and so we will modify the rule for the *left hand*. The left-hand rule states: *If the fingers on the left hand encircle the wire and the thumb of the left hand points in the direction of electron flow in the wire, the fingers point in the direction of the magnetic lines of force* (Fig. 27-2). The right-hand rule is identical except that the direction of conventional current is opposite to that of electron current, and the right hand is used.

Magnetic Field about a Solenoid

We observed in the previous experiment that when two magnetic fields were brought near each other, they interacted and the lines of force were distorted. We can extend this observation to note that the lines of force of two magnetic fields *aid each other* when their lines of force are in the *same direction*,

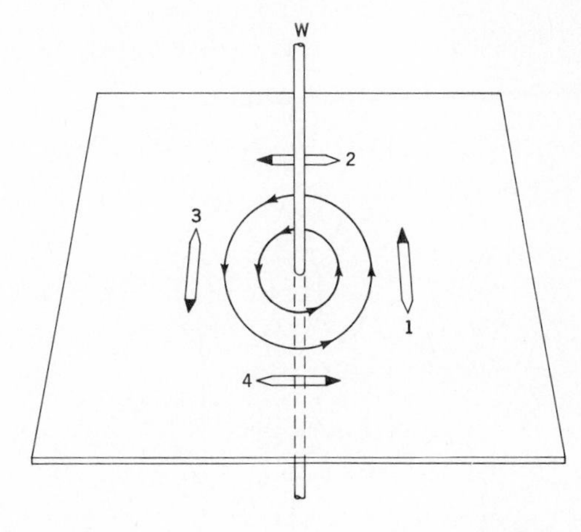

Fig. 27-1. Deflection of a magnetic compass in the vicinity of a wire carrying current.

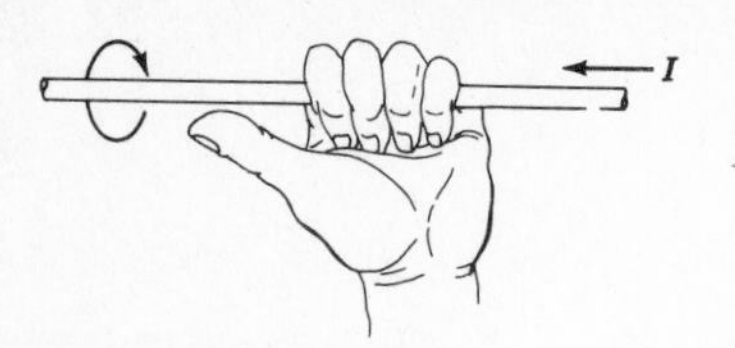

Fig. 27-2. Left-hand rule showing the direction of magnetic lines of force about a wire carrying electron current.

and *oppose (cancel) each other* when their lines of force are in opposite directions. Wire shaped in the form of a solenoid coil (Fig. 27-3) utilizes this characteristic to form an electromagnet when current is sent through the solenoid. The circular lines of magnetic force about the wire in the solenoid combine when they are in the same direction, and cancel when oppositely directed, to form a field pattern which is very much like that of a bar magnet. Note that the magnetic lines of force are most heavily concentrated inside the coil, just as they are in a bar magnet. Moreover the distribution of the field lines outside the coil is very similar to the field distribution about a bar magnet.

The location of the N and S poles of a solenoid may be determined by another left-hand rule. If you grasp the solenoid so that the fingers point in the direction of electron current, then the thumb points in the direction of the N pole (Fig. 27-4). Observe that the position of the poles may be reversed by changing the direction of current or changing the direction of the winding.

The form about which a solenoid is wound may consist of magnetic or nonmagnetic material. However, soft iron is generally used, because the iron core, having a higher permeability than nonmagnetic materials, increases the flux density inside the core. When current is turned on, the iron core is magnetized and acts like a bar magnet. The magnet loses its "strength" when current is turned off. This is another reason for using a soft-iron core: to create a temporary magnet when it is needed and eliminate it when it is not needed.

The strength of a solenoid magnet may be increased by increasing the number of turns of wire in the coil or by increasing the current in the coil.

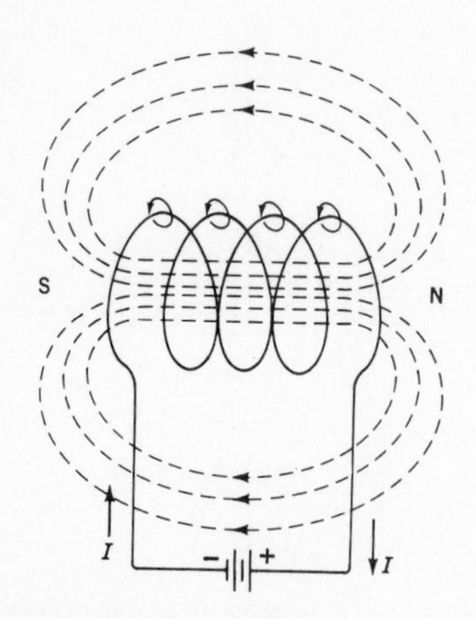

Fig. 27-3. Magnetic field about a solenoid.

We can understand magnetic effects better if we compare these effects with those in an electric circuit. In an electric circuit the variables are V, electromotive force; I, the resulting current; and R, the opposition to current in the circuit. In a magnetic circuit the equivalent variables are *mmf*, magnetomotive force; ϕ, the lines of magnetic force produced; and $\mathcal{R}$ (reluctance), the opposition of the circuit to producing lines of force. Algebraically

$$\text{mmf} = At \tag{27-1}$$

where A equals the amperes in the circuit (coil), and t equals the number of turns in the coil. The formula which relates the three variables in a magnetic circuit is

$$\phi = \frac{\text{mmf}}{\mathcal{R}} \tag{27-2}$$

This is comparable to Ohm's law for electric circuits.

Reasoning from formulas (27-1) and (27-2) it becomes apparent that the strength (ϕ) of a solenoid magnet is directly proportional to the number of ampere-turns (At); that if the current (A) in the coil is changing, the strength of the magnet is changing in direct proportion to A. That is, an increase or decrease in current will result in a proportionate increase or decrease in the strength of the resulting magnetic field.

Industrially, solenoids in the form of relays are used as electric switches to turn a process on and off. Solenoids are also used as valves to permit a liquid to circulate in a system or to turn the liquid off. Bells, buzzers, and many other devices are solenoid-operated.

SUMMARY

1. A magnetic field is developed by a moving electric charge.
2. Circular magnetic lines of force appear around a wire carrying current. The magnetic field is at right angles to the wire and may be visualized as a cylindrical "sleeve" surrounding the wire.
3. The direction of the lines of force depends on the direction of current. If the wire is grasped in the left hand with the thumb pointing in the direction of electron current, the fingers point in the direction of the circular magnetic lines of force (Fig. 27-2).

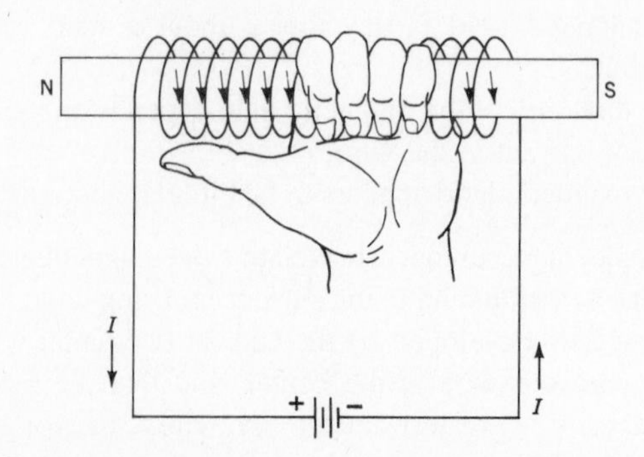

Fig. 27-4. Left-hand rule for locating the N pole of a solenoid magnet.

4. If the wire is wound in the form of a coil and current is permitted in this solenoid coil, the solenoid exhibits a magnetic field pattern very much like the field about a bar magnet (Fig. 27-3).
5. The poles of this electromagnet may be located by grasping the coil with the left hand, the fingers pointing in the direction of current in the windings. The thumb then points to the N pole of the magnet (Fig. 27-4).
6. The position of the poles of a solenoid magnet may be reversed by (*a*) reversing the current or (*b*) reversing the direction of the winding.
7. Solenoids are frequently wound on soft-iron cores. When current is turned on, the iron core acts like a temporary bar magnet, but it loses its magnetism when current is turned off.
8. The strength of a solenoid magnet may be increased by increasing the current or the number of turns, or both.
9. A magnetic circuit may be compared to an electric circuit in which:
   (*a*) mmf (magnetomotive forces) is the equivalent of V (electromotive force).
   (*b*) ϕ, the resulting lines of magnetic flux, is the equivalent of I.
   (*c*) $\mathcal{R}$, the opposition to ϕ, is the equivalent of R.
10. In a magnetic circuit

$$\phi = \frac{\text{mmf}}{\mathcal{R}}$$

11. In a magnetic circuit mmf $= At$.
12. In a solenoid magnet, the strength of the magnet (ϕ) is directly proportional to A, the current in the coil; that is, ϕ increases as A increases, and ϕ decreases as A decreases.
13. Solenoid magnets are used to actuate the switching mechanism in a relay, a bell, buzzer, and other electromagnetic devices.

SELF-TEST

Check your understanding by answering these questions:

1. An electric current creates a magnetic field. __________ (true/false)
2. The magnetic field at any one point about a wire carrying current lies in a plane __________ to the wire at that point.
3. If an observer looks along a wire carrying electron current, in the direction of electron flow, the direction of the magnetic field about the wire appears __________ (clockwise/counterclockwise) to the observer.
4. Soft iron is more permeable than air. Therefore a solenoid with a soft-iron core has a more intense field in the core than does a solenoid with an air core. __________ (true/false)
5. The left-hand rule can be used to determine the identity of the __________ in a solenoid.
6. The strength of an electromagnet can be increased by increasing (*a*) __________ (*b*) __________.

MATERIALS REQUIRED

- Power supply: Variable dc source, low-voltage, high-current
- Equipment: 0–1-A ammeter; Polaroid camera and holder, if available
- Resistors: 25-W 15-Ω
- Miscellaneous: SPST switch, magnetic compass, iron filings, cardboard (8½ × 11 in), 16 ft of #18 varnish-insulated copper wire, 2-in cylindrical (½-in diameter) hollow coil form (cardboard, plastic, or copper tubing), 2-in round soft iron core for coil form

PROCEDURE

Solenoid without Core

1. Use #18 varnish-insulated copper wire to construct a solenoid coil. On a hollow cylindrical form, cardboard, plastic or copper tubing, 2 in long by ½ in in diameter, wind clockwise in two or three layers a 100-turn coil. Scrape the ends of the wire free of insulation and secure one end at conductive terminal A, mounted at one end of the coil form. Similarly secure the other scraped end at conductive terminal B, mounted at the other end of the coil form. Mark ends A and B.
2. Connect the coil constructed in step 1 in the circuit of Fig. 27-5. End A of the coil is connected to the positive terminal of the supply, end B, through a resistor R and ammeter M to the negative terminal of the supply. Power is **off**.

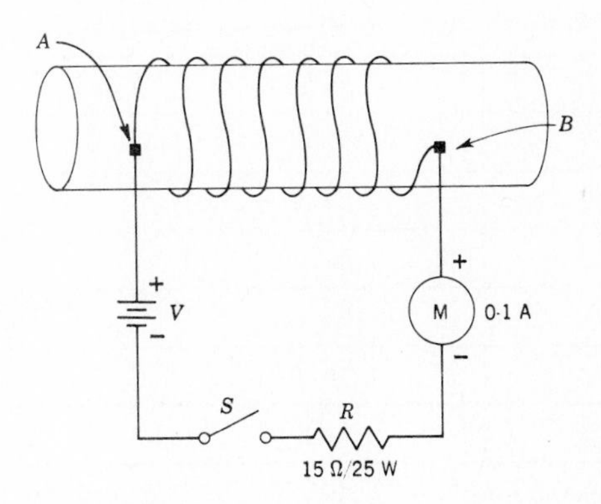

Fig. 27-5. Experimental circuit to determine characteristics of the magnetic field about a solenoid.

3. Set a magnetic compass down on the lab table so that it is not near any magnets or magnetic materials. Orient the compass case so that the compass pointer is set on north N, as in Fig. 27-6. Place end *A* of the coil about 2 in away from the point marked *E* on the compass. Orient the solenoid coil so that its axis is in line with the line of the compass drawn from W to E, as in Fig. 27-6. The solenoid coil, with no current in it, should have no effect on the compass pointer.
4. Turn the power supply **on** and adjust its output so that there is ¾ A in the coil, as measured on *M*. Record in Table 27-1 the direction of the compass pointer. *Data for steps 4 through 14 should be recorded in Table 27-1.*
5. Maintaining the W–E orientation of the axis of the solenoid coil, slowly push the coil so that end *A* is ½ in away from point E on the compass. The effect of the solenoid on the compass pointer should be more evident than in step 4. Record the direction of the compass pointer.
6. Reduce current in the solenoid to 100 mA. The solenoid is still ½ in from point E on the compass, as it was in step 5. Record the direction of the compass pointer.

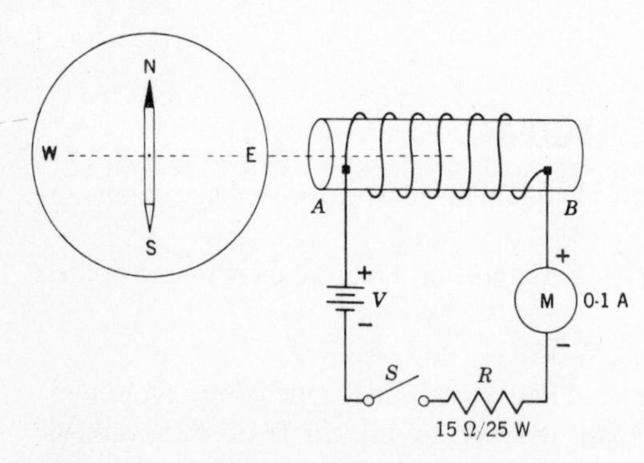

Fig. 27-6. Axis of the solenoid is in line with the line drawn from W to E. When there is no current in the solenoid, the compass points to the north (N).

7. Reduce solenoid current to zero. Record the direction of the compass pointer. **Power off.**

Solenoid with Soft-Iron Core

8. Place a 2-in-long soft-iron core inside the solenoid coil. Move the solenoid so that end *A* is again 2 in from the point on the compass marked E. The solenoid and compass are oriented as in step 3.
9. **Power on.** Adjust the power-supply output until there is ¾ A current in the solenoid. Record the direction of the compass pointer.
10. Move the solenoid coil until it is ½ in away from point E on the compass, maintaining the W–E orientation of its axis. Record the direction of the compass pointer.
11. Reduce solenoid current to 100 mA. Record the direction of the compass pointer.
12. Reduce solenoid current to zero. Record the direction of the compass pointer.
13. **Power off.** Reverse the polarity of current meter leads. Reverse also polarity of power supply leads. **Power on.** Adjust the supply for ¾ A current in the solenoid coil. The solenoid is still ½ in away from the point marked E on the compass, as in step 10. Record the direction of the compass pointer.
14. **Power off.** Record the direction of the compass pointer.

Magnetic Field about a Solenoid

15. The soft-iron core is still inside the solenoid coil, as in steps 8 through 14. The solenoid is connected as in step 13. **Power on.** Set the current in the solenoid to ¾ A.
16. Position a cardboard plane over the solenoid. Sprinkle iron filings on the cardboard and gently tap the cardboard until the filings settle into an identifiable pattern. Photograph or draw this pattern.

TABLE 27-1. Magnetic Field About a Solenoid

Step	*Current in Solenoid,* A	*Solenoid Distance from E, in*	*Current Polarity*		*Direction of Pointer*
			A	*B*	
4	¾	2	+	–	
5	¾	½	+	–	
6	⅒	½	+	–	
7	0	½	+	–	
9	¾	2	+	–	
10	¾	½	+	–	
11	⅒	½	+	–	
12	0	½	+	–	
13	¾	½	–	+	
14	0	½	–	+	

QUESTIONS

Confirm your answers to the following questions by referring specifically to your experimental data.

1. What verification have you of the fact that there is a magnetic field associated with current in a coil?
2. In step 10, which end of the electromagnet, *A* or *B*, is the N pole? Why?
3. In step 13, which end of the electromagnet is the N pole? Why?
4. Is the magnetic polarity of the electromagnet affected by the direction of current in the electromagnet? How?
5. Do the results of your experiment confirm the left-hand rule for determining the poles of an electromagnet? Explain?
6. Is the strength of the electromagnet affected by the amount of current in the solenoid? How?
7. What does the magnetic field pattern about a solenoid coil electromagnet resemble?

Answers to Self-Test

1. true
2. perpendicular
3. counterclockwise
4. true
5. poles
6. (*a*) number of turns; (*b*) amount of current

28

EXPERIMENT

INDUCING VOLTAGE IN A COIL

OBJECTIVES

1. To verify experimentally that a voltage is induced in a coil when the lines of force of a magnet cut across its windings
2. To verify experimentally that the polarity of the induced voltage depends on the direction in which magnetic lines of force cut the coil windings

INTRODUCTORY INFORMATION

Electromagnetic Induction

In the preceding experiment we established that current, which is defined as the *movement* of electric charges, generates a magnetic field. In this experiment we will demonstrate that a moving magnetic field which cuts across a conductor will generate a movement of electric charges in the conductor; that is, it will produce a voltage in the conductor, and if the conductor is part of a complete circuit, current will flow in the conductor.

If a zero-center galvanometer is connected to the ends of a metal rod and the rod is placed in the air gap between the poles of a permanent magnet, there will be no movement of the galvanometer pointer when the rod has come to rest in the magnetic field. However, if the rod is moved up, *cutting* the lines of magnetic force, the galvanometer will deflect, showing current, as in Fig. 28-1. If the rod is then moved down, the galvanometer will again deflect, but in the opposite direction. Once the rod has come to rest, the galvanometer will register zero current. Now, if the rod is permitted to rest and the *magnet* is moved up, away from the rod, there is again current flow in the rod. Then if the magnet is brought down toward the rod, there is current flow in the opposite direction. Note that in each case the rod was perpendicular to the lines of force of the magnet, and the effect of a relative motion, that is, a motion of the rod or the magnet, was to *cut* the magnetic lines of force. The result was to induce a voltage in the conductive rod, causing current, as registered by the galvanometer. If the rod is now moved parallel to the lines of force between the poles of the magnet, there is no deflection of the galvanometer. *The conclusion we reach is that a relative motion between the conductive rod and the magnetic lines of force will induce a voltage in the rod only when the lines of force are cut by the rod.*

Polarity of Induced Voltage

The fact that the galvanometer pointer deflected either to the left or to the right of zero, depending on the direction of cutting of the lines of force, is evidence that the polarity of the voltage induced in the conductive rod depends on the *direction* of cutting of the lines of force. The polarity of the induced voltage can be established by Lenz' law.

Lenz' law states that the induced voltage must be of such a polarity that the direction of the resulting current in the conductor sets up a magnetic field about the conductor which will oppose the *motion* of the inducing field. An example will illustrate the meaning of Lenz' law.

Consider the solenoid wound about an air core (Fig. 28-2). The ends of the solenoid are connected to a galvanometer, as shown. If a bar magnet is pushed into the coil so that the S pole of the magnet enters the left side of the coil, the current resulting in the solenoid should set up a magnetic field whose S pole will be at the left side of the solenoid. Since like magnetic poles repel, Lenz' law would be satisfied, for the induced current in the solenoid would have set up a magnetic field which opposes the inducing field. To verify that a S pole is in fact produced at the left end of the solenoid, observe the direction in which the galvanometer pointer deflects. Note that it moves to the right, indicating a voltage whose polarity is shown in Fig. 28-2. Applying the left-hand rule, grasp the

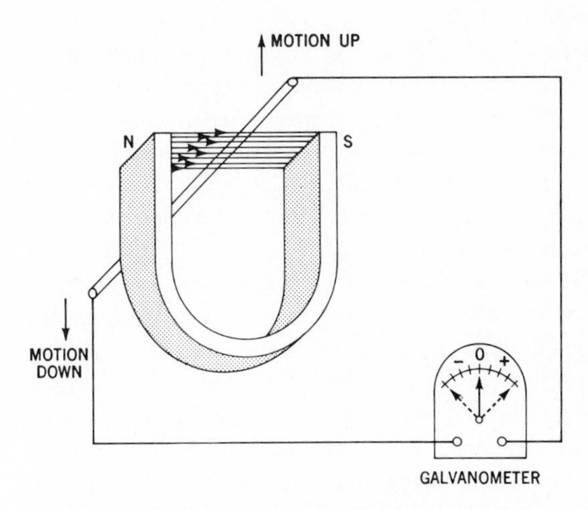

Fig. 28-1. The galvanometer registers current when there is a motion up or down, that is, when the lines of force are cut by the rod.

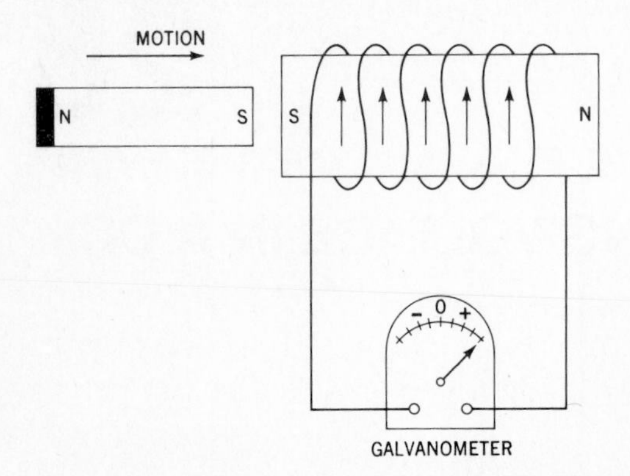

Fig. 28-2. Verifying Lenz' law. The polarity of the magnetic field created by the induced current in the solenoid opposes the polarity of the inducing field.

solenoid so that the fingers point in the direction of electron current, as shown by the arrows on the solenoid core. The thumb then points to the N pole. It is apparent, then, that at the left end of the solenoid winding there is a S pole and at the right end a N pole, and we have confirmation of Lenz' law.

If the solenoid is next moved out of the core, the left end of the solenoid will become a N pole, thus attempting to restrain the bar magnet from moving out of the core. The voltage induced in the solenoid will be of the opposite polarity and the galvanometer will deflect to the left.

Counter EMF

The discussion to this point has dealt with the observation that a voltage is induced in a wire or coil when there is relative motion between the wire or coil and the magnetic field of a permanent magnet, that is, when the wire or coil cuts the lines of force of the magnet. The facts are more general, however. The effect of induction in a coil or conductor is achieved when there is relative motion between the coil or conductor and the field of *any* magnet, including an electromagnet.

In a previous experiment it was demonstrated that a magnetic field exists about a coil carrying current. The coil carrying current is in fact an electromagnet and may be substituted for the permanent magnet in the discussion earlier in this experiment.

The term *relative motion* is not limited merely to the physical movement of a magnet near a conductor or of a conductor in a magnetic field. Relative motion may exist without any physical movement whatsoever. Consider a coil through which an increasing or decreasing (not a steady-state) current is flowing. In the case of an increasing current, the magnetic field about the coil is increasing or expanding; that is, it is a *moving field*. When the current in the coil is decreasing, the magnetic field about the coil is decreasing or collapsing, again a moving magnetic field. A conductor held stationary in this moving magnetic field is in effect *cutting* the lines of magnetic force. Therefore a voltage will be induced in the conductor. Expanding and collapsing magnetic fields will cause voltages of opposite polarity to be induced in the conductor.

Consider again a coil through which an increasing or decreasing current is flowing. The expanding or collapsing (moving) lines of magnetic force about the coil *cut the windings of the coil itself*. Accordingly, a voltage is induced in the coil, which by Lenz' law will oppose the inducing force. That is, the voltage induced in the coil will have a polarity opposite to the voltage which initially caused current to flow in the coil. The voltage induced in the coil may therefore be termed a *counter emf*.

Magnitude of Induced Voltage

The amplitude of the voltage induced in a solenoid depends directly on (*a*) the *number* of turns *N* of wire in the solenoid and (*b*) the *rate* at which the lines of flux are cut by the windings of the solenoid. Thus the more windings, the higher will be the induced voltage across the ends of the coil. Similarly, the faster the magnetic flux is cut by the windings, the higher will be the voltage produced across the solenoid. A mathematical formulation of this law was first given by the physicist Michael Faraday. It is suggested that the student refer to a standard text on electromagnetism for the exact formula, which is the mathematical statement of Faraday's law.

SUMMARY

1. A voltage is induced in a conductor when the conductor *cuts* the lines of force in a magnetic field. Cutting the lines of force can be achieved by either moving the conductor or moving the magnetic field.
2. The polarity of the voltage induced in the conductor is determined by the direction of cutting of the lines of force. Thus if a positive voltage is induced, say, by a conductor cutting the lines of force in a *downward* direction, a negative voltage will be induced by an *upward* cutting of these same lines of force.
3. The polarity of the induced voltage can be predicted by Lenz' law, which states: The polarity of the voltage induced in a conductor must be such that the *magnetic field set up by the resulting current* in the conductor will oppose the motion of the inducing magnetic field.
4. The amplitude of voltage induced in a solenoid when its windings cut magnetic lines of force will depend directly on the number of turns in the winding and on the rate at which the magnetic flux lines are cut by the windings (Faraday's law).

SELF-TEST

Check your understanding by answering these questions:

1. A moving magnetic field will ____________ a ____________ in a copper wire, if the lines of magnetic force ____________ the wire.

2. The polarity of voltage induced in a conductor will depend on the ____________ of cutting of the lines of force by the conductor.
3. A conductor moving parallel to the lines of force in a magnetic field ____________ (will/will not) have a voltage induced in it.
4. The more turns there are in a solenoid whose windings cut the lines of force in a magnetic field, the greater will be the voltage induced in it, all other things being equal. ____________ (true/false)
5. By means of ____________ law, it is possible to predict the polarity of voltage induced in a coil.

MATERIALS REQUIRED

- Equipment: Galvanometer or very sensitive electronic dc microammeter which can be zero centered
- Miscellaneous: Solenoid—100 turns of #18 varnish-insulated copper wire wound in three layers on a 3 × 1 in (diam.) hollow cardboard or plastic cylindrical form

PROCEDURE

1. Connect a galvanometer or sensitive zero-centered dc microammeter, set on its lowest range, to the terminals of the air-core solenoid as in Fig. 28-2.
2. Position a bar magnet near the solenoid, with the S pole closest to the left end of the solenoid, as in Fig. 28-2.
3. With the bar magnet stationary, does the meter show any induced voltage? Record your observation in Table 28-1.
4. Now push the bar magnet into the solenoid so that the south pole of the magnet enters the air core first. Does the meter show any induced voltage? What is the polarity of the voltage? What is the current amplitude? Record these observations in Table 28-1.
5. Now pull the bar magnet out of the solenoid. Does the meter show any induced voltage? What is the polarity of the voltage? Current amplitude? Record these observations in Table 28-1.
6. Now reverse the bar magnet so that the N pole is closest to the left end of the solenoid. Insert the bar magnet into the air core. Does the meter show any induced voltage? What is the polarity? Current amplitude? Record your observations in Table 28-1.
7. Repeat step 5.
8. Again plunge the magnet into the air core, but more rapidly than in step 6. Observe and record in Table 28-1 the amplitude of current in the circuit.
9. Remove the magnet and again plunge it into the air core, but more rapidly than in step 8. Observe and record in Table 28-1 the amplitude of current in the circuit.
10. Now position the bar magnet vertically on the table, with the N pole on the table and the S pole in the air above the table. The magnet will remain stationary.
11. With the meter still connected to the solenoid, plunge the solenoid down over the magnet. Observe and record in Table 28-1 the polarity of the voltage induced.
12. Pull the solenoid out of the magnet. Observe and record in Table 28-1 the polarity of the voltage induced.

TABLE 28-1. Inducing Voltage in a Coil

Step	Condition	*Voltage Polarity*	*Current Amplitude, μA*
3	Magnet stationary		
4	S pole of magnet near left end of solenoid. Magnet is pushed into solenoid.		
5	Magnet pulled out of solenoid.		
6	N pole of magnet near left end of solenoid. Magnet is pushed into solenoid.		
7	Magnet pulled out of solenoid.		
8	N pole of magnet near left end of solenoid. Magnet is pushed into solenoid more rapidly than in step 6.	X	
9	Same as step 8 but more rapidly.	X	
11	Solenoid plunged down over magnet.		X
12	Solenoid pulled up, away from magnet.		X

QUESTIONS

1. What are the conditions for electromagnetic induction of a voltage in a solenoid?
2. What verification have you that a voltage is in fact induced in a solenoid when a bar magnet is plunged into it or pulled out of it? Refer specifically to your data to support your answer.
3. What verification have you that it is the *relative* motion involved in cutting of lines of force which induces a voltage in the solenoid? Refer specifically to your data to support your answer.
4. What is Lenz' law?
5. What verification have you that Lenz' law is true? Refer specifically to your data to support your answer.
6. What is Faraday's law?
7. Do you have any verification of all or part of Faraday's law? Explain.

Answers to Self-Test

1. induce; voltage; cut
2. direction
3. will not
4. true
5. Lenz'

29 EXPERIMENT

APPLICATIONS OF MAGNETISM: THE DC RELAY

OBJECTIVES

1. To become acquainted with some applications of magnetism
2. To study the characteristics and operation of a dc relay

INTRODUCTORY INFORMATION

Applications of Magnetism

One of the earliest applications of magnetism was the magnetic compass. It can be said that the compass ushered in the age of modern marine navigation and exploration, for now mariners were no longer dependent for direction on sighting the stars. With the compass they could find their way at any time of day or night, and in any weather.

The next dramatic development in the application of magnetism was in the generation of electricity. The generator was designed to take advantage of the fact that an emf is created when a conductor cuts lines of magnetic force. Thus the armature of a generator contains many windings of copper wire rotating in the magnetic field associated with its field windings (the field windings act as an electromagnet). The emf induced in the armature is transmitted for use in the home and in industry. Our world could not exist as it is without electricity. Witness how everything comes to a halt when power fails. Life in the developed countries is totally dependent on electricity and electrical technology.

And of course there is the electric motor, which works because of the interaction between the magnetic fields associated with its armature and field windings. There are electric motors in every home: the motor which circulates the furnace heat, the air-conditioner motor, the washing machine and dryer motors, the electric clock motor, etc.

And then there is electronics. Without electricity (which is dependent on magnetism) there would be no radio, no televison, no computer, no telephone, none of the other conveniences which characterize life today.

The Relay

The *relay*, another electromagnetic device, performs countless tasks in industry and in the home. Relays are used for switching, for indicating, for transmission, and for protective devices.

Protective relays remove from service any element of an electrical system when that element short circuits or operates in such a way as to damage other elements of the system.

Indicating or signaling relays may be used together with protective relays to show the location of a component which has failed. Bells, buzzers, and alarms are all relay-type devices.

Relays are electromagnetically operated, remotely controlled switches, with one or more sets of contacts. When energized, the relay operates, opening or closing its contacts or opening some contacts and closing others. Contacts which are open when the relay is not energized are called *normally open* (NO) or simply *open* contacts. Contacts which are closed when the relay is not energized are called *normally closed* (NC) contacts. Normally open contacts are sometimes referred to as *a* contacts. Normally closed contacts are known as *b* contacts. Figure 29-1 shows the schematic representation for open and closed contacts.

There are certain terms usually associated with relays which the technician should understand. A relay is said to "pick up" when it is energized and trips, and the pickup value is the smallest value of actuating current required to close an *a* contact or open a *b* contact. When a relay is deenergized, opening an *a* contact or closing a *b* contact, it is said to "reset" or "drop out."

Relay contacts are held in their normal position either by springs or by some gravity-actuated mechanism. An adjustment or adjustments are usually provided to set the restraining force to cause the relay to operate within predetermined circuit conditions.

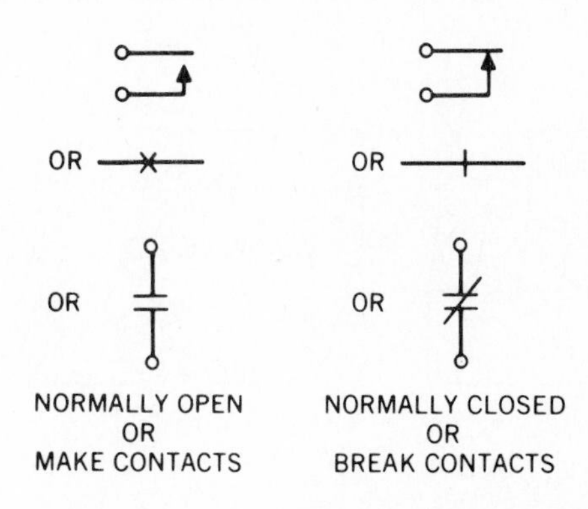

Fig. 29-1. Schematic symbols for relay contacts.

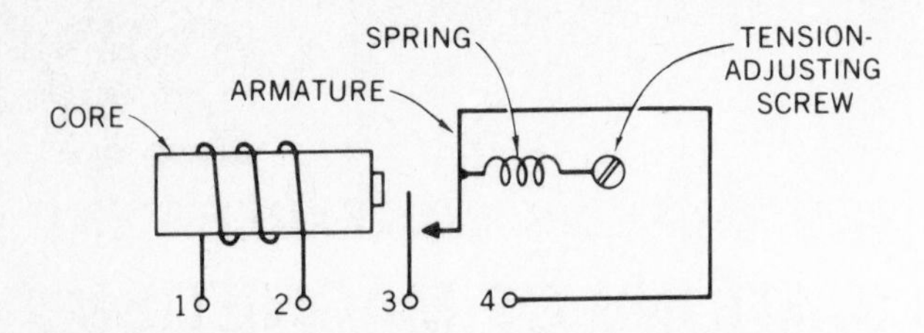

Fig. 29-2. Arrangement of parts of an armature-type relay.

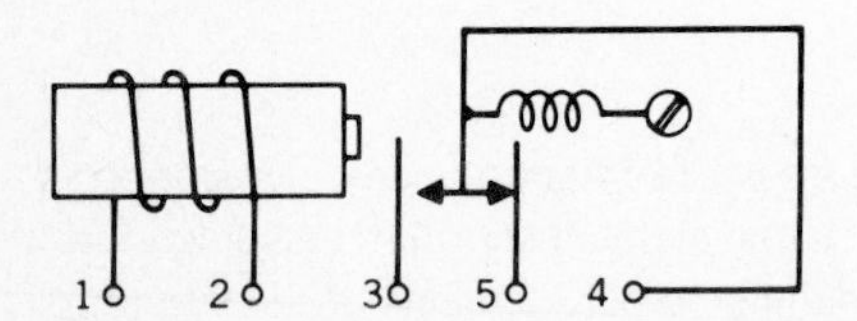

Fig. 29-3. Single-pole double-throw armature relay.

Relays operate on one of two different principles: electromagnetic attraction or electromagnetic induction.

Electromagnetic attraction-type relays, which may be either ac- or dc-actuated, consist of (1) an electromagnet, an armature, and contacts or (2) a solenoid, a plunger, and contacts. The electromagnet consists of a core and a winding. The core, armature, and plunger are made of magnetic materials such as iron, silicon steel, or permalloy (an alloy of nickel and steel). The arrangement of parts in an attraction armature-type relay is shown in Fig. 29-2. Connections to the winding of the electromagnet are brought to terminals 1 and 2. The movable contact is fixed on the armature. A spring, whose tension is adjustable, restrains the armature from closing the gap between the stationary and movable contacts. The relay shown has normally open contacts. Connections to the two contacts are made at terminals 3 and 4 of the relay.

When terminals 1 and 2 are connected to a source of electric current, an electromagnet is formed, and the armature is attracted to the core. If there is sufficient current to overcome the restraining force of the spring, the relay contacts close. The armature will be attracted whether the pole of the electromagnet adjacent to the armature is north or south.

Another arrangement of contacts is shown in Fig. 29-3. Contacts 4 and 5 are normally closed, and contacts 4 and 3 are normally open. When the relay is energized, the armature causes contacts 3 and 4 to close, leaving 5 open. This is comparable to a single-pole double-throw (SPDT) switch. There are many different contact arrangements, but in all cases the principle of operation is the same.

Open- and Closed-Circuit Relay System

It is not our intention here to discuss the countless relay circuits used in industry. We will consider the two basic systems which serve as the building blocks for more complex circuits.

The circuit in Fig. 29-4*a* shows a low-voltage, low-power, normally open circuit relay system, used to control a circuit whose load consumes *more power at a higher voltage* than the relay circuit. When control switch *S* is closed, the relay is energized, closing its contacts. The closed contacts complete the load circuit, pemitting current to flow through the load. Figure 29-4*b* is a schematic representation of the circuit in Fig. 29-4*a*.

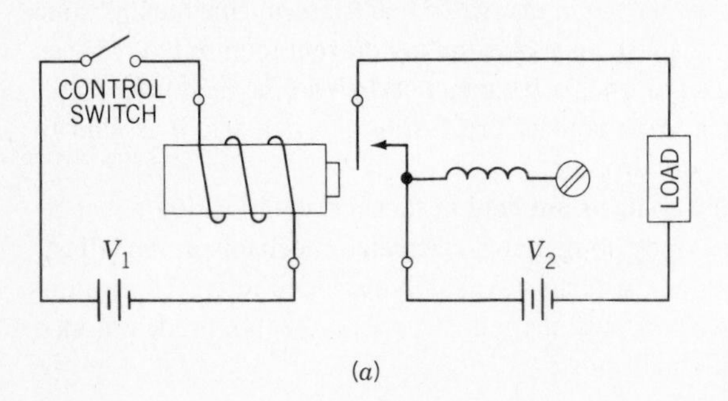

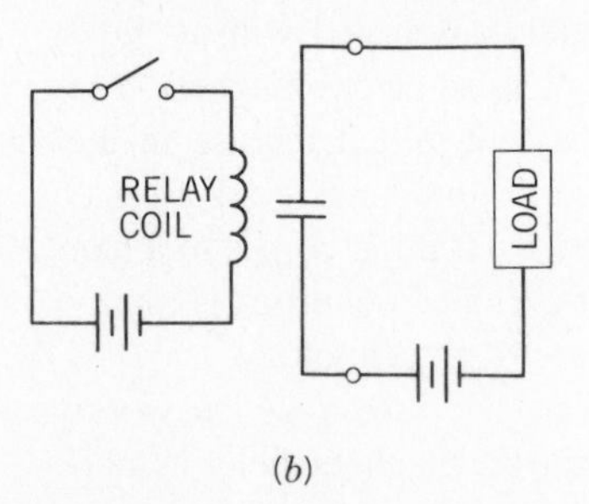

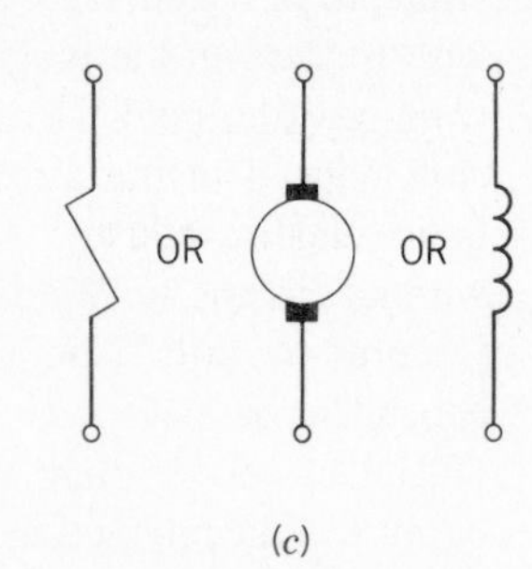

Fig. 29-4. (*a*) Open-circuit system; (*b*) schematic open-circuit system; (*c*) relay coil symbols.

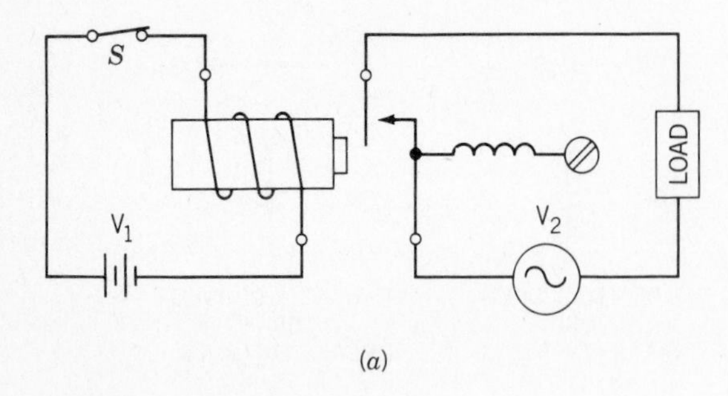

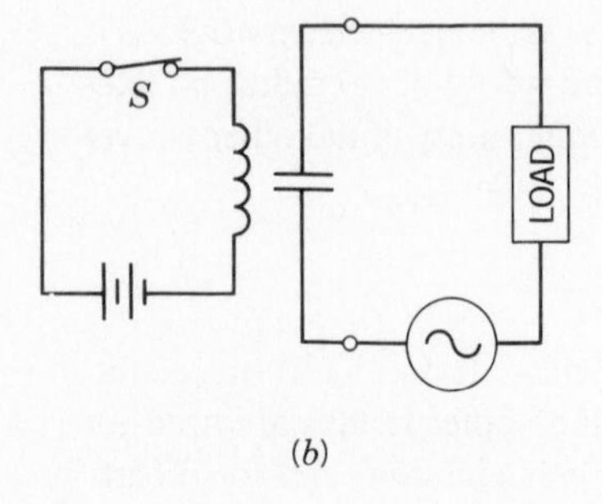

Fig. 29-5. Closed-circuit system.

Relay control of a large-power circuit may also be obtained by the closed-circuit relay system in Fig. 29-5. Here the relay is energized by a switch S, which is normally closed. Current flowing in the relay circuit opens the relay contacts, as shown, and keeps the load circuit open, so that no load current is drawn. When it is desired to operate the load, switch S is opened, deenergizing the relay circuit, thus closing the relay contacts. This completes the load circuit. Figure 29-5*b* is a schematic representation of the circuit shown in Fig. 29-5*a*. We have used the conventional symbol for a coil in designating the relay coil. Other symbols, as in Fig. 29-4*c*, are also used to designate the relay coil.

Closed-circuit systems have one advantage over open-circuit systems which makes them more desirable in certain applications. In the circuit shown in Fig. 29-5, any defect in the control system, such as an open circuit or dead battery V_1 will immediately become apparent, because the relay will become deenergized and the load circuit will be made operative. If the load is an alarm bell, the bell will sound continuously, signaling a defect in the control system. However, a defect in the open-circuit system of Fig. 29-4 will not become known until an attempt is made to operate the system.

Relay Specification and Identification of Relay Contacts

Relay manufacturers supply a specification sheet with each of their relays. This "spec" sheet contains relay ratings, designates whether the relay is dc or ac, and specifies the location and the ratings of the contacts. For example, the relay which you will use in this experiment is a dc SPDT relay. The resistance of the relay coil is 400 Ω. The contacts of the relay are designed to carry a 5-A current.

The relay coil and contact terminals can usually be located by inspection. If a relay is enclosed in a sealed unit, this may not be possible. An ohmmeter may then be used to identify the terminals. The ohmmeter is used to measure the resistance between any two terminals. There are only a limited number of two-terminal combinations possible. The resistance of the coil will be measured between the two terminals which are connected to the relay coil. Relay coils will vary from very low to very high values of resistance. Your relay coil should read approximately 400 Ω. In the process of finding the coil terminals, you will also locate contact terminals. Normally closed contacts will measure zero resistance. Normally open contacts will read infinite resistance. For example, in the relay in Fig. 29-6, the ohmmeter will read the coil resistance between terminals 4 and 5. There will be zero resistance between terminals 2 and 1 and infinite resistance between terminals 2 and 3. The resistance between terminals 1 and 3 will also be infinite. If you cannot determine by inspection whether contact 2 or 1 is the movable contact, the relay should be tripped. A resistance check will then show, as in the case of Fig. 29-6, that there is now zero resistance between contacts 2 and 3, and infinite resistance between contacts 2 and 1 and contacts 3 and 1. It is therefore evident that contact 2 is the movable one.

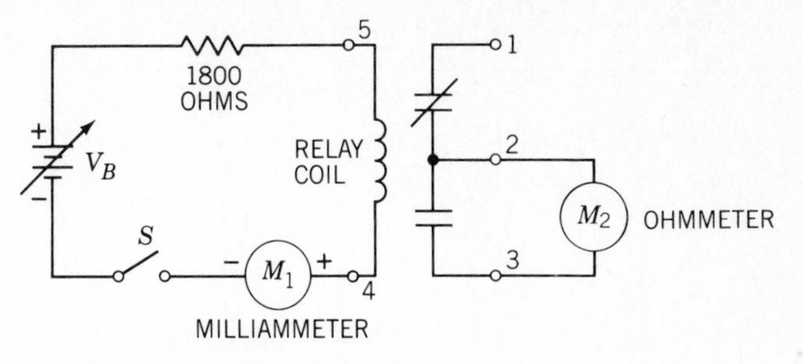

Fig. 29-6. Circuit for determining pickup value of a dc relay.

CAUTION: In making resistance checks of relay contacts, be certain the power (to the load) is *off*; otherwise you may damage the ohmmeter.

Advantages of a Relay System for Remote Control

One of the main advantages of relays is their ability to switch in high-power loads with relatively low-power input. Thus a relay whose input is rated at 100 mA, 5 V (50 Ω) uses 0.5 W of input power. The contacts of the relay may be rated at, say 10 A, 120 V, and accordingly can handle 10 A of load current and can accommodate loads up to 1200 W. If the same load had to be turned on by a mechanical switch, that switch would have to be able to handle 1200 W of power.

In remote-control applications, a relay can control a large load at some distance away more efficiently than a mechanical switch can. The reason becomes apparent when the arithmetic of line-power losses is considered. Figure 29-7*a* is the circuit of a system in which a switch is used to turn on a 10-A, 120-V load which is 1000 ft away from the switch. The length of line used to connect the power source and switch to the load is 2000 ft, 1000 going to the load and 1000 returning to the switch. If #12 wire is used (whose resistance is 1.62 Ω/1000 ft), the total resistance of the line is 3.34 Ω. The power loss W in the line carrying 10 A of load current would be

$$W = I^2R = 10^2 \times 3.34 = 334 \text{ W}$$

The useful power W_L left for the load would be

$$W_L = 1200 - 334 = 866 \text{ W}$$

If we define efficiency as the percentage of useful power to total power expended, then the efficiency of the system just described is,

$$\text{Eff} = \frac{866}{1200} \times 100\% = 72\%$$

Figure 29-7*b* is the circuit of a *relay* controlling the 10-A, 120-V load 1000 ft away. The input to the relay is rated at 100 mA, 5 V. Here the switch S is used to turn on the *relay*, which in turn switches in the load. S is located 1000 ft from the relay coil. However, in this situation, the 2000 ft of line carry the 0.1 A of *relay* current, not the 10 A of load current. If the

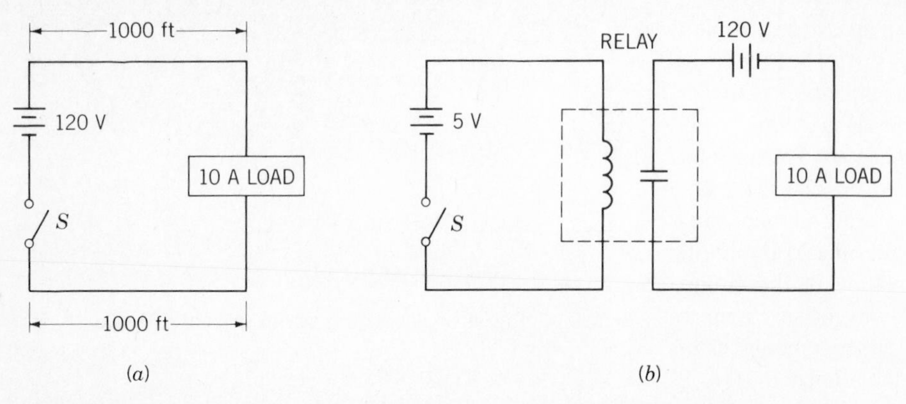

Fig. 29-7. (*a*) Mechanical switch used to control a remote load; (*b*) relay used for remote control of a load.

same 2000 ft of #12 wire were used, the power W dissipated in the line would be:

$$W = I^2R = (0.1)^2 \times 3.24 = 0.0324 \text{ W}$$

The total input power consumed by the relay and line is therefore,

$$W_T = VI + I^2R = 5(0.1) + 0.0324 = 0.5324 \text{ W}$$

The efficiency of this system is

$$\text{Eff} = \frac{1200}{1200.5324} = 99.6\%$$

And so it is clear that the relay circuit in this example is more efficient.

SUMMARY

1. Relays are electromagnetic switches used as protective devices, as indicating devices, and as transmitting devices.
2. Protective relays protect good components from the effects of circuit components that have failed.
3. Transmission relays are used in communication systems.
4. Indicating relays may be used to identify a component which has failed, or they may be used with attention-getting devices such as bells, and buzzers.
5. Relays may be simple SPST switch-type devices, or they may have complex switching arrangements.
6. The switch contacts of relays may be normally open (NO) or normally closed (NC). The contacts are held in their normal positions by springs or by some gravity-actuated mechanism.
7. Attraction-type relays are either ac- or dc-operated and consist of either (*a*) an electromagnet, a movable armature, and contacts or (*b*) a solenoid, a plunger, and contacts.
8. Terminals are provided on a relay for the winding of the electromagnet and for the relay switch contacts.
9. An advantage of a relay over an ordinary switch is that a low-power source may be used to turn a relay **on** and **off**, while in turn, heavy-duty relay contacts open and close the circuit for a high-power load.
10. A relay system may be designed so that its load circuit is open when the relay control switch is open, called an *open-circuit system*. Or a relay may be designed with a *closed-circuit system* in which the load circuit is open when the control switch is closed.

SELF-TEST

Check your understanding by answering these questions:

1. A dc relay is one in which __________ current in the relay coil actuates the relay mechanism.
2. The movable arm of an attraction-type relay is called the __________.
3. A relay which "picks up" at 10 mA is one which is turned __________ (ON/OFF) when 10 mA flows in the __________ __________.
4. Some important electrical specifications of an attraction-type relay are the:
   (*a*) __________ and __________ of the coil;
   (*b*) pickup __________ of the coil;
   (*c*) current-handling capacity of the __________.
5. In Fig. 29-3, switch contacts __________ and __________ complete the load circuit when the relay is turned ON.
6. A switch is more efficient than a relay in turning a remotely located, high-power load on and off. __________ (true/false)

MATERIALS REQUIRED

- Power supply: Variable dc source; 120 V/60 Hz source
- Equipment: 20,000 Ω/V VOM; EVM
- Resistors: ½-W 1800-Ω
- Relay: DC, SPDT, 400-Ω field, 7-mA pickup, 1-A contacts, RBM type 10730-8 or the equivalent
- Miscellaneous: 60-W wired test lamp and socket, fused line cord

PROCEDURE

1. You will receive a relay whose terminals will be numbered. Determine by inspection, if possible, the terminal connections of the coil, the number of contacts, the movable contacts, and whether the contacts are open or closed. If necessary, use an ohmmeter to identify the terminals. Record the relay coil and contact terminal numbers in Table 29-1.
2. Measure and record in Table 29-1 the coil resistance.
3. Connect the circuit of Fig. 29-6. V_B is a source of variable direct current. M_1 is a 0–20-mA milliammeter or a VOM set on an equivalent milliampere range. M_2 is an ohmmeter connected across normally open terminals 2 and 3. Note: Your relay terminals may not be numbered like those in Fig. 29-6. Set V_B for zero output. What will happen when switch S is closed? Will the relay be tripped? Why?
4. Close switch S and slowly increase the dc voltage output of V_B, observing M_2. When the relay trips, M_2 will read zero resistance (you will also be able to hear the relay trip). Observe and record in Table 29-1 the minimum current, read on M_1, required to trip the relay. This is the pickup-current value.

   Do you think that the relay will reset when the current through the coil is reduced just below the pickup value?
5. Slowly reduce the voltage of V_B, observing M_2. When the relay resets, M_2 will read infinite resistance. Observe and record in Table 29-1 the value of current at which the relay resets.
6. Repeat steps 4 and 5, checking your previous pickup and reset readings. Repeat if necessary until your readings are constant.
7. Turn up V_B until the relay turns **on**, then increase the current in the relay coil about 2 mA beyond the pickup value. Open S. The relay is now **off**.
8. Replace the ohmmeter by a 60-W lamp which will act as a load, as in Fig. 29-8. Plug the line cord into a 120-V/60-Hz outlet. Note that the load of the dc relay may be alternating current.
9. Close S. Does the lamp light? ___________
10. Open S. **Power off**.

TABLE 29-1. Relay Characteristics

Function	*Terminal Connection*	*Relay Characteristics*	
Relay coil		Relay coil resistance, Ω	
Normally open contacts		Pickup current, mA	
Normally closed contacts		Reset current, mA	

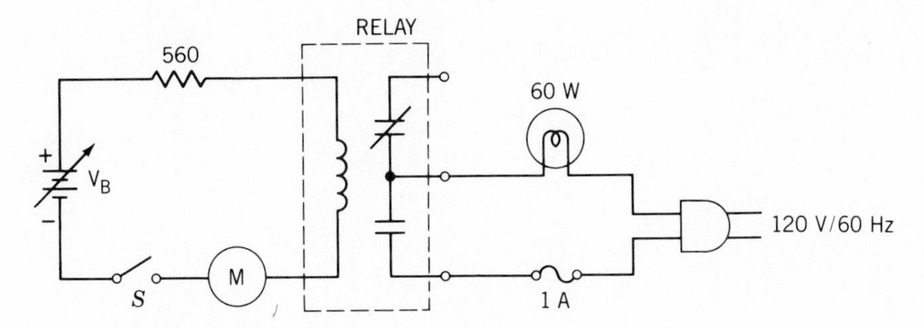

Fig. 29-8. Relay used to turn on a 60-W load.

QUESTIONS

1. Define pickup value; reset value. Refer to Table 29-1: What is the pickup value of the relay you used? The reset value?
2. Is it possible to change the pickup value of the relay you used? Explain.
3. How can you determine how much current is required to trip a relay whose pickup value you do not know?
4. What is the minimum value of voltage which must be placed across the relay coil in this experiment to trip the coil? Show your computations.
5. Would it be possible to employ the relay you used in this experiment to turn on a 120-V dc, 2000-W load? Why?

Answers to Self-Test

1. direct
2. armature
3. ON; relay coil
4. resistance, voltage, current, contacts
5. 3, 4
6. false

30

EXPERIMENT

CHARACTERISTICS OF A DC METER MOVEMENT

OBJECTIVES

1. To determine experimentally the sensitivity I_m of a meter movement
2. To determine experimentally the internal resistance R_m of a meter movement
3. To determine experimentally the linearity of the meter movement

INTRODUCTORY INFORMATION

The Moving-Coil Meter Movement

The measurement of dc voltage and current in electric and electronic circuits is usually effected by meters employing a moving-coil meter movement. This type of movement is based on the principle of the galvanometer developed by the French physicist Arsene d'Arsonval. D'Arsonval's galvanometer was a basic meter which measured very small currents. Though it was substantially modified later by Edward Weston and others, the basic moving-coil meter movement in use today is still known as the d'Arsonval movement, and is the backbone of *non-digital* meters.

Construction

The moving-coil meter consists of a horseshoe magnet with semicircular pole pieces made of soft iron added at the ends. A circular soft-iron core is positioned in a uniform magnetic field provided by the magnet and the pole pieces (Fig. 30-1*a*). Very fine phosphor bronze nonmagnetic wire wound on a rectangular aluminum frame (Fig. 30-1*b*) constitutes the moving coil. The coil assembly is centered in the air gap between the soft-iron core and the magnet (Fig. 30-1*c*). Each end of the coil is connected to a spring. The two springs are symmetrically mounted on the coil, one at the top, the other at the bottom. At the center of the springs are pivots (Fig. 30-1*d*) which fit into jeweled bearings to reduce friction, thus permitting the coil to move freely in the air gap. The free ends of the two springs are brought to the two meter terminals. These serve as the terminals of the coil. A pointer is attached to the coil assembly and moves with it. The pointer can be positioned so that its rest or zero position is at the center or at the left side. A zero adjust screw is used to compensate for minor changes in coil position. Stops are provided at the left and right side of the magnet to limit the degree of movement of the pointer.

Operation

When a potential difference is placed across the meter terminals from an external circuit, current flows through the meter coil, setting up a magnetic field around it. This magnetic field interacts with the field of the permanent magnet, and the torque developed causes the coil to turn on its pivots. The number of degrees of rotation of the moving coil is determined by the strength of the magnetic field about the coil, which in turn is proportional to the amount of current in the coil. The greater the current, the greater the rotation, and therefore the greater the deflection of the pointer attached to the moving-coil assembly. A scale, calibrated in suitable units, is used to read the measurement made in the circuit.

The direction of current through the coil determines the polarity of its magnetic field and the direction of rotation of the coil assembly. If current direction is reversed in a meter whose zero position is at the left, the pointer attempts to move downward to the left, off the scale. It is restrained by the *stop* on the left. Since the system requires a clockwise movement of the pointer, improper direction of current makes measurement impossible. It is apparent, therefore, that the moving-coil meter is a polarized device, and polarity must be observed in measuring voltage or current with it.

The springs play an important role in the operation of the meter. First, they act as a restraining force on the rotation of the coil. Hence, they must be precision-wound to permit accurate measurements. The springs also act as conductors to carry current to and from the coil. Their action brings the pointer back to zero rest after the measurement is completed and there is no longer current in the coil. Finally, they are wound in opposite directions to compensate for temperature changes, because a temperature change affects both springs in the same manner. Since they are oppositely wound, the effect of an expansion or contraction of one is counteracted by an expansion or contraction of the other acting in the opposite direction. Hence, the net change is zero.

The aluminum frame on which the coil is wound prevents the tendency of the pointer to oscillate or "hunt" before it comes to rest. This is accomplished by the eddy currents

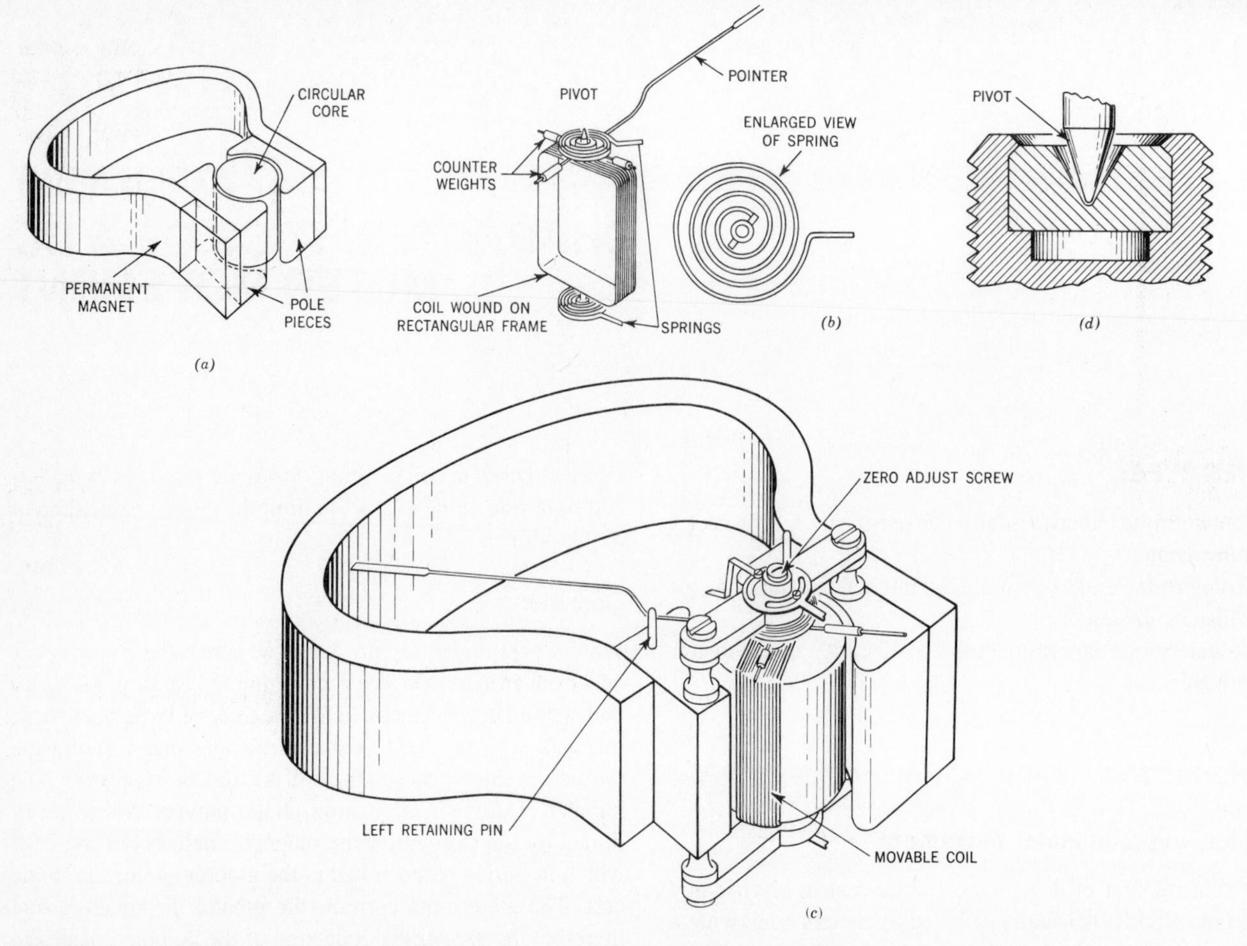

Fig. 30-1. Construction and arrangement of parts in a moving-coil meter movement (*reprinted from Army Technical Manual TM11-664*).

induced in the frame as it moves in the magnetic field of the meter. These eddy currents set up a magnetic field about the frame which, by Lenz' law, opposes the action of the inducing field of the permanent magnet. This slows down or damps the action of the moving coil when its current is either increasing or decreasing, thus minimizing its tendency to hunt.

Sensitivity of the Meter Movement

The amount of current (I_m) required to produce full-scale deflection of the pointer is called the *sensitivity* of the meter movement. Meters have sensitivities ranging from a few microamperes to many milliamperes. The smaller the current required to produce full-scale deflection, the greater the sensitivity of the movement.

The sensitivity of the movement depends on the strength of the permanent magnet and the number of ampere-turns of the coil. The more turns on the coil, the smaller the coil current required for full-scale deflection. Of course, the more turns, the greater the internal or dc resistance of the coil.

The characteristics which are therefore associated with a meter movement are its sensitivity and internal resistance. Another characteristic derived from these two is the ohms/volt rating. The significance of this characteristic will be more evident when the loading effects of voltmeters are considered.

Meter Linearity

In the design of the meter movement great pains are taken to provide a uniform magnetic field in the air gap where the coil is located. The purpose is to assure linear changes in deflection with linear changes in current through the coil. In a linear meter the number of degrees of rotation is directly proportional to the current in the coil. Thus, full-scale deflection results when 100 percent of the rated current is present in the coil. Half-scale deflection occurs when there is 50 percent of the rated current, etc. A linear meter movement makes possible a linear scale.

Meter Accuracy

Meter movements can be built with a high degree of precision and hence can be made accurate within a fraction of a percent. These are intended for very precise laboratory measurements. General-purpose meters, however, are usually rated at 2 percent accuracy.

The percentage of accuracy is based on the full-scale reading of the meter. Hence, for readings lower than full scale, the percentage of error is greater than the rated value. The significance of this fact is more apparent when the movement is converted into a voltmeter. Thus, a meter rated at 2 percent on the 100-V range is accurate within 2 V for any value measured on this range. The percent of error in a 10-V reading on this range could conceivably be as high as 20 percent. It is evident, therefore, that meter readings close to full-scale deflection are the most accurate, if circuit loading, or the effects of the meter on the circuit, are ignored.

Circuit loading will be discussed in detail later in the sections dealing with voltage and current measurement.

Determining the Characteristics of a Meter Movement

The instrument engineer must know the characteristics (internal resistance and sensitivity) of the meter movement in designing voltmeters, current meters, and ohmmeters based on this movement. These characteristics are specified by the manufacturer of the movement. They may also be measured by the technician in the laboratory.

Sensitivity

One method of measuring the sensitivity (I_m) of a meter movement is shown in the circuit in Fig. 30-2. The meter movement under test is placed in series with a rheostat R, a limiting resistor R_1, a "standard" current meter M, and a battery V. Rheostat R is adjusted until the meter-movement pointer of the meter under test reaches full-scale deflection. The current required to effect full-scale deflection is read from the "standard" current meter. This is the required *sensitivity* (I_m) of the meter movement under test.

The accuracy of the "standard" current meter limits the accuracy of the measurement.

Internal Resistance

A calibrated rheostat R, or decade box, connected in series with a linear movement under test, receiving voltage from a voltage-regulated dc source V, as in Fig. 30-3, may be used to measure the internal resistance R_m of the meter movement. The technique is based on the assumption that deflection of the meter movement varies linearly with current in the movement. (Linearity of the movement may be verified by using the circuit of Fig. 30-2.)

Measurement of R_m is achieved in the following manner:

First, R is adjusted to the value R_1 for full-scale deflection of the movement pointer. The value of R_1 is read from the calibrated dial(s) of the rheostat.

Next, R is adjusted to the value R_2 for half-scale deflection of the pointer. R_2 is read from the dial(s) of the rheostat. Then the internal resistance of the movement may be calculated from the relationship

$$R_m = R_2 - 2R_1 \qquad (30\text{-}1)$$

An analysis of the circuit of Fig. 30-3 will show why the method described yields the correct value of R_m. In measurement 1,

$$V = I_m(R_1 + R_m) \qquad (30\text{-}2)$$

In the second measurement,

$$V = \frac{I_m}{2}(R_2 + R_m) \qquad (30\text{-}3)$$

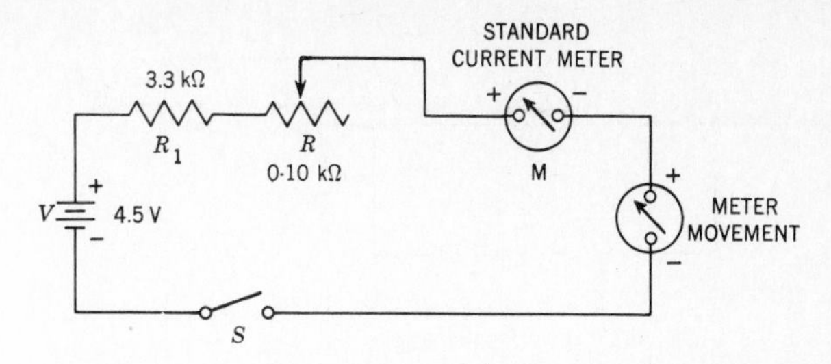

Fig. 30-2. Circuit for measuring sensitivity I_m of a meter movement.

Eliminating V from Eqs. (30-1) and (30-2), we have

$$I_m \times (R_1 + R_m) = \frac{I_m}{2}(R_2 + R_m) \qquad (30\text{-}4)$$

Solution of Eq. (30-4) leads to

$$2R_1 + 2R_m = R_2 + R_m$$

and

$$R_m = R_2 - 2R_1 \qquad (30\text{-}5)$$

If a laboratory-type resistance decade box is used as the rheostat R, together with a low-voltage source, an accurate measurement of R_m may be made (that is, accurate within the tolerance of the decade box).

Equations (30-2) and (30-3) set one requirement for the dc source. The voltage V delivered for both measurements must be the same. Therefore, a well-regulated power source is required, or V must be closely monitored and maintained at the same level for the two load conditions, that is, for I_m and $I_m/2$.

Another requirement is that R remain low in value. A concrete example will explain why this is necessary. Suppose V is a 5-V source and the meter is a 0–1-mA milliammeter with an internal resistance of 50 Ω. For full-scale deflection

$$R + R_m = \frac{V}{I_m} = \frac{5}{0.001} = 5000\ \Omega$$

Since $R_m = 50\ \Omega$, R must be set at $R_1 = 4950\ \Omega$ for full-scale deflection. Similarly, at half-scale deflection, R must be set at $R_2 = 9950\ \Omega$. At this resistance level it is very possible to make a measurement error. For if R were set at $R_2 = 9900\ \Omega$ instead of 9950, the current in the meter circuit would increase from 500 microamperes (μA) (½ mA) to 502 μA. On a 1000-μA (1-mA) scale, a 2-μA change is hardly noticeable. It is therefore possible that the erroneous setting of R_2 could be selected. Now, substituting the values $R_1 = 4950$ and $R_2 = 9900$ in Eq. (30-5), we discover that $R_m = 9900 - 2\ (4950) = 0$, a totally meaningless conclusion.

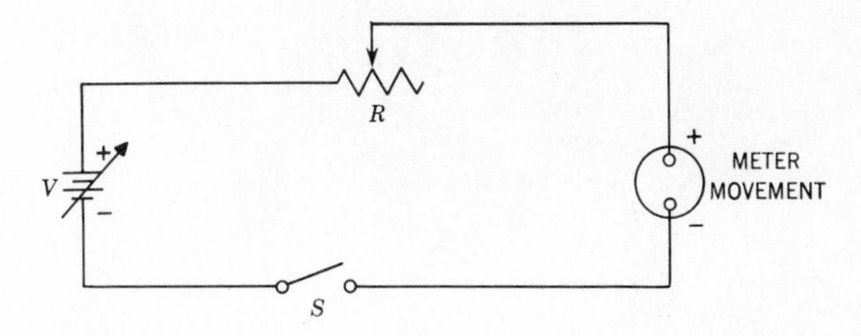

Fig. 30-3. Circuit for determining the internal resistance R_m of a meter movement.

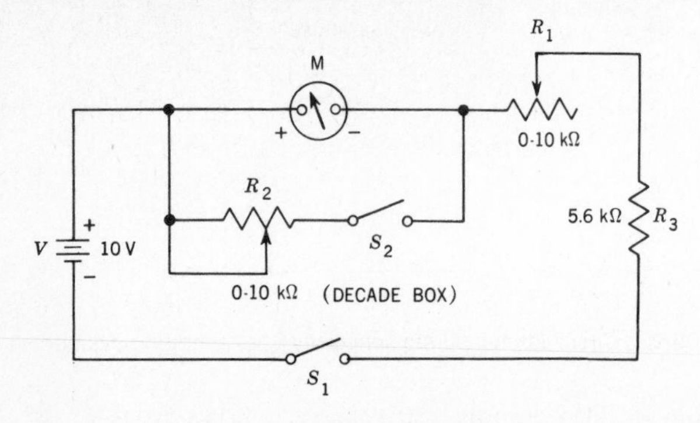

Fig. 30-4. Experimental circuit for measuring the internal resistance R_m of a meter movement.

To overcome this difficulty, the applied voltage V must be very low so that the limiting resistance R can be low. To illustrate, consider again the 0–1-mA movement whose $R_m = 50\ \Omega$. If, in the circuit of Fig. 30-3, $R = R_1 = 25\ \Omega$ for full-scale deflection, then $V = I_m(R_1 + R_m) = 0.001 \times (75)$, and $V = 75 \times 10^{-3}$ V = 75 millivolts (mV). With the 75-mV source, a value of $R = R_2 = 100\ \Omega$ is required for half-scale deflection. In this case, a change of 75 Ω in R causes a change of 500 μA (½ mA). Thus, an error in setting R can readily be observed. And so, for effective measurement of R_m by the method of Fig. 30-3, a very low-voltage regulated dc voltage source is required.

A Preferred Procedure for Measuring R_m

The measurement of R_m for a low-sensitivity movement can be made with a good degree of accuracy, even if a very low regulated dc source is not available, using the circuit of Fig. 30-4. The output V of a variable voltage-regulated supply is set at 10 V. R_1, a 10,000-Ω potentiometer, is connected as a rheostat in series with R_3, a 5600-Ω resistor. R_2 is a calibrated decade rheostat whose resistance may be varied from 0 to 10,000 Ω. (A low-value potentiometer will serve if a decade rheostat is not available.)

Measurement of R_m begins with switch S_2 open and S_1 closed. R_1 is adjusted for full-scale deflection of the movement. Next, switch S_2 is closed, throwing R_2 in parallel with R_m. R_2 is now adjusted for half-scale deflection of the movement. The resistance of R_2, read from the calibrated scale of the rheostat, is the required R_m, the internal resistance of the movement.

NOTE: If a potentiometer is used for R_2, the resistance of R_2 can be measured with an ohmmeter after half-scale adjustment is made. Open S_2 before measuring R_2.

This procedure introduces a small error because there is a slight increase in circuit current when R_2 is placed in parallel with R_m. This increase in circuit current actually appears in R_2, and therefore the current I_2 in R_2 is slightly higher than $I_m/2$, the current in the meter movement. However, if R_m is small, the increase in current is negligible and may be ignored.

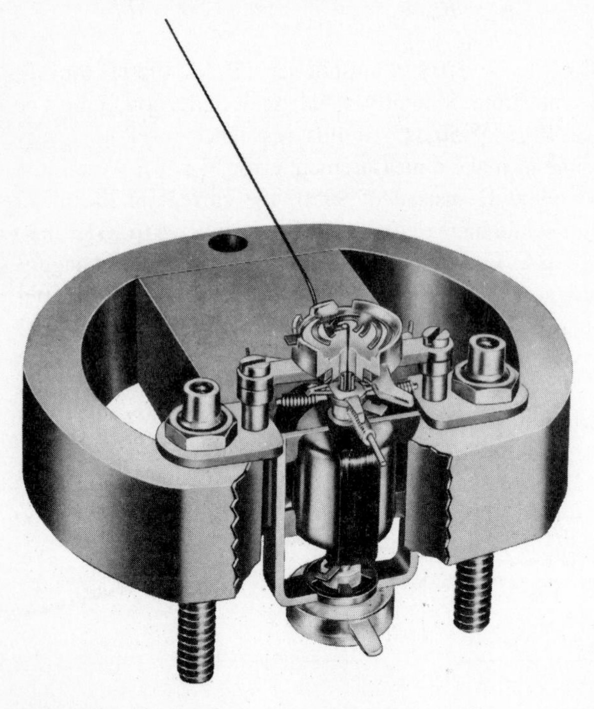

Fig. 30-5. Annular taut-band meter movement (*Simpson Electric Company*).

Taut-Band Meters

Another development in meter design features *taut-band construction*, which is a change from the conventional pivot-and-jewel type just described. The moving element of the taut-band movement is supported by two thin metal ribbons, kept under tension by elastic members at each end of the assembly (see Fig. 30-5). This type of construction eliminates the friction associated with the moving parts in the pivot-and-jewel arrangement.

The main advantage of the taut-band movement is its sensitivity; it can operate with very limited power—less than 1 microwatt. A disadvantage of the taut-band movement is that it does not have the fast response of the pivot-and-jewel construction, because it does not develop as much torque. Moreover, rocking movements can cause pointer vibration, and the instrument is more subject to breakdown under shock than is the pivot-and-jewel type.

Taut-band meters are used where highly sensitive movements are required.

SUMMARY

1. Nondigital meters used to measure dc voltage and current usually employ a moving-coil meter movement called a d'Arsonval meter.
2. The actuating elements of a d'Arsonval movement consist of a permanent horseshoe magnet and a moving-coil assembly freely suspended on a frame in the air gap between the semicircular pole pieces of the permanent magnet and a circular soft-iron core.
3. A very fine nonmagnetic wire is used for winding the moving coil on a rectangular aluminum frame.
4. Direct current is brought to the coil from the external circuit which is to be measured. The current in the coil

sets up a magnetic field which interacts with the permanent magnetic field of the horseshoe magnet and creates a torque, causing the freely suspended moving-coil assembly to rotate. A pointer attached to the moving coil moves along a calibrated scale. Readings are made from the scale.

5. The amount of current required to cause full-scale deflection of the pointer is called the sensitivity of the meter movement (I_m).
6. The sensitivity of the movement depends directly on the strength of the horseshoe magnet and on the number of ampere-turns of the coil.
7. The direction of rotation of the coil depends on the direction of current in the coil. The dc meter movement is therefore a polarized device.
8. The resistance of a movement (R_m) is the resistance of the coil winding. The more turns of winding, the greater the resistance of the movement. A very sensitive meter movement normally has a higher internal resistance than a less-sensitive movement.
9. The electrical characteristics of a meter movement of concern to the technician and engineer are sensitivity I_m and internal resistance R_m.
10. In the design of a meter movement, pains are taken to ensure a uniform magnetic field in the air gap where the coil is located. A uniform field ensures equal changes in deflection of the meter pointer for equal increments of current in the coil. This characteristic is called the *linearity* of the meter movement.
11. The percentage of accuracy of a d'Arsonval movement is based on the full-scale reading of the meter. Therefore, for readings lower than full-scale, the percentage of error is greater than the rated value.
12. It is possible to determine the sensitivity I_m of a meter movement experimentally by connecting the movement in series with a standard direct current meter, a rheostat or resistance decade box, a current-limiting resistor, and a power source. The power source and/or decade box are adjusted for full-scale deflection of the meter movement. The current (I_m) required for full-scale deflection of the movement under test is then read from the standard meter (Fig. 30-2).
13. A preferred method for determining the internal resistance R_m of a meter movement is to connect the meter movement in series with a rheostat, a limiting resistor, and a power source (Fig. 30-4). The rheostat is then adjusted for full-scale deflection of the meter movement. Now a resistance decade box is connected in parallel with the meter movement. The decade box is adjusted for half-scale deflection of the meter pointer. The resistance read on the decade box at this point is the required R_m. Of course, it is essential that the output voltage of the source V remain the same during this process.
14. A method for determining the linearity of a meter movement is to connect the meter movement in series with a standard direct current meter, a rheostat (or resistance decade box), a current-limiting resistor, and a voltage source (Fig. 30-2). The rheostat is adjusted to effect fixed percentage changes of full-scale deflection of the pointer of the meter movement under test. The increments of current required to effect these changes are then read from the standard current meter. If equal deflections of the pointer (read along the linear scale of the meter movement) are effected by equal increments of current (read on the standard current meter), the movement under test is linear.

SELF-TEST

Check your understanding by answering these questions:

1. A galvanometer is a ___________.
2. The coil assembly of a d'Arsonval meter movement rotates on its pivots because of the ___________ developed by the interaction of the magnetic field of the horseshoe magnet and the magnetic field set up in the coil when there is ___________ in it.
3. The amount of current required to cause full-scale deflection of the pointer is called the ___________ of the movement and is designated ___________.
4. The smaller the current required to cause full-scale deflection of the pointer, the ___________ ___________ is the movement.
5. The sensitivity of the meter movement depends on the strength of the ___________ ___________ and on the number of ___________ ___________ of the coil.
6. If equal increments of current in the moving coil cause equal increments of angular deflection of the meter pointer, the movement is said to be ___________.
7. The internal resistance R_m of a meter movement is the resistance of the ___________.
8. Percentage of meter accuracy is based on ___________-___________ ___________ of the meter.
9. One method of measuring the sensitivity of a meter movement is to connect a series circuit consisting of the meter movement, a ___________ ___________ meter, and a ___________ ___________ ___________. The current in the circuit is then adjusted for ___________ ___________ ___________ of the movement, and the current required to do this is read from the ___________-___________ ___________.
10. The preferred method of measuring the internal resistance R_m of a meter movement requires the use of a ___________ ___________ ___________ connected in ___________ (series/parallel) with the meter movement.

MATERIALS REQUIRED

- Power supply: A variable regulated dc supply
- Equipment: "Standard" multirange milliammeter whose lowest range is 0–1 mA (approximately); rheostat

(resistance decade box) 0–10 kΩ adjustable in 1-Ω steps; meter movement whose I_m = 1 mA
- Resistors: ½-W 3300-, 5600-, 15,000-Ω
- Miscellaneous: Two SPST switches; 10,000-Ω 2-W potentiometer; 10,000-Ω 2-W potentiometer (if decade box is not available)

PROCEDURE

Measuring Sensitivity of a Meter Movement

1. Connect the meter movement whose sensitivity you wish to measure in the circuit of Fig. 30-2. S is open. V is a 4½-V battery or power supply set to 4.5 V. Rheostat R is set to 10,000 Ω, at the start.

CAUTION: Check to see that the "standard" current meter (an accurate commercial current meter may be used as the "standard") and the meter movement under test are connected with the proper polarity.

2. Close S and slowly reduce the resistance of R until full-scale deflection is indicated on the meter movement. Read the value I_m from the "standard" meter and record in Table 30-1.

Linearity Check

3. Increase the resistance R until the meter-movement (under test) pointer is deflected three-fourths of the way on its scale. Measure the current (using the standard meter) at this level, and record in Table 30-1 under the column labeled "I (¾)."

   Measure also the current required to cause half-scale deflection and quarter-scale deflection of the meter-movement pointer, and record in Table 30-1, under columns "I (½)" and "I (¼)," respectively. For quarter-scale deflection it will be necessary to replace R_1 with a 15,000-Ω resistor.

   Draw a graph of measured current (vertical axis) versus pointer deflection (horizontal axis). If the graph is a straight line, then deflection of the meter movement varies linearly with meter current.

Measuring R_m

4. Connect the circuit (Fig. 30-4). Switches S_1 and S_2 are open. Set V at 10 V and R_1 at maximum resistance.
5. S_2 is still open. Close S_1. Gradually decrease the resistance of R_1 until M reads full-scale deflection.
6. Close switch S_2. Adjust R_2 until M reads half-scale deflection. Read the value of R_2 from the calibrated dial and record in Table 30-1, under column "Measured Value R_m."

TABLE 30-1. Meter Movement Characteristics

Measured Values				*Measured Value*
I_m	I (¾)	I (½)	I (¼)	R_m

QUESTIONS

1. What determines the accuracy of your measurement of I_m?
2. Is the scale on the meter movement under test linear or nonlinear? What do your measurements indicate concerning meter-movement linearity?
3. What is the value of voltage across the meter movement required to cause full-scale deflection of the pointer?
4. What two characteristics describe a dc meter movement?
5. What error is introduced in measuring the R_m of a meter movement using the experimental circuit (Fig. 30-4)? Why was this error ignored in this experiment?
6. Why does the circuit of Fig. 30-3 require a constant source voltage V to measure R_m?

Extra Credit

7. Explain how you would measure the R_m of a high-sensitivity meter movement, as for example, a movement whose I_m = 50 μA and whose R_m = 2500 Ω. Show the circuit design you would use, the value of dc voltage required, and the values of all resistors and rheostats used. Discuss the factors relating to the accuracy of your procedure.

Answers to Self-Test

1. very sensitive moving-coil dc meter movement
2. torque; current
3. sensitivity; I_m
4. more sensitive
5. permanent magnet; ampere-turns
6. linear
7. coil
8. full-scale deflection
9. standard current; resistance decade box; regulated voltage supply; full-scale deflection; standard current meter
10. resistance decade box; parallel

31

EXPERIMENT

VOLTMETER MULTIPLIERS

OBJECTIVES

1. To develop by analysis a formula for determining the value of the multiplier required to convert a meter movement into a specified range voltmeter
2. To verify, experimentally, the value of the multiplier required to convert the meter movement into a voltmeter of a specified range
3. To calibrate the scale of the experimental voltmeter

INTRODUCTORY INFORMATION

Nondigital meters used in electronics consist of basic meter movements connected in suitable circuit arrangements. The basic movement has a moving element to which a pointer is attached. Current in the meter movement results in a force which acts on the moving element. The pointer is deflected, and moves in an arc along a scale suitably calibrated in the units being measured. These units may be volts, ohms, amperes, etc., as in a VOM which performs all these functions of measurement.

In Experiment 30 you learned that a meter movement has two characteristics associated with it:

1. Sensitivity—the current required for full-scale deflection. For example, a 0–1-mA milliammeter movement requires 1 mA of current to give full-scale deflection of the pointer.
2. The resistance of the movement.

Voltmeter Multipliers

The circuit components of a *voltmeter* depend on these meter-movement characteristics. Thus, suppose it is required to construct a 0–100-V voltmeter from a basic 0–1-mA milliammeter movement whose resistance is 200 Ω.

Figure 31-1 shows the basic circuit arrangement for this voltmeter. *A* and *B* are the terminals of the meter where the test leads are inserted. *A* and *B* are coded (−) and (+), respectively, from the polarity of the movement. When measuring voltage, *A* is placed on the negative side of the voltage source and *B* on the positive terminal. Otherwise, current through *M* is reversed and the movement may be damaged. *M* is the meter movement. It is shown as a 200-Ω resistance R_m, with a pointer, connected in series with a resistor R_s called the "multiplier." The purpose of R_s is to limit the current in this circuit to 1 mA when 100 V is applied across the meter test leads (*A* and *B*). This will cause full-scale deflection of the pointer, and the scale can be calibrated for 100 V at this point.

If a linear meter movement is used, deflection of the pointer is directly proportional to the current in the movement. The current (see Fig. 31-1) depends on the voltage across the terminals *AB* of the meter. The meter scale can therefore be calibrated in volts, though the movement itself measures (or responds to) current.

Ohm's law is used to find the value of the multiplier R_s. Consider the experimental voltmeter of Fig. 31-1. Since we wish to construct a 100-V meter, that is, a meter whose scale may be calibrated 100 V at full-scale deflection,

$$V_{\text{range}} = V_{AB} = 100 \text{ V},$$

and

$$I = I_m = 0.001 \text{ A}$$

Substituting these values and the value $R_m = 200$ in the equation

$$V_{\text{range}} = I_m(R_s + R_m) \quad (31\text{-}1)$$

we get

$$100 = 0.001(R_s + 200) \quad (31\text{-}2)$$

Solving Eq. (31-2), we find that

$$R_s = 100{,}000 - 200 = 99{,}800 \ \Omega \quad (31\text{-}3)$$

The total resistance R_T of this 100-V meter is 100,000 Ω, since it equals $R_s + R_m$.

Formula for Determining Multiplier

A general formula for determining the value of the multiplier R_s required to convert a meter movement (I_m, R_m) into a voltmeter whose range is V_{range} can now be easily derived.

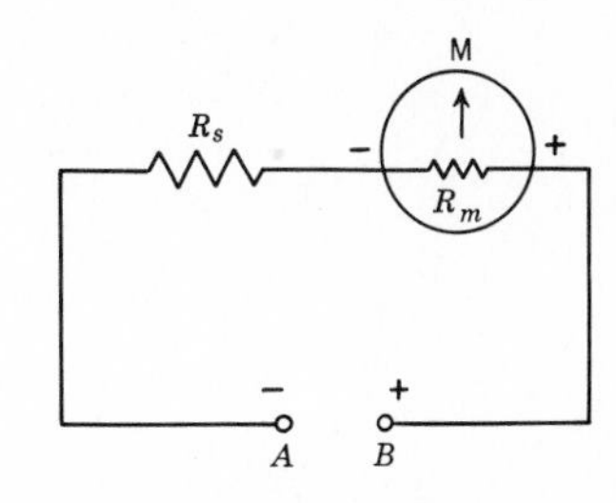

Fig. 31-1. Constructing a voltmeter.

Refer to Fig. 31-1. R_T is the total resistance of this meter and is equal to $R_s + R_m$. Thus:

$$R_T = R_s + R_m \qquad (31\text{-}4)$$

and

$$R_s = R_T - R_m \qquad (31\text{-}5)$$

But

$$R_T = \frac{V_{AB}}{I_m} = \frac{V_{\text{range}}}{I_m} \qquad (31\text{-}6)$$

Therefore, from Eq. (31-5) and Eq. (31-6),

$$R_s = \frac{V_{\text{range}}}{I_m} - R_m \qquad (31\text{-}7)$$

Simplifying Eq. (31-7) leads to

$$R_s = \frac{V_{\text{range}} - I_m \times R_m}{I_m} \qquad (31\text{-}8)$$

Either Eq. (31-7) or (31-8) may be considered a general formula for determining the value of a multiplier R_s for a voltmeter of range V_{range}. In these formulas R_s is the resistance of the multiplier in ohms; V_{range} is the range of the meter in volts; R_m is the resistance of the meter movement in ohms; and I_m is the sensitivity of the meter movement in amperes. Several examples will demonstrate how the formula may be applied.

Example 1. Convert a meter movement whose sensitivity $I_m = 0.001$ A and whose internal resistance $R_m = 200\ \Omega$ into a voltmeter of 100 V.

NOTE: This is the same movement as that in Fig. 31-1 and is the same problem as that solved in Eqs. (31-1) through (31-3).

Solution. Substituting the given values in Eq. (31-7) gives

$$R_s = \frac{100}{0.001} - 200 = 100{,}000 - 200 = 99{,}800\ \Omega$$

Or substituting the given values in Eq. (31-8) gives

$$R_s = \frac{100 - 0.001(200)}{0.001} = \frac{99.8}{0.001} = 99{,}800\ \Omega$$

The results are the same, whichever formula is used.

Example 2. Convert the meter movement in Fig. 31-1 into a voltmeter whose range is 10 V.

Solution. Substituting the given values ($I_m = 0.001$, $R_m = 200$, $V_{\text{range}} = 10$) in Eq. (31-7) gives

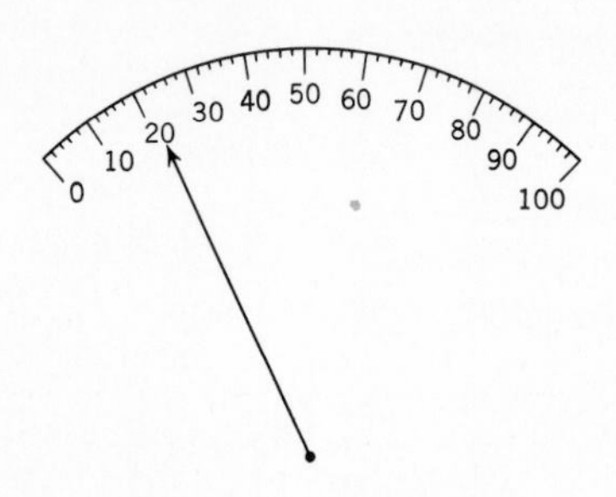

Fig. 31-2. Voltmeter scale.

$$R_s = \frac{10}{0.001} - 200 = 10{,}000 - 200 = 9800\ \Omega$$

Similarly, it can be shown that the multipliers required to convert the meter movement of Fig. 31-1 into voltmeters of ranges 50 and 300 V, are respectively 49,800 Ω and 299,800 Ω. These exact values are difficult to obtain. Hence 1 percent resistors are generally employed as multipliers in general-purpose voltmeters.

Calibrating the Voltmeter

Suppose Fig. 31-1 is the experimental 100-V meter. It is required to calibrate the scale of this voltmeter. If zero volts is applied ($V_{AB} = 0$) across the voltmeter, there is no current in the meter movement and hence no deflection of the pointer. This is the 0-V position of the pointer on the scale. If 50 V is applied, there will be ½ mA of current in the movement, and therefore there will be half-scale deflection of the pointer. This will be the 50-V position on the scale. If 25 V is applied, there will be ¼ mA of current, and therefore there will be quarter-scale deflection of the pointer. This will be the 25-V position on the scale, and so on.

The entire 100-V scale is calibrated in this manner and appears as in Fig. 31-2. It is a linear scale because the meter movement used is linear.

SUMMARY

1. A basic dc meter movement may be converted into a voltmeter by adding a multiplier resistor R_s in series with the meter movement, as in Fig. 31-1.
2. The value of the multiplier R_s for a specified range V_{range} must be such that it will permit full-scale deflection of the pointer when the voltmeter terminals are placed across the voltage value V_{range}. This means that for full-scale deflection the total resistance of the voltmeter, $R_s + R_m$, must limit the current through the movement to I_m, the sensitivity of the movement.
3. Though the scale of a voltmeter is calibrated in volts, the meter itself is responding to current in the movement, that is, it is a current-actuated device.
4. A general formula for determining the value of multiplier R_s required to convert a meter movement whose sensitivity is I_m and internal resistance is R_m into a voltmeter of range V_{range} is

$$R_s = \frac{V_{\text{range}}}{I_m} - R_m$$

or

$$R_s = \frac{V_{\text{range}} - I_m \times R_m}{I_m}$$

where R_s and R_m are in ohms, V_{range} is in volts, and I_m is in amperes.
5. If a linear meter movement is used for a voltmeter, the scale of the voltmeter will also be linear.

SELF-TEST

Check your comprehension by answering these questions:

1. A dc voltmeter may be made by connecting a precision resistor called a ___________ in ___________ with a dc meter movement.
2. If the sensitivity of the meter movement used to build a dc voltmeter is 50 μA, then the current in a 50-V meter at full-scale deflection is ___________.
3. The total resistance of a dc voltmeter consists of the ___________ of the multiplier resistance and the resistance of the ___________.
4. A 100-V meter is made from a meter movement whose sensitivity is 50 μA and whose internal resistance is 2000 Ω.
   (*a*) The total resistance of the meter is ___________ Ω.
   (*b*) The resistance of the multiplier is ___________ Ω.
5. In the voltmeter of question 4, half-scale deflection of the pointer corresponds to ___________ V; quarter-scale deflection of the pointer corresponds to ___________ V.

MATERIALS REQUIRED

- Power supply: Variable low-voltage dc source
- Equipment: Electronic voltmeter, meter movement (same as in Experiment 30), resistance decade box as in Experiment 30
- Resistors: ½-W as required
- Miscellaneous: One SPST switch, potentiometers as required in step 11 (optional)

PROCEDURE

(Show all your computations.)

Constructing a Voltmeter of Specified Range (30 V)

1. The meter movement you will use will be the same as that whose internal resistance R_m and sensitivity I_m were measured in Experiment 30. Record these in Table 31-1.
2. Compute and record in Table 31-1 the multiplier resistance R_s required to convert your meter movement into a 30-V meter. Choose a resistor (or combination of resistors) whose total resistance R_s is equal or close to the computed value. With an ohmmeter measure and record in Table 31-1 the resistance of the multiplier R_s.

NOTE: A resistance decade box may be used to select the proper value of R_s. When this is used, read the value of R_s from the calibrated dial.

   Using the selected value of R_s, construct a voltmeter as in Fig. 31-1.

3. Connect the electronic voltmeter V_1 and the *experimental voltmeter* V_2 in parallel across the low-voltage supply V, and adjust the voltage V for a 30-V output, as measured on the electronic voltmeter V_1 used as a standard (Fig. 31-3). Mark "30" the point on the *uncalibrated* scale of the *experimental* voltmeter (V_2) at which the pointer has stopped. If deflection is greater than full-scale, so state.
4. If exact full-scale deflection is not reached on experimental voltmeter V_2, adjust voltage supply V to obtain it and record under "Measured Range" the voltage V_1 required to obtain full-scale deflection of V_2, using the electronic voltmeter V_1 as the standard.

Calibrating the Experimental Voltmeter

5. Using V_1 as the standard of reference, calibrate the linear scale of experimental voltmeter V_2 in intervals of 5 V, from zero volts to full-scale deflection.
6. Compute the input resistance R_{in} of the experimental voltmeter V_2 and record in Table 31-1. Use the equation

$$R_{in} = R_T = \frac{V_{range}}{I_m}$$

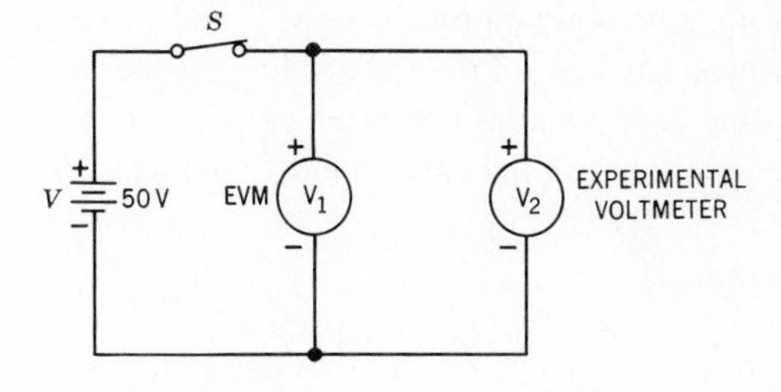

Fig. 31-3. Calibrating the experimental voltmeter.

TABLE 31-1. Experimental Voltmeter Measurements

Meter-Movement Measured Values		*Required Range V,*	*R_s (multiplier), Ω*		*Measured Range V_1,*	*Resistance (R_{in}) of Experimental Voltmeter*	
R_m	I_m	V	*Computed*	*Measured*	V	*Computed*	*Measured*
		30					
		10					

7. Disconnect V_2 from the circuit of Fig. 31-3. Connect an ohmmeter across the terminals of experimental voltmeter V_2. Measure and record in Table 31-1 the input resistance of V_2. Does the voltmeter pointer move? If it does, is it upscale or downscale?
8. Reverse the ohmmeter leads and repeat step 7.

Experimental Voltmeter (10 V)

9. Repeat steps 2 through 4, constructing a 10-V-range voltmeter. Record your data in Table 31-1.
10. Repeat steps 5 through 8, calibrating the 10-V meter and measuring its input resistance.

Design Problem (Extra Credit)

11. Design a three-range voltmeter, 5-V, 25-V, and 40-V, using the meter movement assigned to you in this experiment. Draw the circuit diagram. Include the design value for each multiplier R_s and the values of R_m and I_m. Show all your computations. (You may not use a switch to change range.) Secure the necessary values of R_s and breadboard the three-range meter (you may use a resistance decade box or a potentiometer connected as a rheostat for R_s). Experimentally determine and record in Table 31-2 the voltage required for full-scale deflection on each of the ranges.

TABLE 31-2. Three-Range Voltmeter Characteristics

Meter-Movement Characteristics		*Multiplier R,* Ω			*Full-Scale Voltage,* V		
R_m, Ω	I_m, A	*5 V*	*25 V*	*40 V*	*5 V*	*25 V*	*40 V*

QUESTIONS

1. In Table 31-1, how do the required ranges V compare with the measured ranges V_1 for the experimental voltmeter on (*a*) 30-V range, (*b*) 10-V range? Explain any differences between V and V_1 for each specific range.
2. How do the computed and measured values of input resistance in Table 31-1 compare? If they are not the same, explain why.
3. Without using specially designed precision resistors, explain how you obtained a required value R_s for use as a multiplier in a single-range voltmeter.
4. In calibrating your experimental voltmeter, what factor(s) limit(s) the accuracy of calibration?
5. From your observations in steps 7 and 8, what conclusions, if any, can you draw about the effect of the ohmmeter on the experimental voltmeter pointer, when measuring the resistance of the experimental voltmeter? Explain.

Extra Credit

6. Explain in detail how you secured the proper values of R_s in the design of the three-range voltmeter. Explain how you determined full-scale deflection on each range, and comment on the accuracy of your design.

Answers to Self-Test

1. multiplier; series
2. 50 μA
3. sum; meter movement
4. (*a*) 2,000,000; (*b*) 1,998,000
5. 50, 25

32

EXPERIMENT

LOADING EFFECTS OF A VOLTMETER

OBJECTIVES

1. To find by analysis the relationship between I_m and the ohms/volt rating of a voltmeter and to confirm this rating experimentally
2. To determine analytically and confirm experimentally the loading effects of a voltmeter
3. To consider analytically a method for eliminating the loading effects of a nonelectronic voltmeter

INTRODUCTORY INFORMATION

Ohms/volt Characteristic and Input Resistance of a Voltmeter

In Experiment 31 it was shown that the total input resistance that a voltmeter presents to the circuit under test, on any voltage range, is equal to $R_s + R_m$, where R_s is the multiplier for the specified range and R_m is the resistance of the movement. Thus, the input resistances of the experimental voltmeter in Fig. 32-1 on the 10-, 50-, 100-, and 300-V ranges are shown in Table 32-1.

Observe, in Table 32-1, that the ratio of input resistance to voltage on each range is 1000. This constant ratio is called the ohms per volt (ohms/volt) characteristic of the meter; in the case of the meter under consideration, it is 1000 Ω/V. This characteristic is also known as the *voltmeter sensitivity*.

Refer to the voltmeter in Fig. 32-1. If the range of this meter is V volts, that is, if there is full-scale deflection of the meter pointer when V volts is applied across the meter terminals, then

$$V = I_m(R_s + R_m) \qquad (32\text{-}1)$$

where I_m is the sensitivity of the meter movement and R_m is the internal resistance of the movement.

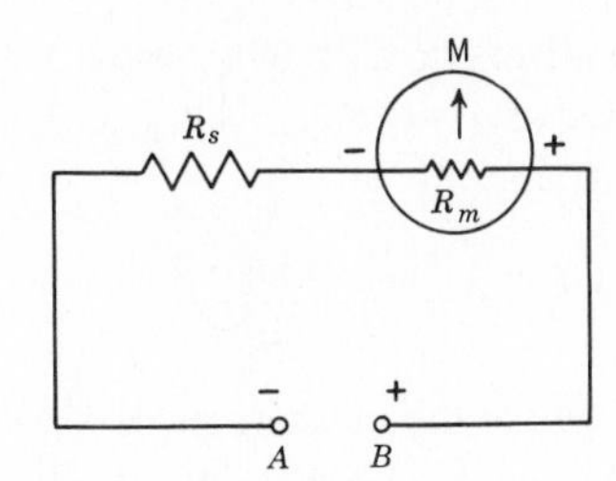

Fig. 32-1. Equivalent circuit of a voltmeter.

From Eq. (32-1) it follows that

$$\frac{R_s + R_m}{V} = \frac{1}{I_m} \qquad (32\text{-}2)$$

The ratio $(R_s + R_m)/V$ is the ohms/volt characteristic according to our definition. It is therefore apparent that the ohms/volt characteristic of a voltmeter is the reciprocal of the sensitivity $(1/I_m)$ of the meter movement.

$$\text{Ohms/volt} = \frac{1}{I_m} \qquad (32\text{-}3)$$

where I_m (in amperes) = sensitivity of meter movement.

For example, using Eq. (32-3), we can show that the ohms/volt characteristic of a voltmeter using a 1-mA movement is 1000, that of a voltmeter using a 50-μA movement is 20,000, etc.

The ohms/volt rating of a voltmeter multiplied by the set voltage range of the meter is the *input resistance* of the meter on that range; that is, it is the resistance which the circuit "sees" when the voltmeter is inserted in the circuit to measure voltage.

DC Voltmeter Circuit Loading

The ohms/volt rating is significant to the technician because it indicates the extent to which the voltmeter will "load" the circuit when it is used in voltage measurement. Consider the circuit of Fig. 32-2. Suppose it is desired to measure the voltage from A to B, using the 100-V range of a 1000 Ω/V meter. Here the meter places 100,000 Ω resistance in parallel with the 4 megohms (MΩ) of the divider, and the total resistance R_{T2} of the parallel circuit from A to B now becomes slightly less than 100,000 Ω. Suppose, however, that the

TABLE 32-1. Input Resistance of Voltmeter Using a 0-1-mA Movement

Voltage Range	*Input Resistance*	*Ohms/volt: Input Resistance / Voltage*
10	10,000	1000
50	50,000	1000
100	100,000	1000
300	300,000	1000

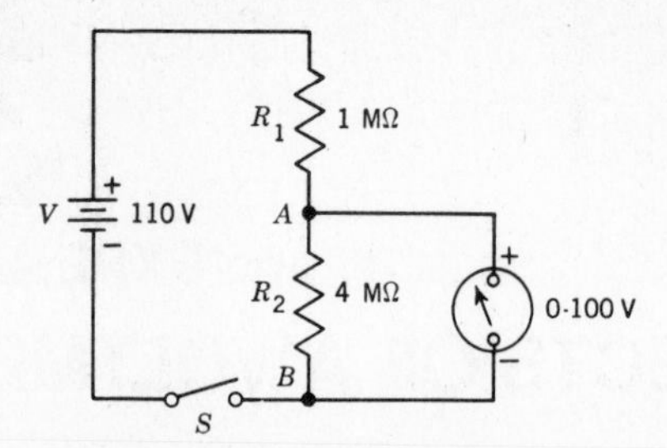

Fig. 32-2. Voltmeter loading.

figure 100,000 Ω is used. The equivalent circuit (Fig. 32-3) shows what happens when the meter is placed across the circuit.

The voltage across R_{T2}, and hence across the voltmeter, is $\frac{1}{11} \times 110 = 10$ V, and the meter would read 10 V. But this would not be the true voltage from A to B in the absence of the meter, since a computation shows that the voltage from A to B before the meter was connected was $\frac{4}{5} \times 110 = 88$ V. The voltmeter has "loaded" the circuit. Obviously this is undesirable because it can lead to serious errors. To minimize this loading, use a voltmeter with a much *higher* ohms/volt rating which is set on the highest range sufficient to make a voltage reading possible.

The effect of voltmeter "loading" in a series circuit may be generalized. Thus, in the circuit of Fig. 32-2, the computed voltage V_{AB} across R_2 is

$$V_{AB} = \frac{R_2}{R_1 + R_2} \times V \tag{32-4}$$

When the leads of a voltmeter with input resistance R_{in} are placed across R_2, the voltage V'_{AB} measured between points A and B is

$$V'_{AB} = \frac{R_2 R_{\text{in}}}{R_1 R_2 + R_{\text{in}}(R_1 + R_2)} \times V \tag{32-5}$$

The ratio V_{AB}/V'_{AB} is the ratio of the voltage across AB before the voltmeter is inserted in the circuit to the voltage after it is placed across AB. And it can be shown that

$$\frac{V_{AB}}{V'_{AB}} = \frac{R_1 R_2 + R_{\text{in}}(R_1 + R_2)}{R_{\text{in}}(R_1 + R_2)} = \frac{R_1 R_2}{R_{\text{in}}(R_1 + R_2)} + 1 \tag{32-6}$$

From Eq. (32-6) it is apparent that the voltages V_{AB} and V'_{AB} are not the same. It can be seen further that V'_{AB} equals V_{AB} only when

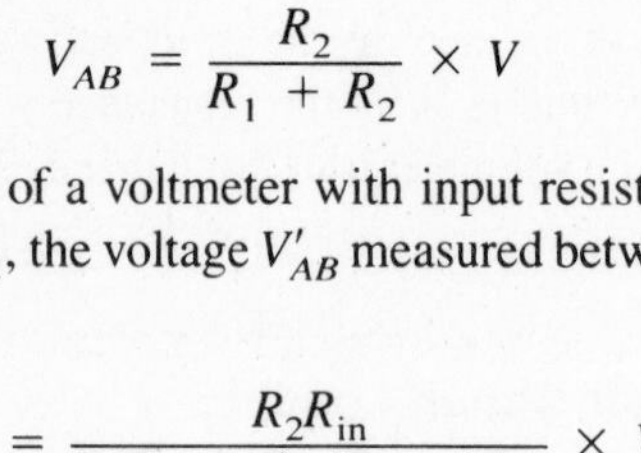

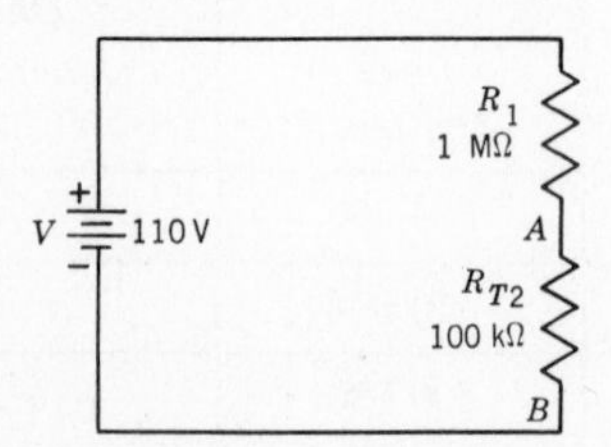

Fig. 32-3. Effect of voltmeter in the circuit.

$$\frac{R_1 R_2}{R_{\text{in}}(R_1 + R_2)} = 0 \tag{32-7}$$

Since R_1 and R_2 are finite values, Eq. (32-7) is true only when R_{in} is infinitely large. It follows, then, that the input resistance of a voltmeter loads the circuit by changing the effective resistance across which the voltage is measured, but that the higher the value of R_{in}, the less is the loading of the circuit because as R_{in} increases, the value of

$$\frac{R_1 R_2}{R_{\text{in}}(R_1 + R_2)}$$

decreases.

Eliminating Loading Effects of a Nonelectronic Voltmeter

The voltmeter in Fig. 32-1 is not an electronic voltmeter, that is, it contains no electronic amplifiers. The input resistance of this nonelectronic voltmeter changes from range to range, as is evident from Table 32-1. It is theoretically possible to eliminate the loading effects of such a voltmeter by making two measurements with it, on different ranges of the meter.

Refer to Fig. 32-2. Let the voltage measured across AB on range 1 be V_1, and that measured across AB on range 2 be V_2, and let the ratio of voltage range 1 to voltage range 2 be a. (Assume voltage range 1 is a higher range than range 2. Therefore a is a number greater than 1.) It can be shown that the voltage V_{AB} which exists across AB before the voltmeter is inserted, that is, before the circuit is loaded, may be computed from the formula

$$V_{AB} = \frac{(a - 1)V_1 V_2}{aV_2 - V_1} \tag{32-8}$$

An example will show how Eq. (32-8) may be applied. Suppose range 1 is the 100-V range of a 1000-Ω/V meter and $V_1 = 50$ V, and suppose range 2 is the 25-V range of the same meter and V_2 is 20 V. Then $a = (100/25) = 4$, and by substituting the values $a = 4$, $V_1 = 50$, $V_2 = 20$ in Eq. (32-8) we get

$$V_{AB} = \frac{(4 - 1) \times 50 \times 20}{4 \times 20 - 50} = \frac{3 \times 50 \times 20}{30} = 100 \text{ V}$$

That is, the unloaded voltage across AB is actually 100 V.

NOTES: (1) It is an interesting fact that this technique may be employed to find the true value of V_{AB} without knowing the values of $R_1 R_2$ or the voltage V applied to the circuit under test.
(2) This technique can be used only with a voltmeter whose input resistance differs on each of the ranges, for the ratio a, previously defined as $a = (V_{\text{range 1}}/V_{\text{range 2}})$, is also the ratio $R_{\text{in}}(\text{range 1})/R_{\text{in}}(\text{range 2})$. That is

$$a = \frac{R_{\text{in}}(\text{range 1})}{R_{\text{in}}(\text{range 2})}$$

(3) The more accurate the measurements V_1 and V_2 are, the closer is the value of V_{AB} computed from Eq. (32-8) to the actual value of V_{AB} in the circuit.

(4) The input resistance of electronic voltmeters is usually the same for all ranges of the meter. Hence this technique cannot be used with such a meter. However, the input resistance of an electronic voltmeter is usually very high, so that its loading effect in a circuit is relatively low.

SUMMARY

1. The input resistance R_{in} of a nonelectronic voltmeter such as that in Fig. 32-1 differs for every range of the meter.
2. If a meter movement of sensitivity I_m is used to construct a voltmeter of range V, then the input resistance of the meter on that range is

$$R_{in} = \frac{V}{I_m}$$

3. The ohms/volt characteristic of a voltmeter is defined as the ratio of the input resistance of the meter on a particular range (R_{in}) to the voltage of that range (V):

$$\text{Ohms/volt} = \frac{R_{in}}{V}$$

4. The ohms/volt characteristic of a nonelectronic voltmeter using a movement whose sensitivity is I_m is constant on each range. Moreover,

$$\text{Ohms/volt} = \frac{1}{I_m}$$

5. When a voltmeter is placed across a resistance R_{AB} connected between two points A and B in a circuit, it "loads" the circuit by changing the effective resistance between A and B. The "new" resistance which the circuit sees is the parallel combination of R_{AB} and R_{in}, the input resistance of the voltmeter on the specific range to which it is set.
6. As a result of voltmeter loading, the voltage measured may be very much lower than the nonloaded voltage which existed across the points A and B before the voltmeter was placed in the circuit.
7. Loading may be minimized by using high-sensitivity meter movements, that is, meter movements whose I_m is so low that the ohms/volt rating of the voltmeter, $1/I_m$, is very high. For electronic work, 20,000 Ω/V meters are usually used.
8. Loading may also be minimized by using the higher rather than the lower ranges of a voltmeter, where a choice is possible.
9. The loading effects of a nonelectronic voltmeter may be eliminated by making two measurements with the same voltmeter, each on a *different range*. If the measurements on range 1 and range 2 are respectively V_1 and V_2, and if the ratio Voltage range 1/Voltage range 2 $= a$, then the actual voltage may be found by substituting V_1, V_2, and a in the formula

$$V_{actual} = \frac{(a - 1)V_1 \times V_2}{aV_2 - V_1}$$

SELF-TEST

Check your comprehension by answering these questions:

1. The input resistance of a 20,000 Ω/V meter on the 50-V range is ________ Ω; on the 300-V range its input resistance is ________ Ω.
2. A meter movement whose sensitivity $I_m = 10\ \mu A$ is used to construct a nonelectronic voltmeter. The ohms/volt characteristic of that meter is ________ Ω/V.
3. The input resistance of the voltmeter in question 2 on the 10-V range is ________ Ω.
4. In the circuit of Fig. 32-2, $R_1 = 1$ MΩ, $R_2 = 1$ MΩ, and $V = 100$ V. If a 1000 Ω/V meter on the 100-V range is placed across R_2, the effective resistance R_{AB} between points A and B becomes ________ Ω.
5. The voltmeter in question 4 would read ________ volts instead of the unloaded value of ________ V.
6. If the meter in question 4 were set on the 50-V range, the voltage across R_2 would measure ________ volts.
7. Using the measurements in 5 and 6 to eliminate the loading effects of the voltmeter, the unloaded voltage $V_{AB} =$ ________ V.

MATERIALS REQUIRED

- Power supply: Variable low-voltage dc supply
- Equipment: Electronic voltmeter, meter movement (same as in Experiment 31), resistance decade box as in preceding experiment
- Resistors: ½-W, two 100,000-Ω; others as required
- Miscellaneous: One SPST switch, potentiometers as required

PROCEDURE

1. The voltmeters constructed in Experiment 31 will be used again in this experiment. Transfer from Table 31-1 to Table 32-2 the following *measured* values.
   a. Sensitivity I_m of the meter movement
   b. Voltage ranges of the experimental voltmeter, that is, V_1 on range 1 (required 30-V range) and V_2 on range 2 (required 10-V range)
   c. Multiplier R_s for range 1; for range 2
   d. Input resistance of the voltmeter on range 1; on range 2

TABLE 32-2. Experimental Voltmeter Characteristics

	Measured Values			Computed Values	
				Ohms/Volt	
Meter Movement Sensitivity I_m, A	*Voltmeter Range,* V	*Multiplier* (R_s), Ω	R_{in}, Ω	R_{in}/V	$\frac{1}{I_m}$
	V_1(range 1)				
	V_2(range 2)				

TABLE 32-3. Voltage Across R_1

Measured With			*Computed*
Experimental Voltmeter (V_2)		*Electronic Voltmeter* (V_1)	$V \times \frac{R_1}{R_1 + R_2}$
Range 1			
Range 2			

2. Compute and record in Table 32-2 the ohms/volt rating of the voltmeter on range 1. Use the measured value V_1 and R_{in} for range 1 and substitute in the formula

$$\text{Ohms/volt} = \frac{R_{in}}{V_1}$$

3. Repeat step 2 for range 2.
4. Using the measured value I_m, compute and record in Table 32-2 the ohms/volt rating of the voltmeter by substituting I_m in the formula

$$\text{Ohms/volt} = \frac{1}{I_m} \text{ (where } I_m \text{ is in amperes)}$$

Voltmeter Loading

5. Construct the experimental voltmeter in Fig. 32-1 range 1 (from Table 32-2). Construct also the circuit in Fig. 32-4. The electronic voltmeter V_1 is used to measure the voltage V of the voltage supply. The *experimental voltmeter* V_2 is used to measure voltage across R_1.
6. With both meters connected in the circuit, adjust V to read 40 V on V_1 (electronic voltmeter). Measure the voltage across R_1 as read on the experimental voltmeter V_2 and record in Table 32-3.
7. *Remove the experimental voltmeter* V_2 from the circuit. Readjust the voltage source V, if necessary, to read 40 V on the electronic voltmeter V_1. Now connect the electronic voltmeter V_1 across R_1, and measure and record the voltage across R_1 as read on V_1.
8. Repeat the procedure in step 5 for range 2 of the experimental voltmeter.
9. Repeat steps 6 and 7, recording your measurements in Table 32-3.
10. Compute and record in Table 32-3 the voltage which should appear across R_1 if no voltmeters were connected in the circuit, using the circuit values of R_1, R_2, and V and substituting in the formula:

$$V_{(\text{across } R_1)} = V \times \frac{R_1}{R_1 + R_2}$$

Extra Credit

NOTE: The readings with the experimental voltmeter are too crude (since the scale is not fully calibrated) to apply the formula in Eq. (32-8) to eliminate the loading effects of the experimental voltmeter. Hence this will not be attempted. It is possible to check the formula by repeating steps 5 through 9 using a commercial 1000 Ω/V voltmeter instead of the experimental voltmeter.

11. Verify Eq. (32-8) using a commercial 1000 Ω/V meter and the circuit of Fig. 32-4. Construct a special table for the data and substitute the values V_1, V_2, and a in Eq. (32-8). If a 20,000 Ω/V meter is used, let R_1 and R_2 be 1-MΩ resistors instead of 100-kΩ resistors.

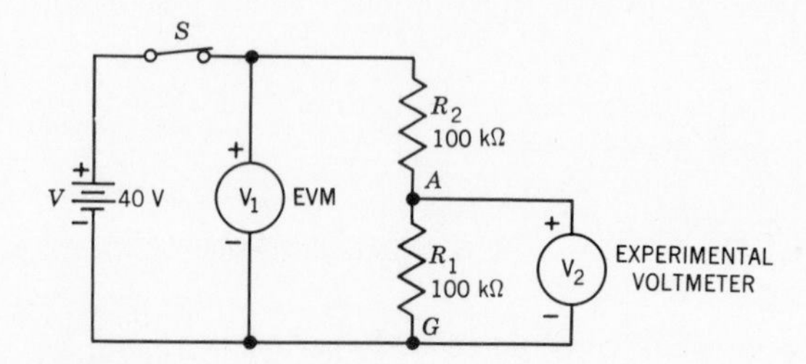

Fig. 32-4. Circuit to determine loading effects of experimental voltmeter.

QUESTIONS

1. Which meter loaded the circuit more, the electronic voltmeter or the experimental voltmeter? How do you know? Refer to the data in your tables to support your answer.
2. Which range of the experimental voltmeter loaded the circuit more and why? Refer specifically to the data in your tables to support your answer.
3. How did the two ohms/volt ratings (computed values) in Table 32-2 compare? If there was any difference, explain why.
4. What is the ohms/volt rating of a commercial VOM using
   (*a*) 10-mA movement?
   (*b*) 50-μA movement?
   (*c*) 1-mA movement?
   (*d*) 10-μA movement?
5. Of four voltmeters using, in turn, meter movements (*a*) through (*d*) (question 4), which one would cause the least amount of circuit loading in measuring voltage across R_1 in Fig. 32-4? Most amount of loading? Why?
6. In measuring an unknown dc voltage, which range of a multirange voltmeter should you use? Why?

Answers to Self-Test

1. 1,000,000; 6,000,000
2. 100,000
3. 1,000,000
4. 90,900 (approx.)
5. 8.33 (approx.); 50
6. 4.55 (approx.)
7. 49.22 (approx.)

33

EXPERIMENT

CURRENT-METER SHUNTS

OBJECTIVES

1. To determine by analysis how a basic meter movement is converted into a specified higher-range current meter by the addition of a shunt
2. To verify experimentally the value of the shunt required to convert the meter movement into a current meter of a specified range

INTRODUCTORY INFORMATION

The basic meter movement is a current-reading meter. Thus, a 0–10-mA milliammeter movement requires 10 mA of current for full-scale deflection, 5 mA for half-scale deflection, etc. This movement could therefore be used without change as a 0–10-mA milliammeter to measure current in circuits where there is 10 mA or less.

Where it is necessary to measure more than 10 mA of current, it is possible to convert the meter into a higher-range current meter. Thus, if it is desired to measure the current in a circuit where there is 40 mA, a 0–50-mA milliammeter could be used.

To extend the range of the 0–10-mA milliammeter movement to read 50 mA of current at maximum deflection would require a "shunt." This is a resistor connected in parallel with the movement to provide another path for current. In Fig. 33-1, R_{sh} is the shunt.

Consider the movement which registers full-scale deflection with 10 mA of current. To convert it into a 0–50-mA milliammeter, it is necessary to send 40 mA through the shunt and 10 mA through the movement.

If the resistance R_m of the movement is known, it is a simple matter to determine the size of the shunt R_{sh}. Since the shunt and movement are connected in parallel, the voltage across each must be the same. Hence this voltage:

$$V_m = I_m \times R_m = I_{sh} \times R_{sh} \tag{33-1}$$

where I_m is the current in amperes in the movement and I_{sh} is the current in amperes in the shunt.
Therefore,

$$\frac{R_{sh}}{R_m} = \frac{I_m}{I_{sh}} \tag{33-2}$$

and

$$R_{sh} = R_m \times \frac{I_m}{I_{sh}} \tag{33-3}$$

If it is assumed that the resistance R_m of the movement is 200 Ω, and at full-scale deflection I_m = 10 mA and I_{sh} = 40 mA, the size of the required shunt is

$$R_{sh} = 200 \times \frac{10}{40} = 50\ \Omega$$

As in the voltmeter application, R_{sh} must be a precision resistor.

To change the basic 0–10-mA movement into a 500-mA meter, a shunt R_{sh} would be used, whose size would be

$$R_{sh} = 200 \times \frac{10}{490} = 4.08\ \Omega$$

A meter capable of measuring small currents accurately would require a more sensitive movement. Thus a 0–50-μA movement could be used as a current meter giving full-scale deflection at 50 μA. Shunts could then extend the range of the meter upward as required.

Equation (33-3) is used to determine the value of the shunt resistor R_{sh} when the meter current I_m, the shunt current I_{sh}, and the meter-movement resistance R_m are known. If I_m represents the meter-movement sensitivity (i.e., current required for full-scale deflection), then $I_m + I_{sh} = I_t$ represents the maximum current which the meter can measure on that range. I_t, then, is the current range for the particular shunt R_{sh}. Equation (33-3) may now be written as

$$R_{sh} = R_m \times \frac{I_m}{I_t - I_m} \tag{33-4}$$

In the case of the 10-mA meter movement which was converted into a 50-mA meter,

$$I_m = 10 \times 10^{-3}\ \text{A, and } I_t = 50 \times 10^{-3}\ \text{A}$$

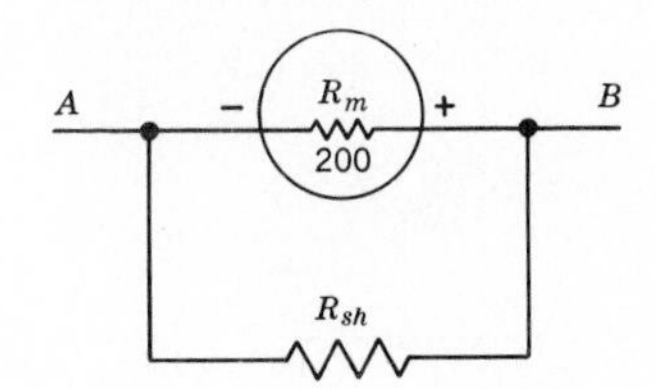

Fig. 33-1. Milliammeter shunt.

It may sometimes be necessary to determine the current range I_t of a meter when the shunt resistor R_{sh}, meter-movement resistance R_m, and meter-movement sensitivity I_m are given. The following formula can be used to find I_t:

$$I_t = I_m \times \frac{R_m + R_{sh}}{R_{sh}} \quad (33\text{-}5)$$

SUMMARY

1. A dc meter movement is a direct-current–reading device.
2. A dc meter movement whose sensitivity is I_m can be used as a direct-current meter which can measure currents equal to or less than I_m.
3. To measure direct currents higher than I_m, the meter movement must be converted into a higher-range current meter by the addition of a *shunt*.
4. A shunt is a precision resistor (R_{sh}) connected in parallel with the meter movement.
5. At full-scale deflection, the current in the movement is I_m. If the current in the shunt is I_{sh}, then the range of the meter is I_t, where

$$I_t = I_m + I_{sh}$$

6. If R_m, I_m, and I_{sh} are known, then the resistance of the shunt may be computed from the formula

$$R_{sh} = R_m \times \frac{I_m}{I_{sh}}$$

7. If the range current I_t, I_m, and R_m are known, the resistance of the shunt may be computed from the formula

$$R_{sh} = R_m \times \frac{I_m}{I_t - I_m}$$

8. If it is required to measure currents very much smaller than I_m, a meter movement of higher sensitivity than I_m must be used.

SELF-TEST

Check your understanding by answering these questions:

1. If a meter movement whose sensitivity is 1 mA is used as a current meter, it ________ (is/is not) possible to measure current greater than 1 mA with this movement alone.
2. It is required to convert a 1-mA (I_m) 50-Ω (R_m) dc meter movement into a 25-mA meter. The value of the shunt resistor (R_{sh}) must be ________ Ω.
3. At full-scale deflection a meter movement takes 50 μA of current. If the shunt in parallel with this movement takes 450 μA at full-scale deflection, the range of the meter is ________ μA or ________ mA.
4. If the resistance of the movement in question 3 is 2500 ohms, the value of shunt resistance for the current meter above must be ________ Ω.

MATERIALS REQUIRED

- Power supply: Variable low-voltage direct current
- Equipment: EVM, 0–1-mA meter movement, the same as in Experiment 31, resistance decade box, same as in Experiment 32, 0–10-mA milliammeter
- Resistors: ½-W 10,000-, 5600-Ω
- Miscellaneous: Hand tools, 100-Ω 2-W potentiometer (if a decade box is not available); other components as required in step 16 (optional); one SPST switch

PROCEDURE

Current-Meter Shunt

1. Connect the circuit of Fig. 33-2, using a 0–1-mA milliammeter movement as the current meter *M* and an EVM as the voltmeter *V*. *Observe meter polarity.*
2. Adjust the variable voltage source for full-scale deflection on *M*. Measure, and record in Table 33-1, the voltage *V* as measured on the EVM. The current *I* in the circuit is 1 mA.
3. **Power off.** Connect a rheostat R_{sh} in shunt with *M* as in Fig. 33-3. Set R_{sh} for maximum resistance. A resistance decade box R_{sh} set at 100 Ω is preferable to a 100-Ω rheostat.
4. **Power on.** Adjust R_{sh} until *M* reads half-scale deflection. Monitor the voltage source and hold the level at *V*, as in step 2. Ignoring the loading effect of the meter, has the current in the circuit of Fig. 33-3 changed value? *M* and R_{sh} now constitute a current meter whose range is

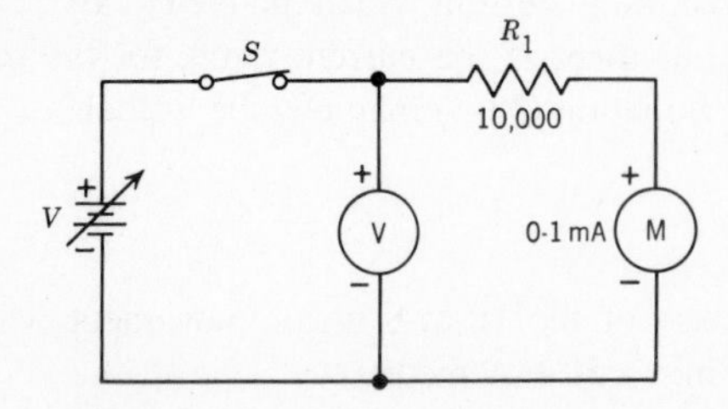

Fig. 33-2. Basic meter movement used as a 0–1-mA milliammeter.

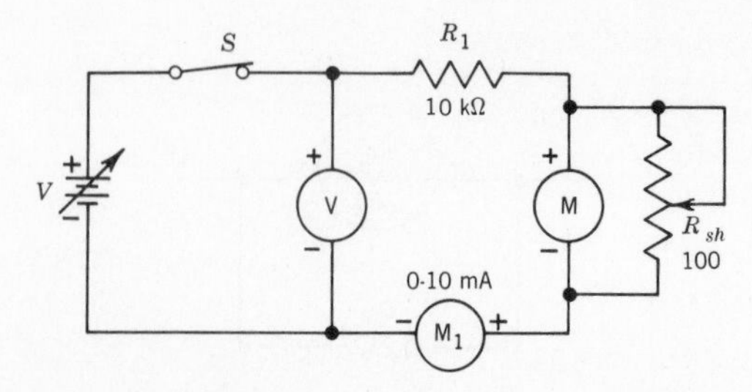

Fig. 33-3. Using a shunt to extend the range of a milliammeter.

TABLE 33-1. Experimental 2-mA Meter

Step	V, volts	I, mA	R_{sh}, Ω
2		1 mA	X
4			X
5			X
6	X	X	
11, 12	X		

TABLE 33-2. Experimental 5-mA Meter

Step	R_{sh}, Ω	I, mA
13		X
15	X	

____________ mA. Measure the voltage V and current I in the circuit and record in Table 33-1.

5. Increase the voltage V until there is full-scale deflection of M. Measure and record this voltage and the current I in R_1 as measured by the experimental current meter. **Power off.**
6. Remove shunt R_{sh} from the circuit. Do not change the setting of R_{sh}. Read the value of R_{sh} from the decade box dials or measure and record the resistance of R_{sh}.
7. How does the resistance R_m of the meter movement (see Experiment 31) compare with R_{sh} above?
8. Without changing the value of R_{sh}, place it in parallel with M as in Fig. 33-3. **Power is still off.**
9. Add a 0–1-mA milliammeter (M_1) to be used as a standard in the circuit, as in Fig. 33-4.
10. Close switch S and adjust the voltage V, if necessary, until there is full-scale deflection of M.
11. Record in Table 33-1 the current I read on M_1, the standard current meter. **Power off.**
12. Again remove R_{sh} from the circuit and measure its resistance. Record in Table 33-1.

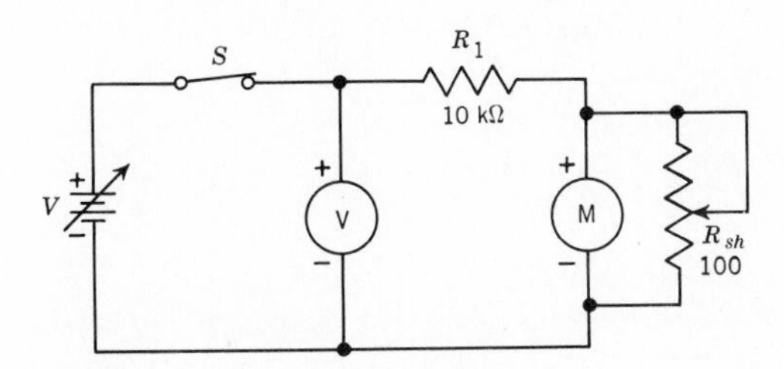

Fig. 33-4. A standard current meter M_1 is used to check the range of the experimental current meter M.

NOTE: The value of R_{sh} in steps 6 and 11 should be exactly the same. If it is not, reset R_{sh} until its resistance is the same as in step 6 and repeat steps 10 and 11.

13. **Power off.** Analytically determine the value of R_{sh}, required to convert the meter movement M to a 0–5-mA current meter. Show your computations. Record this value in Table 33-2.
14. Remove R_{sh} from the circuit and set it to the value determined in step 13. Place it in the circuit of Fig. 33-4. Replace the 10-kΩ limiting resistor R_1 with a 5.6-kΩ resistor.
15. **Power on.** Adjust V until meter movement M reads full-scale deflection. Record in Table 33-2 the current I measured by standard current meter M_1.

Design Problem (Extra Credit)

16. Design a milliammeter with the following ranges: 2 mA, 5 mA, 10 mA. Use the 0–1-mA meter movement assigned to you in this experiment. Draw the circuit diagram. In Table 33-3 include the design values for each shunt resistor R_{sh}. Show all your computations. You may use a switch to change range. Secure the necessary values of R_{sh} and breadboard the three-range current meter.

NOTE: A resistance decade box will serve for selecting the proper shunt value. Using a commercial milliammeter as standard, experimentally measure the amount of current required to cause full-scale deflection of the experimental current meter, on each range.

TABLE 33-3. Experimental Three-Range Milliammeter

Meter-Movement Characteristics		*Shunt Resistor R_{sh}, Ω for*			*Current for Full-Scale Deflection*		
R_m, Ω	I_m, A	2 mA	5 mA	10 mA	2 mA	5 mA	10 mA
	0.001						

QUESTIONS

1. In step 5 how much current is there in the meter movement? In the shunt?
2. What is the range of the experimental current meter in step 5?
3. What is the computed (formula) value of R_{sh} required to convert the meter movement into a current meter of the same range as in step 5? Show your computations.
4. How does the computed value of R_{sh} compare with the measured value in step 6? Explain any difference.
5. How does the measured value of R_m (see Experiment 30) compare with the measured value of R_{sh} in procedural step 6? Should they be the same? Explain.
6. Explain in detail the experimental procedure you followed to test the accuracy of the experimental current meter. Use a circuit diagram.
7. In connection with the procedure in question 6, what factors limit the accuracy of this procedure?

Answers to Self-Test

1. is not
2. 2.08
3. 500; 0.5
4. 278

34

EXPERIMENT

INSERTION (LOADING) EFFECTS OF A CURRENT METER

OBJECTIVES

1. To determine analytically the effects of inserting a current meter in the circuit
2. To confirm experimentally the insertion (loading) effects of a current meter

INTRODUCTORY INFORMATION

Circuit Insertion Effects of a Current Meter

The effect of inserting a current meter in a series circuit is to add the resistance of the meter to the total resistance of the circuit. The circuit effect of inserting a current meter, then, depends on the resistance of the meter and the resistance of the circuit under test. For example, to measure current in the circuit of Fig. 34-1, the meter M would, as usual, be placed in series with the circuit, as in Fig. 34-2.

If it is assumed that the meter is a 0–10-mA milliammeter with 200 Ω resistance, then it would add 200 Ω to R_1 and R_2 and the battery would see a total resistance 330 + 470 + 200 = 1000 Ω. M would then read 10 mA in the circuit of Fig. 34-2.

However, computing the current in Fig. 34-1, we find

$$I = \frac{V}{R_T} = \frac{10}{800} = 12.5 \text{ mA}$$

There is actually a 2.5-mA difference between the calculated value and the measured value, and the difference is obviously due to the resistance of the meter.

Of course, if R_1 = 4700 Ω and R_2 = 3300 Ω, the "insertion" effect of the milliammeter would be negligible.

In dc voltage measurement, minimum circuit loading requires the use of a *high-resistance voltmeter.* In current measurement, on the other hand, the *ammeter* should have *low resistance* to reduce circuit errors.

The error introduced by a current meter M connected in the series circuit (Fig. 34-2) may be readily determined for the general case. Let I_B be the current computed in the circuit before the meter is inserted, and I_A the current measured in the circuit, after the meter is connected. By Ohm's law,

$$I_B = \frac{V}{R_T}$$

where
$$R_T = R_1 + R_2 \tag{34-1}$$

and
$$I_A = \frac{V}{R_T + R_m} \tag{34-2}$$

where R_m is the resistance of the meter.

The error I_{error} caused by meter insertion is

$$I_{\text{error}} = I_B - I_A \tag{34-3}$$

By substituting in Eq. (34-3) the values of I_A and I_B from Eqs. (34-1) and (34-2), and solving, the value of I_{error} is found to be

$$I_{\text{error}} = I_A \times \frac{R_m}{R_T} \tag{34-4}$$

From Eq. (34-4) it is evident that I_{error} is zero when R_m = 0. It is clear, therefore, that no error is introduced when the meter resistance is zero. Since the meter resistance is never zero, some error is introduced in measurement, but this error can be made relatively small by using an ammeter with low internal resistance.

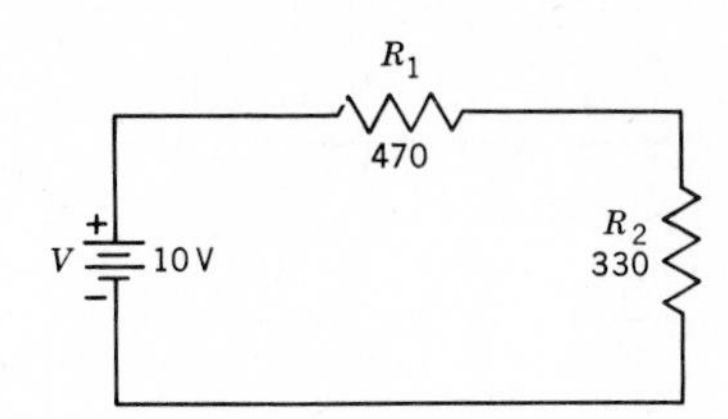

Fig. 34-1. Circuit where current is to be measured.

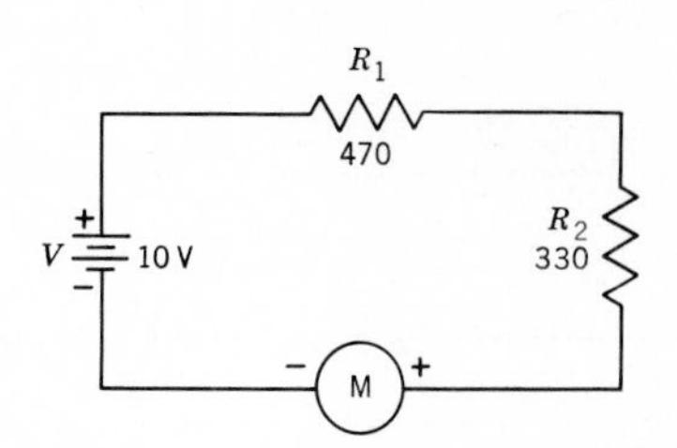

Fig. 34-2. Connecting milliammeter to measure current.

SUMMARY

1. When a current meter is inserted in a circuit, the meter adds its resistance in series with the circuit resistance. The total resistance of the circuit is therefore higher with the current meter in than it was before the meter was added.
2. This increase in circuit resistance is called the loading effect of a current meter.
3. The total current in the circuit, with the current meter in, is less than the current in the circuit with the meter out.
4. To minimize the loading effect of a current meter, use a *low-resistance* meter.
5. The current error introduced in the circuit may be calculated from the formula

$$I_{error} = I_A \times \frac{R_m}{R_T}$$

where I_A is the current measured in the circuit, R_m is the resistance of the current meter, and R_T is the total resistance in the circuit before the meter is added.

SELF-TEST

Check your understanding by answering these questions:

1. An ammeter is connected in ____________ (series/parallel) with the circuit to measure current in the circuit.
2. The insertion effect of a current meter is to ____________ resistance to the circuit.
3. The effect of connecting a current meter in a circuit, then, is to ____________ the circuit.
4. The current measured by a current meter is actually ____________ than the current in the circuit before the meter was added.
5. In the circuit of Fig. 34-1, $R_1 = 120\ \Omega$, $R_2 = 280\ \Omega$, and $V = 10$ V. The computed circuit current I_B = ____________ mA.
6. When a current meter M whose resistance R_m is 100 Ω is connected in the circuit of question 5, the meter reads ____________ mA (I_A).
7. From Eq. (34-4), the error current I_{error} = ____________ mA.
8. The error current determined by subtracting I_A from I_B (that is, $I_B - I_A$) = ____________ mA.

MATERIALS REQUIRED

- Power supply: Variable low-voltage dc source
- Equipment: Electronic voltmeter, 0–1-mA meter movement and resistance decade box (the same as in Experiments 30 through 33)
- Resistors: ½-W 47-Ω, 100-Ω
- Miscellaneous: 500-Ω 2-W potentiometer; one SPST switch

PROCEDURE

1. Connect the circuit of Fig. 34-3. **Power off.** V is an electronic voltmeter; M is the 0–1-mA meter movement acting as a 1-mA current meter. Switch S is *open.*

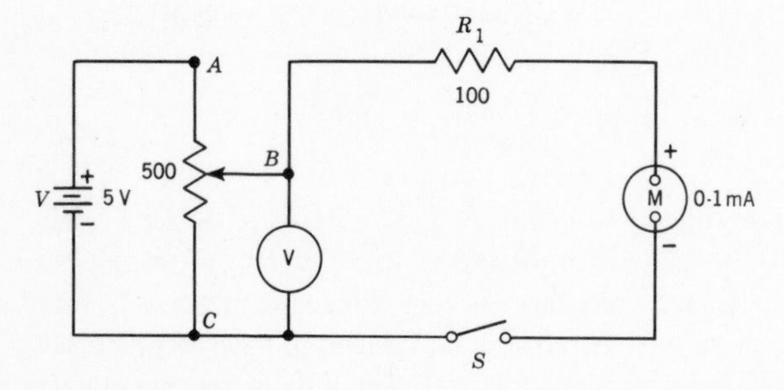

Fig. 34-3. Experimental circuit to determine loading effects of a current meter.

2. Set the arm of the 500-Ω potentiometer, acting as a circuit voltage control, at A. The voltmeter V will then measure the voltage V delivered by the power supply.
3. Turn the power supply **on** (switch S is still *open*) and adjust V for 5 V as measured by the voltmeter V. Keep power-supply output at this level. Turn the arm of the voltage-control potentiometer down to C until V reads *zero* volts. *Close switch S.*
4. Slowly turn the arm of the voltage control up until M shows full-scale deflection (1 mA). This is the current I_A in Table 34-1. Record in Table 34-1 the voltage V_{BC} as measured on the voltmeter. **Power off**, switch S *open*.
5. Remove R_1 from the circuit. Measure its resistance and record in Table 34-1. Record also the value of R_m taken from Experiment 30, Table 30-1.

TABLE 34-1. Current-Meter Loading

Step	V_{BC}	I (A)			R_1, Ω	R_m, Ω	I_{error}, A
		I_A	I_B	$I_B - I_A$			
4 to 8		0.001					
10		0.001					

6. Compute, and record in Table 34-1, the current I_B in the circuit of Fig. 34-3, with the meter M removed, but with voltage V_{BC} the same as in step 4:

$$I_B = \frac{V_{BC}}{R_{1(\text{measured})}}$$

7. Compute and record in Table 34-1, $I_B - I_A$.
8. Using Eq. (34-4), compute I_{error} and record in Table 34-1, using the values I_A, $R_{1(\text{measured})} = R_T$.
9. Using a 47-Ω resistor as R_1 again, connect the circuit of Fig. 34-3. Switch S is *open*.
10. Repeat steps 2 through 8, recording your results in Table 34-1.

QUESTIONS

1. What effect on circuit current, if any, results from inserting a milliammeter to measure current in a circuit? What is this effect called?
2. For steps 4 to 8 in Table 34-1, compare the values I_{error} and $I_B - I_A$. Explain the relationship, if any, between these values.
3. Which value of $I_B - I_A$ (that is, which error current), if any, was greater, that in step 7 or that in step 10? Why?
4. Which current meter will load a circuit more, a high-resistance meter or a low-resistance meter? Why?

Answers to Self-Test

1. series
2. add
3. load
4. less
5. 25
6. 20
7. 5
8. 5

35

EXPERIMENT

THE SERIES OHMMETER

OBJECTIVES

1. To determine analytically and confirm experimentally how a basic meter movement may be converted into a series ohmmeter
2. To calibrate experimentally the scale of a series ohmmeter

INTRODUCTORY INFORMATION

The ohmmeter is an instrument used to measure the resistance of resistors and of other circuit components. In addition, it is used to locate open or shorted components and to determine circuit continuity. Resistance is read on a meter scale calibrated in ohms.

Like the nonelectronic voltmeter and ammeter, this basic test instrument utilizes a direct-current-actuated meter movement. However, unlike the voltmeter and ammeter, the ohmmeter requires a self-contained source of voltage.

The circuit in Fig. 35-1 illustrates a series ohmmeter. M, a 0–1-mA movement with internal resistance R_m equal to 200 Ω, is connected in series with a current-limiting resistor R_1, a 500-Ω rheostat R_2, and a 3-V battery V. A and B are jacks where test leads are inserted. These leads are connected across resistor R_X, whose resistance we wish to measure. The circuit values in Fig. 35-1 were assigned to simplify the discussion of the operation of this instrument.

When the test leads at A and B are shorted together, the circuit is complete and there is current in M. R_2 can be adjusted so that the total resistance that the battery "sees," including R_m, is 3000 Ω. It is apparent, by Ohm's law, with R_T = 3000 Ω and V = 3 V, that I = 1 mA is the current in this circuit. The meter will therefore read full-scale deflection. This is the zero position of the ohmmeter, at the right-hand side of the scale, and it occurs when A and B are shorted and R_2 is adjusted for full-scale deflection. R_2 is called the "zero-ohms adjust."

After the meter has been adjusted for zero ohms, the test leads are separated. The circuit is broken and the meter pointer returns to the open circuit position on the left side of the scale. When the leads are connected across a resistor, the circuit is again complete and there is current in the meter. However, the circuit current is now less, than that required for full-scale deflection (1 mA), because the total series resistance R_T has increased. Thus, if a 1000-Ω resistor is inserted between A and B (i.e., if a 1000-Ω resistor is being measured with the experimental ohmmeter), $\frac{3}{4}$ mA will flow through M and the pointer will be deflected three-quarters of the way. This position on the ohmmeter scale (Fig. 35-2) can now be calibrated as 1000 Ω.

If a 2000-Ω resistor is measured, $\frac{3}{5}$ mA of current will flow through M. The pointer will deflect three-fifths of the scale, and this will be the 2000-Ω point on the scale.

It can be shown similarly that in this ohmmeter $\frac{1}{2}$ deflection corresponds to 3000 Ω, $\frac{3}{8}$ deflection corresponds to 5000 Ω, $\frac{3}{13}$ deflection to 10,000 Ω, $\frac{3}{53}$ deflection to 50,000 Ω, etc. No deflection corresponds to an infinite resistance, that is, an open circuit between A and B. The symbol for infinity is ∞.

Several facts are apparent from the ohmmeter scale. It is evident that it is not linear, but crowds toward the left-hand side. The need for other ohmmeter ranges is also apparent. Thus, the scale of Fig. 35-2 would not be adequate for measuring a 10-Ω resistor. However, a range in which the

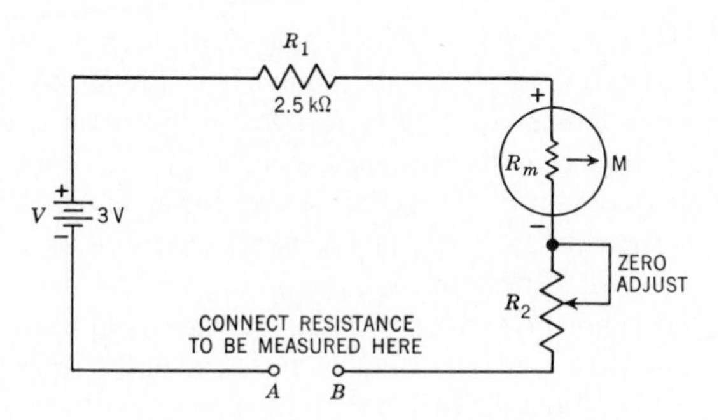

Fig. 35-1. The basic series ohmmeter circuit.

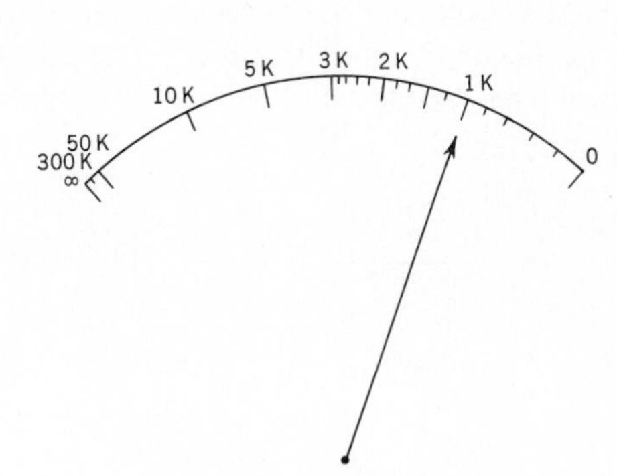

Fig. 35-2. Ohms scale.

center of the scale is, say, 30 Ω could be used readily for this measurement. Another ohmmeter circuit arrangement would be required for a range with a 30-Ω calibration at the center of the scale.

It should also be evident why an ohmmeter must not be used in a circuit to which power is applied. Consider the circuit of Fig. 35-3. If it is necessary to measure R, power must be removed before the ohmmeter leads are placed across it. Otherwise, 100 V would be applied to the ohmmeter and the meter would burn out.

The calibration of a series ohmmeter scale can be generalized. Any point on an ohmmeter scale can be found from the equation

$$R_X = R_T \frac{(I_m - I_X)}{I_X} \qquad (35\text{-}1)$$

where R_X = unknown resistance
R_T = total circuit resistance, with test leads shorted, giving full-scale deflection of the meter movement
I_m = meter-movement sensitivity, that is, current for full-scale deflection (ohmmeter current with test leads shorted)
I_X = ohmmeter current with test leads connected across R_X

The results of our previous discussion on scale calibration can now be verified by means of Eq. (35-1). For example, in the circuit of Fig. 35-1, if a resistor R_X inserted across AB causes the meter current I_X to decrease to ¾ mA, we may find the value of R_X by substituting in Eq. (35-1) these values:

$$R_T = 3000\ \Omega$$
$$I_m = 1 \times 10^{-3}\ \text{A}$$
$$I_X = \tfrac{3}{4} \times 10^{-3}\ \text{A}$$

$$R_X = 3000 \times \frac{(1 - \frac{3}{4}) \times 10^{-3}}{\frac{3}{4} \times 10^{-3}} = 1000\ \Omega$$

The center-scale value of a series ohmmeter, that is, the value of R_X which will give half-scale deflection of the pointer, may readily be found from Eq. (35-1) by letting $I_X = I_m/2$. It follows, then, that

$$R_X = R_T \frac{I_m - I_m/2}{I_m/2} = R_T \qquad (35\text{-}2)$$

Equation (35-2) states that the center-scale value of a series ohmmeter is equal to R_T, the total circuit resistance, with test leads shorted, required for full-scale deflection of the meter pointer.

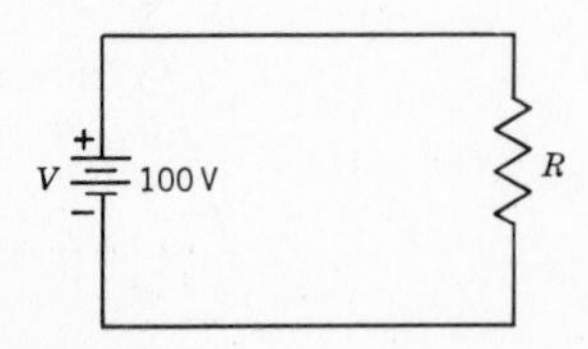

Fig. 35-3. Why must power be removed before measuring R?

The series ohmmeter in Fig. 35-1 cannot be used to measure low-valued resistances accurately because these resistances are crowded together on the low end of the scale, on the right. Consider the range of resistances on a tenth of the scale at the extreme right end. The meter current, corresponding to these end points, is $I_X = I_m$, and $I_X = \frac{9}{10} I_m$. The corresponding values of R_X, from Eq. (35-1), are

$$R_X = 0 \qquad \text{and} \qquad R_X = 333\tfrac{1}{3}\ \Omega$$

It is evident, therefore, that in the last tenth of meter scale arc, the resistances measured are between 0 and 333⅓ Ω. It would be impossible to distinguish an accurate measurement of 5 Ω or 10 Ω, etc., in this part of the arc. To measure such low-valued resistors, an ohmmeter with a much lower center-scale reading must be used. A series-type ohmmeter would not be employed in such an application; a shunt- or series-shunt type ohmmeter would.

SUMMARY

1. The ohmmeter is used to measure resistance.
2. The series ohmmeter utilizes a basic meter movement in series with circuit resistors, including a zero-ohms adjust and a self-contained voltage source. This is a nonelectronic meter.
3. The resistance we wish to measure is connected in series with the circuit of the series ohmmeter. In Fig. 35-1 the points for connecting the resistor are shown as AB.
4. The meter is set on zero by shorting the test leads at A and B and setting the zero-adjust control for full-scale deflection. This position is zero ohms, on the right-hand side of the scale.
5. Infinite resistance is marked on the left-hand end of the scale, when there is *zero* current in the meter, that is, when the circuit of the meter is not complete. See Fig. 35-2 for the 0 and ∞ points on the scale.
6. If the total resistance required for full-scale deflection of the series ohmmeter is R_T, then half-scale deflection is the calibration point for resistance R_T. In Fig. 35-2, half-scale deflection is 3 kΩ.
7. The calibration of the scale of a series ohmmeter can be achieved by use of the formula

$$R_X = R_T \frac{I_m - I_X}{I_X}$$

where R_X is the resistance value in ohms on the scale corresponding to I_X, the ohmmeter current with test leads connected across R_X; R_T is the total circuit resistance, with test leads shorted, giving full-scale deflection of the meter movement; and I_m is the meter-movement sensitivity, that is, meter current required for full-scale deflection.
8. The ohmmeter scale is a nonlinear scale, with the high-value resistances crowded at the ∞ end of the scale and the low-value resistances crowded together at the *zero* end of the scale.

9. The purpose of the zero-ohms control (rheostat) is to compensate for deterioration of the battery used to power the ohmmeter. Thus, as the internal resistance of the battery *increases* with use, the resistance of the zero-ohms adjust must be *decreased* to permit full-scale deflection of the pointer at zero ohms.
10. An ohmmeter must *never* be used in a circuit to which power is applied to prevent the ohmmeter from being damaged and to ensure a correct resistance reading. Also, to conserve the battery, the ohmmeter test leads should remain open (not shorted) when the ohmmeter is not in use.

SELF-TEST

Check your understanding by answering these questions:

1. In a series nonelectronic ohmmeter the resistance to be measured is connected in ____________ (series/parallel) with the meter movement.
2. Zero resistance on the series ohmmeter scale occurs when there is ____________ (zero/full-scale) deflection of the meter pointer.
3. As the self-contained battery in an ohmmeter is used, its internal resistance ____________ (increases/decreases).
4. The zero ohms control is used to ____________ for changes in the ____________ ____________ of the ohmmeter battery.
5. A series ohmmeter utilizes a 1½-V battery, a 50-μA meter movement and the appropriate resistors, and zero ohms rheostat. Assuming the meter has been properly zeroed, half-scale deflection of the pointer occurs when a resistance of ____________ Ω is being measured.
6. The 10-μA point on the scale of this meter would be calibrated ____________ Ω.
7. The 48-μA point on this meter (questions 5 and 6) would correspond to ____________ Ω.
8. This meter ____________ (would/would not) be useful for measuring a resistance of 30 Ω.

MATERIALS REQUIRED

- Power supply: 1.5-V battery
- Equipment: 0–1-mA milliammeter movement (same as in Experiment 33), EVM, resistance decade box
- Resistors: ½-W 330-, 1000-, 1200-, 1500-, 2200-, 3300-, and 56,000-Ω
- Miscellaneous: 500-Ω potentiometer; alligator-type test leads

PROCEDURE

1. Connect the circuit of Fig. 35-4. Alligator clip leads should be connected at *A* and *B* to act as ohmmeter test leads.
2. Short the test leads together and adjust R_2 (Zero Adj.) for full-scale deflection of the pointer. Calibrate the linear scale (Fig. 35-5) for zero resistance at this point. Do not disturb the setting of R_2.

   With the test leads shorted, measure V, the battery voltage (use an EVM), and record in Table 35-1. Open the test leads.

   Using the measured value of V and the known deflection sensitivity I_m of the meter movement (1 mA), compute, and record in Table 35-1, the total resistance R_T, which will give full-scale deflection of the pointer in the experimental ohmmeter (Fig. 35-4). Record also the known values I_m and R_m.
3. Open the test leads when the ohmmeter is not in use. To conserve the battery, the circuit should remain open.

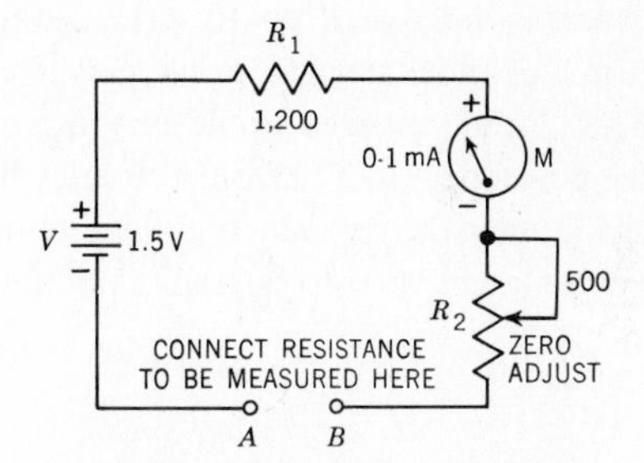

Fig. 35-4. Experimental ohmmeter.

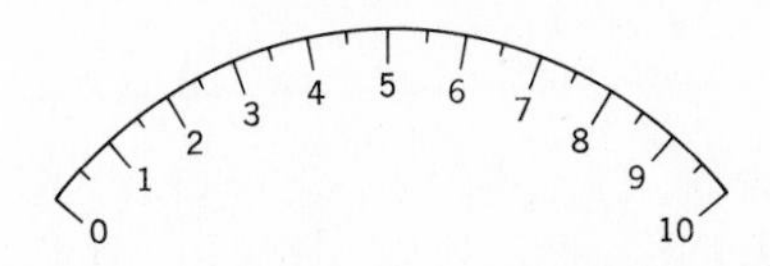

Fig. 35-5. Calibrating the ohmmeter scale.

TABLE 35-1. Calibrating Ohmmeter Scale

V, Volts	I_m, mA	R_m, Ω	R_T, Ω	R_X (Ω), *when* I_X (mA) *is:* 1	0.9	0.8	0.7	0.6	0.5	0.4	0.3	0.2	0.1	0
Rated value R_X				330	1000		1500		2200		3300		56,000	
Measured value R_X														

4. Calibrate the scale ∞ at this point, that is, with test leads open.
5. a. Compute and record the size resistor R_X which, when measured, will give half-scale deflection of the meter. Show the computation.
   b. Verify this computation by measuring with the experimental ohmmeter a resistor equal in value to the computed value above. Set the resistance decade box to the proper value of R_X. Calibrate the scale (Fig. 35-5) for this value.
6. Compute, and record in Table 35-1, the value of R_X required to limit the meter current to the following values: $I_X = 0.9I_m$; $0.8I_m$; $0.7I_m$; $0.6I_m$; $0.5I_m$; $0.4I_m$; $0.3I_m$; $0.2I_m$; $0.1I_m$; $0I_m$.
7. Calibrate the ohmmeter scale at these points.
8. Measure, and record in Table 35-1, the following resistors using the experimental ohmmeter; 330, 1000, 1500, 2200, 3300, and 56,000 Ω.

Design Problem (Extra Credit)

9. Design an ohmmeter whose center-scale value is 50 Ω, using the same meter movement and battery as in the experimental ohmmeter (Fig. 35-4). Draw the circuit diagram, showing all values. Show your computations. Explain how you would calibrate the scale of this meter.

HINT: Try a series shunt, or shunt-type ohmmeter.

QUESTIONS

1. On what does the center-scale reading of a series ohmmeter (connected as in Fig. 35-4) depend?
2. If the 1.5-V battery in the ohmmeter of Fig. 35-4 deteriorated and developed only 1 V output under load, would it be possible to zero the ohmmeter? Explain.
3. In applying Eq. (35-1) to calibrate a series ohmmeter, what is assumed about the relationship between pointer deflection and meter current?
4. Why must the test leads of an ohmmeter never be left shorted for any length of time?
5. If we wished to make an $R \times 10$ series ohmmeter using the same meter movement as in the experimental ohmmeter, what circuit changes would be required?
6. Why must power be removed from a circuit when using an ohmmeter to measure resistance in the circuit?
7. In your experiment, how accurately was the scale calibrated? Refer to the results of your measurements in Table 35-1.
8. A circuit contains two resistors connected in parallel, and their individual values must be measured with an ohmmeter. Explain the procedure.
9. Why is an ohmmeter scale nonlinear? What type of scale is it?

Answers to Self-Test

1. series
2. full-scale
3. increases
4. compensate; internal resistance
5. 30,000
6. 120,000
7. 1250
8. would not

36

EXPERIMENT

DESIGN OF A VOLT-OHM-MILLIAMMETER

OBJECTIVES

1. To design a dc VOM which will meet specified range requirements
2. To calibrate the scales of the VOM

NOTE: This may be assigned as a research and development term project for the student to complete independently. It is not intended as a regular laboratory experiment.

INTRODUCTORY INFORMATION

In Experiments 30 to 35 you studied the characteristics of a meter movement and its application as a circuit element in the measurement of voltage, current, and resistance. You learned how to determine the characteristics of a meter movement; how to design multipliers and shunts for specified voltage and current ranges; how to calculate the value of the series resistance required to convert the movement into a series ohmmeter of specified range. You can now apply this knowledge to the design and construction of a multirange dc volt-ohm-milliammeter (VOM).

Simple DC VOM

Several techniques may be employed in the design of a VOM. The simplest utilizes pin jacks for the selection of circuits and ranges, as in Fig. 36-1. Here a ½-mA meter movement, together with the appropriate circuit, serves as a three-range milliammeter, a three-range voltmeter, and a two-range ohmmeter. For all measurements the black test lead is inserted into the Common jack. The red test lead is then plugged into one of the remaining jacks to select the proper function (current, voltage, or resistance) and range. For example, for current measurements of 1 mA or less, the red lead is inserted in the 1-mA jack. For currents between 1 mA and 10 mA, the 10-mA jack is used. For voltages less than 10 V, the 10-V jack is used, etc.

The resistance ranges are marked 100,000 Ω and 1,000,000 Ω. These markings represent maximum readable resistance on each respective range.

The milliammeter ranges employ a ring or Ayrton shunt, R_1, R_2, and R_3, whose operation will be explained. The voltmeter multipliers are R_4, R_5, and R_6. The values of these multipliers are computed in the usual manner. It should be noted that in computing the values of the multipliers, both the current drawn by the meter movement M and the current in the series shunt arrangement $R_1 + R_2 + R_3$ must be considered. Thus, M and its shunt act like a 1-mA movement. For example, if the resistance R_{AB} between points A and B is 50 Ω, then

$$1 \times 10^{-3}(R_4 + 50) = 10 \qquad (36\text{-}1)$$

and $$R_4 = 10{,}000 - 50 = 9950\ \Omega$$

(The current in R_4 for full-scale deflection is 1 mA.)

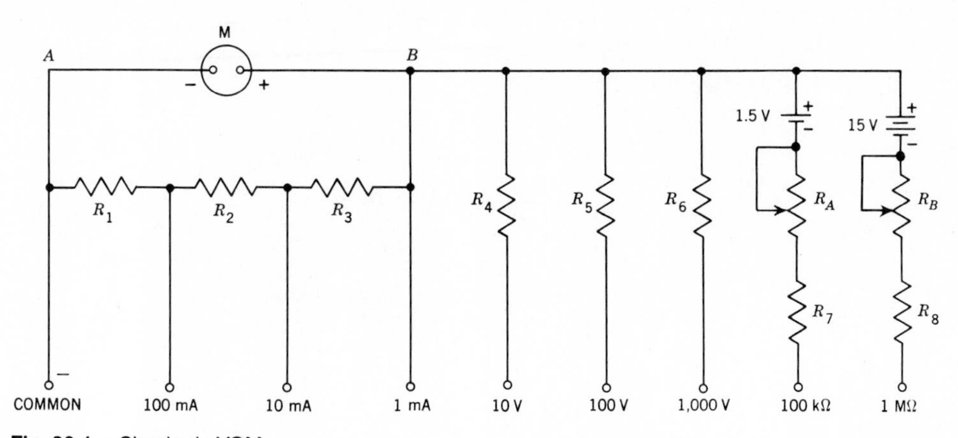

Fig. 36-1. Simple dc VOM.

R_A and R_B are, respectively, ohm-adjust rheostats for the 100,000-Ω and 1,000,000-Ω resistance ranges. On the 100,000-Ω range a 1.5-V battery is used, while on the 1,000,000-Ω range a 15-V battery is required. In computing the series range resistors, for example $R_A + R_7$, the meter movement and its shunt are again treated like a 1-mA movement. Therefore, when the meter is zeroed on this range, we can find $R_A + R_7$ by applying Ohm's law.

$$1 \times 10^{-3}(R_A + R_7 + 50) = 1.5 \qquad (36\text{-}2)$$

and $$R_A + R_7 = 1500 - 50 = 1450\ \Omega$$

We can select a 1200-Ω resistor for R_7 and a 500-Ω rheostat for R_A. Note that the center-scale reading on this range is 1500 Ω.

Ayrton Shunt

The three-range current meter circuit, by itself, is shown in Fig. 36-2. M is a meter movement whose sensitivity $I_m = ½\ \text{mA} = 0.0005\ \text{A}$, and whose internal resistance $R_m = 100\ \Omega$. We wish to determine the values of the ring shunt resistors R_1, R_2, and R_3 which will convert the meter movement into the current meter with 1-mA, 10-mA, and 100-mA ranges.

On the 1-mA range, the shunt consists of the sum of the three resistors. That is,

$$R_{sh} = R_1 + R_2 + R_3 \qquad (36\text{-}3)$$

At full-scale deflection there will be ½ mA in the movement, and ½ mA in the shunt. Therefore, the shunt and meter resistance are equal and

$$R_1 + R_2 + R_3 = 100\ \Omega \qquad (36\text{-}4)$$

On the 10-mA range, the circuit will appear as in Fig. 36-3. The shunt now consists of $R_1 + R_2$. These are in parallel with M which is in series with R_3. Since $R_1 + R_2 + R_3 = 100\ \Omega$,

$$R_1 + R_2 = 100 - R_3 \qquad (36\text{-}5)$$

We can now solve for R_3. At full-scale deflection, the current in the meter movement is 0.0005 A, while the current in the shunt is 0.0095 A. Since the voltage across $R_1 + R_2$ is equal to the voltage across $R_m + R_3$,

$$(0.0095)(100 - R_3) = 0.0005(100 + R_3) \qquad (36\text{-}6)$$

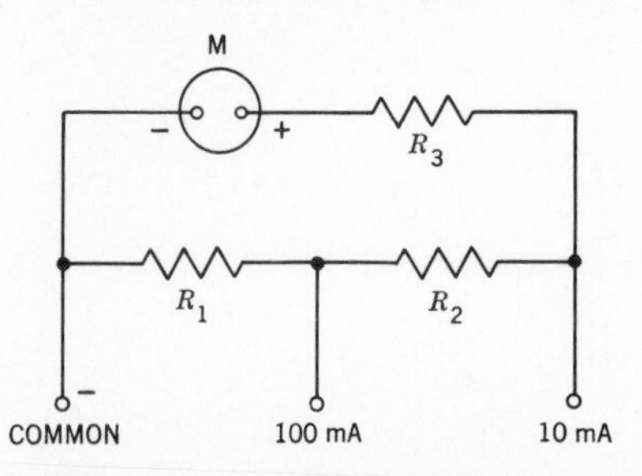

Fig. 36-3. Circuit on 10-mA range.

Solving Eq. (36-6), we find that

$$R_3 = 90\ \Omega \qquad (36\text{-}7)$$

Substituting $R_3 = 90$ in Eq. (36-5), it is evident that

$$R_1 + R_2 = 10 \qquad (36\text{-}8)$$

Now consider the 100-mA range, Fig. 36-4. R_1 is the shunt, in parallel with the series combination of the meter movement, R_3 and R_2. Since

$$\begin{aligned} R_1 + R_2 &= 10 \\ R_1 &= 10 - R_2 \end{aligned} \qquad (36\text{-}9)$$

We can now solve for R_2. The current in the meter movement at full-scale deflection is again 0.0005 A, while that in the shunt is 0.0995 A. Therefore

$$0.0005(100 + 90 + R_2) = 0.0995(10 - R_2) \qquad (36\text{-}10)$$

Solving Eq. (36-10) we find

$$R_2 = 9\ \Omega \qquad (36\text{-}11)$$

and $$R_1 = 1\ \Omega \qquad (36\text{-}12)$$

This method may be extended to find the values of any number of resistors connected in a ring shunt.

Switch-type DC VOM

The dc VOM (Fig. 36-1) may also be designed using only two jacks for the common (−) and hot (+) meter leads, employing a somewhat complex switching arrangement (Fig. 36-5). The test leads are plugged into the (−) and (+) jacks. Two switches are used, S_1 and S_2. Each is a three-position switch. S_2 is the function switch. In the position shown, the meter is set to measure current. In the center position of S_2, the meter will measure volts, and in the third position, it will measure ohms.

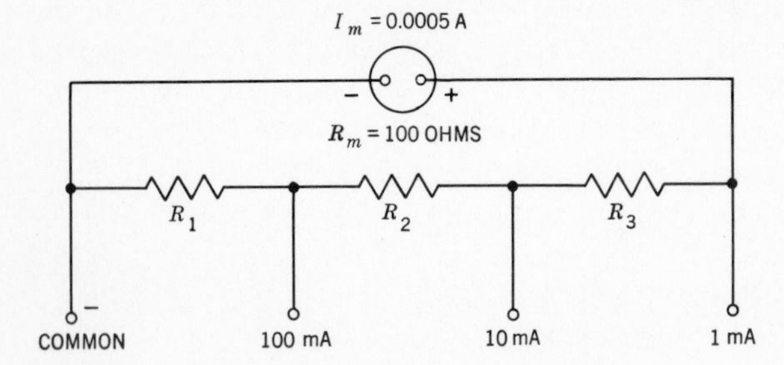

Fig. 36-2. Three-range milliammeter using a ring shunt.

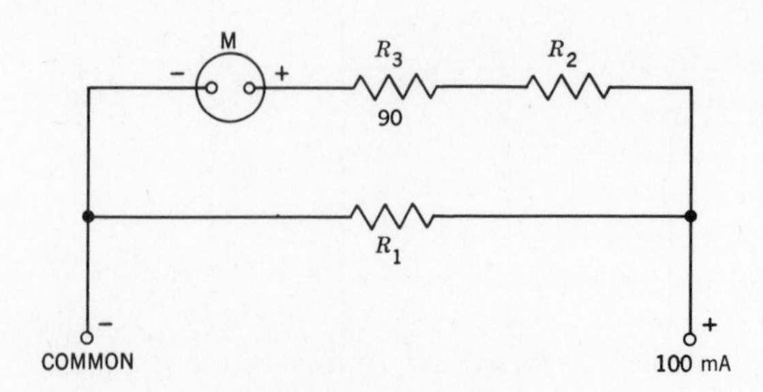

Fig. 36-4. Circuit on 100-mA range.

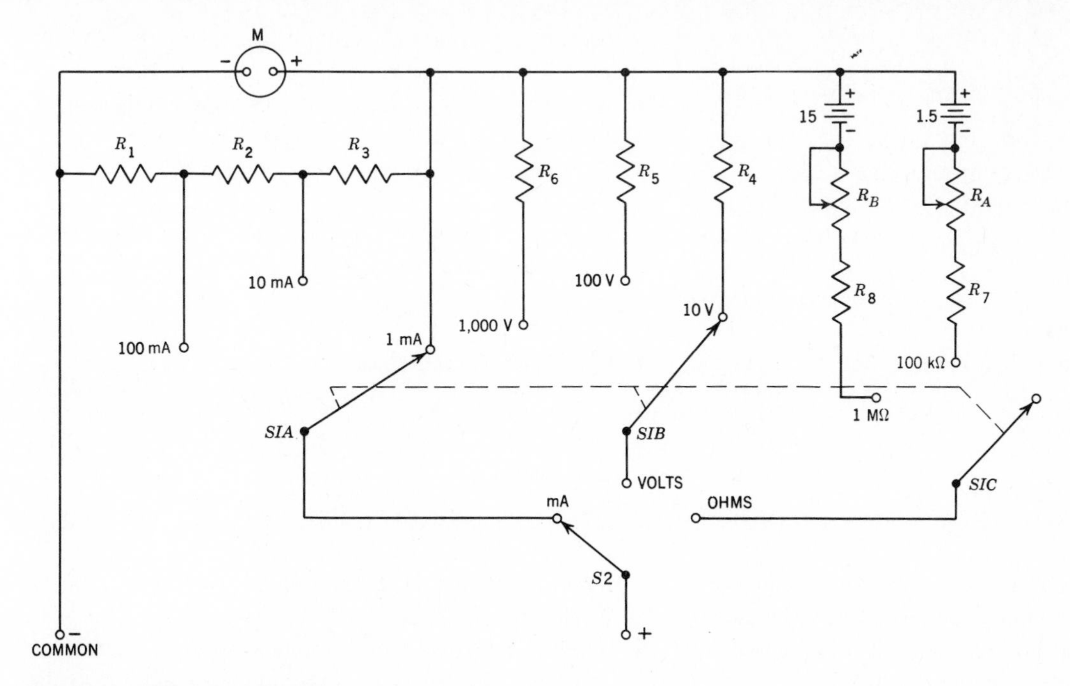

Fig. 36-5. Switch-type dc VOM.

S_1 acts as the range selector. S_1 really consists of three independent switches, S_{1A}, S_{1B}, and S_{1C}. These switches are on separate wafers, ganged together, so that they are actuated simultaneously. S_2 is shown in the current (mA) position. Note that the common arm of S_2 is connected to the + terminal of the meter. The + terminal of the meter in Fig. 36-5 is connected to the common arm of S_{1A}, which is shown in the 1-mA range. These, then, are the settings of S_1 and S_2 for measuring current on the 1-mA range. By keeping S_2 in the mA position and rotating S_1, we may also select the 10-mA or 100-mA range.

When S_2 is in the volts position, the common arm of S_{1B} is connected to the + terminal of the meter. The 10-, 100-, or 1000-V range may then be selected by S_{1B}. A similar arrangement is used for measuring resistance. However, since we assumed only two resistance ranges, switch S_{1C} has one open terminal, the position shown in Fig. 36-5.

Figures 36-1 and 36-5 illustrate two design arrangements for a simple dc VOM. For other designs the student is referred to the technical literature on VOMs.

SUMMARY

1. A nonelectronic multirange volt-ohm-milliammeter (VOM) provides, in one instrument, facilities for measuring voltage, resistance, and current. The number of ranges included for each function depends on the switch and/or circuit arrangement.
2. A dc meter movement is the common element employed for each function and each range of a VOM.
3. A simple VOM may eliminate a switching arrangement and provide jacks where the test leads are inserted to select function (voltage, resistance, or current) and range (Fig. 36-1).
4. A switch-type VOM eliminates the need to move the positive test lead from jack to jack. In this type of instrument (Fig. 36-5) selection of function and range is accomplished by a rather complex switching arrangement.
5. In Fig. 36-5 the function switch S_2 is a single-pole three-position switch. Its terminal positions are marked mA, Volts, and Ohms.
6. In Fig. 36-5 the range switch S_1 is a three-pole triple-throw switch. The common arms of the switch S_{1A}, S_{1B}, and S_{1C} are ganged, and they work together for the selection of the proper range for current (S_{1A}), voltage (S_{1B}), and resistance (S_{1C}).
7. The current-meter shunts are frequently arranged in a ring with the meter movement—an arrangement called an Ayrton shunt.
8. The lowest range of the current meter is that which utilizes all the series-connected ring resistors as a shunt for the meter movement (Fig. 36-2).
9. In succeeding switch positions of the Ayrton shunt (Figs. 36-3, 36-4), some resistors in the ring are thrown in series with the meter movement, giving the effect of a meter movement with increased internal resistance. For example, in Fig. 36-3 the meter movement acts like a movement whose resistance equals $R_m + R_3$. In Fig. 36-4 the meter movement acts like a movement whose resistance is equal to $R_m + R_3 + R_2$.
10. Computation of the resistors in the Ayrton shunt follows the system employed in computing a simple shunt resistor and a meter movement (Experiment 33).

SELF-TEST

Check your understanding by answering these questions:

1. The measurement of ____________, ____________, and ____________ may be made with a VOM.
2. The circuit arrangement of a VOM uses the same ____________ ____________ for each function and range of the meter.
3. One or more ____________ are used to select function and range of the meter.
4. A ring shunt, normally used in the design of the current shunts of a VOM, is called an ____________ shunt.
5. In the circuit of Fig. 36-5, if M is a 50-μA, 2000-Ω movement, and the lowest current range of the meter is 60 μA, then $R_1 + R_2 + R_3 =$ ____________ Ω.
6. In the meter of question 5, if the next range of the meter is 0.5 mA, then $R_3 =$ ____________ Ω.
7. In the meter of question 5, the value of R_5 on the 100-V range is ____________ Ω.
8. Center-scale deflection on the highest resistance range of the VOM in question 5 is ____________ Ω.

MATERIALS REQUIRED

As required in the design of the VOM (see Procedure)

PROCEDURE

1. You will experimentally determine the characteristics of a meter movement (I_m and R_m) assigned by the instructor and check the linearity of the movement at intervals of $I_m/10$. In the comprehensive report, which is required, all circuits and the details of the measurement process should be clearly explained.
2. You will design a VOM using this movement with the following specifications, or as specified by your instructor:
   a. Three voltage ranges: 5, 20, 40 V
   b. Three current ranges: 2, 10, 50 mA
   c. One resistance range: 1500-Ω center scale
   In the accompanying report, show all computations and the circuit values determined for each component.
3. Submit the circuit design (showing all values, including voltage and tolerance) to your instructor.
4. After your instructor has approved your design, secure the proper components and breadboard the circuit. (For the multipliers, shunts, etc., use a resistance decade box.) Test it on each of the operating ranges. Include in your report the procedure employed in each check, and the results of your checks.
5. (Optional) Design a case for the instrument. Show all material, construction, and mounting details in the accompanying report.
6. Construct the case and mount the components. Again test your meter on each range. In the accompanying report, list and describe all electrical and mechanical components.
7. Calibrate the meter scales for voltage, current, and resistance on each range. Draw the meter scales on the accompanying report.
8. Conclude your report by listing these specifications for your instrument:
   a. Voltage ranges
   b. Current ranges
   c. Center-scale resistance
   d. Ohms/volt rating of the meter
   e. Accuracy
   Explain how you determined instrument accuracy.

Answers to Self-Test

1. voltage; resistance; current
2. meter movement
3. switches
4. Ayrton
5. 10,000
6. 8800
7. 1,665,000
8. 250 kΩ

37 EXPERIMENT

OSCILLOSCOPE OPERATION—TRIGGERED SCOPE

OBJECTIVES

1. To identify the operating controls of a triggered oscilloscope
2. To set up the oscilloscope and adjust the controls properly to observe an ac voltage waveform

INTRODUCTORY INFORMATION

The cathode-ray oscilloscope (CRO) or "scope," as it is familiarly known, is the most versatile instrument in electronics. The technician must therefore be able to operate this instrument and understand how and where it is used.

For purposes of this book, oscilloscopes will be classified as triggered or nontriggered. Triggered oscilloscopes are the more sophisticated of the two, can do more, and generally are used in industrial laboratories and plants, in engineering and technical school laboratories, and in any application requiring the study of low- and high-frequency waveforms, precise measurement of time, and timing relationships. At this point this very brief statement of the applications of triggered oscilloscopes must suffice. As you progress in the study of electronics, additional uses of this laboratory instrument will become evident.

The nontriggered scope is historically important because it was the first oscilloscope developed. It was used to view ac waveforms at test points in a circuit where knowledge of the existence or nonexistence of a waveform was usually all that was required. Precise time measurements could not be made with this elementary instrument. Nontriggered scopes were popular with TV service technicians. However, they have almost all been replaced with triggered oscilloscopes.

What an Oscilloscope Does

An oscilloscope automatically graphs a time-varying voltage; that is, it displays the instantaneous amplitude of an ac voltage waveform versus time. Most triggered scopes also measure dc voltages. The indicator in an oscilloscope is a cathode-ray tube (CRT). Inside the cathode-ray tube are an electron gun assembly, vertical and horizontal deflection plates, and a fluorescent screen.

The electron gun emits a high-velocity low-inertia beam of electrons, which strikes the fluorescent screen and causes the screen to emit light. The intensity of the light given off by the screen is determined by the voltage relationships between the elements in the electron gun assembly. The manual control of brightness is effected by a control located on the oscilloscope panel.

The motion of the beam over the CRT screen is controlled by a deflection system which includes deflection voltages generated in electronic circuits outside of the CRT and the deflection plates inside the CRT to which the deflection voltages are applied.

Figure 37-1 is an elementary block diagram of an oscilloscope. The CRT serves as the indicator on which electrical waveforms are viewed. These "signal" waveforms are applied to the vertical input on the oscilloscope and are processed by "vertical" amplifiers in circuitry external to the CRT. Since the oscilloscope must handle a wide range of signal-voltage amplitudes, a vertical attenuator, a variable voltage divider, acts to set up the proper signal level for viewing. The signal voltage applied to the *vertical* deflection plates causes the electron beam of the CRT to be deflected vertically. The resulting up-and-down trace is significant in that *the extent of vertical deflection is directly proportional to the amplitude of signal voltage applied to the V input.*

To make it possible for the oscilloscope to graph a time-varying voltage, a linearly changing (time-base) deflection voltage is applied to the horizontal deflection plates. This voltage is developed, in electronic circuits external to the CRT, by a time-base or sweep generator. It is this sweep generator which is either triggered or nontriggered.

Dual-Trace Oscilloscopes

Triggered oscilloscopes with *two* traces are in common use. By means of an electronc switching arrangement two traces are developed on the screen of the scope. Dual-trace oscilloscopes make it possible to observe simultaneously *two time-related waveforms* at different points in an electronic circuit. Familiarity with the operation of dual-trace oscilloscopes will be helpful to the student in the study of electricity and electronics.

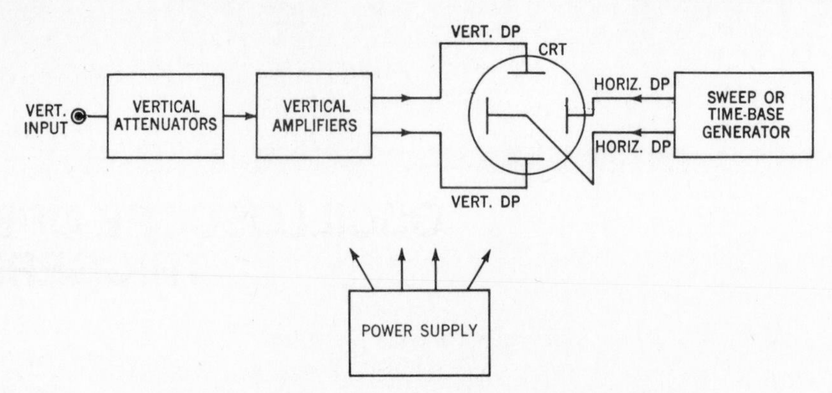

Fig. 37-1. Elementary block diagram of an oscilloscope.

Triggered Oscilloscope—Elementary Considerations

Some triggered scopes utilize a single-frame construction; that is, the electronic circuits external to the CRT are mounted on a chassis. This chassis, together with the CRT mount and the front panel, constitute a single assembly. The sweep generator, the vertical amplifier, and all the other electronic circuitry which make up the oscilloscope are self-contained in this single unit. Other triggered oscilloscopes utilize a multiframe construction. These have separate assemblies for the sweep generator and the vertical amplifier which plug in to the main frame. The main frame holds the CRT, its associated circuits, and the remaining oscilloscope circuits.

Manual Operating Controls

Intensity. This control sets the level of brightness or intensity of the light trace on the CRT. Rotation in a clockwise (CW) direction increases the brightness. Intensity should not be set too high to prevent damage to the CRT screen.

Focus. This control is adjusted in conjunction with the intensity control to give the sharpest trace on the screen. There is interaction between these two controls, so adjustment of one may require readjustment of the other.

Astigmatism. This is another beam-focusing control found on some oscilloscopes which operates in conjunction with the focus control for the sharpest trace. The astigmatism control is sometimes a screwdriver adjustment rather than a manual control.

Horizontal and Vertical Positioning or Centering. These are trace-positioning controls. They are adjusted so that the trace is positioned or centered both vertically and horizontally on the screen. In front of the CRT screen is an etched faceplate called the *graticule*. The etchings take the form of horizontal and vertical graph lines. Calibration markings are usually placed on the center vertical and horizontal lines on this faceplate.

Volts/Div. (also called Volts/cm). There are two concentric controls which act as attenuators of the vertical input signal waveform (which is to be viewed on the screen). The center control marked *Variable* is continuously variable for setting the height (vertical amplitude) of the signal on the screen. Its completely clockwise position is *calibrated* for making peak-to-peak voltage measurements of the vertical input signal. Volts/Div. is the outer of the two concentric vertical attenuators. It is a switched control. A dot on the control can be thrown to the calibrated voltage markings on the panel around the control. Thus when the variable control is set to its calibrated position, the setting of the Volts/Div. control determines the voltage which is equivalent to every division of vertical signal deflection on the screen.

Time/Div. (also called Time/cm). There are two concentric controls which affect the timing of the sweep or time-base generator. The inner control is marked *Variable* and is continuously adjustable over each range of the Time/Div. control. The complete clockwise (CW) position of the variable control is calibrated for making time measurements of waveforms displayed on the screen. Time/Div. is the outer of the two concentric time-base generator controls. It is a switched control. A dot on the control can be thrown to the calibrated time markings on the panel around the control. When the variable control is in its calibrated position, the settings of the Time/Div. control determine the time it takes the trace to move horizontally across one division of the graticule.

Triggering Controls. A simple calibrated time base usually has four triggering controls associated with it. Thus one oscilloscope has a:

1. *Level control.* There is a switch position, associated with this control labeled *Auto*(-matic). In this position the trigger circuit is free-running and on each cycle triggers the sweep generator. Hence a trace always appears on the screen. The oscilloscope is frequently used in this mode of operation. When the oscilloscope is not in the automatic mode, triggering depends on some external or internal signal, and the setting of the level control determines the stability or synchronization of the sweep. In the nonautomatic mode there will be *no trace* on the screen in the absence of a triggering signal.
2. *Slope.* This switch is marked + and −, and its setting determines whether triggering of the sweep is effected by the positive or negative portion of the triggering signal.

3. *Coupling*. This selects the manner in which trigger coupling is achieved. The particular oscilloscope we are describing has three coupling modes: AC slow, AC fast, and DC.
4. *Source*. The trigger signal may be *Ext*(ernal), *Int*(ernal), or *Line*. In this experiment this switched control will be set on Int.

NOTES:

(1) The controls described may have other names, depending on the manufacturer and oscilloscope model. However, once you understand the operation of a triggered oscilloscope, it will be relatively easy to operate other triggered oscilloscopes.

(2) A triggered oscilloscope usually has facilities for switching off the internal sweep generator. A horizontal input jack will then receive some external sweep voltage, apply it to the horizontal processing circuits, and thus cause a horizontal trace. Of course the calibrated Time/Div. controls of the oscilloscope do not operate for this external sweep, and the time base is uncalibrated.

(3) There is a vertical signal input jack on the panel which receives the input signal, via a shielded coaxial cable terminated in a probe. There are various types of oscilloscope probes, for example, direct probes and low-capacitance probes. We will be using a *direct* probe in this experiment.

(4) Additional features and controls are found on many oscilloscopes. However, knowledge of their operation is not needed at this point.

SUMMARY

1. A triggered oscilloscope can be used to measure dc, as well as low- and high-frequency ac, waveforms and time.
2. A nontriggered scope is normally used to observe low-frequency waveforms but cannot measure time directly.
3. An oscilloscope displays a graph of the amplitude of an ac waveform versus time.
4. A cathode-ray tube is the indicator or *screen* of an oscilloscope.
5. The purpose of the electron gun in a CRT is to emit an electron beam which strikes the screen and causes it to give off light.
6. The signal voltages applied to the vertical deflection plates of a CRT cause the beam to be deflected up and down.
7. The horizontal deflection plates receive the linearly changing deflection voltage which generates the time base.
8. The intensity control is used to set the brightness of the trace.
9. The focus control is used to narrow the beam into the sharpest trace. There may be an auxiliary astigmatism control for focusing.
10. The horizontal and vertical centering or positioning controls are used to position the trace on the CRT screen.
11. The etched faceplate in front of the CRT face which appears as vertical and horizontal graph lines is called the graticule. Linear calibration markers (height and width) are frequently etched on the graticule.
12. The Volts/Div. or Volts/cm control is calibrated for the measurement of the amplitude of signal waveforms along the vertical axis.
13. The Time/Div. or Time/cm control is calibrated for the measurement of time along the horizontal axis.
14. The triggering controls determine the manner in which a trigger pulse is initiated to start the sweep generator.
15. The trigger can be run automatically, the mode which is frequently used.
16. The trigger circuit can be actuated on Int. by the signal waveform from within the oscilloscope circuits.
17. The trigger circuit can also be actuated on Ext. by an external signal voltage applied to an input jack labelled Ext. trigger.
18. Also, the trigger circuit can be actuated on Line position of the trigger switch by a power line-derived voltage from within the oscilloscope circuits.

SELF-TEST

Check your understanding by answering these questions:

1. Oscilloscopes are used for the observation of ac waveforms. ___________ (true/false)
2. Precise time measurements can be made with both triggered and nontriggered scopes. ___________ (true/false)
3. The waveform seen on the screen of a CRT is a ___________ of amplitude versus ___________.
4. The ___________ deflection plates of a CRT are the signal plates; the ___________ deflection plates are for the time-base voltage.
5. Of the controls on an oscilloscope, those that affect the height of the signal are called ___________.
6. Those controls that affect the sharpness of the trace are called ___________ and ___________.
7. The etched faceplate in front of the face of the CRT is called the ___________.
8. A frequently used triggering mode of a triggered oscilloscope is ___________.
9. The controls that affect the up-and-down movement of the trace are called ___________ ___________.
10. The height of the waveform displayed on the oscilloscope screen is directly proportional to the ___________ of the waveform.

MATERIALS REQUIRED

- Power supply: Source of 120 V ac
- Equipment: Triggered-type oscilloscope with calibrated time base, with calibrated vertical amplifier, with internal voltage calibrator, and with direct probe

PROCEDURE

NOTE. Before attempting the experiment, the student should read and become thoroughly familiar with the operating instructions of the oscilloscope.

CAUTION. Do not operate the scope with trace intensity too high.

Operating the Controls Which Affect the Trace

1. List each manual control and switch on your oscilloscope and state its function in Table 37-1. Include also the input jacks.
2. Turn the oscilloscope **on**. The scope normally contains a protective time-delay relay. Wait until you hear the relay click in. This will occur after the oscilloscope is warmed up and ready for operation. Set the Time/Div. control to 1 ms.
3. If a trace does not appear on the screen, check to see that the triggering switch is on Auto(-matic). If it is not, set it on Auto.
4. If there is still no trace, turn the Intensity control completely clockwise.
5. If there is still no trace, try the Positioning (centering) controls until a trace does appear.
6. Adjust the focus, astigmatism, and brightness controls for a clear, sharp trace. Center the beam both vertically and horizontally. Set the triggering *Slope* to +, triggering *Coupling* to AC or AC Fast, triggering *Source* to Int. The oscilloscope is now ready for viewing ac waveforms.
7. Throw off the controls you adjusted in steps 2 through 6 and repeat the entire procedure for setting up a trace. When you are satisfied that you can operate the controls properly, notify your instructor.

Viewing a Waveform

8. Connect the vertical input leads of the oscilloscope to the output of the voltage calibrator on the scope. Set the calibrator to 2 V output (approx.). The calibration waveform will be seen when the vertical attenuators (Volts/Div.) and Time/Div. controls are properly set.
9. Set the variable vertical attenuator on *Calibrated* and vary the Volts/Div. control for about three divisions of signal height.
10. Now set the Variable Time/Div. control to Calibrated and vary the Time/Div. switch until three cycles (approx.) of the calibration waveform appear on the screen.

TABLE 37-1. Manual Controls and Switches and Their Functions

Control or Switch	*Function*

11. Now, leaving the Height/Div. controls as set, change the setting of the calibrator output to ½ V (approx.) if possible. What happens to the height of the waveform on the screen?
12. Reset the calibrator output to 5 V, if possible. What happens to the height of the waveform?
13. Reset the calibrator output to 2 V, as in step 8. The waveform should now have a height of three divisions as in step 9.
14. *Increase* the sweep speed (Time/Div.) one setting on the Time/Div. switch (for example, if the switch was set to 0.5 ms/div., set it to the next faster time calibration, say 0.2 ms/div.). What happens to the number of cycles on the screen?
15. *Decrease* the sweep speed one setting past its position in step 10. What happens to the number of cycles on the screen?
16. When you are satisfied that you understand the operation of all the manual controls on the scope, have your instructor throw them off. Readjust the operating controls until you can see the calibrator waveform, properly centered, on the screen, as in step 10.
17. Repeat step 16 until you are completely satisfied that you can operate the oscilloscope controls properly.

QUESTIONS

1. After the Volts/Div. controls were set to view the calibrator waveform, what happened to the height of the waveform when the calibrator voltage output was increased? Why?
2. After the Time/Div. controls were set to view three cycles of the calibration waveform, what happened to the number of cycles seen when the sweep speed was increased? decreased? Why?
3. Refer to Procedure steps 14 and 15. Was the frequency of the calibration waveform affected when the sweep speed was increased or decreased? What was changed?
4. What is the relationship, if any, between the number of cycles of waveform and the setting of the Time/Div. control?
5. List the controls on your scope which affect the (*a*) height of the waveform; (*b*) brightness of the trace; (*c*) sharpness of the trace; (*d*) position of the trace; (*e*) triggering of the sweep generator.

Answers to Self-Test

1. true
2. false
3. graph; time
4. vertical; horizontal
5. Volts/Div. or Volts/cm
6. focus; astigmatism
7. graticule
8. automatic
9. vertical positioning or centering
10. amplitude (voltage)

38

EXPERIMENT

OPERATION OF AN AUDIO-FREQUENCY GENERATOR

OBJECTIVES

1. To identify the operating controls of an audio-frequency (AF) generator
2. To observe on an oscilloscope the signal delivered by an AF generator

INTRODUCTORY INFORMATION

Audio Oscillator (AF Signal Generator)

A dc power supply provides dc voltages for dc circuits. Similarly, the "signal generator" supplies ac voltages for ac circuits. These generators develop signals whose frequency is variable over a specified range.

Alternating-current (ac) frequencies cover a very wide spectrum, from a fraction of a cycle per second to thousands of millions of cycles per second. No one instrument has been designed that can cover this extensive range. Many generators are available commercially to provide the various frequency needs in electronics.

One of the characteristics by which signal generators are identified is the frequency range which the generator covers. The audio-frequency (AF) generator supplies frequencies from several hertz up to 20,000 Hz approximately. It is called "audio" (which means sound) because this is the range of frequencies (approximately) to which the ear responds. This is not to say that the ear can hear the direct electrical signal delivered by the generator. But the electrical signal can be converted to a sound signal (a vibration in the air) by a suitable device such as a loudspeaker.

An AF generator often covers a range much wider than the audio frequencies. Thus, one "audio" generator supplies a signal whose frequency can be varied from several hertz up to 600 kHz. However, this coverage is achieved in several ranges.

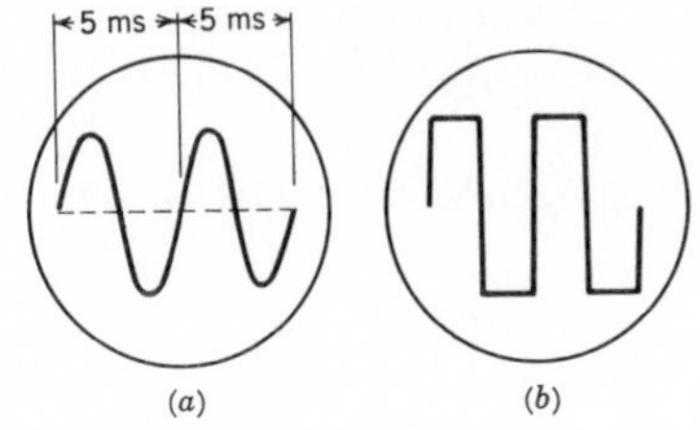

Fig. 38-1. AF generator develops (*a*) sine wave; (*b*) square wave.

Another characteristic which identifies a signal generator is the shape of the waveform it develops. Thus, there is a sine-wave generator whose output is sinusoidal (Fig. 38-1*a*), a square-wave generator whose output is the square waveform (Fig. 38-1*b*), and other waveform generators with which you need not be concerned at the moment.

The usual controls and switches found on the front panel of a signal generator are:

1. On-off switch for applying power to the generator.
2. *Range*. A coarse control to select a specified frequency coverage (range); for example, 10 to 1000 Hz or 1000 to 100,000 Hz, etc.
3. *Frequency*. This is a continuously variable control to select a specified frequency within any range. Associated with this control is a calibrated frequency scale.
4. *Level (output) control*. This is used to set the voltage of the output signal. There may be two types of level controls on the panel, a coarse control and a continuously variable output control. Both types use voltage-divider networks. The coarse control is usually a decade attenuator which reduces the signal level in multiples of 10.

The output of a general-purpose sine-wave generator is not metered. Its level must be measured by an external ac voltmeter or by an oscilloscope. Some laboratory-standard signal generators include a metering circuit which measures the signal delivered at the output terminals. General-purpose AF generators can deliver a signal from a fraction of a volt to many volts. Thus, the approximate output range of one instrument is from a millivolt to 20 V.

A shielded cable is used to deliver the generator signal to the external circuit. This cable is detachable from the generator.

Characteristics of an AC Signal Voltage

Alternating-current voltages are identified by certain characteristics. These are:

The Waveform or Shape of the Voltage

This characteristic pertains to the manner in which the voltage varies between maximum and minimum. Thus, if the voltage is varying in a sinusoidal fashion, the scope will display the voltage as a sine wave (Fig. 38-2).

The Amplitude of the Voltage

This characteristic describes the difference between the positive and negative peaks of the voltage and is expressed in peak-to-peak volts (Fig. 38-3). The amplitude of a periodic waveform is usually measured by using an oscilloscope.

The Frequency of the Voltage

An ac periodic wave completes a number of cycles every second. The time interval for each of these cycles is called the period t. Thus the period or time in seconds of one complete cycle of an ac voltage is found from the relationship

$$t = \frac{1}{F}$$

where F is the frequency of the voltage in hertz. Thus, if the voltage of Fig. 38-3 has a frequency of 60 Hz, the period is $\frac{1}{60}$ s. Hence, the time required for the scope to "graph" one cycle of the waveform is $\frac{1}{60}$ s.

Measuring the Period of an AC Voltage with an Oscilloscope

Since the period t and frequency F of an ac signal waveform are inversely related; that is, since

$$t = \frac{1}{F}$$

and

$$F = \frac{1}{t}$$

it is possible to calculate the frequency of a waveform if its period is known. If its frequency is known, its period may be computed.

The triggered oscilloscope has facilities for direct measurement of the period of an ac signal voltage. The characteristic which makes this possible is the calibrated sweep or time base of this type of oscilloscope.

In Experiment 37 you noted that to display a signal waveform on the screen, it was necessary to set the Time/cm control properly. It is this control which determines the rate at which the cathode-ray-tube beam is moving across the screen to generate the trace. An example will show how the screen can be used to measure the period of the waveform displayed on the CRT.

Suppose the Time/cm control is set to 1 ms, the Variable/Time/cm to *calibrated*, and suppose the calibrated face

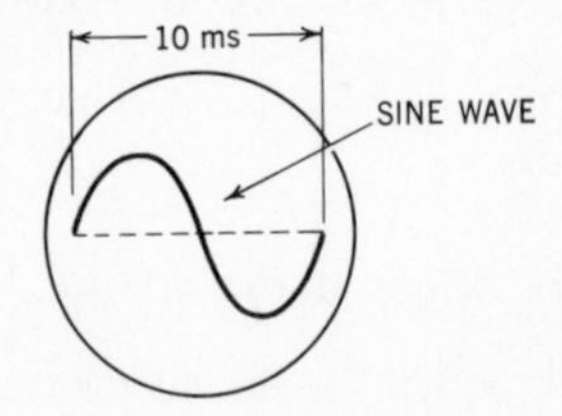

Fig. 38-2. Sine wave seen on scope screen.

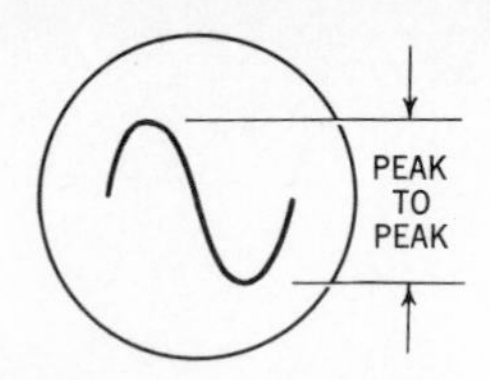

Fig. 38-3. Measuring peak-to-peak voltage.

of the oscilloscope, called the graticule, is 10 cm wide. Then it takes 10 ms for the electron beam to move across the 10-cm screen. Now, if the width of a waveform is exactly 10 cm, as in Fig. 38-2, then the period of the waveform is 10 ms. That is, $t = 10 \times 10^{-3}$ s. The frequency of the displayed waveform can now be calculated, for

$$F = \frac{1}{t} = \frac{1}{10 \times 10^{-3}} = 100 \text{ Hz}$$

If two waveforms are displayed on the 10-ms trace (Fig. 38-1), then the period of each cycle is 5 ms and the frequency is

$$F = \frac{1}{5 \times 10^{-3}} = 200 \text{ Hz}$$

SUMMARY

1. A signal generator supplies ac voltages for ac circuits.
2. An AF signal generator provides ac signals in the audio range of frequencies, that is, 10 to 20,000 Hz (approximately).
3. There are other generators which develop signal voltages over other frequency ranges.
4. An AF generator may provide a sinusoidal voltage waveform, a square wave, or some other waveshape, depending on the type and purpose of the generator.
5. The controls normally found on a general-purpose AF generator are: *on-off* (for power), *range* (coarse frequency), *frequency* (continuously variable within each range), *level* (output).
6. The output of a general-purpose sine-wave generator is normally not metered, so it must be measured by an external voltmeter or oscilloscope.
7. Laboratory standard AF generators do have metered outputs.
8. An ac voltage waveform may be identified by
   (*a*) Its waveform—is it a sine wave, square wave, triangular wave, etc.?
   (*b*) The amplitude of the wave—usually measured in peak-to-peak volts.
   (*c*) Its frequency (F) in hertz.
   (*d*) The period or time (t) in seconds of one cycle of the waveform.
9. Frequency and period are related by the formulas

$$F = \frac{1}{t}$$

$$t = \frac{1}{F}$$

10. A triggered oscilloscope has a calibrated time base which may be used for measuring the period t of a cycle of a periodic waveform. Once t has been measured, F can be calculated.

SELF-TEST

Check your understanding by answering these questions.

1. The instrument which supplies an ac test signal is called a ____________ ____________.
2. The audio frequencies are in the range ____________ Hz to ____________ Hz (approximately).
3. The general-purpose AF signal generator has a metered output. ____________ (true/false)
4. The characteristics which identify a periodic ac signal are:
   (*a*) ____________
   (*b*) ____________
   (*c*) ____________
   (*d*) ____________
5. If the period of an ac waveform is known, the frequency of that waveform may be calculated from the formula: ____________ = ____________.
6. A triggered oscilloscope has a ____________ time base, which can be used to measure the ____________ of an ac waveform.
7. An oscilloscope may also be used to measure the ____________ of an ac waveform.

MATERIALS REQUIRED

- Power supply: ac voltage source
- Equipment: AF signal generator; triggered oscilloscope with direct probe

PROCEDURE

Familiarization with AF Signal Generator

1. Familiarize yourself with the signal generator issued to you. Study the instruction manual and learn the function and use of the operating controls on the front panel.
2. **Power on.** Connect the output leads of the signal generator to the vertical input of the oscilloscope. Set the output controls of the generator in the middle of their range. Adjust the frequency control to 100 Hz.
3. Set the oscilloscope Volts/div. selector until the waveform is deflected 4 divisions vertically (approximately). Set scope on automatic triggering and sync on Int. +. Adjust the oscilloscope Time/div. control for a display of two sine waves (approximately).
4. Reduce the output level of the generator signal and observe the effect on the height of the signal displayed on the oscilloscope. Is the stability of the presentation affected as the signal is reduced below a certain level?
5. Increase the level of the generator signal to maximum. Readjust, if necessary, the Volts/div. control until the entire waveform is displayed on the screen. Is the Volts/div. factor now higher or lower than in step 3?
6. Reset the frequency of the signal generator to 200 Hz. There are now ____________ cycles on the screen.
7. Reset the frequency of the signal generator to 300 Hz. There are now ____________ cycles on the screen.
8. Reset the oscilloscope Time/div. control until one or two cycles are displayed on the screen. Measure the width of one sine wave. ____________ div.
9. Multiply the Time/div. setting for this display by the width (number of div.) of the waveform. This is the period of the sine wave. ____________ s
10. Using the formula $F = 1/t$, compute the frequency of the sine wave. (NOTE: F is in hertz when t is in seconds.) $F =$ ____________ Hz.

QUESTIONS

1. What is the range of frequencies which your AF generator can deliver?
2. Does there seem to be any relationship between the number of cycles displayed on the screen and the frequency setting of the generator, assuming there is no change in Time/div. (or sweep) and sync settings of the oscilloscope? If so, what is the relationship?
3. For the triggered oscilloscope, how did the signal frequency, computed in step 10, compare with the frequency setting of the generator? If they were not the same, explain why.
4. What is the purpose of each of the following controls on the signal generator?
   (*a*) Range
   (*b*) Frequency
   (*c*) Output or level

Answers to Self-Test

1. signal generator
2. 10; 20,000
3. false
4. (*a*) waveform; (*b*) amplitude; (*c*) frequency; (*d*) period
5. $F = 1/t$
6. calibrated; period
7. amplitude

39 EXPERIMENT

OSCILLOSCOPE VOLTAGE MEASUREMENTS

OBJECTIVES

1. To check the calibration accuracy of the oscilloscope for use as an ac voltmeter
2. To make peak-to-peak ac voltage measurements with the scope
3. To make dc voltage measurements with the scope

INTRODUCTORY INFORMATION

An oscilloscope is used to view and measure ac waveforms. It has many other functions, but in this experiment we will be concerned with its voltage-measuring capabilities.

Measuring Voltages with the Scope

The height of the voltage waveform displayed on the oscilloscope screen is directly proportional to the peak-to-peak amplitude of the voltage. Thus, for the same settings of Volts/div., or vertical-gain controls, a 100-V signal will have twice the height of a 50-V signal. A 25-V signal will have one-fourth the height of a 100-V signal. Moreover, if we suppose that for this same fixed setting of the vertical controls, every division of vertical deflection corresponds to 25 V of input, then a 25-V signal will have 1 division of height, a 50-V signal will have 2 divisions of height, a 75-V signal will have 3 divisions of height, and so on. Two and one half (2½) divisions of vertical deflection (height) will correspond to 62½ V, etc. This characteristic of the signal circuits makes it possible to use the oscilloscope for measuring ac voltages.

To measure dc, the method described above is also used. The ac-dc switch is placed in the dc position and the probe is connected to the point in the circuit where the dc voltage is. The ground lead of the scope is connected to the ground of the circuit. The number of divisions the trace rises above or falls below the zero base setting is a measure of the + or − dc voltage.

NOTE: In this experiment we shall refer to the number of centimeters (cm) or the number of divisions of height of the voltage waveform, because these two markings are found on the graticules of oscilloscopes. The triggered oscilloscope uses a graticule whose division markings are usually in centimeters (Fig. 39-1).

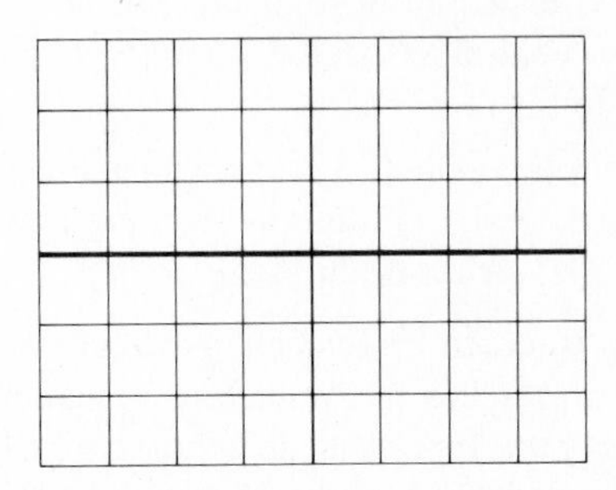

Fig. 39-1. Graticule of a triggered-type oscilloscope. Markings are in centimeters.

In measuring voltages with the oscilloscope, the procedure is to apply the signal voltage to the calibrated vertical input of the oscilloscope and to measure on the etched faceplate or graticule the number of divisions of height of the waveform. This number is then multiplied by the calibration factor of the oscilloscope, giving the peak-to-peak voltage of the waveform. For example, if the oscilloscope is set for 20 V/div., 3 div. of deflection (height) measures a 60-V signal. Similarly, if the scope is calibrated for 10 V/div., 6 div. of height also measures a 60-V signal (Fig. 39-2).

It should be noted that there are no voltage scales on the graticule. There are simply linear markings whose value is determined by the calibration factor to which the Volts/div. vertical-gain controls of the oscilloscope are calibrated. Thus the Volts/div. or vertical-gain controls correspond to the range controls of a voltmeter; the graticule markings correspond to the scale of the voltmeter.

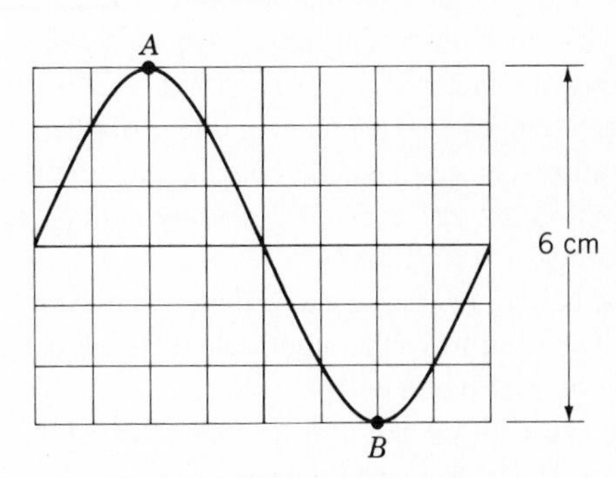

Fig. 39-2. The maximum height of the sine wave (between points *A* and *B*) is 6 cm. If the scope is set at 10 V/cm, then the sine-wave signal voltage is 60 V peak-to-peak.

Checking Calibration Accuracy

Triggered Oscilloscope. The triggered oscilloscope has a calibrated vertical amplifier. There are two concentric input controls which affect, by attenuation, the height of the waveform displayed on the screen. The inner control is continuously variable. The switch-type Volts/div. control changes the signal attentuation by a precalibrated factor. *The inner control must be turned to its "calibrated" position, which is completely clockwise. Then, the marking to which the switch-type Volts/div. control is set is the calibration factor of the signal-measuring vertical amplifier of the oscilloscope.* Thus, with the inner control set to its calibrated position, and the Volts/div. control set to 0.5 V, every 1 div. of height corresponds to 0.5 V.

NOTE: These rules apply to a scope with a direct probe. If a low-capacitance probe is used, the reading must be multiplied by the attenuation factor of the probe.

The Volts/div. control is frequently a decade-type attenuator; that is, it switches the voltage-range calibration by a factor of 10. For this type of attenuator the calibration marker settings might read: 0.01 V, 0.1 V, 1.0 V, 10 V, and so on.

Many triggered oscilloscopes contain a warning light which goes on when the inner control is not set to its calibrated position; that is, when the scope is uncalibrated. Measurements cannot be accurately made when this light is on. Before measurement, the inner control must be turned completely clockwise to its calibrated position. The warning light will then go out and the oscilloscope is again calibrated.

Triggered oscilloscopes have a self-contained voltage calibrator for checking the calibration of the Volts/cm selector. The output of this calibrator may be a single voltage, as for example 1 V peak-to-peak, or the scope may have a range of calibration voltages selected by a switch. For this switch-type calibrator, the measured outputs usually correspond to the positions of the Volts/cm control. Thus, for the Volts/cm vertical-gain selector described above, the calibrator would deliver 0.01 V, 0.1 V, 1.0 V, and so on, depending on the switch setting of the calibrator.

If the scope calibrator delivers just a single voltage, say 1 V, the calibration of the vertical amplifiers may be checked on several, but not all, settings. Thus, the 1 V/cm vertical selector setting of the scope may be checked, as may the 0.5 V/cm setting. But a setting of, say, 20 V/cm cannot be checked with any degree of accuracy. We must either assume that the other attenuator settings are correct, that is, that there has been no change in the attenuator voltage divider, or use an external calibrator, with variable outputs. Accurate, external voltage calibrators are available, although they are rarely used for triggered oscilloscopes.

External calibrators provide line-derived adjustable signal voltages. The peak-to-peak output-voltage level may be read on a self-contained meter or from calibrated dials. The output of the voltage calibrator is fed to the vertical input of the oscilloscope to check its calibration.

If an external voltage calibrator is not available, a sine-wave generator and an accurately calibrated peak-to-peak-reading ac voltmeter can be used. The generator frequency control is set at 1000 Hz, and the output control is set for a specified voltage as read by the ac voltmeter used here as a standard. The measured signal acts as the calibration source and is fed to the vertical input leads of the scope.

To check the calibration accuracy of the Volts/cm selector, the technician would apply the oscilloscope probe to the calibrator output jack and note the height of the waveform displayed. Thus, if the calibrator were set for 1.0 V output, the calibrator signal should be deflected 1 cm in height when the Volts/cm selector is set to 1 V/cm.

SUMMARY

1. An oscilloscope can be used for ac voltage measurements.
2. Triggered oscilloscopes contain calibrated signal amplifiers. These are used directly in the measurement of ac voltage waveforms and dc voltages.
3. Before a triggered oscilloscope is used to measure voltage, its calibration factors, Volts/div. setting, should be checked by applying the measured calibration voltages available on the oscilloscope to the input terminals of the scope. If they check, it is ready to make measurements.
4. An accurate, known voltage source, whether provided by the oscilloscope calibrator or by an external calibrator, may be used to check the accuracy of calibration of the vertical amplifiers.
5. Oscilloscope signal amplifiers are normally calibrated to make peak-to-peak measurements.
6. There are instruments called voltage calibrators which may be used as external calibration sources. They deliver a measured signal voltage for calibration purposes.

SELF-TEST

Check your understanding by answering these questions:

1. AC voltages __________ (can/cannot) be measured with an oscilloscope.
2. For peak-to-peak waveform measurements, the oscilloscope __________ amplifiers must be __________.
3. Triggered-type oscilloscopes __________ (do/do not) have calibrated vertical amplifiers.
4. The Volts/cm selector setting on a triggered scope determines the __________ factor by which the signal __________ of the waveform must be multiplied to get the voltage of the waveform.
5. The Volts/cm vertical-gain selector is calibrated and set at 2 V/cm. At this setting the height of a waveform is 2.5 cm. The voltage of this waveform is __________ V.
6. If the same signal voltage as in question 5 is applied to the calibrated vertical input of the oscilloscope but the vertical-gain selector is set at 1 V/cm, the height of the waveform will be __________ cm.
7. Calibration voltage(s) is(are) available on triggered oscilloscopes. __________ (true/false)

8. If an oscilloscope does not have self-contained facilities for checking voltage calibration, an ____________ ____________ may be used.

MATERIALS REQUIRED

- Power supply: Known ac voltage source of 18 V peak-to-peak; fused line cord
- Equipment: Triggered oscilloscope with direct probe; AF signal generator; EVM or VOM

  NOTE: In this and future experiments when the term oscilloscope is used, a *triggered* oscilloscope will be required.

- Resistors: ½-W 5100-, 10,000-, and 15,000-Ω

PROCEDURE

Checking Accuracy of Calibrated Vertical Amplifiers

1. Locate the calibration voltage jack on your oscilloscope. Set the vertical amplifier-gain control on Calibrated.
2. Connect the direct probe of the oscilloscope to the calibrator voltage jack and check the accuracy of the Volts/cm selector on every range you can possibly check. Record your results in Table 39-1. Identify the Volts/cm range, the calibration voltage used to check that range, and the height of the calibration waveform on the screen of your oscilloscope. If the scope is not properly calibrated, notify your instructor. If calibration is OK, proceed to the next step.

Measuring AC Voltages with Oscilloscope

3. Remove the vertical input cable from the calibration source and apply 18 V p-p (as measured with an ac voltmeter) across the circuit of Fig. 39-3. With the oscilloscope, measure and record in Table 39-2 the peak-to-peak voltages form *A* to *G*, *B* to *G*, and *C* to *G*. Compute and record in Table 39-2 the peak-to-peak voltages which should appear at these points. Show your computations.
4. Measure and record in Table 39-2 the maximum signal voltage delivered by the AF signal generator set at 1000 Hz. Also measure and record the minimum measurable output of the AF generator at 1000 Hz.

TABLE 39-1. Checking Calibration Accuracy

Volts/cm Setting	*Calibration Voltage, V p-p*	*Height of Calibration Voltage on Screen*

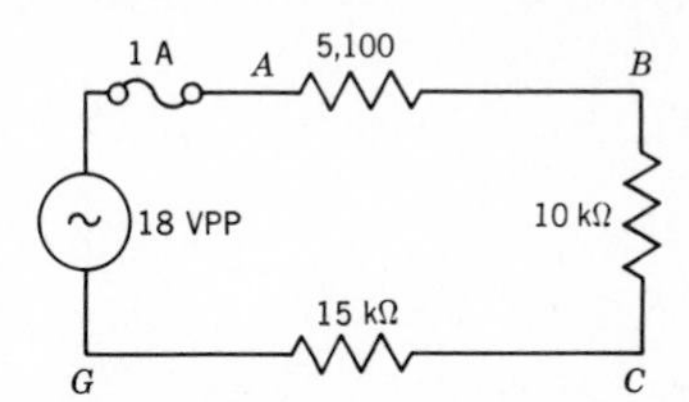

Fig. 39-3. AC voltage across divider network.

Measuring DC Voltages with Oscilloscope

5. If you are using an ac-dc oscilloscope, switch to dc. Remove the 18 V ac source from Fig. 39-3 and connect a dc voltage source across points *AG* of Fig. 39-3. Set the output of the dc supply at 30 V, as measured with an EVM. You now have a dc voltage divider.
6. Center the trace vertically on the screen. This is the zero reference line. With the oscilloscope measure and record in Table 39-3 the dc voltage across each set of test points listed. Now measure the same voltages with an EVM, and record your measurements in the appropriate column of Table 39-3. Compute and record the dc voltage across each set of points.

TABLE 39-2. AC Voltage Measurement

Test Points	*Computed, V p-p*	*Measured, V p-p*
A to *G*		
B to *G*		
C to *G*		
AF Generator Output	Maximum, V p-p	Minimum, V p-p

TABLE 39-3. DC Voltage Measurement

Test Points	*Voltage, V*		
	Oscilloscope	*Voltmeter*	*Computed*
A to *G*			
B to *G*			
C to *G*			

QUESTIONS

1. (*a*) Does your oscilloscope have calibrated or noncalibrated vertical amplifiers?
   (*b*) Is it a triggered scope?
   (*c*) Does it have dc input?
2. What are the vertical-gain calibrations on your scope? Which is the most sensitive, that is, which requires the least voltage for one division of vertical deflection?
3. How do the computed and measured values in Table 39-2 compare? Refer specifically to your data.
4. List the advantages, if any, in using the scope as an ac voltmeter.
5. What are the disadvantages of using the scope as an ac voltmeter?
6. Which provided more accurate dc voltage measurements, the scope or the voltmeter? Justify your answer by referring to your readings in Table 39-3.

Answers to Self-Test

1. can
2. vertical or signal; calibrated
3. do
4. calibration; height
5. 5
6. 5
7. true
8. external calibrator

40 EXPERIMENT

PEAK, RMS, AND AVERAGE VALUES OF AC

OBJECTIVES

1. To learn the relationships between peak, rms, and average values of alternating voltage and current
2. To confirm the relationship between peak and rms values of an ac voltage

INTRODUCTORY INFORMATION

Generating an AC Voltage

When a conductor cuts lines of magnetic force, a voltage is induced in the conductor. This is the principle of the electric generator. We will apply this principle to the generation of an ac voltage.

Consider a single loop of wire so arranged on a shaft that it can be rotated through the magnetic field which exists between the N and S poles of a magnet, Fig. 40-1. The loop of wire is part of the armature of an ac generator. Any mechanical means, such as a hand crank, a waterfall, or a gasoline engine, may be used to turn the armature. When the armature is rotated, the long sides of the loop move past the magnetic poles, and a voltage is induced in each side of the loop as it cuts the lines of force between the poles. The direction (polarity) of the induced voltage is determined by the direction of the magnetic field and the direction of motion of the conductor.

Assume a CCW rotation of the loop. Figure 40-1 shows that in position 1-3 of the loop, side 3 is cutting the field *up*, while side 1 is cutting the lines of force in a *downward* direction. The polarity of voltage induced in each is therefore opposite. But the total effect is *additive*, and the generated voltage can be measured at the ends of the loop.

An ac generator consists of many such loops. They constitute the armature winding. The loop ends terminate in two slip rings (Fig. 40-2). As the loops rotate, the slip rings make contact with carbon brushes, which act as the output terminals of the induced voltage.

The amplitude of voltage induced in the armature coil is proportional to the rate at which the lines of force are cut by each loop. In position 1-3 (Fig. 40-1) the loop cuts the lines of force perpendicularly, thus cutting the largest number of lines of force. The *maximum* voltage is therefore induced in this position. As the coil rotates toward position 2-4, fewer lines of force are cut. Finally, in position 2-4, the loop is moving parallel to the lines of force; *no* lines of force are cut and the induced voltage is zero. When the coil rotates past the zero-voltage position, the number of lines of force cut increases from zero to maximum (90° later) and back to zero. Note that past the zero point the direction of cutting of lines of force

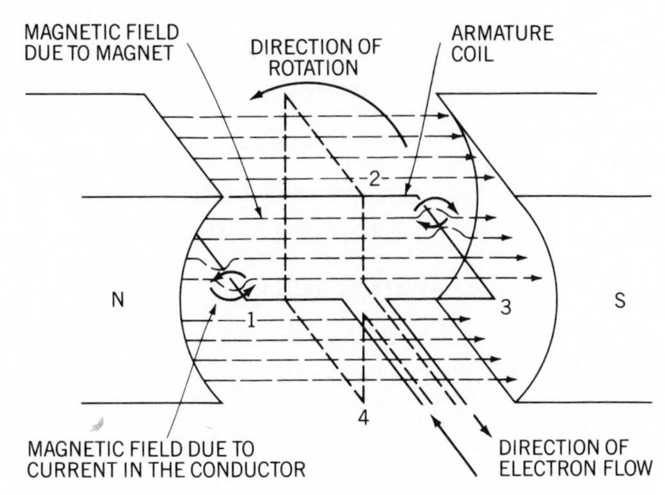

Fig. 40-1. Simple electric generator (*reprinted from Army Technical Manual TM 11-466*).

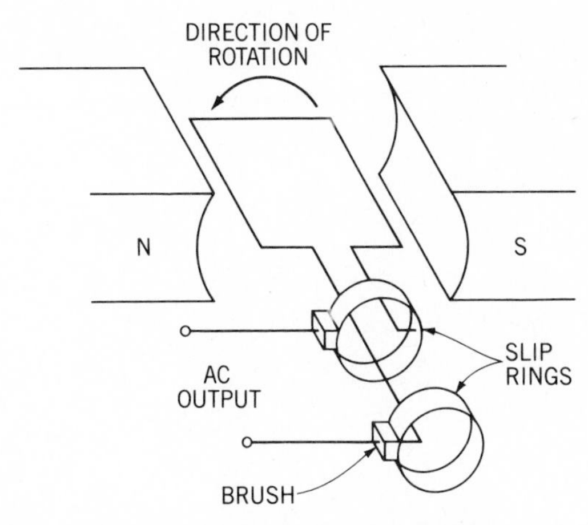

Fig. 40-2. Slip rings are used to take power from an ac generator (*reprinted from Army Technical Manual TM 11-466*).

changes, and the polarity of voltage induced in the loop changes. In a complete rotation, the induced voltage goes from zero to maximum positive (at 90°) back to zero (at 180°), to maximum negative (at 270°), down to zero again (at 360°).

A complete rotation of the loop, 360°, is called a *cycle*. The positive or negative half of each cycle is called an *alternation*.

Now consider radius OA, Fig. 40-3. As it rotates CCW around the center through 360°, it completes a circle. If the motion of OA is stopped at some point, say A', and a perpendicular is dropped from A' to line OA, it will intersect line OA at B'. As OA moves past point A', the height of $A'B'$ increases continuously to OA''' (at 90°). Past OA''' the height of $A'B'$ decreases continuously until it becomes zero again (after 180° of rotation). This occurs when OA reaches and coincides with OB. Now as OA continues to move CCW past OB to OA''''(at 270°), the height of $A'B'$increases from 0 to maximum. Finally, as the rotating radius moves CCW from OA'''' back to point A (360°), the height of $A'B'$ decreases from maximum back to 0.

Consider the height $A'B'$, positive above the diameter AB and negative below AB. If we now draw a graph of $A'B'$ over the 360° of rotation of OA, we obtain the waveform in Fig. 40-4. In this graph, the horizontal axis shows the degrees of rotation of the radius OA, and the vertical axis shows the height of $A'B'$. From Fig. 40-3 it is clear that $A'B'/O'A' = \sin\theta$, where θ is the angle of rotation from the position OA to OA'. If we consider OA as a radius of unit length (1), then it is evident that $A'B' = \sin\theta$. So the graph of $A'B'$ versus θ is a sine wave.

If Fig. 40-1 is rotated 90° CW, and loop 2-4 is the starting position of a uniform CCW rotation, the voltage induced in the loop will vary in exactly the same manner as does $A'B'$ in Fig. 40-4. That is, the voltage induced in the loop will be a sine wave. Thus, the voltage developed in the output of our ac generator of Fig. 40-1 is sinusoidal, in which a 360° rotation of the loop (in the armature) generates *one* cycle of the sine wave. If we call the peak of the instantaneous value of this

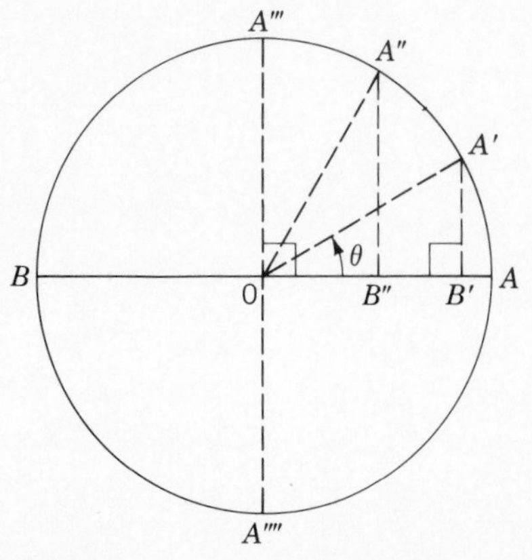

Fig. 40-3. Height $A'B'$ of the rotating radius OA from diameter AB varies as a function of the angle of rotation θ.

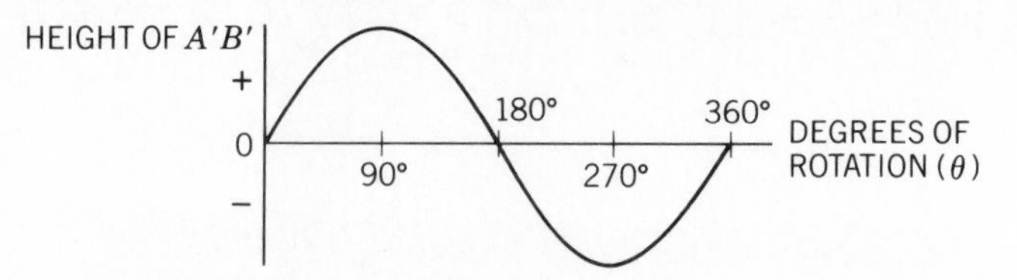

Fig. 40-4. $A'B'$ varies sinusoidally with the angle of rotation θ (theta) of OA. If $OA = 1$, then $A'B' = \sin\theta$.

sine wave V_M, then the voltage V at any instant of time, when the armature has rotated θ degrees, is

$$V = V_M \sin\theta \tag{40-1}$$

AC Voltage and Alternating Current

The generator in Fig. 40-1 produces an ac voltage. When an ac voltage is impressed across a resistive circuit, *alternating current* flows in the circuit. Since a cycle of ac voltage consists of one *positive* and one *negative* alternation, current in the circuit will flow in one direction during the positive alternation, and in the opposite direction during the negative alternation. Ohm's and Kirchhoff's laws apply to a resistive ac circuit in the same way that they apply to dc circuits. However, in a dc circuit the applied voltage has a *single* value. In ac circuits the voltage is constantly changing in amplitude, and periodically its polarity.

The effect of an ac voltage on current in a resistive circuit is shown in Fig. 40-5. As the voltage $V_M \sin\theta$ increases from 0 to V_M, the current increases from 0 to I_M; that is, a maximum current occurs at the same time as a maximum voltage is reached. Now as the voltage drops from V_M back to 0, the current drops from I_M to 0. During the second (negative) alternation of V, the current through R reverses direction. Now as the voltage reaches its maximum negative value ($-V_M$), the current also reaches its maximum value ($-I_M$), then decreases to 0 as the voltage decreases to 0. Thus Fig. 40-5 shows that current variations follow exactly the voltage variation in a resistive ac circuit. It is also evident that the shape of the current waveform is also a sine wave. The instantaneous values of current in an alternating-current waveform can be given by the formula

$$I = I_M \sin\theta \tag{40-2}$$

We can also say that in a resistive ac circuit, current and voltage are *in phase*. There are ac circuits which contain capacitors and/or inductors in which current and voltage are not in phase. These will be considered in later experiments.

Peak, RMS, and Average Values

The amplitude of a dc voltage can be identified by a single value. Can we specify an ac voltage by a single value? The answer is yes. There are actually three different values which can be used. There is the *peak*, the *rms*, and the *average* value. Each identifies a different characteristic of the voltage, but they are all related. Given *one* value, the other two can be easily computed.

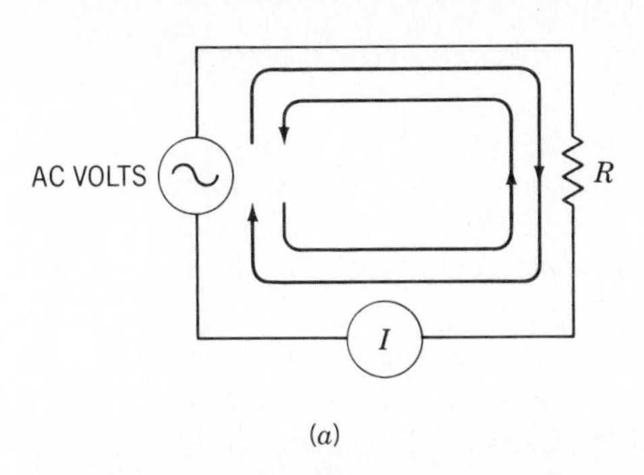

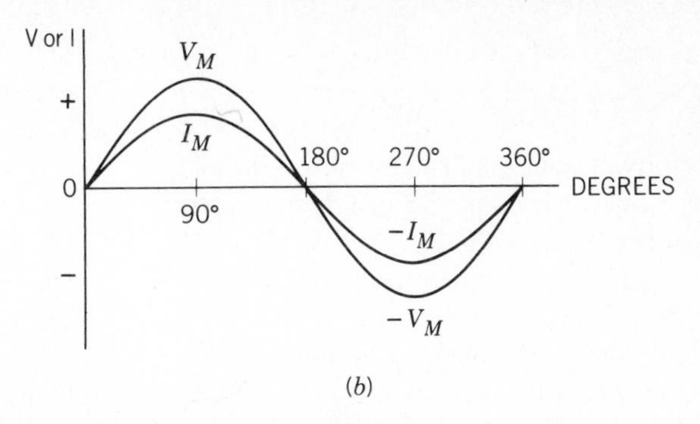

Fig. 40-5. (*a*) A sinusoidal voltage impressed on a resistive circuit; (*b*) graph of current and voltage in the resistive circuit (*a*).

An ac voltage can be described by its *peak* value, V_M. Thus, if we say that the peak value of a sinusoidal voltage is 100 V, we mean that the voltage reaches a maximum of +100 V on the positive alternation and −100 V on the negative alternation. From the peak value we can compute the instantaneous level of voltage at any angle θ in the cycle using Formula (40-1).

Does an ac voltage whose peak value is 100 V produce the same power as 100 V dc? It does *not*, because the ac voltage varies constantly in amplitude, whereas the dc voltage maintains a constant level. But there is a value of ac voltage which will produce the same power as the equivalent dc level. This is the rms, or root mean square, value. That is, if we say that the rms value of an ac voltage is 100 V, we mean that it will deliver the same power as 100 V dc. What is the rms value of an ac voltage? It is the (square) *root* of the *mean* (average) of the sum of the *squares* of the instantaneous values of voltage in an ac alternation. That is,

$$V_{rms} = \sqrt{\frac{V_1^2 + V_2^2 + \cdots + V_n^2}{n}}$$

where V_1, V_2, etc., are succeeding instantaneous values of $V \sin \theta$. It can be shown that

$$V_{rms} = 0.707V_M \qquad (40\text{-}3)$$

and that

$$V_M = 1.414V_{rms} \qquad (40\text{-}4)$$

And so we have two unique, but related, values of V which describe the amplitude of an ac voltage. RMS values are more popularly used than peak values in giving the amplitude of an ac voltage. That is why ac voltmeters are calibrated in rms rather than peak values.

Electric power is a square function of voltage (V^2/R) or current ($I^2 \times R$). That is why the rms value, which is derived from a square value (V_1^2, etc.), is comparable to the equivalent value of dc voltage.

There is still another method of identifying an ac voltage, and that is by using the *average* value of the waveform. V_{ave} is also dependent on the peak value and the rms value, and may be determined from these values by the formulas

$$V_{ave} = 0.636V_M = 0.899V_{rms} \qquad (40\text{-}5)$$

V_M can also be specified in terms of V_{rms} and V_{ave}. Thus,

$$V_M = 1.414\ V_{rms} = 1.572\ V_{ave} \qquad (40\text{-}6)$$

It should be noted that alternating current is also identified in any one of three ways, peak (I_M), rms (I_{rms}), and average (I_{ave}). The formulas for current are identical to the formulas for voltage except that I replaces V.

Measurement of AC Voltages and Currents

In the design of ac circuits, voltage and current measurements are usually made in rms values. A d'Arsonval meter movement serves as the basic element in a VOM used for measuring ac voltages. This movement responds to average values, but the ac scale is calibrated in rms values. Normally a VOM does not have an alternating-current function. Current measurements can be made with a digital VOM which has both ac voltage and current scales. These are also calibrated in rms values. Some meters have both rms and peak value scales. But these may be used only in the measurement of *sinusoidal* voltages or currents.

An oscilloscope is normally used for measuring peak ac voltages, or peak-to-peak values.

SUMMARY

1. An ac generator produces a sinusoidal voltage.
2. The equation for the instantaneous value V of a sinusoidal voltage is

   $$V = V_M \sin \theta$$

   where V_M is the peak or maximum voltage of the waveform and θ is the angle the waveform has reached at the instant it is measured.
3. The amplitude of a *sinusoidal* ac waveform may be specified in three ways. These are: (*a*) its *peak* value, V_M; (*b*) its *root mean square* value, V_{rms}; and (*c*) its *average* value, V_{ave}.
4. The rms value is comparable to the equivalent value dc voltage because the *power* that each can produce is the same.
5. The three values of a sinusoidal voltage are interdependent, as their formulas indicate. Thus

   $$V_{rms} = 0.707V_M$$

   $$V_{ave} = 0.636V_M = 0.899V_{rms}$$

6. These formulas may also be written as

$$V_M = 1.414V_{\text{rms}} = 1.572V_{\text{ave}}$$

7. The ac scale of a voltmeter (and current meter) is calibrated in rms values, although the meter movement responds to average values.
8. When an ac voltage is given without specifying whether it is rms, peak, or average, the *rms* value is meant.
9. The relationships among rms, peak, and average values of alternating *current* are the same as those for ac voltage. Just substitute *I* for *V* in the voltage formulas.
10. Ohm's and Kirchhoff's laws apply to *resistive* ac circuits just as they do to dc circuits. Thus the rms value of current in a resistor of $R\Omega$, across which V_{rms} is applied, may be calculated from the formula

$$I_{\text{rms}} = \frac{V_{\text{rms}}}{R}$$

Similarly, I_M may be calculated for the resistor R by substituting in the formula

$$I_M = \frac{V_M}{R}$$

SELF-TEST

Check your understanding by answering these questions:

1. An ac voltage changes constantly in ____________ and periodically in ____________.
2. The shape of the voltage waveform deliverd by an ac generator is a ____________.
3. The rms and average values of a sinusoidal voltage or current may be computed from its peak value. ____________ (true/false)
4. The V_M of a sinusoidal voltage is 100 V. The rms value of that voltage is ____________ V.
5. The rms value of a sinusoidal current is 10 mA. The peak value of that current is ____________ mA.
6. In question 5, the average value of current is ____________ mA.
7. The rms voltage measured across a 56-Ω resistor is 10 V. The rms current in that resistor is ____________ mA.
8. The peak current in the resistor in question 7 is ____________ mA.
9. AC voltmeters are usually calibrated in ____________ (peak/rms/average) values.
10. The peak value of an ac voltage may be measured with an ____________.

MATERIALS REQUIRED

- Power source: 120 V/60 Hz; isolation transformer; variac
- Equipment: Variable af sine-wave generator; digital EVM with ac voltage and current ranges; oscilloscope
- Resistors: ½-W two 33-, 47-, two 1000-, 1500-Ω
- Miscellaneous: SPST switch; fused line cord

PROCEDURE

60-Hz AC Voltage and Current

1. Connect the circuit of Fig. 40-6. **Power on.** Close switch S. Adjust the output of the variac so that the voltage V_{AD} is 35 V rms, as measured by the digital ac voltmeter V.
2. With the voltmeter, measure and record in Table 40-1 the rms voltages V_{AB} across R_1, V_{BC} across R_2, and V_{CD} across R_3.
3. With an oscilloscope whose vertical amplifiers are voltage calibrated, measure and record in Table 40-1 the peak voltage across *AD*, *AB*, *BC*, and *CD*.
4. **Power off.** Break the circuit at point *B* and insert into the circuit the digital EVM set on the 20-mA ac range.
5. **Power on.** Measure and record in Table 40-1 the alternating current in the circuit.
6. Open S. Restore the circuit to its original condition. Now break the circuit at point *C* and insert the ac milliammeter into the circuit at that point.
7. Close S. Measure and record in Table 40-1 the current at this point. Open S. **Power off.**
8. Compute and record in Table 40-1 the alternating current which should be in the circuit. Show your computations.
9. Compute and record in Table 40-1 the rms voltage across *AB*, *BC*, and *CD*. Show your computations.
10. Compute and record in Table 40-1 the peak voltage across *AD*, *AB*, *BC*, and *CD*.

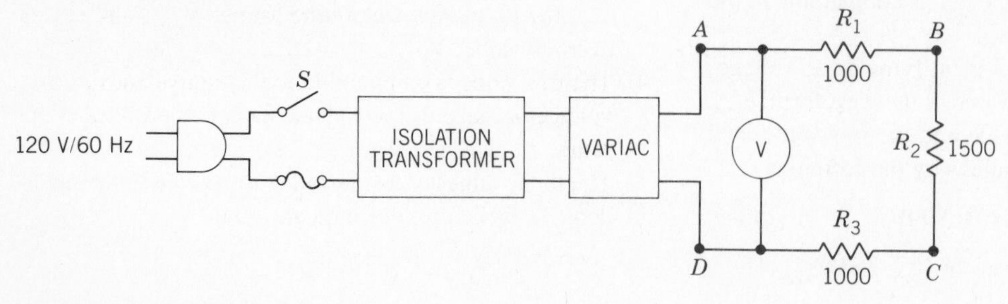

Fig. 40-6. 60-Hz experimental circuit.

1000-Hz AC Voltage and Current

11. Connect the circuit of Fig. 40-7. An af signal generator acts as the voltage source for the circuit. Close S. **Power on.**
12. Set the frequency of the generator at 1000 Hz and the output at 5 V rms, as measured by the ac voltmeter across AD. If the generator cannot deliver 5 V, set the output at 4 V or 3 V rms.
13. Measure and record in Table 40-2 the rms voltage across AD, AB, BC, and CD.
14. With the oscilloscope, measure and record in Table 40-2 the peak voltage across AD, AB, BC, and CD. Open S.
15. Break the circuit at B. Insert the milliammeter into the circuit at point B. Close S. Measure and record in Table 40-2 the current in the circuit at this point. Open S.
16. Restore the circuit to its original condition. Now break the circuit at C. Insert the milliammeter at C. Close S. Measure and record the current in Table 40-2.
17. **Power off.** Compute and record in Table 40-2 the rms voltage across AB, BC, and CD.
18. Compute and record in Table 40-2 the peak voltage across AD, AB, BC, and CD. Also, compute and record the current in the circuit.

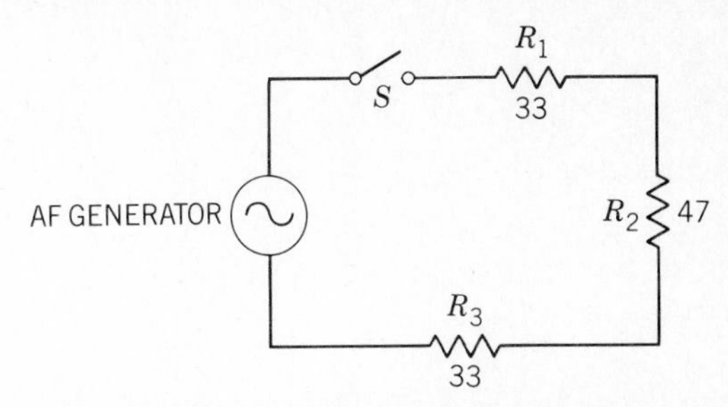

Fig. 40-7. 1000-Hz experimental circuit.

TABLE 40-1. 60-Hz Voltage and Current

V rms				
	AD	*AB*	*BC*	*CD*
Measured				
Computed	X			
V peak				
	AD	*AB*	*BC*	*CD*
Measured				
Computed				
***Current*, mA**				
	Point *B*		Point *C*	
Measured				
Computed				

TABLE 40-2. 1000-Hz Voltage and Current

V rms				
	AD	*AB*	*BC*	*CD*
Measured				
Computed	X			
V peak				
	AD	*AB*	*BC*	*CD*
Measured				
Computed				
***Current*, mA**				
	Point *B*		Point *C*	
Measured				
Computed				

QUESTIONS

1. How do the rms measurements in Table 40-1 compare with the computed rms values? Explain any discrepancies.
2. How do the peak measurements in Table 40-1 compare with the computed peak values? Explain any differences.
3. From each set of measurements in Table 40-1, what is the *average* ratio of V_M/V_{rms}? Show your computations.
4. Does the answer to question 3 substantiate the formula $V_M = 1.414V_{\text{rms}}$? Explain.
5. From each set of measurements and computed values in Table 40-2, what is the average value of V_{rms}/V_M? Show your computations.
6. Does the answer to question 5 substantiate the formula $V_{\text{rms}} = 0.707V_M$? Explain.
7. Do the current data in Tables 40-1 and 40-2 substantiate the application of Ohm's law to resistive ac circuits? Explain.
8. What can you say about the amplitude of alternating current in a resistive circuit? Refer to your data.

Answers to Self-Test

1. amplitude; polarity
2. sine wave
3. true
4. 70.7
5. 14.14
6. 8.99
7. 179
8. 253
9. rms
10. oscilloscope

41

EXPERIMENT

CHARACTERISTICS OF AN INDUCTANCE

OBJECTIVES

1. To observe experimentally the effect of an inductance on current in a dc and ac circuit
2. To verify experimentally the formula for inductive reactance

$$X_L = 2\pi FL$$

INTRODUCTORY INFORMATION

Inductance and Reactance of a Coil

In dc circuits, resistors limit the amount (amplitude) of current. Resistors also oppose current in ac circuits. In addition to resistors, reactive components, namely, inductors and capacitors, impede currents in ac circuits.

In our experiments with magnetism we observed that a magnetic field exists about a coil carrying current. We observed also that a voltage is induced in a conductor when the conductor is *cut* by a magnetic field. By Lenz' law, the polarity of the induced voltage in the conductor is such that any current resulting from it will produce its own magnetic field which will oppose the effect of the original magnetic field.

In the circuit of Fig. 41-1 a sine wave of voltage V causes current through the coil to increase and decrease sinusoidally. This changing current creates a moving magnetic field about the coil which therefore *cuts* the windings of the coil, inducing a voltage V' in these windings. By Lenz' law this voltage is in opposition to the source voltage V, and can therefore be termed a *counter emf*. When V is positive, V' is negative but smaller than V. The numerical value of the resultant emf which powers the circuit is $|V| - |V'|$, where

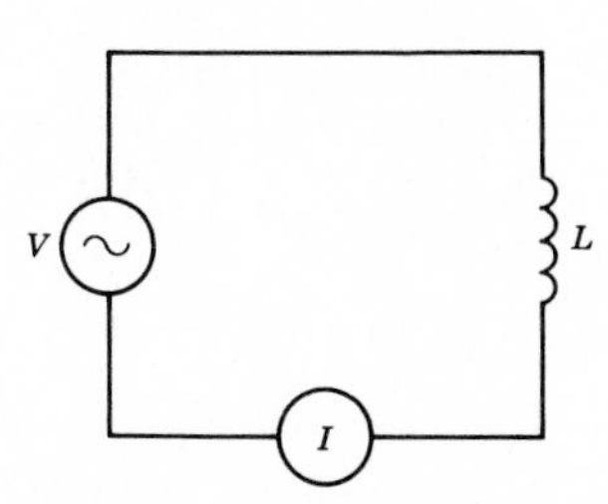

Fig. 41-1. Effect of an inductance on current in an ac circuit.

$|V|$ and $|V'|$ are the absolute values of V and V' respectively, ignoring their *signs*. It is evident therefore that because of the counter emf the alternating current in the coil is less than it would have been if V were a dc source and a steady-state current were flowing. This property of the coil which opposes a *changing* current is termed the *inductance* of the coil.

The unit of inductance is the henry (H), named in honor of the scientist Joseph Henry. The number of henrys in an inductor can be measured by means of an inductance bridge, an instrument with which you are not yet familiar. Inductance (L) is that characteristic of a coil which *opposes* a change in current. The *amount* of opposition offered by an inductor is called inductive reactance (X_L). Inductive reactance is measured in ohms.

The inductive reactance of a coil is not constant but is a variable quantity. X_L depends on the inductance L and on the frequency F of an alternating current. X_L may be computed from the formula

$$X_L = 2\pi FL \tag{41-1}$$

where π is the constant 3.14, F is the frequency in hertz of the alternating current in the inductance, and L is the inductance in henrys.

Some important facts may be deduced from Eq. (41-1). The first is that X_L varies directly with frequency, that is, it is directly proportional to F. This variation may be shown graphically by a specific example.

Problem. Draw a graph of X_L versus F for an inductor L = 1.59 H.

Solution. Find X_L at each of a series of frequencies. Thus, when

$$F = 0$$
$$X_L = 2\pi FL = 2(3.14)(0)(1.59) = 0$$

when F = 100 Hz,

$$X_L = 6.28(100)(1.59) = 1000\ \Omega$$

Similarly, it can be shown that when

$$F = 200\text{ Hz}, X_L = 2000\ \Omega$$
$$F = 300\text{ Hz}, X_L = 3000\ \Omega$$
$$F = 400\text{ Hz}, X_L = 4000\ \Omega$$
$$F = 500\text{ Hz}, X_L = 5000\ \Omega\text{, etc.}$$

Figure 41-2 shows the variation of X_L versus F for $L = 1.59$ H. It is a straight-line graph. Hence we say that the relationship between X_L and F is a *linear one*.

A second fact which may be deduced from Eq. (41-1) is that for direct current, that is, for $F = 0, X_L = 0$. This fact is consistent with the definition of inductance, as that characteristic of a coil which opposes a *change* in current. A steady-state direct current produces no change. Hence an inductance has no effect on a direct current which is constant in amplitude, that is, $X_L = 0$ for such a direct current.

Equation (41-1) also defines the relationship between X_L and L. If F is kept constant, X_L increases or decreases as L is increased or decreased, respectively. Again, if X_L were plotted as a function of L, a linear graph would result.

Resistance of a Coil

An inductor consists of a number of turns of insulated wire wound around a core. There may be only several turns, or there may be thousands of turns. The more turns there are on a particular core, the greater is the inductance of the coil. The diameter of the wire used in winding a coil depends on the maximum current which that coil will be required to pass. The larger the diameter of the wire, the greater the current capabilities of the coil. A smaller-diameter wire will sustain less current. If more current is permitted to flow in a coil than that for which it is rated, the coil will overheat and may burn away the insulation around the windings, thus shorting windings together. Or an overheated coil may burn out, that is, the wire may open. In either event the coil will become defective.

In electronics we generally deal with relatively small currents. Hence, very fine diameter wire is normally used in the construction of coils. Since the resistance of a wire is directly proportional to its length, the resistance of inductors consisting of many turns of fine wire will be relatively high. Thus, the characteristics of inductance *and resistance* are associated with an inductor. Though the resistance of a coil cannot be separated from the inductance, a schematic representation of a coil with inductance L and resistance R (Fig. 41-3) shows these two characteristics as though they were lumped constants, acting in series. The resistance can be measured with an ohmmeter. A coil may therefore be represented by its series equivalent form (Fig. 41-3).

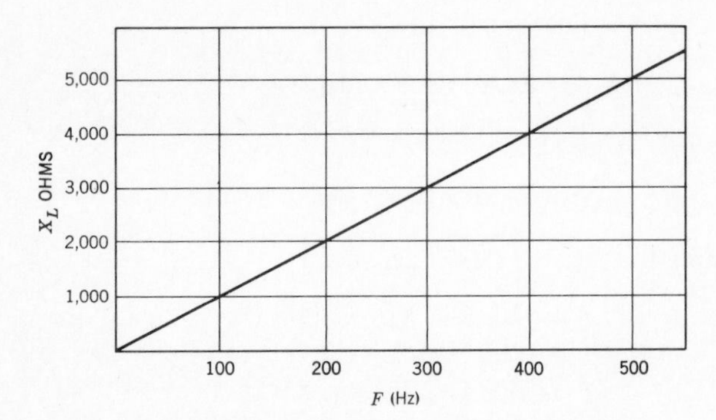

Fig. 41-2. Variation of X_L versus F for a 1.59-H inductor.

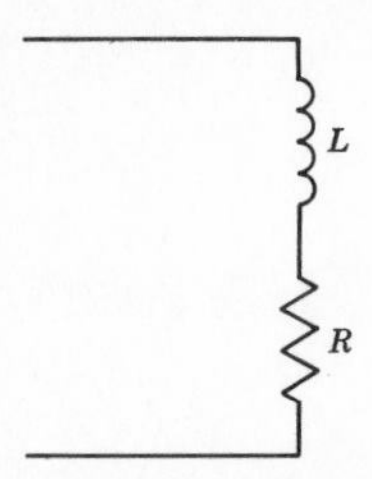

Fig. 41-3. Series-equivalent form of an inductance.

The resistance associated with the length of wire in the winding suggests that one method of testing an inductor is to place an ohmmeter across the two terminals and compare the measured with the rated resistance. If they are the same, it may be assumed that the winding is continuous, that is, it is neither open nor shorted. If the resistance is infinite, the inductor is open. If the resistance is very much smaller than the rated value, say 20 Ω instead of 500 Ω, it may be assumed that a large part of the coil winding is shorted. In practice, resistance checks of an inductor are made to determine if it is continuous or open. Special instruments are required to identify coils with shorted turns.

Measurement of X_L

The reactance of a coil (X_L) may be determined by measurement. In the circuit of Fig. 41-4, a sinusoidal voltage V causes a current I to flow. Ohm's law extended to ac circuits states that the voltage V_L across the coil is $V_L = I \times Z$. Where the resistance of the coil is small compared to X_L, $Z = X_L$ and

$$V_L = I \times X_L \tag{41-2}$$

This equation may be written in the form

$$X_L = \frac{V_L}{I} \tag{41-3}$$

Equation (41-3) can be used to determine the X_L of a coil at a given frequency F. This is accomplished by using an ac voltmeter to measure the voltage V_L across L, and measuring the current I with an ac ammeter. The measured values of V_L and I are then substituted in Eq. (41-3), and X_L is computed.

Note that a slight error is introduced by this procedure, for L has resistance in addition to inductance. However, if the value of X_L is appreciably larger than the resistance R of the coil, R may be ignored.

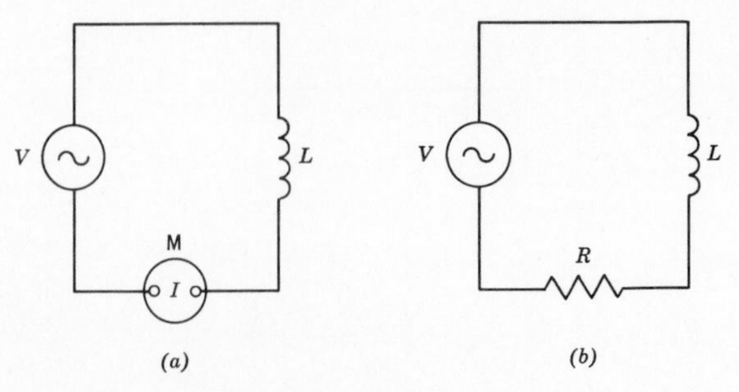

Fig. 41-4. Circuits used to determine X_L.

An alternate method for measuring the current I does not require the use of an ammeter. Instead of the meter M, a resistor R of known value is placed in series with L, as in Fig. 41-4*b*. In a series circuit current is the same everywhere. Hence the current in R is the same as the current in L. The voltage V_R is then measured across R. Current is next determined by substituting the measured value of V_R and the known value of R in the equation

$$I = \frac{V_R}{R} \qquad (41\text{-}4)$$

Next, V_L is measured and X_L computed by substituting the values of V_L and I in Eq. (41-3).

SUMMARY

1. Inductance (L) is that characteristic of a coil which opposes a *change* in current. The unit of inductance is the *henry* (H).
2. The amount of opposition that an inductance offers to a changing current is called inductive reactance (X_L), or simply reactance. Reactance is measured in ohms.
3. The inductive reactance of a coil is not a constant, but is directly proportional to the frequency (F) of the changing current. It is also directly proportional to the value of the inductance.
4. Inductive reactance may be computed from the formula

$$X_L = 2\pi FL$$

where X_L is in ohms, F in hertz, and L in henrys.
5. Every coil has some resistance R associated with it. The continuity of a coil winding may be determined by measuring the resistance of the coil. If the resistance measured is infinite, the coil winding is open.
6. Only the resistance of an inductance will limit direct current in a coil. The inductance of a coil will not affect direct current.
7. The ac voltage across a coil V_L is equal to the product of the alternating current in the coil and the X_L of the coil, that is,

$$V_L = I \times X_L$$

8. If the ac voltage V_L across a coil and the alternating current I in the coil are known, the inductive reactance of the coil may be computed from the equation

$$X_L = \frac{V_L}{I}$$

This equation assumes that the resistance R of the coil is small compared with the reactance of the coil.

SELF-TEST

Check your understanding by answering these questions:

1. Inductance of a coil is measured in ohms. ______ (true/false)
2. The ______ of a coil is directly ______ to the ______ of the current in the coil, if the L of the coil remains constant.
3. If L = 2.5 H and F = 120 Hz, X_L = ______ Ω.
4. If the frequency F is doubled while L is held constant, X_L is ______.
5. The graph of X_L versus F is a ______ ______ if L is held constant.
6. The resistance R of a coil is associated with the resistance of the wire in the ______ of the coil.
7. If the ac voltage V_L across a coil is 4.5 V and the current I in the coil is 0.0225 A, then X_L = ______ Ω.
8. If F = 200 Hz, and X_L = 15,100 Ω, then L = ______ H.

MATERIALS REQUIRED

- Power supply: Line-isolation transformer and variable autotransformer operating from 120-V ac line
- Equipment: Variable regulated dc source; 0–25-mA ac meter or a multirange ac milliammeter covering this range; oscilloscope; EVM; 0–25-mA dc meter; variable-frequency AF sine-wave generator
- Resistor: ½-W 1000-Ω; R as required for step 2
- Inductors: 8 H at 50 mA whose R = 250 Ω; 30 mH
- Miscellaneous: SPDT switch, SPST switch; fused line cord

PROCEDURE

Effect of an Inductance on DC and AC Current

1. Measure, and record in Table 41-1, the resistance R of the 8-H choke.
2. Connect the circuit of Fig. 41-5*a*. V is a variable regulated dc source. L is the 8-H choke; R a resistor equal in value to the resistance of the 8-H choke; S_2 a single-pole double-throw switch (which selects either L or R to complete the circuit); M a 0–25-mA dc meter; V_1 an EVM set on dc voltage; S_1 a switch to apply voltage to the circuit. S_1 is open. V is set for *zero* volts.
3. Throw S_2 to position 1. R is in the circuit. Close S_1. **Power on.** Adjust V until M measures 20 mA. Measure and record the voltage V across the resistor. *Do not vary the supply voltage.*
4. Throw S_2 to position 2. L is in the circuit. The voltage across L should be the same as across R. Measure the current I in L and record in Table 41-1. **Power off.** Disconnect and remove the dc source, the milliammeter M, and the electronic voltmeter.
5. Connect the circuit of Fig. 41-5*b*. V is an ac source consisting of a variable autotransformer plugged into an

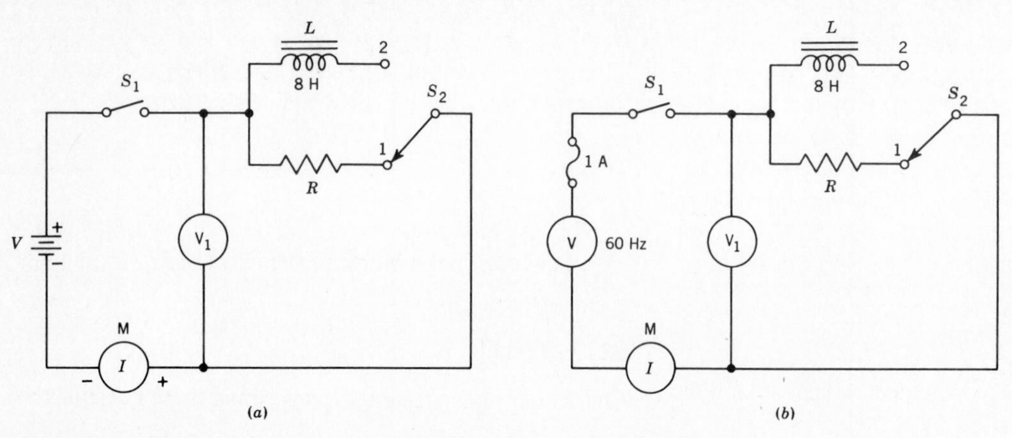

Fig. 41-5. Experimental circuit to demonstrate the effects of an inductance on direct and alternating current.

TABLE 41-1. Effect of an Inductance on Direct and Alternating Current

	Resistance, Ω	*S_2 Position*	*Component in Circuit*	*dc* V, volts	*dc* I, A	*ac* V, volts	*ac* I, A
L		1	*R*		0.02		0.02
R		2	*L*				

TABLE 41-2. X_L of L at 60 Hz

I, A	0.025	0.020	0.015	0.010	*Average X_L at Line Frequency,* Ω
V, volts					
$X_L = \frac{V}{I}$ (ohms)					

isolation transformer. M is a 0–25-mA ac meter, V_1 is an EVM set on ac voltage. The other components are the same as in the preceding circuit. S_1 is *off*. V is set at 0 V.

6. Throw S_2 to position 1, placing R in the circuit. **Power on.** Adjust V until M measures 20 mA. Measure and record the voltage V across the resistor. *Do not vary the supply voltage*.
7. Throw S_2 to position 2, placing L in the circuit. Readjust V if necessary to place the same voltage across L as there was across R in step 6. Measure and record the current I in L.

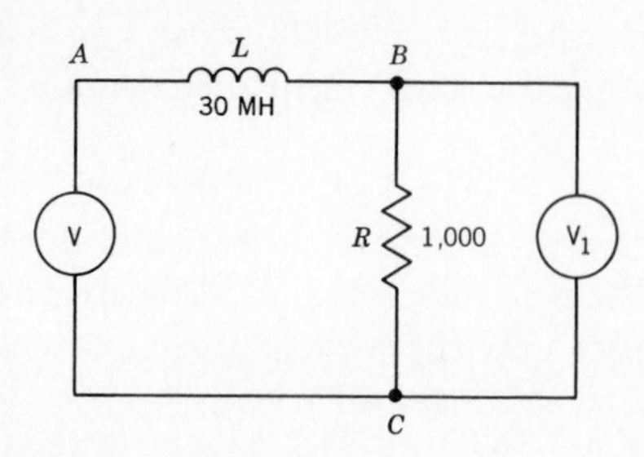

Fig. 41-6. Circuit to determine X_L and the variation of X_L versus F.

Reactance (X_L) of L at Line Frequency

8. The circuit of Fig. 41-5*b* is still connected. S_2 is in position 2 and L is in the circuit. M is the 0–25-mA range of the meter. Adjust V in turn for each of the currents shown in Table 41-2. Measure and record the voltage V at each current level. Compute the V/I ratio at each of the current levels and record as X_L in Table 41-2.

NOTE: The ratio of V/I is really the impedance Z. In this case we will ignore the resistance of L and consider the impedance at line frequency as a pure reactance, X_L. The error introduced by this procedure is small enough to be ignored.

Average the four values of X_L and record. Draw a graph of the voltampere characteristic of the inductance at line frequency. (Let V be the horizontal, I the vertical axis.)

Inductive Reactance (X_L) versus Frequency (F)

In this part of the experiment you will determine how X_L varies with frequency. X_L will not be measured directly, but will be computed as the ratio of V_L/I. The current I

TABLE 41-3. X_L versus F

F, Hz	V, V p-p	V_L, V p-p	V_R, V p-p	I, A $\left(\frac{V_R}{R}\right)$	X_L, Ω $\frac{V_L}{V_R} \times R$	$2\pi FL$
2,000						
3,000						
5,000						
7,000						
8,000						
9,000						
10,000						

will not be measured directly, but will be computed from the voltage V_R measured across R, where $I = V_R/R$. The calculation for X_L can be made directly form V_R (the voltage across R), V_L (the voltage across L), and R, by substituting these measured values in Eq. (41-5).

For
$$X_L = \frac{V_L}{I} = \frac{V_L}{\frac{V_R}{R}}$$

Therefore
$$X_L = \frac{V_L}{V_R} \times R \qquad (41\text{-}5)$$

9. Connect the circuit of Fig. 41-6. V is a variable-frequency sine-wave generator capable of delivering a 20-V peak-to-peak ac voltage, variable from 10 Hz to 100 kHz. V_1 is an oscilloscope calibrated to measure peak-to-peak volts. Set the signal generator in turn to each of the frequencies shown in Table 41-3. Maintain the output V of the generator at 15 V peak-to-peak. Measure and record the voltage V_L and V_R for each frequency. Compute and record the value of I (V_R/R) in amperes, and the value of X_L [use Eq. (41-5)] for each frequency. Show sample computations. Compute also, and record in Table 41-3, the formula value ($2\pi FL$) of X_L. Show a sample computation. Draw a graph of X_L (vertical axis) versus F (horizontal axis) from the data.

QUESTIONS

1. From the data in Table 41-1, is there any significant difference between alternating and direct current in L and R? Identify the difference and explain the reason for it.
2. Refer to the data in Table 41-2. Is X_L the same for each of the currents developed in the circuit? Should it be? Explain any unexpected result.
3. What does the voltampere characteristic (graph) tell you about X_L?
4. How does X_L vary with frequency? Substantiate your answer by referring to the data in Table 41-3. Refer also to the graph (Fig. 41-2).
5. Did your experiment verify the formula $X_L = 2\pi FL$? Substantiate your answer by reference to the data.
6. If X_L of a certain inductance is 5000 Ω at frequency F, what is the X_L of the coil at $2F$? $5F$? nF?
7. Refer to Table 41-3. Are the computed values of I, rms or peak-to-peak? Explain.

Answers to Self-Test

1. false
2. reactance; proportional; frequency
3. 1885
4. doubled
5. straight line
6. winding
7. 200
8. 12 (approx.)

42

EXPERIMENT

MEASURING FREQUENCY AND PHASE WITH AN OSCILLOSCOPE

OBJECTIVES

1. To measure the frequency of an ac waveform using an oscilloscope with a calibrated horizontal sweep
2. To measure the phase angle between voltage and current in an inductance, using a dual-trace oscilloscope

INTRODUCTORY INFORMATION

Measuring Frequency

Oscilloscopes are used to measure ac voltages and to observe waveforms. They may also be used, if they have a calibrated horizontal sweep (time base), to measure the period (time of 1 cycle) of a waveform. Once the period t is known, the frequency F of the waveform may be calculated using the formula

$$F = \frac{1}{t} \tag{42-1}$$

where F is in hertz and t is in seconds.

Let v be a sinusoidal voltage of unknown frequency F, Fig. 42-1, displayed on an oscilloscope with calibrated time base. The waveform is centered with respect to the vertical and horizontal axes. Assume that the calibrated sweep of the scope is set at 0.5 ms/cm. This means that it takes the sweep 0.5 ms to travel 1 cm horizontally across the oscilloscope face. Note in Fig. 42-1 that one complete cycle covers the sweep from A to C, a distance of 8 cm on this scope. Since the sweep speed is 0.5 ms/cm, the period of the sine wave is 4 ms. That is, $t = 4 \times 10^{-3}$ s. Substituting this value of t in Eq. (42-1), we obtain

$$F = \frac{1}{t} = \frac{1}{4 \times 10^{-3}} = \frac{10^3}{4} = 250 \text{ Hz}$$

This value, 250 Hz, is the required frequency of the waveform.

And so it is a relatively simple matter to determine the frequency of an ac waveform. How accurate the measurement is depends on the accuracy of the sweep-speed calibrations on the oscilloscope, and on how accurately the width of one cycle is measured on the scope. Reading accuracy is improved by setting the sweep speed on the scope so that one cycle of waveform occupies as far as possible the width of the graticule.

More accurate measurements can be made with a frequency meter. Where a high degree of accuracy is required, frequency meters are used. In this experiment an oscilloscope will be employed.

Phase Relations between Voltage and Current in a Coil

In a dc circuit, as the voltage across a resistor increases, the current in R increases. This is also true in a resistive ac circuit, for if a sinusoidal voltage v is applied across R, the instantaneous variations of current i in R follow exactly the instantaneous changes in voltage v. Thus, at the instant v is going through zero, i is going through zero. When v is at maximum, i is at maximum. Voltage and current in R are said to be in phase. This relationship is shown graphically in Fig. 42-2b.

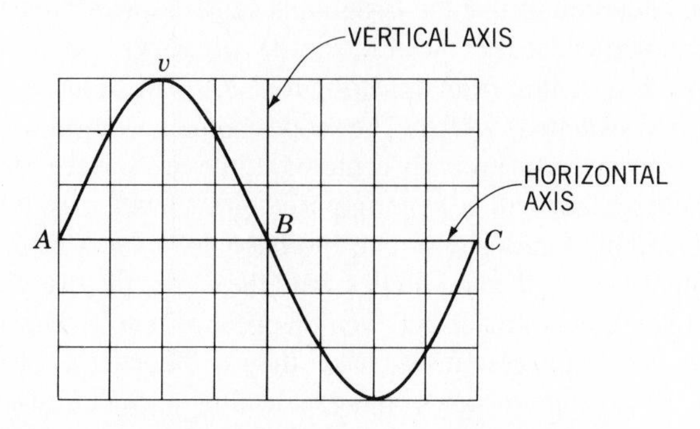

Fig. 42-1. Sine wave displayed on an oscilloscope is centered with respect to the vertical and horizontal axes of the graticule.

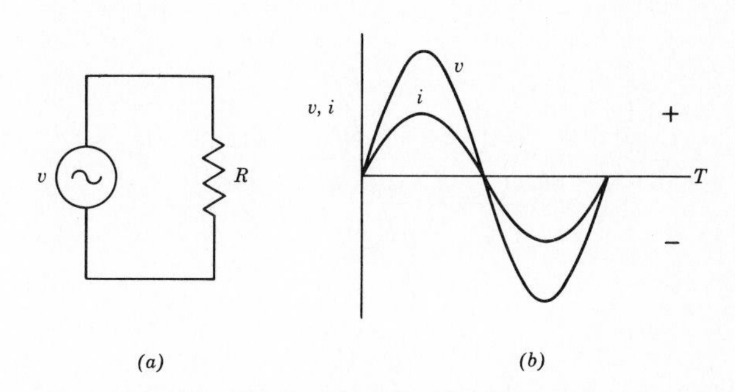

Fig. 42-2. In a resistive circuit, voltage and current are in phase.

In a pure inductive circuit voltage and current are *not* in phase, but current lags voltage by 90°. A graph of voltage and current (Fig. 42-3) shows this relationship. A phase meter or an oscilloscope acting as a phase indicator may be used to verify the phase relationship between voltage and current in an inductor. In this experiment an oscilloscope will be employed.

Dual-Trace Oscilloscope Used to Measure Phase

It is possible to measure the phase difference between ac voltage and current in an inductance by using a dual-trace oscilloscope. This type of scope can simultaneously display two time bases on the screen of a single cathode-ray tube, Fig. 42-4*a*. On each trace may be imposed a signal voltage, and the characteristics of the signals can then be compared, Fig. 42-4*b*. The signals, however, must be of the same frequency, or they must be derived from the same frequency source. In this type of application, a dual-trace oscilloscope acts like two oscilloscopes, with the convenience of a single screen.

Figure 42-5 shows a dual-trace scope used in electronics. The two separate vertical channels, *A* and *B*, have separate inputs, identified respectively by the numbers 23 and 15. Associated with the channel *A* input are the Variable and Volts/cm controls, 21 and 20; a vertical positioning (centering) control for time trace *A*, 25, and other operating controls. Channel *B* has equivalent controls which affect the signal voltage applied to the *B* input circuits. There is also a Mode control, 19, which makes it possible to operate the instrument as a single-trace oscilloscope on channel *A* or channel *B*. In the *A and B* mode, both traces appear simultaneously. In the $A + B$ mode, the *A* and *B* signals are added algebraically and displayed as a single signal on one trace. In the $A - B$ mode, the *A* and *B* signals are subtracted algebraically and displayed as a single signal.

The Variable (6) and Sweep Time/cm (5) controls are the sweep speed controls for the oscilloscope. In the "calibrated" position of the Variable control, the setting of the Sweep-Time/cm control determines the length of time it takes the sweep to trace through 1 cm horizontally. Controls 5 and 6 are common to all positions of the Mode control.

There is one other switch-type control whose operation is important in using a dual-trace oscilloscope. This three-

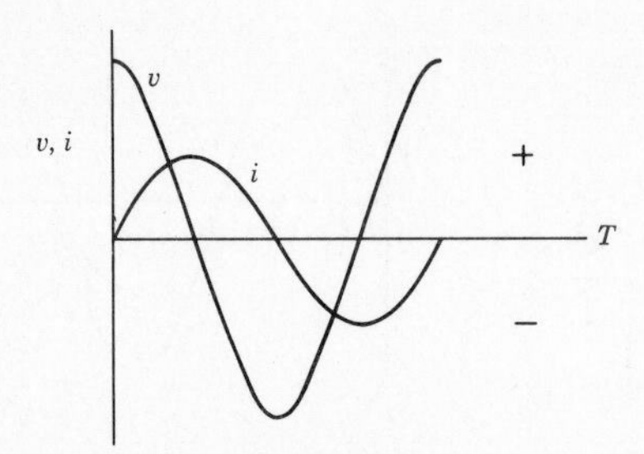

Fig. 42-3. In a pure inductive circuit, voltage and current are out of phase, and current lags voltage by 90°.

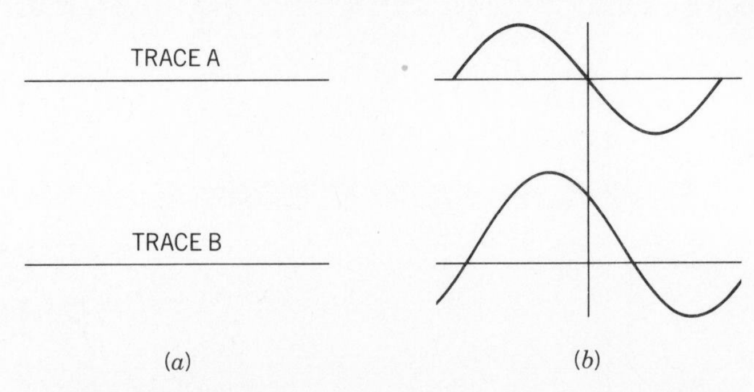

Fig. 42-4. (*a*) Two traces on a dual-trace oscilloscope; (*b*) separate signals are displayed on each trace.

position switch, 12, called the Source control, selects a triggering source for the sweep. In mode *A&B* (the dual trace mode), both sweeps are triggered by the same source. When the Source switch is set to ChA, the *A* sweep is triggered by the channel *A* signal. In the ChB position, the channel *B* sweep is triggered by the channel *B* signal. In the Ext. position, the sweep is triggered by an external signal applied to the Ext. Sync jack, 10. To measure phase difference between two time-related signals, the Source switch is set on Ext. and a time-related reference pulse triggers the sweep.

Phase Measurement Using a Dual-trace Oscilloscope

To compare the phase of two signals, the reference signal is fed to the channel *A* input, and the second signal is connected to the channel *B* input. Operating the oscilloscope in mode *A*, the proper controls are adjusted so that one cycle of the channel *A* reference signal occupies the width of the graticule, 10 cm for the oscilloscope in Fig. 42-5. The ChA centering control and the horizontal centering control are adjusted so that the waveform is centered with respect to the vertical and horizontal axes on the graticule, as in Fig. 42-1. The ChA Variable and Volts/cm controls are set so that the waveform is about 3 cm high and still properly centered. The oscilloscope controls, which were set for viewing the reference waveform on the *A* channel, *must not be disturbed* until the phase measurement is completed. The oscilloscope is then switched to mode *B*, and the second signal, already fed to the ChB input, is displayed on the screen. The channel *B* vertical-gain controls are set so that the *B* signal occupies about 2 cm in height. The ChB vertical centering control is then adjusted so that the *B* signal is centered with respect to the horizontal axis of the graticule. *Do not vary the horizontal centering* control from its setting for ChA. The Mode control is then switched to *A&B*. The scope should now display both waveforms, each properly centered. If the two waveforms are in phase, they will be superimposed on each other, as in Fig. 42-2. This figure shows two in-phase voltages of different amplitudes. If the amplitudes were the same, the two waveforms would coincide and would appear as one. However, if the waveforms are out of phase, they will appear as in Fig. 42-6. This figure shows voltage v leading current i by some angle called the *phase angle*. Here it is assumed that the amplitudes of both signals are the same.

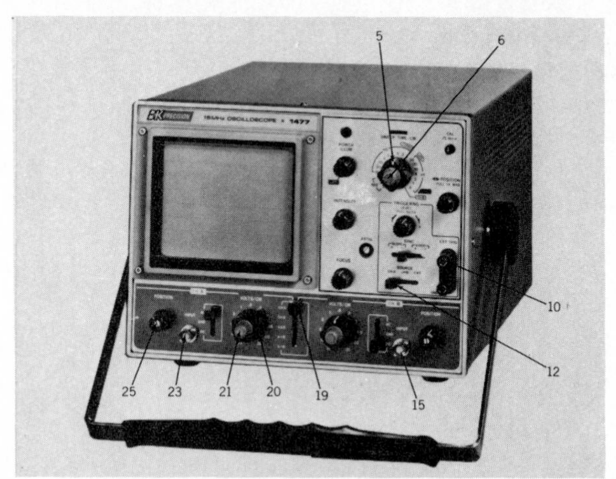

Fig. 42-5. Controls and inputs on a dual-trace oscilloscope (*Dynascan Corporation*).

At this point the ChB Volts/cm control should be adjusted to equalize, approximately, the A and B signal amplitudes, so that the phase difference between the two waveforms can be determined. The angular time-base distance between the zero crossover points, A and B, of the waveforms in Fig. 42-6 shows the phase difference between v and i. Let us see how this phase difference can be calculated in degrees. The time base (horizontal axis) of the graph is in degrees. As shown, 8 cm correspond to 360° for one complete cycle. Therefore each centimeter, horizontally, represents 45°. The distance between the crossover points, A and B, is 2 cm. Therefore the phase difference between the two waveforms is 90°. There is a variation of this method for calculating the angular phase difference. As before, the distance between the zero cross-over points of the two waveforms is 2 cm. The width of one cycle of the waveform is 8 cm. Therefore the phase difference equals $2/8 \times 360° = 90°$.

It is possible to measure the phase difference between voltage and current in an inductance by connecting an external resistor R in series with L, as in the circuit of Fig. 42-7. Be careful to select a resistor of such a value that the ratio X_L/R is greater than 10. This circuit will act almost like a pure inductance. Current i in the series circuit of Fig. 42-7 is the same throughout the circuit. And the voltage v_R across R is in phase with current i in R. Hence, when we measure the phase angle between the voltage v_R across R and the applied voltage v, we are in fact measuring the phase angle between voltage and current in the coil.

NOTE: If the ratio of X_L/R is equal to 10, the error in phase angle introduced by R is approximately 6°. As X_L/R becomes greater than 10, the error decreases.

SUMMARY

1. Triggered oscilloscopes can be used to measure the frequency of an ac waveform.
2. To determine the frequency of an ac waveform, the time-base controls are set on "calibrated." A single waveform is displayed on the screen, and the width of the waveform in centimeters is measured. The setting of the Sweep Time/cm control is then multiplied by the number of centimeters in the width of the waveform, yielding a figure t. The frequency F of the waveform is found from the formula $F = 1/t$, where F is in hertz when t is in seconds.

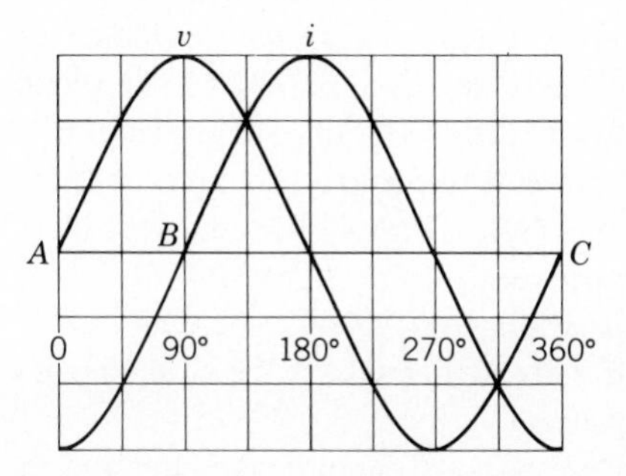

Fig. 42-6. Calculating phase difference between v and i.

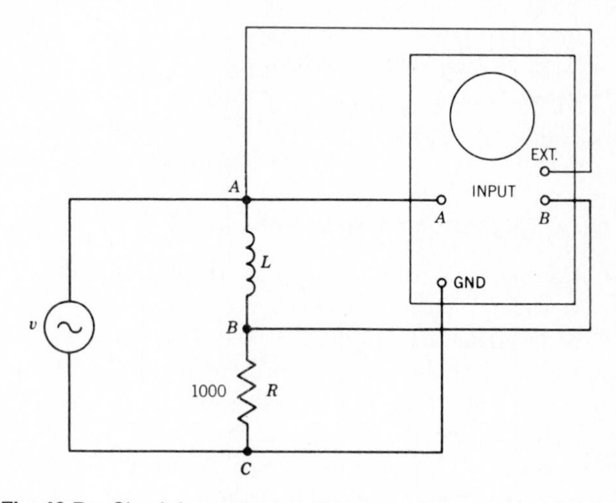

Fig. 42-7. Circuit for measuring phase angle between v and i in an inductive circuit.

3. In a resistive circuit current and voltage are in phase. This means that variations in current follow variations in voltage. Thus as voltage increases, current increases. As voltage decreases, current decreases. When the voltage is zero, current is zero.
4. In an inductive circuit alternating current and voltage are not in phase. In a pure inductive circuit, that is, one without resistance, current lags voltage by 90°.
5. We can demonstrate the phase difference between ac voltage and current in an inductance using a dual-trace oscilloscope.
6. A dual-trace oscilloscope is one which displays two time bases on a single CRT screen.
7. A dual-trace oscilloscope has two separate vertical (signal) inputs. One input, say *A*, places its signal on one trace, call it the *A* trace. The other, the *B* input, places its signal on the second trace, the *B* trace.
8. Each signal is centered on the graticule (etched face plate) of the scope, as in Fig. 42-6. The scope sweep is set to "calibrated." The time-base distance at the zero crossover points, measured in degrees, between the two waveforms (*AB* in Fig. 42-6) represents the phase difference between the two signals.
9. To measure phase difference, the two waveforms must have the same frequency, or their frequencies must be time-related.
10. To measure phase difference, the sweep of the oscilloscope must be triggered from the same source. This can be the signal from either the *A* input or the *B* input, or from an external source which is time-related to the two signals (EXT).

SELF-TEST

Check your understanding by answering these questions:

1. A sinusoidal voltage is 5 cm wide on the calibrated time base of an oscilloscope whose Sweep Time/cm control is set at 2 ms. The frequency of the waveform is ____________ Hz.
2. In a ____________ circuit, ac voltage and current are in phase.
3. In an inductive circuit whose ____________ is at least ten times greater than its ____________, current lags voltage by approximately 90°.
4. In Fig. 42-6, current ____________ voltage by 90°.
5. In Fig. 42-6, the peaks of the two waveforms are ____________ degrees apart.
6. A dual-trace oscilloscope may be used to determine the phase difference between two ____________ ____________ waveforms.
7. The phase difference between a 60-Hz signal and a 100-Hz signal may be determined by the method described in Introductory Information of this experiment. ____________ (true/false)

MATERIALS REQUIRED

- Equipment: Dual-trace oscilloscope; AF sine-wave generator
- Resistors: ½-W two 1000-Ω
- Inductor: 30 mH

PROCEDURE

Measuring Frequency

1. Your instructor will set the dial of an AF sine-wave generator at some frequency, and will mask the dial setting.
2. Using a dual-trace triggered oscilloscope, determine the frequency setting of the generator. ____________ Hz. Show any required computations.
3. If you found the frequency correctly, you may go on to step 4. Otherwise, repeat steps 1 and 2 until you have determined the unknown frequency accurately.

Phase Relations in a Resistive Circuit

4. Connect the circuit of Fig. 42-8. The output of the AF generator is fed to the channel *A* input. Connect point *B* of the circuit to the channel *B* input. Connect point *A* of the circuit to the EXT Sync/trigger jack of the oscilloscope, and set the Source switch to EXT. Set the frequency of the AF generator at 50 kHz and the output voltage at 15 V peak-to-peak. Channel *A* is the reference (applied) voltage channel; channel *B* is the current channel.
5. Switch to channel *A*. Adjust your oscilloscope controls until a single sine wave about 3 cm high is stationary on the screen, covering the width of the graticule. Adjust the channel *A* vertical centering control and the horizontal centering control until the waveform is properly centered with respect to the horizontal and vertical graticule axes.

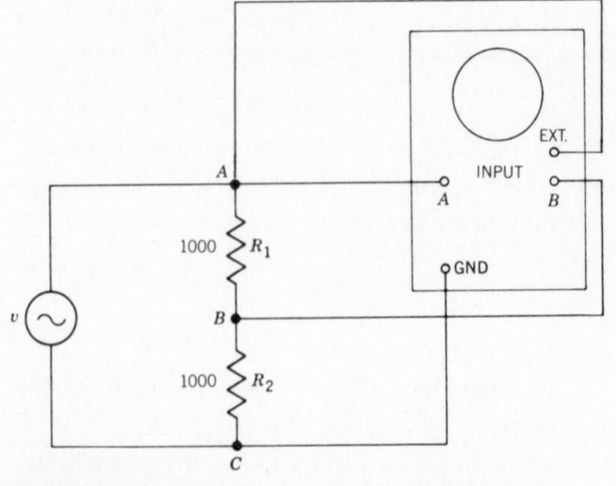

Fig. 42-8. Circuit for measuring phase angle between *v* and *i* in a resistive circuit.

TABLE 42-1. Waveforms in a Resistive Circuit

(*a*) (*b*)

TABLE 42-2. Waveforms in an Inductive Circuit

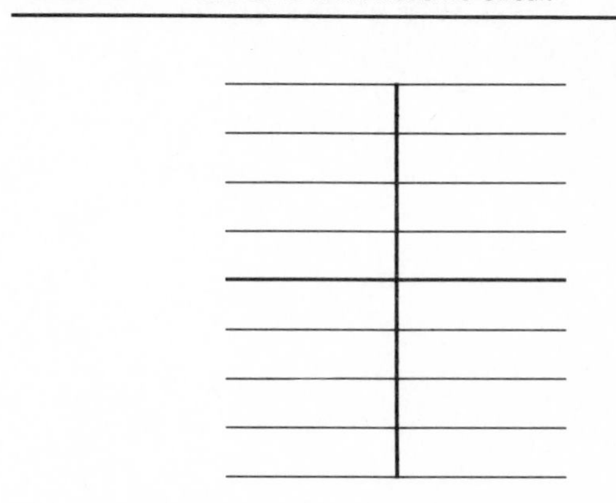

6. Switch the oscilloscope to channel *B* and adjust the Volts/cm control until the sine wave on the screen is approximately 2 cm high. Center it properly with the channel *B vertical* centering control. *Do not touch the horizontal centering control.*
7. Switch your oscilloscope to the *A and B* mode. Both signals should appear on the screen. Draw these waveforms in Table 42-1*a*. Identify them by labeling *v* and *i*.
8. Reset the variable Volts/cm control of channel *B* until the height of each waveform is approximately the same, 3 cm. Draw the waveforms in Table 42-1*b*.

Phase Relations in an Inductive Circuit

9. In Fig. 42-8, replace R_1 with a 30-mH choke, as in Fig. 42-7. The AF generator is now set at 100 kHz, 15 V peak-to-peak. The oscilloscope controls are still set as in step 8.
10. Recheck the waveform *v* in channel *A*. It should still be 3 cm high and properly centered with respect to the horizontal and vertical axes of the graticule. If not, readjust the scope controls as in step 5.
11. Switch the scope to channel *B* and adjust the ChB Volts/cm controls until the waveform *i* is 3 cm high, approximately, and properly centered vertically. *Do not vary the horizontal centering control from its position in step 10.* Observe the zero volts points of the waveform, that is, note where waveform *B* cuts the horizontal graticule axis.
12. Switch the oscilloscope to the *A and B* mode. Waveform *v* should cross the horizontal axis on the extreme left, at the start of the horizontal sweep, and should be going in a positive direction during the first alternation. The crossover point of waveform *i* should be to the right of the zero crossover point of waveform *v*. At the start of the horizontal sweep, waveform *i* should be negative.
13. Draw and identify the two waveforms in Table 42-2.
14. Measure and record in Table 42-2 the width w_1 in centimeters of waveform *v*. Also measure and record in Table 42-2 the distance w_2 in centimeters between the zero crossover points of waveforms *v* and *i*.
15. Determine and record the phase difference between waveforms *v* and *i*. Show the formula you used and show your computations. **Power off.**

QUESTIONS

1. (*a*) Is it necessary to use a dual-trace oscilloscope to measure the frequency of a periodic waveform?
   (*b*) What must be true about the sweep of an oscilloscope used to measure the frequency of a waveform?
2. In step 7, Table 42-1*a*, do waveforms *v* and *i* coincide? Are they in phase? Explain.
3. Refer to step 8. How can you tell if both waveforms were properly centered? Explain.
4. Did you establish in this experiment that *v* and *i* are in phase in a resistive circuit? Explain.
5. Is the ratio of X_L/R in this experiment greater than 10? Show your computations.
6. In this experiment, how did the measured phase difference between *v* and *i* in the inductive circuit compare with the theoretical phase difference in a pure inductance? Explain any discrepancy.

Answers to Self-Test

1. 100
2. resistive
3. X_L; resistance
4. lags
5. 90
6. time-related
7. false

43

EXPERIMENT

INDUCTANCES IN SERIES AND IN PARALLEL

OBJECTIVES

1. To verify experimentally that the total inductance L_T for two inductors L_1 and L_2 connected in series, where no mutual coupling exists, is

$$L_T = L_1 + L_2$$

2. To verify experimentally that L_T for two inductors connected in parallel, where no mutual coupling exists, is

$$\frac{1}{L_T} = \frac{1}{L_1} + \frac{1}{L_2}$$

INTRODUCTORY INFORMATION

Just as in the case of resistors, inductors may be connected in series or in parallel in a circuit. The effect of series-connected inductors is to offer greater opposition to alternating currents than either inductor can, acting alone. Measurement shows that series-connected inductors, where no mutual coupling exists, act like an equivalent single inductor whose inductance L_T is the sum of the individual inductances. That is,

$$L_T = L_1 + L_2 + L_3 + \cdots \tag{43-1}$$

Two or more inductors connected in parallel, where no mutual coupling exists, act like an equivalent single inductor whose inductance L_T is given by the equation

$$\frac{1}{L_T} = \frac{1}{L_1} + \frac{1}{L_2} + \frac{1}{L_3} + \cdots \tag{43-2}$$

It is apparent that the total inductance of parallel-connected inductors is less than the smallest inductor in the circuit.

Verification of the total inductance of series- and parallel-connected inductors, as given by Eqs. (43-1) and (43-2), respectively, may be obtained by measuring L_T directly on an inductance bridge. Or L_T may be measured indirectly, as in this experiment.

Consider the circuit of Fig. 43-1. V is an ac source sending current through R and L. The reactance X_L of L may be determined, as in the preceding experiment, by measuring I and the voltage V_L across L, and substituting these values in the equation

$$X_L = \frac{V_L}{I} \tag{43-3}$$

After X_L has been computed, L may be found from the equation

$$L = \frac{X_L}{2\pi F} \tag{43-4}$$

In this equation, X_L is the value in ohms just obtained, F is the frequency in hertz of the alternating current, and L is the required inductance in henrys.

Example. What is the inductance of a coil whose reactance at 100 Hz is 3000 Ω?

Solution. $L = \frac{X_L}{2\pi F} = \frac{3000}{6.28(100)} = 4.78$ H

In the circuit of Fig. 43-1, the current I may be measured directly with an ac ammeter. Another method is to measure the resistance of R and the voltage V_R across R. The current may then be computed from the equation

$$I = \frac{V_R}{R} \tag{43-5}$$

Effect of Direct Current on the Inductance of an Iron-Core Choke

In electronics iron-core coils, called *chokes*, are used as current-smoothing elements in power supplies. The inductance of these chokes is dependent on the direct current present in them. A filter choke rated 8 H at 50 mA direct current means that L is 8 H when a 50-mA direct current is in the choke. In the absence of the direct current, the choke will not measure 8 H. If there is less than 50 mA dc, the choke will measure more than 8 H; when there is more than 50 mA dc, the choke will measure less than 8 H.

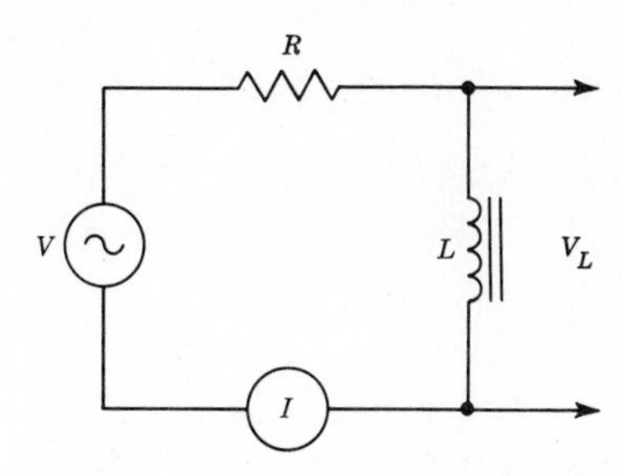

Fig. 43-1. A series *LR* circuit may be used to determine the X_L of an inductor.

In this experiment you will verify the effect of direct current on the inductance of an iron-core choke by means of the experimental circuit (Fig. 43-2). V_{rms} is an isolated line-derived ac voltage. With switch S in position 1, current flows in the ac circuit consisting of the inductor L connected in series with R. There is no direct current in L. The inductance of L is determined by finding the alternating current I in L, and the ac voltage V_{BC} across L. Now,

$$X_L = \frac{V_L}{I} \quad \text{and} \quad L = \frac{X_L}{2\pi F}$$

Next S is switched to position 2, and the output of the dc supply V_{dc}, connected in series with V_{rms}, is adjusted to the rated dc level of L. The inductance of L is again determined for comparison.

SUMMARY

1. The total inductance L_T of series-connected inductors, where no mutual coupling exists, is given by the equation

$$L_T = L_1 + L_2 + L_3 + \cdots$$

where L_1, L_2, L_3, etc., are inductors connected in series.
2. The total inductance L_T of parallel-connected inductors, where no mutual coupling exists, is given by the formula

$$\frac{1}{L_T} = \frac{1}{L_1} + \frac{1}{L_2} + \frac{1}{L_3} + \cdots$$

3. It is apparent from summary statements 1 and 2 that the laws for L_T of inductors connected in combination (series or parallel), where no mutual coupling exists, are the same as the laws for R_T of resistors connected in combination.
4. The inductance of iron-core coils, or chokes, is rated for a particular level of direct current. Thus a choke rated 8 H at 50 mA dc will measure more than 8 H when there is less than 50 mA dc, and less than 8 H when there is more than 50 mA dc.
5. The inductance of coils and chokes may be measured directly with an inductance bridge.
6. The inductance of coils and chokes may also be determined indirectly by measuring the ac voltage V_L across the coil and the alternating current I in the coil. From these measurements X_L can be computed, for $X_L = (V_L/I)$. Knowing X_L, it is possible to calculate L by the equation

$$L = \frac{X_L}{2\pi F}$$

SELF-TEST

Check your understanding by answering these questions:

1. Three chokes, 11, 2.5, and 8 H, are connected in series. No mutual coupling exists. Their total inductance is ____________ H.
2. The same three chokes (question 1) are connected in parallel. Their total inductance is ____________ H.
3. The ac voltage V_L across a coil L is 22 V. The frequency is 60 Hz. The alternating current in the coil is 0.025 A. What is
   (*a*) X_L = ____________ Ω
   (*b*) L = ____________ H
4. A choke (L) and a 1-kΩ resistor are connected in series as in Fig. 43-3. The voltage across the resistor is 50 V, and the voltage across the choke is 40 V. The frequency of the applied voltage is 60 Hz. What is the
   (*a*) Current I in the circuit? ____________ A
   (*b*) X_L of the choke? ____________ Ω
   (*c*) L of the choke? ____________ H

MATERIALS REQUIRED

- Power supply: Variable voltage-regulated dc source, isolation transformer, variable autotransformer
- Equipment: Oscilloscope, EVM, 0–100-mA dc milliammeter
- Resistors: ½-W 10,000-Ω; 5-W 500-Ω
- Inductors: Two iron-core chokes, 8 H at 50 mA dc, whose R = 250 Ω (as in Experiment 41)
- Miscellaneous: SPST and SPDT switches; additional parts as required for step 8; fused line cord

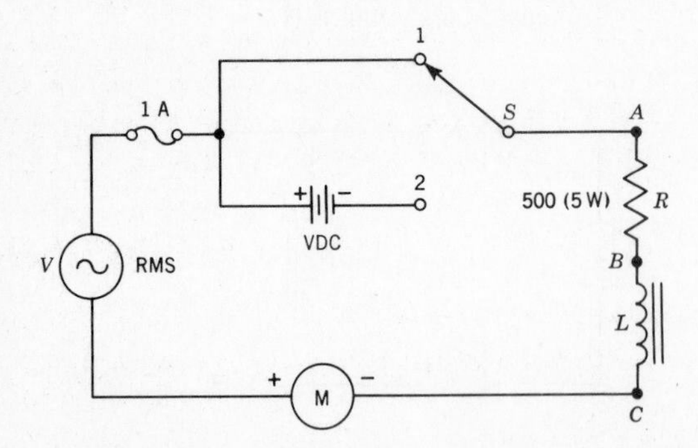

Fig. 43-2. Experimental circuit to determine the inductance of *L*.

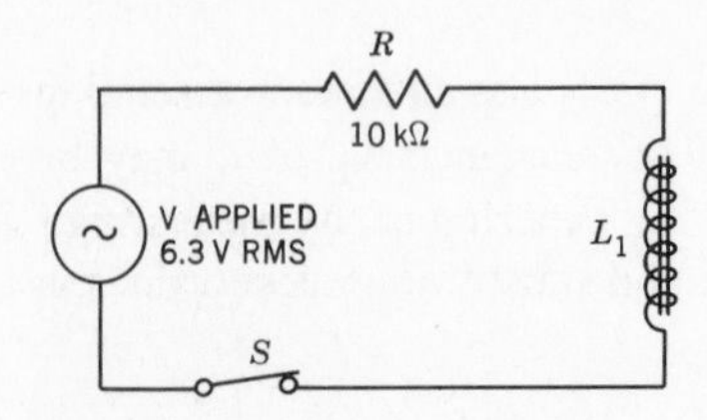

Fig. 43-3. *LR* series circuit.

PROCEDURE

Determining the Effect of Direct Current on the X_L of an Iron-Core Choke

All ac voltages should be measured with an EVM and checked with an oscilloscope calibrated to measure rms values.

1. Connect the circuit of Fig. 43-2. S is in position 1. V_{rms} is the output of a variable autotransformer plugged into an isolation transformer deriving its power from the 120-V ac line. V_{rms} is set at *zero* volts. V_{dc} is set at *zero* volts. M is a 0–100-mA dc meter to measure direct current in the inductor.
2. With the EVM and oscilloscope connected across R, adjust the output of the ac supply until V_{AB}, the voltage across R, is 1 V. Determine the alternating current I in R and record in Table 43-1. Measure and record the voltage V_{BC} across L. Compute and record the X_L of L. Show your computations. Compute and record the inductance L. Show your computations. Record also the rated value in henrys and the dc rating of L.
3. Set switch S to position 2. Adjust V_{dc} until there is 15 mA direct current in L. Measure the ac voltage across R. If it is not precisely 1 V rms, readjust V_{rms} until this voltage is obtained. Measure and record V_{BC}, the voltage across L. Determine and record I (alternating current), X_L, and L.
4. Repeat step 3 for 30 mA of direct current and finally for the rated value of direct current in L. *Be certain that the voltage V_{AB} across R is precisely 1 V. If it is not, readjust V_{rms} until there is precisely 1 V across R.*

Inductors in Series

All ac voltage measurements should be made with an EVM. These readings should be confirmed by measurement with an oscilloscope set on the ac voltage range and calibrated in rms values.

5. Determine the inductance of each inductor L_1 and L_2 using the following procedure:
   a. Connect the circuit of Fig. 43-3.
   b. Measure, and record in Table 43-2, the voltage V_{L_1} across the inductance L_1, and V_R across the resistor.
   c. Replace L_1 by L_2 and repeat step **b**.
   d. Compute the values of I_1 and I_2 and record them in Table 43-2. Show formulas used and computations.
   e. Compute the values of L_1 and L_2 and record them in Table 43-2. Also show formulas used and computations.
6. Determine the total inductance L_T of L_1 and L_2 connected in series by the following method:
   a. Connect the circuit of Fig. 43-4. Separate L_1 and L_2 as far as possible so that there is no coupling between them.
   b. Measure, and record in Table 43-3, the voltages V_{LT} across L_1 and L_2 (i.e., from point A to B), and V_R across R.
   c. Compute and record the current I and L_T.

TABLE 43-2. Determining the Inductance of a Choke

V_{L_1}, V	V_R, V	I_1, A	L_1, H
V_{L_2}, V	V_R, V	I_2, A	L_2, H

TABLE 43-3. Series-connected Chokes

V_{LT}	V_R	I	L_T

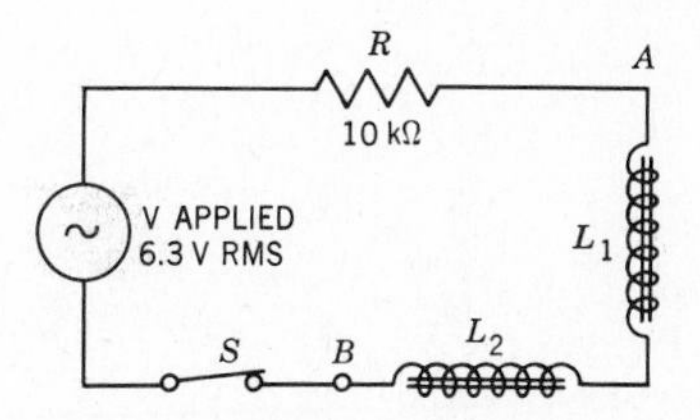

Fig. 43-4. Effect of adding inductors in series.

TABLE 43-1. X_L of an Iron-Core Choke, with and without DC

Direct current, mA	V_{AB}, V	*I(ac)*, A	V_{BC}, V	X_L, Ω	*L*, H	*L*, Rated value, H
0	1					
15	1					
30	1					direct current
Rated value	1					

Inductors in Parallel

7. Determine the total inductance L_T of L_1 and L_2 connected in parallel by the following method:
   a. Connect the circuit of Fig. 43-5. Separate L_1 and L_2 as far as possible so that there is no coupling between them.
   b. Measure, and record in Table 43-4, the voltages V_{LT} across L_1 and L_2 (A to B) and V_R across R.
   c. Compute and record the current I and L_T.

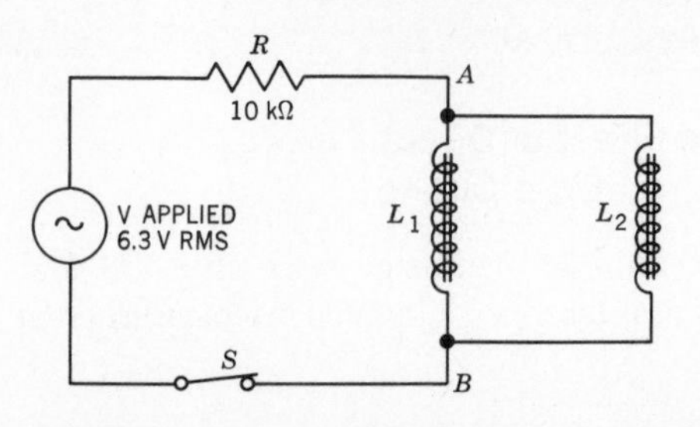

Fig. 43-5. Effect of adding inductors in parallel.

TABLE 43-4. Parallel-connected Chokes

V_{LT}	V_R	I	L_T

QUESTIONS

1. Explain the method you used to determine the inductance of an iron-core choke.
2. (*a*) What is the sum L_T of L_1 and L_2 recorded in Table 43-2?
   (*b*) How does this compare with the measured value of L_T in Table 43-3?
   (*c*) Explain the difference, if any, between these two values.
3. (*a*) Determine L_T from the formula

$$\frac{1}{L_T} = \frac{1}{L_1} + \frac{1}{L_2}$$

   (*b*) How does this computed value of L_T compare with that obtained in Table 43-4?
   (*c*) Explain the difference, if any, between these two values.
4. Show the formula for L_T of three inductors in series, assuming no mutual coupling.
5. Show the formula for L_T of three inductors in parallel, assuming no mutual coupling.
6. Compare the formulas for R_T of resistors with those for L_T of inductors where no mutual coupling exists.
7. Give a simple formula L_T for two inductors in parallel where no mutual coupling exists.
8. Why aren't the measured values of L_1 and L_2 in step 5 the same as their rated values?

Answers to Self-Test

1. 21.5
2. 1.62
3. (*a*) 880; (*b*) 2.34
4. (*a*) 0.05; (*b*) 800; (*c*) 2.12

44

EXPERIMENT

TRANSFORMER CHARACTERISTICS

OBJECTIVES

1. To determine experimentally the turns ratio of a given transformer
2. To measure the effect of an increase of current in the secondary winding (load current) on current in the primary (source current)
3. To resistance-test a transformer

INTRODUCTORY INFORMATION

Ideal Transformer

A transformer is a device for coupling ac power from a source to a load. The source of power is connected to the primary winding, the load to the secondary winding. In the process, certain transformations take place which are related to the construction, and to the primary to secondary turns ratio, of the transformer.

Transformers find many uses in electronics. There are power transformers, filament transformers (for heating the filaments of a vacuum tube), audio transformers, radio-frequency transformers, etc.

A conventional transformer consists of two or more windings on a core, magnetically coupled (Fig. 44-1). An ac voltage applied across the input or primary winding causes current in the primary. This sets up an expanding and collapsing magnetic field which cuts the turns of the secondary winding, inducing an ac voltage in the secondary. When a load is connected across the secondary, current flows in the load.

The core around which the primary and secondary are wound may be iron, as in the case of low-frequency power and audio transformers. An air core may be employed for coupling higher-frequency circuits. The core material and the geometry of the windings determine the characteristics of coupling.

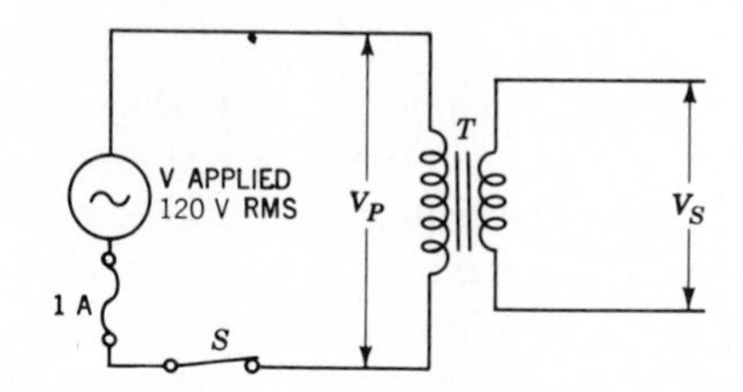

Fig. 44-1. Stepdown transformer without load.

An alternating current in a winding of an iron-core transformer magnetizes the core first in one direction and then in the other. It is this moving magnetic field in the core which cuts the windings of the other coil, inducing a voltage in it.

In an ideal iron-core transformer, that is, one in which there are no power losses, 100 percent of the source (primary) power would be delivered to the load. Equation (44-1) shows this power relationship.

$$V_p \times I_p = V_s \times I_s \qquad (44\text{-}1)$$

Here V_p and I_p are the primary voltage and current, respectively, while V_s and I_s are the secondary voltage and current. In a lossless transformer, the ratio between the primary voltage and the voltage induced in the secondary is the same as the ratio (a) between the number of turns N_p of the primary and number of turns N_s of the secondary.

$$\frac{V_p}{V_s} = \frac{N_p}{N_s} = a \qquad (44\text{-}2)$$

If $a = 1$, there are as many turns in the primary as there are in the secondary, and the voltages appearing across the primary and secondary are equal. This type of 1:1 transformer is called an *isolation transformer*.

If a is greater than 1, a lower voltage appears across the secondary than across the primary. This is called a voltage stepdown transformer.

If a is less than 1, a higher voltage appears across the secondary than across the primary. This is a voltage stepup transformer.

Equation (44-1) for an ideal transformer may also be written as

$$\frac{V_p}{V_s} = \frac{I_s}{I_p} \qquad (44\text{-}3)$$

From Eqs. (44-2) and (44-3) we have

$$\frac{V_p}{V_s} = \frac{I_s}{I_p} = a \qquad (44\text{-}4)$$

The last equation states that current and voltage in the windings of a transformer are inversely related. Therefore, a voltage stepup transformer is also a current stepdown transformer, whereas a voltage stepdown is a current stepup transformer.

Power Losses in a Transformer

The ideal transformer does not exist because there are power losses which do not permit 100 percent transfer of power from the source to the load. One such loss is related to the resistance of the windings and is the I^2R or heating loss. Thus, there are I^2R losses associated with the resistance of the primary winding (primary loss) and also with the resistance of the secondary (secondary loss). When there is no load on the secondary winding, there is no current and hence no power loss in the secondary. However, there is current and hence some I^2R loss in the primary.

Eddy currents are present in the core of an iron-core transformer. A circulating eddy current is induced in the iron core by the changing magnetic field. Eddy currents heat the transformer and act like an I^2R loss. Eddy currents thus rob the source of power and represent another power loss.

Another loss associated with a transformer core is *hysteresis loss*. The hysteresis effect results from the fact that magnetic lines of force lag behind the magnetizing force that causes them. Hysteresis can be understood by considering the fact that when a magnetizing force is removed from an iron-core magnetic circuit, a portion of the flux remains within the iron. This residual magnetism can be removed by applying to the iron a magnetizing force opposite in direction to that of the initial force. The energy required to demagnetize the iron acts as a core loss, which is associated with the reversal of magnetizing current in the winding.

One other loss must be mentioned. This is associated with the magnetic leakage which exists in a transformer. Not all the lines of magnetic force will link the turns of the secondary winding. The lines of force thus lost to the magnetizing circuit constitute magnetic leakage.

Effect of Load Current on Primary Current

Primary current in the transformer of Fig. 44-2 depends on the load current in the secondary, as is evident from the equation

$$\frac{I_s}{I_p} = a$$

or

$$I_p = \frac{I_s}{a} \tag{44-5}$$

Therefore, as load current I_s increases due to a decrease in load resistance R_2, primary current must increase. The increase in primary current is explained by the assumption that the change in load impedance (as is evidenced by an increase in load current) appears also as a reflected impedance in parallel with the primary. Since power in the primary and secondary was assumed to be equal, and since power can be dissipated only in a resistance, the reflected impedance must be the resistance R_1 (Fig. 44-2).

In addition to the primary current I_p resulting from the reflected load impedance, the primary supplies the current for the iron-core losses and the magnetizing current. The vector sum of these two currents (the magnetizing current is out of phase with the voltage) is called the "exciting current." The exciting current is about 3 to 5 percent of the rated output of the transformer. This explains why primary current will normally measure more than that predicted by Eq. (44-5). Note that when a vector is used to represent time-varying voltages and currents the term phasor is used. Phasors are discussed more fully in Experiment 51 and in subsequent experiments.

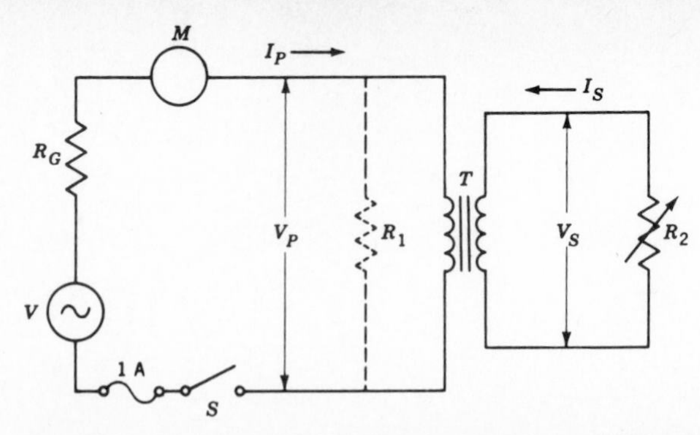

Fig. 44-2. Primary current I_P increases when secondary current I_S increases, due to reflected impedance (R_1) from the secondary to the primary.

Resistance-Testing Transformer Windings

Resistance (ohmmeter) tests of the individual windings of the small transformers used in electronics are used to determine the continuity of the windings. An ohmmeter test also establishes the resistance of each winding. The technician then compares the measured resistance with the rated value to determine if a suspected transformer is in fact defective.

The following considerations may be helpful in analyzing the results of resistance measurements.

Winding Measures Infinite Resistance. This winding is "open." The break may be at the beginning or end of the winding, where the connection is made to the terminal leads. This type of break can be easily repaired by resoldering the leads to the winding. If the discontinuity is elsewhere, the transformer must be replaced.

Winding Resistance "Very" High. A winding whose resistance measures "very" high compared with its rated value may be open, or there may be a cold-solder joint at the terminal connections. If the condition cannot be corrected, the transformer must be replaced.

Winding Resistance "Very" Low Compared with Rated Value. Turns of the winding must be "shorted" somewhere on the transformer, or the winding may be shorted to the frame. However, a small difference in resistance between the rated and measured values may be insignificant. For example, if a primary rated at 120 Ω measures 100 Ω, the difference may be attributable to the inaccuracy of the meter. In cases of doubt, other tests (which will not be discussed here) are indicated.

Resistance between Windings. The windings of transformers, other than autotransformers, are insulated from each other. There should be infinite resistance between insulated windings, as long as the transformer is not connected in a circuit. If the insulation between two windings breaks down, there will be a measurable resistance between these windings, signifying a defective transformer.

Factors which Determine Resistance of a Winding. The resistance of a winding depends on the diameter of the wire and the number of turns. Thus, resistance varies inversely as the square of the diameter and directly as the number of turns. Large diameters are required for high-current windings, smaller diameters for windings which carry less current. The primary of a voltage stepdown transformer has more windings than the secondary. Moreover, a voltage stepdown is also a current stepup transformer. Hence the secondary must carry more current than the primary. Therefore, the resistance of the primary will be higher than that of the secondary. How much higher depends on the transformer ratio and on the diameter of the two wires.

SUMMARY

1. A transformer consists of two or more windings on a core. The core is iron for low-frequency applications or air for higher frequencies.
2. When an ac source is applied across the primary winding, a voltage is induced in the secondary. If there is a load on the secondary, current also flows in that winding.
3. A transformer is therefore a device which couples ac power from a source connected to the primary, to a load connected to the secondary winding.
4. In an ideal transformer all the source power would be delivered to the load. Thus in Fig. 44-2

$$V_p \times I_p = V_s \times I_s$$

   where V_p and V_s are the primary and secondary voltages, respectively; I_p and I_s are the primary and secondary currents, respectively.
5. In an ideal transformer the ratio between the number of turns in the primary N_p and the number of turns in the secondary N_s is the same as the ratio of the voltage in the primary V_p to the voltage in the secondary V_s. Thus

$$\frac{N_p}{N_s} = a = \frac{V_p}{V_s}$$

6. In an ideal transformer the ratio of current in the secondary I_s to current in the primary I_p is also a. Thus

$$\frac{N_p}{N_s} = a = \frac{I_s}{I_p}$$

7. A voltage stepdown transformer is one in which the secondary voltage is lower than the primary. A voltage stepdown transformer is also a current stepup transformer. In this type of transformer $a > 1$.

NOTE: The symbol $>$ means greater than; $<$ means smaller than.

8. If $a < 1$, the transformer is a voltage stepup, current stepdown device.
9. The power losses which occur in a transformer include:
   (*a*) I^2R or heating losses due to the resistance of the windings.
   (*b*) Eddy current losses present in an iron-core transformer.
   (*c*) Hysteresis losses, which result from the lag between the magnetic lines of force and the magnetizing force which induces them.
   (*d*) Magnetic leakage, which occurs because not all the magnetic lines of force link the turns of the secondary winding.
10. An increase in I_s will result in an increase in I_p and a resulting decrease in primary impedance. The decrease in primary impedance is caused by the impedance reflected from the secondary into the primary. The reflected impedance is the resistance R_1 (Fig. 44-2) which acts in parallel with the primary winding.
11. The primary current supplies the power losses associated with a transformer. Therefore, because there are losses associated with every transformer, $I_p > I_s/a$.
12. Ohmmeter tests of the resistance of individual windings of a transformer are used to determine the continuity of the windings.
13. The separate windings of a transformer are insulated from each other and from the transformer frame. Therefore, the resistance between windings should be infinite, as long as the transformer is *not* connected in the circuit.
14. The resistance between an individual winding and the frame of the transformer should also be infinite.
15. The resistance of a transformer winding depends on the wire diameter and the number of turns. In a voltage stepdown, current stepup transformer:
   (*a*) There are more turns in the primary than in the secondary.
   (*b*) The secondary wire diameter is larger than that of the primary wire.
   (*c*) The resistance of the secondary is lower than the primary.

SELF-TEST

Check your understanding by answering these questions:

1. In a particular ideal transformer, the primary has 10 times as many turns as the secondary. The voltage across the primary is 120 V. The voltage across the secondary must be __________ V.
2. In the transformer of question 1, the current in the secondary is 3 A. The current in the primary must therefore be __________ A.
3. In the transformer of question 1, the turns ratio N_p/N_s is __________.
4. There is no *ideal* transformer for all practical purposes. __________ (true/false)
5. The power losses in a transformer include those due to:
   (*a*) __________ of the windings
   (*b*) __________ __________
   (*c*) __________
   (*d*) __________ __________

6. An increase in secondary current causes a(n) ____________ (increase/decrease) in primary current.
7. The impedance reflected from the secondary into the primary is a ____________ (resistance/inductance).
8. The power losses in a transformer are supplied by the ____________ ____________ ____________.
9. In an iron-core power transformer which is not connected in a circuit, the resistance measured between the primary and secondary windings is 10 Ω. This resistance ____________ (is/is not) normal.
10. The primary and secondary windings must be ____________ in the transformer in question 9.

MATERIALS REQUIRED

- Power supply: Source of 120-V 60-Hz
- Equipment: EVM; multirange ac milliammeter; decade-type rheostat; ohmmeter
- Transformer: AF output, 6 K6 to 3-Ω speaker
- Miscellaneous: SPST switch; fused line cord

PROCEDURE

Transformer Turns Ratio, a:

1. Connect the circuit of Fig. 44-1.

CAUTION: Be certain that the line voltage (120 V ac) is applied to the *primary* of the voltage stepdown transformer. The secondary is not loaded.

2. Measure the voltages V_p (across the primary) and V_s (across secondary) and record in Table 44-1.

3. Compute and record the turns ratio a.

Effect of Load on Primary Current

4. Connect an ac milliammeter M in the primary circuit as in Fig. 44-2. The ac milliammeter should be set on the highest range, then adjusted to the proper range. R_G represents the internal resistance of the generator. V is the 120 V of the 60-Hz line. R_2 is a decade rheostat acting as the load resistor.

TABLE 44-1. Transformer Turns Ratio

V_p	V_s	$a = \frac{V_p}{V_s}$

5. Measure and record the primary current I_p, the voltage across the primary V_p, and the voltage across the secondary V_s, in turn for each value of load resistance R_2 shown in Table 44-2. An EVM is used to measure V_p and V_s.

6. Compute and record the secondary current ($I_s = V_s/R_2$). Substitute the value of I_s in the equation $I_{P1} = I_s/a$. Compute and record I_{P1}. (The computed value I_{P1} is the current which would appear in the primary of an ideal transformer.)

7. Subtract I_{P1} from I_p (the current measured in the primary) and record in Table 44-2 under column labeled "ΔI."

8. Compute and record the power W_s dissipated by the load resistor R_2 for each value of R_2.

CAUTION: In all cases, be certain that the wattage rating of R_2 is high enough for the application.

Resistance Measurement of the Windings

9. Remove power and disconnect the transformer primary and secondary from the circuit.

10. Measure the resistance of the primary and secondary with an ohmmeter, and record in Table 44-3. Also measure and record the resistance between windings and between each winding and the frame.

TABLE 44-2. Effect of Load Current on Primary Characteristics

R_2, Ω	I_p, mA	V_p, V	V_s, V	$I_s = \frac{V_s}{R_2}$ mA	$I_{P1} = I_s/a$, mA	$\Delta I = I_p - I_{P1}$, mA	W_s
15							
10							
5							
4							
3							
2							
1							

TABLE 44-3. Resistances in a Transformer

Resistance, Ω				
Primary	*Secondary*	*Primary to Secondary*	*Primary to Frame*	*Secondary to Frame*

QUESTIONS

1. Is the transformer you used a voltage stepup or a current stepup?
2. How is primary current affected by current in the secondary? Why?
3. Does ΔI appear to be relatively constant in Table 44-2? What does ΔI represent?
4. What losses, if any, are incurred in the transformer used in this experiment?
5. Where does the power come from to supply the transformer losses?
6. What is the relationship between the resistance of a winding and the
   (*a*) Current rating of that winding?
   (*b*) Number of turns of the winding?
   Refer to your data to substantiate the answers.
7. What are the sources of error in the experimental data?

Answers to Self-Test

1. 12
2. 0.3
3. 10
4. true
5. (*a*) heating; (*b*) eddy current; (*c*) hysteresis; (*d*) magnetic leakage
6. increase
7. resistance
8. input power source
9. is not
10. shorted

45

EXPERIMENT

CAPACITOR COLOR CODE AND TESTING CAPACITORS

OBJECTIVES

1. To find the capacitance value of a capacitor by means of the EIA color code
2. To use a capacitance checker to determine capacitance, leakage, and power factor
3. To use an EVM for testing shorted or leaky capacitors

INTRODUCTORY INFORMATION

Color Code

At present, capacitance values of molded paper, mica, and disk ceramic capacitors are usually printed on the unit by the manufacturer. However, some equipment in use today may contain components marked according to several obsolescent systems. These obsolescent codes are shown here along with the codes which are currently used for ceramic tubular-style capacitors.

The two types of molded capacitors, tubular and flat, are shown in Fig. 45-1. Color bands are used on the tubular type and color dots on the flat type in accordance with the code shown in Fig. 45-2. Figures 45-1 and 45-2 explain how the code is applied. All values are in picofarads. It should be noted that the MIL code markings are somewhat different from the EIA markings. There are two types of molded mica capacitors, the flat postage-stamp type and the button silver mica, shown in Fig. 45-3. The accompanying obsolescent color code is shown in Figs. 45-3 and 45-4.

Figures 45-5 and 45-6 show an obsolescent system for color-coding disk ceramic capacitors.

Figures 45-7 and 45-8 show the EIA color codes which are presently in use for ceramic tubular-style capacitors.

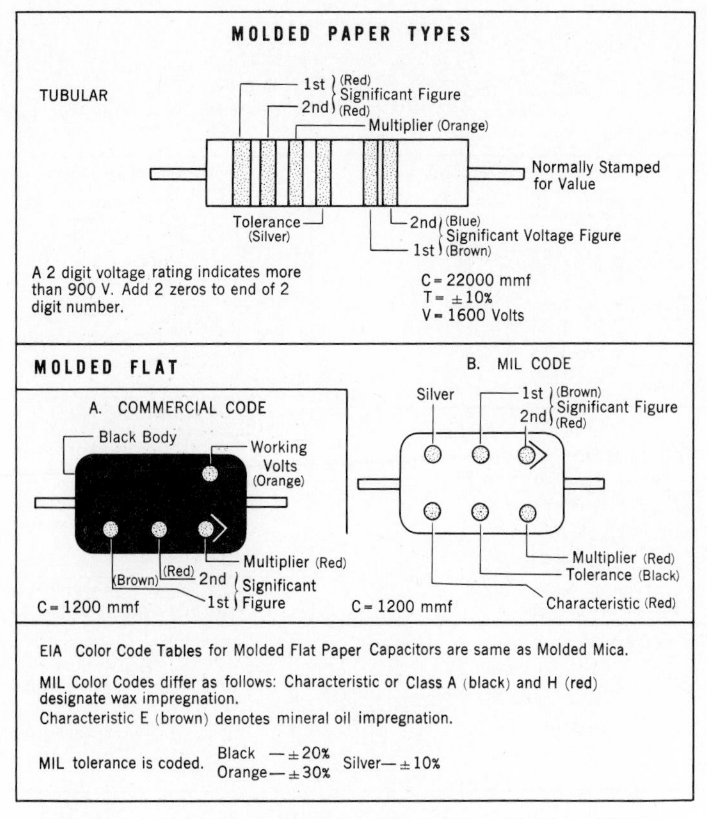

Fig. 45-1. Molded capacitor types (*Centralab, a Division of Globe-Union, Inc.*).

Color	Significant Figures	MOLDED PAPER TUBULAR CAPACITORS		
		Decimal Multiplier	Tolerance ±%	Voltage Volts
Black	0	1	20	—
Brown	1	10	—	100
Red	2	100	—	200
Orange	3	1000	30	300
Yellow	4	10000	40	400
Green	5	10^5	5	500
Blue	6	10^6	—	600
Violet	7	—	—	700
Gray	8	—	—	800
White	9	—	10_{EIA}	900
Gold	—	0.1	—	—
Silver	—	—	10_{EIA}	—
No Color	—	—	—	—

Fig. 45-2. Color code for molded capacitors (*Centralab, a Division of Globe-Union, Inc.*).

Color	Significant Figures	MOLDED MICA EIA Standard—MIL Specification Capacitors			
		Decimal Multiplier	Tolerance ±%	RMA Voltage Rating (All Capacitors)	Class
Black	0	1	20 (JAN RMA 1948)	—	A
Brown	1	10	—	100	B
Red	2	100	2	200	C
Orange	3	1000	3 (EIA)	300	D
Yellow	4	10000	—	400	E
Green	5	—	5 (EIA)	500	F (MIL)
Blue	6	—	—	600	G (MIL)
Violet	7	—	—	700	—
Gray	8	—	—	800	I (EIA)
White	9	—	—	900	J (EIA)
Gold	—	0.1	5 (MIL)	1000	—
Silver	—	0.01	10	2000	—
No Color	—	—	20 (old RMA)	500 (old RMA)	—

Fig. 45-4. Color code for molded mica capacitors (*Centralab, a Division of Globe-Union, Inc.*).

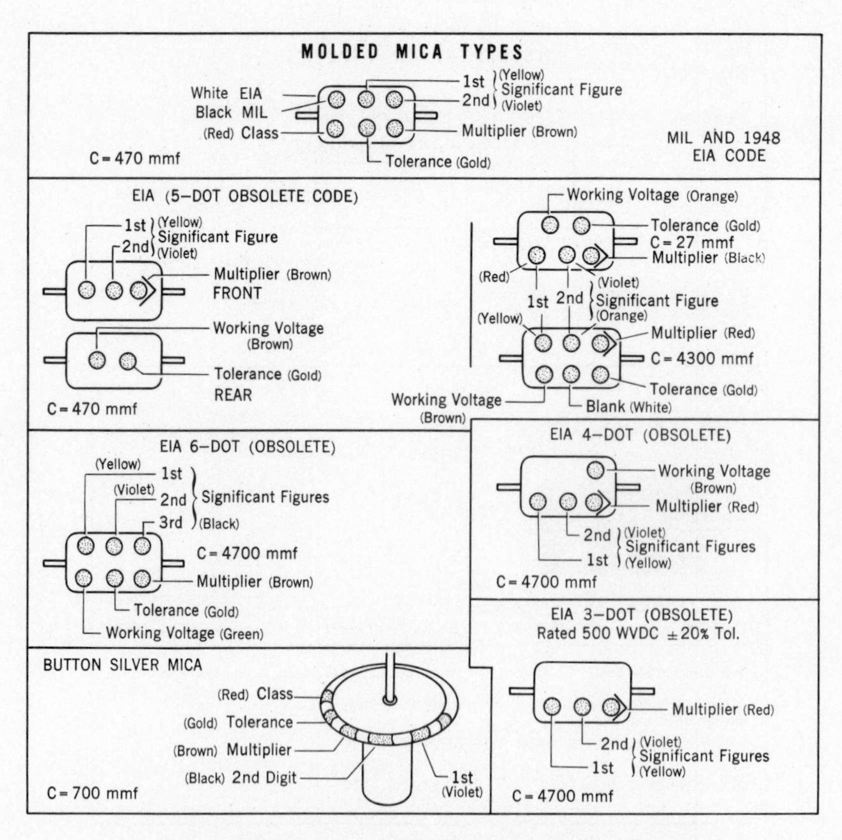

Fig. 45-3. Molded mica capacitor types (*Centralab, a Division of Globe-Union, Inc.*).

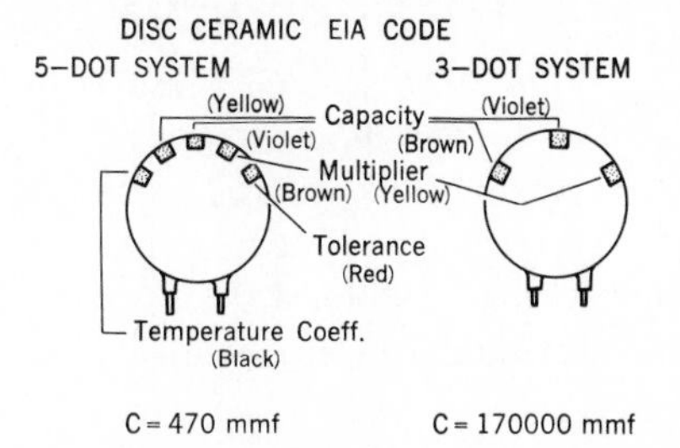

Fig. 45-5. Disk ceramic capacitor types (*Centralab, a Division of Globe-Union, Inc.*).

Color	Significant Figures	Decimal Multiplier	CERAMICS EIA Standard—MIL Specification Capacitors				
			Tolerance		Temperature Co-efficient Parts Per Million Per Degree Centigrade	T.C. for Extended Range TC HiCap	
			Capacity 10 mmf or Less	Capacity More Than 10 mmf		Significant Fig.	Multiplier
Black	0	1	±2.0 mmf(MIL)	±20%	NPO	0	−1
Brown	1	10	±0.1 mmf	±1%	N33	—	−10
Red	2	100	—	±2%	N75	1	−100
Orange	3	1000	—	±2.5%(EIA)	N150	1.5	−1000
Yellow	4	10000(EIA)	—	—	N220	2.2	−10000
Green	5	—	±0.5 mmf	±5%	N330	3.3	+1
Blue	6	—	—	—	N470	4.7	+10
Violet	7	—	—	—	N750	7.5	+100
Gray	8	0.01	±0.25 mmf	—	+30	—	+1000
White	9	0.1	±1.0 mmf	±10%	N330±500	—	+10000
Gold	—	—	—	—	+100 (MIL)	—	—
Silver	—	—	—	—	Bypass & Coupling (EIA)	—	—
No Color	—	—	—	—	—	—	—

Fig. 45-6. Color code for disk ceramic capacitors (*Centralab, a Division of Globe-Union, Inc.*).

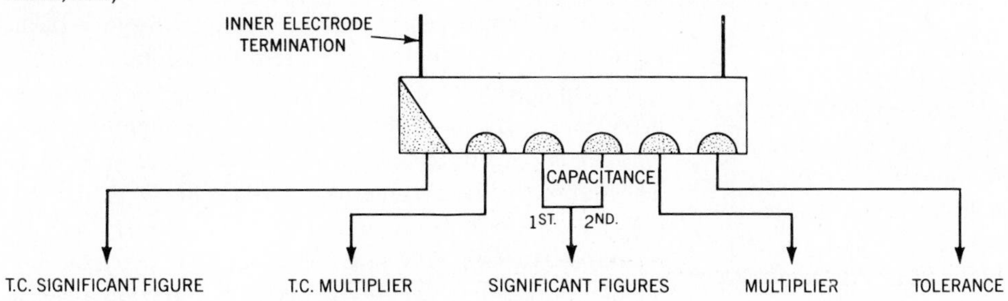

Fig. 45-7. Color coding for general-purpose tubular capacitors (*Centralab, a Division of Globe-Union, Inc.*).

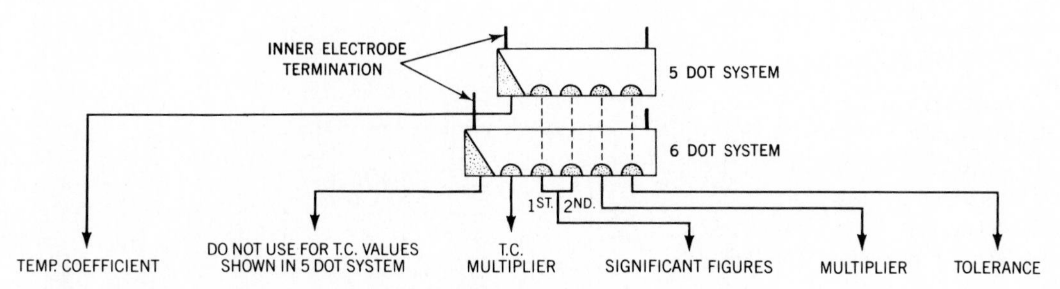

Fig. 45-8. Color coding for temperature-compensating ceramic tubular capacitors (*Centralab, a Division of Globe-Union, Inc.*).

Capacitance Testers

Different types of capacitance testers are available to the technician. There is the tester which checks only the capacitance of a capacitor. This may be used to check the rated value of a capacitor or to determine the capacitance of a capacitor whose value is unknown. This type of tester will also detect open and shorted capacitors.

Other types, in addition to acting as capacitance bridges, also determine some or all of the following capacitor characteristics.

1. *Insulation Resistance for Paper, Mica, and Ceramic Capacitors*. Insulation resistance R_I is effectively a resistance in parallel with the capacitor. The value of insulation resistance considered normal varies with the type of capacitor. This characteristic is considered only for electrostatic and nonelectrolytic capacitors. Approximate values of insulation resistance are shown below.
   (*a*) *Molded Paper Capacitors*. Insulation resistance R_I is inversely proportional to capacitance. The product of $R_I \times \mu F$ should equal about 1000 megohms-microfarads when new, depending on the ambient temperature.
   (*b*) *Mica Capacitors*. When new, these will have a value of R_I greater than 3000 MΩ.
   (*c*) *Ceramic Capacitors*. When new, these will have a value of R_I greater than 7500 MΩ.
2. *Leakage Current of Electrolytic Capacitors*. This test for electrolytic capacitors takes the place of the R_I test for electrostatic capacitors. Leakage current I is directly proportional to capacitance. Maximum leakage current permissible for electrolytics is usually included in the instruction manual of the capacitor checker. Values of leakage current are read directly on the calibrated meter scale of the tester.
3. *Power Factor of Electrolytic Capacitors*. The power factor (PF) of an electrolytic capacitor is a measure of the power losses in that capacitor. The losses occurring in a capacitor may be indicated as a resistance R in series with the capacitor. The phase angle of this RC combination, therefore, indicates the losses of this capacitor. The cosine of this phase angle is defined as the power factor. Acceptable PF values for new electrolytic capacitors are usually given in the instruction manual accompanying the capacitance tester.

NOTE: A capacitor should be replaced if any of its measured characteristics do not come within the range of acceptable norms.

For proper testing, the entire capacitor, or at least one lead of the capacitor, should be disconnected from the circuit. An exception to the rule is a tester on the market which checks shorts and opens of capacitors wired in the circuit.

The residual capacitance of a checker should be determined before it is used to measure the values of very small capacitances. Moreover, for accurate readings the pigtails of the capacitor should be placed directly across the terminals of the checker. Long clip leads should be avoided, since their capacitance will add to that of the capacitor under test.

The student or technician should be fully familiar with the operating instructions of the capacitance checker being used in order to derive full benefit from it. Moreover, the student should perform all the checks which the instrument is capable of making.

Testing Capacitors with an EVM

Completely shorted capacitors or very "leaky" capacitors (capacitors with a very low value of insulation resistance) may frequently be detected with a VOM or EVM.

A simple resistance test in which the ohmmeter is placed across the terminals of the capacitor is used. If the capacitor is shorted, the meter will read zero resistance. Of course, this capacitor must be discarded. A low value of insulation resistance may also be detected by an ohmmeter test.

Occasionally a capacitor will become shorted or leaky only when a voltage is placed across it. This capacitor is said to break down under load. This defect cannot be detected with an ohmmeter, but it can be found with a dc voltmeter. In this "dynamic" test (see Fig. 45-9), a dc voltage is placed across the series combination of the voltmeter and capacitor. A good capacitor will show only a momentary charge on the voltmeter, after which zero voltage will be indicated. However, a capacitor breaking down under the voltage will give a voltage indication on the meter. The higher the leakage current, the higher the voltage will read. When checking for leakage in this fashion, the leakage resistance is in series with the meter and limits the current. If the leakage resistance is much greater than the input resistance of the voltmeter, the maximum current is mainly dependent upon the value of leakage resistance. The rated dc working voltage V of the capacitor should be used for this test if possible.

The input resistance of an *EVM* is much higher than that of a VOM on the low-voltage ranges. It *is* therefore *more effective in measuring leakage of a nonelectrolytic capacitor.* The best procedure to follow is to connect the EVM in series with the capacitor C, as in the circuit of Fig. 45-9. The EVM is first set to a high dc voltage range. If the voltage indication is low, after the capacitor has charged, the voltmeter is switched to the lowest range on which the meter is not overloaded. The voltage measured can be used to determine

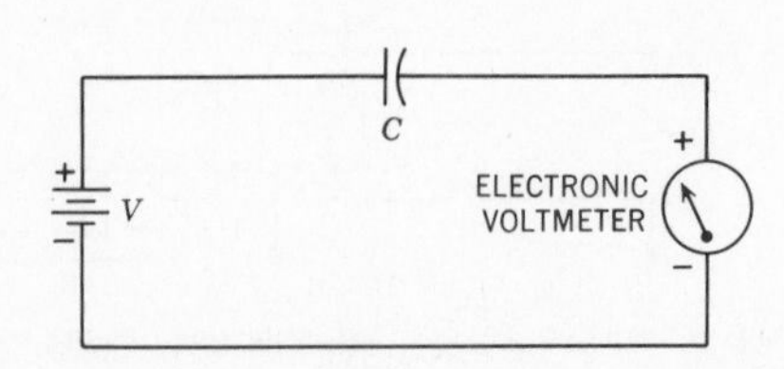

Fig. 45-9. Testing capacitor with an electronic voltmeter.

the insulation resistance R_I of the capacitor by means of the following equation:

$$R_I = R_{in} \frac{V - V_1}{V_1} \quad (45\text{-}1)$$

where V = applied voltage
V_1 = measured voltage
R_{in} = input resistance of the meter. In the case of an EVM it includes both the resistance of the dc probe and the input resistance of the meter on the dc voltage ranges

12. The capacitor is connected in series with the electronic voltmeter, and the rated voltage V of the capacitor is placed across this series combination. The capacitor leakage resistance R_I can be determined from the voltage V_1 read on the voltmeter by means of the formula

$$R_I = R_{in} \times \frac{V - V_1}{V_1}$$

where R_I is the leakage resistance of the capacitor and R_{in} is the input resistance of the electronic voltmeter.

SUMMARY

1. Capacitance values of paper, mica, and disk ceramic capacitors are presently printed on the body of the capacitors.
2. Older nonelectrolytic capacitors used a color code to designate their values. The systems employed and the corresponding values are shown in Figs. 45-1 through 45-8.
3. Whichever system was used, these significant characteristics were normally included in the code or markings: (*a*) value in microfarads or picofarads; (*b*) tolerance in percent; (*c*) voltage rating.
4. Capacitance checkers are used to find the capacitance value of an unknown capacitor.
5. Capacitance testers may also check:
   (*a*) Insulation or leakage resistance of nonelectrolytic paper, mica, and ceramic capacitors.
   (*b*) Leakage current of electrolytic capacitors.
   (*c*) Power factor of electrolytic capacitors.
6. A capacitor should be replaced if any of its measured characteristics are not in the normal range.
7. Some capacitor checkers are especially designed to test for opens and shorts of capacitors connected in the circuit.
8. If the capacitance checker is not of the *go, no go* type used in summary step 7, it is *not* possible to check a capacitor directly *in* the circuit. It *is* necessary to *disconnect* one of the leads of the capacitor from the circuit before testing it.
9. In finding the capacitance value of very small valued capacitors, it is necessary to subtract the residual capacitance of the tester from the measured value. Moreover, long clip leads should be avoided, and the pigtails of the capacitor should be placed directly across the terminals of the checker.
10. In testing for shorted capacitors, an ohmmeter test will reveal the short. For this type of test a VOM or EVM may be used.
11. Some capacitors may break down or display excessive leakage only under load, that is, when their rated voltage appears across their terminals. To locate this type of trouble, an EVM on the voltage ranges must be used.

SELF-TEST

Check your understanding by answering these questions:

1. The characteristics of a nonelectrolytic capacitor which can be checked on a capacitance tester are:
   (*a*) __________ __________
   (*b*) __________ __________
2. The characteristics of an electrolytic capacitor which can be checked on a capacitance tester are:
   (*a*) __________ __________
   (*b*) __________ __________
   (*c*) __________ __________
3. When a color code is used on a nonelectrolytic capacitor, these are the capacitor characteristics which are usually designated:
   (*a*) __________ __________
   (*b*) __________
   (*c*) __________ __________
4. An __________ may be used to check a capacitor for shorts.
5. A dynamic test of a capacitor requires placing a voltage equal to the __________ __________ of the capacitor across a series combination including the capacitor and an __________.
6. Using the test circuit of Fig. 45-9 and Eq. (45-1), the insulation resistance of a capacitor, if V = 600 V, $R_{in} = 13 \times 10^6\ \Omega$, and V_1 = 10 V, is __________ Ω.
7. The insulation resistance of a capacitor is considered to be in __________ (series/parallel) with the capacitor.
8. A capacitor is considered defective if its insulation resistance is very __________ (high/low).
9. A capacitor is considered defective if the voltage measured in the test circuit of Fig. 45-9 is very low. __________ (true/false)

MATERIALS REQUIRED

- Power supply: Variable dc voltage supply
- Equipment: EVM, capacitance checker
- Capacitors: Assortment of paper, mica, ceramic, and electrolytic, including 0.01 μF, 0.1 μF, 0.5 μF

PROCEDURE

1. Determine the value of each capacitor supplied, from its color code. Fill in the information required in Table 45-1.
2. Refer to the instruction manual of the capacitance checker for the procedure for measuring capacitance values.
3. Determine and record the residual capacitance of the checker.
4. Measure each capacitor and record the results in Table 45-1 in the row labeled "Measured value." If the capacitor is very small, compensate for the residual capacitance of the checker by subtracting residual capacitance from measured value.
5. If the checker has facilities, measure the insulation resistance of the molded mica and ceramic capacitors supplied, and record in Table 45-1.
6. If the checker has facilities, measure the leakage current and power factor of electrolytic capacitors supplied, and record in Table 45-1.
7. **a.** Place the leads of an ohmmeter across one of the nonelectrolytic capacitors. Note the charging effect on the meter.
   b. Reverse the ohmmeter leads across the capacitor. Describe the effect.
8. Measure each of the nonelectrolytic capacitors with an ohmmeter. Record the resistance of any whose insulation resistance is measurable.
9. Measure the resistance of each of the electrolytic capacitors supplied and record the results in Table 45-2. Be sure to use the ohmmeter test-lead polarity which gives indications of greater resistance. In most cases this occurs when the positive ohmmeter test lead is connected to the positive capacitor terminal.
10. Connect a dc voltage across the series combination of a 0.01-μF capacitor and a VOM as in Fig. 45-9. Set V at the rated voltage of the capacitor, or to the maximum voltage of the supply if the dc output of the power supply is not as high as the rated voltage of the capacitor. Record V in Table 45-3. Measure the voltage V_1 and record in Table 45-3. Record also the voltage range of the VOM and R_{in}. Compute and record R_l using Eq. (45-1).
11. Now, using the same capacitor, measure V_1 with an EVM and record in Table 45-3. Record also the applied voltage V, the meter range, and R_{in}. Compute and record R_l.
12. Repeat steps 10 and 11 for the other two capacitors in Table 45-3.

TABLE 45-1. Capacitor Measurements

	Capacitor									
	1	2	3	4	5	6	7	8	9	10
Type										
1st color										
2nd color										
3rd color										
4th color										
Coded value, μF										
Tolerance, %										
Measured value										
Insulation resistance										
Leakage current										
Power factor										

TABLE 45-2. Testing Electrolytic Capacitors

	Capacitor			
	1	2	3	4
Value				
DC working voltage				
Resistance				

TABLE 45-3. Voltmeter Measurement of Insulation Resistance

Meter Used	*Capacitance,* μF	*V Applied,* V	V_1 *Measured,* V	*Meter Range*	R_{in}, Ω	R_I *Computed,* MΩ
VOM	0.01					
	0.1					
	0.5					
EVM	0.01					
	0.1					
	0.5					

QUESTIONS

1. What is meant by residual capacitance of a capacitance checker?
2. When is it necessary to compensate for the residual capacitance of a checker?
3. How do the measured values of capacitance compare with the rated values (Table 45-1)? Explain any discrepancies.
4. What indication is there on your capacitance tester of (*a*) A shorted capacitor? (*b*) An open capacitor?
5. Is a resistance check of a capacitor usually a reliable test of capacitor leakage? Why?
6. Assuming the meter in Fig. 45-9 is a 1000 Ω/V voltmeter, answer (*a*) through (*d*).
   (*a*) Replacing the 1000 Ω/V voltmeter of Fig. 45-9 by its equivalent resistance on the 100-V range, draw the equivalent circuit of Fig. 45-9.
   (*b*) What would be the voltage shown on the meter on this range if the insulation resistance of the capacitor under test is 4 MΩ and $V = 600$ V? Compute the percentage deflection in terms of full-scale deflection.
   (*c*) Substitute the equivalent resistance of the 10-V range. What is the percentage of deflection? What would be the voltage shown on the meter?
   (*d*) How do the two percentages of deflection compare? Why?
7. Refer to the data in Table 45-3. Which is more effective in measuring the leakage resistance R_I of a capacitor, the VOM or the electronic voltmeter? Why?

Answers to Self-Test

1. (*a*) capacitance value; (*b*) insulation resistance
2. (*a*) capacitance value; (*b*) leakage current; (*c*) power factor
3. (*a*) capacitance value; (*b*) tolerance; (*c*) voltage rating
4. ohmmeter
5. voltage rating; EVM
6. 767×10^6
7. parallel
8. low
9. false

46

EXPERIMENT

RC TIME CONSTANTS

OBJECTIVES

1. To determine experimentally the time it takes a capacitor to charge through a resistance
2. To determine experimentally the time it takes a capacitor to discharge through a resistance

INTRODUCTORY INFORMATION

Charging and Discharging a Capacitor

In electronic circuits, capacitors are used for many purposes. They are used to store energy, to pass alternating current, and to block direct current. They act as filter elements and as components in "tuned" or resonant circuits, they are used in timing circuits, and they have many other functions.

Capacitors carry out their functions by charging and discharging. A capacitor can store and hold a charge of electrons, a process known as *charging*. To charge a capacitor a voltage source is required, V in Fig. 46-1*a*. When switch S is closed, and the circuit is completed (Fig. 46-1*b*), electrons leave the negative terminal of the battery and enter the lower plate of capacitor C. Simultaneously, electrons leave the upper plate of C and return to the positive terminal of the battery. This process continues until the capacitor is fully charged, which occurs when the voltage across the capacitor equals the applied voltage V. When the capacitor is charged, the lower plate contains an excess of electrons and the upper plate a deficiency of electrons. The quantity of excess electrons on the lower plate is exactly equal to the quantity of electrons which left the upper plate. The voltage built up on the capacitor results from this difference in electron charge between the upper and lower plates. The polarity of charge on the capacitor plates is as shown in Fig. 46-1*b*. If, after charging, C is removed from the voltage source, as in Fig. 46-1*c*, it will continue to hold its charge as long as there is no complete circuit through which it can discharge.

If the charged capacitor is now placed across a resistor R (Fig. 46-2*a*), a discharge path is provided for C. Electrons now leave the negative plate of C, move through R, and enter the positive plate of C. This process continues until the excess electrons on the lower plate have replaced the deficiency of electrons in the upper. When this exchange of electrons has taken place, both plates return to their neutral or uncharged state, and the capacitor is said to be *discharged* (Fig. 46-2*b*). A voltmeter placed across C would now show *zero* voltage across the capacitor. The movement of electrons (charges) during the charge and discharge interval, in effect, constitutes electron current flow.

The relationship between the charge Q on a capacitor, the size C of capacitance, and the voltage V across C is expressed by

$$Q = C \times V \tag{46-1}$$

In this formula Q is in coulombs, C in farads, and V in volts. It is apparent from Eq. (46-1) that the larger C is, the more coulombs it will store when fully charged from a given voltage source V.

Time Required to Charge a Capacitor

Consider the circuit of Fig. 46-3. We see a series circuit consisting of a capacitor C, a resistor R, a dc voltage source V, and an on-off switch S_1. A second switch S_2 is connected across the RC combination. Both S_1 and S_2 are open.

Assume that capacitor C has no charge on it. Now close S_1, leaving S_2 open, as in Fig. 46-4*a*. Closing S_1 completes the charging circuit for C. If a voltmeter were connected across C, the meter would read zero volts at the instant S_1 is closed. As C charges, the meter reads an increasing voltage. This

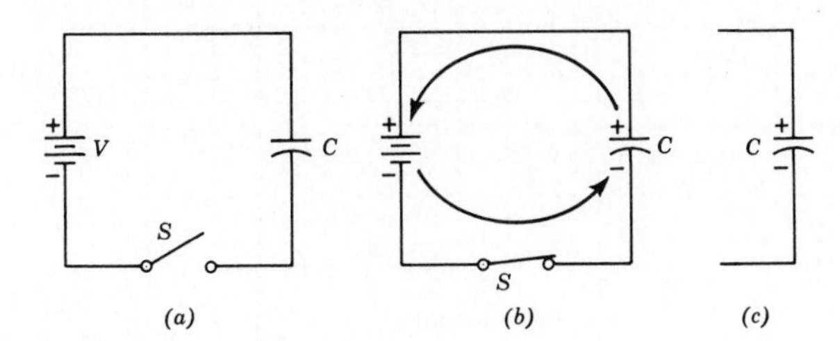

Fig. 46-1. Charging a capacitor.

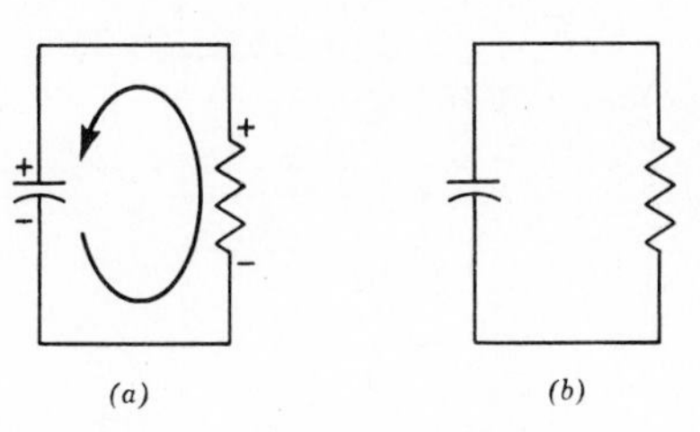

Fig. 46-2. Discharging a capacitor.

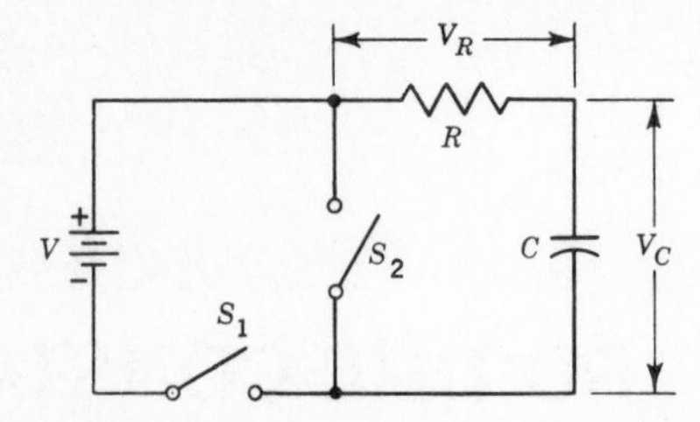

Fig. 46-3. *RC* circuit for charging a capacitor.

experiment simply demonstrates that a capacitor does not charge instantaneously. It takes *time* to charge it. Our concern here is with the question, "How much time does it take to charge a capacitor?"

The manner in which C charges is shown in the accompanying graph (Fig. 46-4*b*). While S_1 was open, C did not charge because the battery voltage was not applied across the *RC* circuit. Thus, with S_1 open, the applied voltage $V = 0$, the charging current $i_c = 0$, the voltage across R, $v_R = 0$, and the voltage across C, $v_c = 0$. At the *instant* switch S_1 is closed, the battery "sees" only the resistance R in the circuit. That is, C acts like a short circuit. Hence, all the applied voltage V appears across R (at the instant S_1 is closed). There is maximum charging current limited only by the size of R, and there is zero voltage across C. But as S_1 remains closed, charging current causes a voltage to build up across C, with the polarity shown in Fig. 46-4*a*. This is a bucking voltage which opposes the battery voltage and acts to reduce the charging current.

At any instant of time after the start of the charging interval, the voltage which causes current to flow in the circuit is the voltage source V minus the voltage v_c to which the capacitor has already charged. If we call this the active voltage V_A, then

$$V_A = V - v_c \tag{46-2}$$

It is apparent that when $v_c = V$, the capacitor is fully charged, $V_A = 0$, and current in the circuit ceases.

The decay in voltage across the resistor follows the decay of charging current. Thus, at the end of the charging interval, though the battery is still applied across the *RC* circuit, the charging current i_c is zero, $v_R = 0$, and $v_c = V$.

Figure 46-4*b* shows that it took time for C to charge. From a mathematical analysis of this circuit, which is beyond the scope of this book, we find that it takes one time constant to charge C to 63.2 percent (approximately) of the applied voltage. The time required to charge a capacitor to 63.2 percent of the applied voltage is known as the *time constant* of the circuit. The value of the time constant t in seconds is equal to the product of the resistance R in ohms and the capacitance of C in farads. Thus

$$t = R \times C \tag{46-3}$$

A time constant is not a fixed unit of time such as a second, but is a *relative* unit of time whose actual value in seconds is determined by the values of R and C in the charging circuit. Thus, in Fig. 46-4, if $R = 1\ \text{M}\Omega$ and $C = 1\ \mu\text{F}$, $t = 1$ s. But if $R = 100{,}000\ \Omega$ and $C = 0.1\ \mu\text{F}$, then $t = 0.01$ s.

The student may find the following relations a useful reference in calculating time constants:

R (in ohms) $\times$ C (in farads) $=$ t (in seconds)
R (in megohms) $\times$ C (in microfarads) $=$ t (in seconds)
R (in ohms) $\times$ C (in microfarads) $=$ t (in microseconds)
R (in megohms) $\times$ C (in picofarads) $=$ t (in microseconds)

Charge Rate of a Capacitor

An accurate graph of the rise of voltage across a capacitor which is charging is given in the "charge" graph of Fig. 46-5. This shows how the voltage rises across a capacitor C charging through a resistance R toward a voltage source V. The horizontal axis is the time axis set off in time constants (*RC*). The vertical axis is a percentage axis, with 100 percent representing the total voltage to which a capacitor can charge.

Examination of this graph shows that capacitor C charges approximately 63 percent in one time constant, 86 percent in two time constants, 95 percent in three time constants, 98 percent in four time constants, and 99 percent (practically 100 percent) in five time constants. Let us consider these figures further. We note that at the end of the first time constant the "active voltage" is 37 percent of the original applied voltage. Also, during the second time constant the increase in charge of capacitor C is 23 percent (86 − 63 =

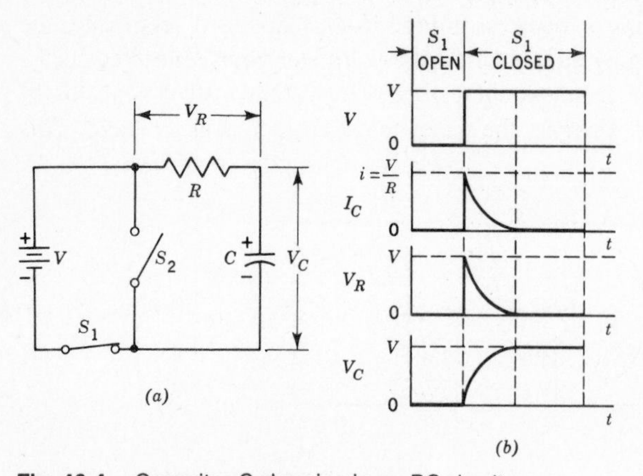

Fig. 46-4. Capacitor C charging in an *RC* circuit.

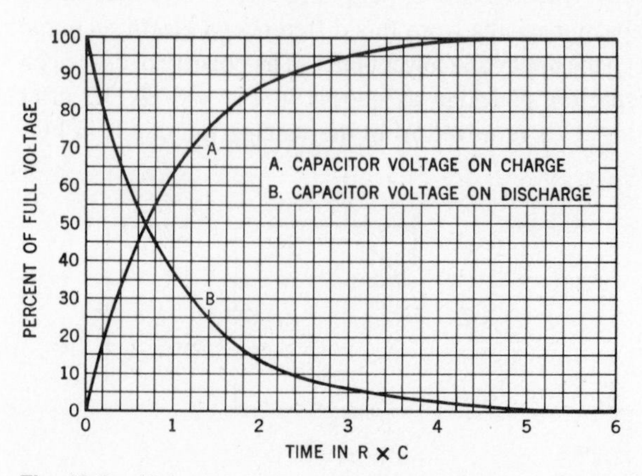

Fig. 46-5. Universal chart for charge and discharge of a capacitor.

23). Now, this 23 percent rise during the second RC is 63 percent of 37. This suggests the rule which actually holds true for capacitor charge: In every time constant, a capacitor charges 63 (more accurately, 63.2) percent of the active voltage. The active voltage can, of course, be computed using Eq. (46-2).

Though a capacitor can theoretically "never" charge fully to the applied voltage, since there is always a remainder of 37 percent of the active voltage, to all intents and purposes we assume that C is fully charged at the end of five time constants.

Capacitor Discharge—Time Required

Now how does a capacitor discharge? Assume that C in Fig. 46-4 has charged to the full applied voltage. We now open S_1 and close S_2, as in Fig. 46-6*a*, permitting C to discharge through R. The variation of discharge current i_d, the change in voltage v_R across R, and the decay in voltage v_c across C are shown in the graphs of Fig. 46-6*b*. We note that the discharge current i_d is opposite in direction to the charge current i_c. Therefore, the voltage v_R across R has the opposite polarity to that in Fig. 46-4*b*.

Figure 46-5 shows the discharge curve for C. We note the following significant similarities between the discharge of C and the charge of C:

In one RC, C discharges 63 percent of its total voltage. In two RCs, C discharges 86 percent of its total voltage, etc. The rule for discharge is therefore similar to the rule for charge: A capacitor discharges 63 percent of its remaining voltage in one time constant.

If we refer to the graphs in Figs. 46-4*b* and 46-6*b*, we note that Kirchhoff's law applies to both charge and discharge circuits. In each case, at any instant of time, the applied voltage V is equal to the sum of the voltage across the capacitor and that across the resistor. That is,

$$V = v_c + v_R \tag{46-4}$$

Before we conclude our discussion of time constants, we should mention that we refer to the curves in Fig. 46-5 as *exponential curves*. That is, the equation which describes how v varies as a function of time involves t (time) as a variable exponent. Thus

$$v_c = V(1 - \epsilon^{-t/RC}) \tag{46-5}$$

where ϵ is a number (it is the base of the natural logs) equal to 2.71828.

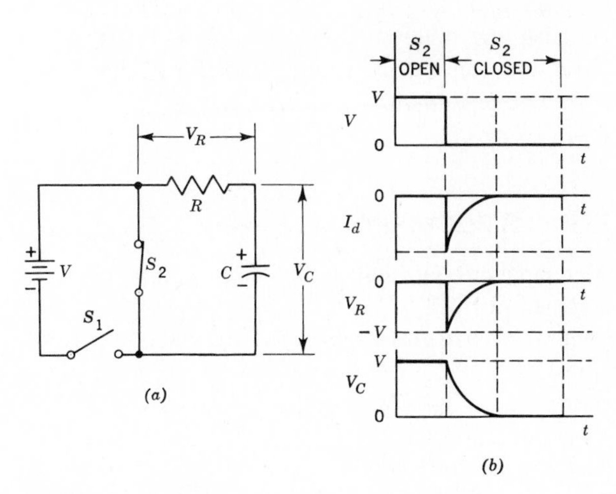

Fig. 46-6. Capacitor discharging in an RC circuit.

Experimental Techniques in Observing Capacitor Charge and Discharge

We can set up an experiment using simple equipment to demonstrate how long it takes a capacitor to charge and discharge. However, our results will be approximate because we will not be using laboratory-precision timing or measuring equipment.

Refer to Fig. 46-7*a*. By selecting the values of R and C, we can find how long it takes C to charge and discharge, using an ordinary watch with a seconds hand. In determining the charge time for C, we use a large nonelectrolytic capacitor, say 1 μF, a 1-MΩ resistor, an EVM, a variable dc voltage source, and a watch. When we close switches S_1 and S_2, C will start charging toward the dc supply voltage. The EVM will read v_c, the voltage across C. We will find that in approximately 5 RCs (5 s in this case) the capacitor will be fully charged.

We should note here that C will not actually charge up to the total applied voltage, but to a value determined by the voltage divider, consisting of the internal resistance of the EVM and the value of R. Thus, if we use an EVM with an input resistance of 11 MΩ and if R is a 1-MΩ resistor, C will charge up to 11/12 of the applied dc voltage. Thus if we wish to have C charge to 100 V, we would set the output of our dc supply to 109 V (approximately).

To observe how long it takes C to discharge, we open S_1. Now C will discharge through the input resistance of the meter, which we assumed to be 11 MΩ. The discharge time constant is $C \times$ resistance of the meter, in this case 11 s. We can verify the discharge voltage at RC intervals, 11-s intervals in this case. Our results should roughly conform to the capacitor discharge curve (Fig. 46-5).

If, in the circuit of Fig. 46-7*a*, R is replaced by the EVM, as in Fig. 46-7*b*, this provides another means for determining the charge time of C. When switch S_1 is closed, C will charge through the input resistance R_{in} of the EVM. At the moment

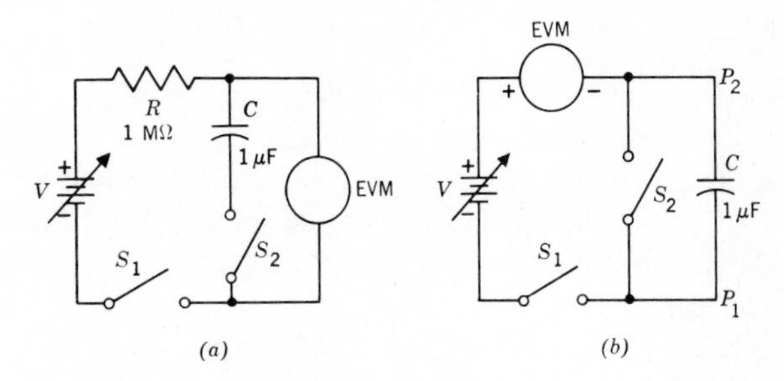

Fig. 46-7. Experimental circuit for charge and discharge of a capacitor.

S_1 is closed, the full-battery voltage V appears across R_{in}, and the EVM measures the full voltage V. There is, of course, *zero* voltage across C at that instant. As C charges, the voltage v_c developed across C increases, while the voltage v_R developed across the input resistance of the EVM decreases. At the end of one RC, only 37 percent (100 − 63) of the applied voltage V will be measured on the meter. At the end of two RCs, 14 percent (100 − 86) will be measured by the EVM. At the end of the third, fourth, and fifth RCs, the EVM will show respectively, 5, 2, and 1 percent (approximately) of the applied voltage V. Of course, the charging time constant in this circuit is $R_{in}C$.

Using this method to determine the length of time it takes C to charge, we would connect the circuit of Fig. 46-7*b* and set V to some value, say 100 V. The timed interval would begin when S_1 is closed and would end when the meter voltage dropped to 1 V (approximately). If the meter reading stabilizes at a measurable value of voltage, say 3 V, it does so because of capacitor leakage. In that event, readings should be discontinued just when the voltage reaches this stable level.

NOTE: As the voltage measured on the EVM falls, we can switch the meter to a lower range without disturbing the charging time constant.

By this method we can measure the charging time of C somewhat more accurately than by the preceding means. Hence it will be used in this experiment.

SUMMARY

1. The charge in coulombs Q on a capacitor is equal to the capacitance C in farads times the voltage V in volts to which the capacitor has charged. Thus, $Q = C \times V$.
2. The time required to charge a capacitor to 63.2 percent (approximately) of the applied voltage is called a *time constant*.
3. When a capacitor of C farads is charging through a resistance of R ohms, the time constant t in seconds of the charging circuit is $t = R \times C$.
4. If at any instant of time while a capacitor is charging, the active voltage V_A equals the difference between the charging source voltage V and the voltage on the capacitor, that is,

$$V_A = V - V_C$$

then the added voltage to which the capacitor charges in one time constant is 63.2 percent of V_A.
5. It takes five time constants for a capacitor to charge to 99+ percent (approximately) of the applied voltage. We can say, therefore, that a capacitor charges to the applied voltage in five time constants.
6. When a capacitor of C farads is discharging through a resistor of R ohms, the discharge time constant t in seconds is $t = R \times C$.
7. When discharging, a capacitor will lose 63.2 percent of its charge in *one* time constant.
8. In each succeeding time constant the capacitor will lose an additional 63.2 percent of the voltage still across it.
9. It takes five time constants, approximately, to discharge a capacitor.
10. The graph of the charge and discharge of a capacitor is exponential and is shown in Fig. 46-5, the universal chart for the charge and discharge of a capacitor.
11. A time constant is a *relative*, not an absolute, measure of time.
12. The formula which gives the voltage V_C to which a capacitor charges at any instant of time t after the start of the charging cycle is

$$V_C = V(1 - \epsilon^{-t/RC})$$

where V is the applied voltage and ϵ is the number 2.71828 (the base of the natural logs).

SELF-TEST

Check your understanding by answering these questions:

1. A 0.25-μF capacitor is charging through a 2.2-MΩ resistor toward an applied voltage of 50 V. In one time constant the capacitor will have charged to ______ V.
2. The capacitor in question 1 has a time constant of ______ s.
3. In ______ s, the capacitor in question 1 will have charged to 43 V.
4. A 0.05-μF capacitor charged to 100 V is permitted to discharge through a 220-kΩ resistor. At the end of ______ s the voltage across the capacitor will have dropped to 37 V (approximately).
5. It will take ______ time constants to discharge the capacitor almost completely.
6. The discharge time constant in question 4 is ______ s.
7. The charging or discharging curve shown in the universal time-constant chart (Fig. 46-5) is called an ______ curve.

MATERIALS REQUIRED

- Power supply: Regulated variable direct current
- Equipment: EVM
  Optional for special credit: Square-wave generator, oscilloscope
- Resistors: ½-W 1-MΩ; 10,000-Ω
- Capacitors: 1 μF; 0.005 μF
- Miscellaneous: Two SPST switches; watch with a seconds hand

PROCEDURE

Input Resistance of EVM

1. Connect the circuit of Fig. 46-7*a*. S_1 and S_2 are open. Set $V = 33$ V (approximately).
2. Close S_1. If your EVM has an input resistance of 11 MΩ, the meter will read 30 V. If it does not, readjust V until the voltmeter reads 30 V.
3. With the EVM, measure the voltage V across the dc supply and record in Table 46-1.
4. Compute and record the input resistance R_{in} in megohms of your EVM by substituting the measured value V in the formula

$$R_{in} = \frac{30}{V - 30} \times 1\ \text{M}\Omega \qquad (46\text{-}6)$$

   Record also the rated value for R_{in}, including the resistance in the EVM probe. The computed and rated values should be the same (approximately).

5. Close S_2. Observe that the voltage indicated on the EVM falls to zero the moment S_2 is closed and then rises as C charges. After 10 s, approximately, C should be fully charged and the voltmeter should measure 30 V.

NOTE: Some meters are heavily damped and should be avoided in this experiment where instantaneous time is a factor.

Discharge Time Constant

6. Compute the discharge time constant $R_{in}C$ and record in Table 46-1. Use the rated value of R_{in}.
7. Open S_1. Observe and record v_c, the voltage across C, at every discharge time constant interval shown in Table 46-1. For example, if $R_{in}C = 11$ s, readings should be made at 11-s intervals and recorded in the column labeled "v_c, 1st trial," until C is completely discharged.

NOTE: When the voltage read on the meter stabilizes at some level, you are reading a voltage produced by the dielectric strain within the capacitor. This may occur at about the tenth time constant.

8. Close S_1 and permit C to charge fully (30 V).
9. Repeat step 7. Record your readings in the column labeled "v_c, 2d trial."
10. Repeat steps 8 and 9, recording your readings in the column labeled "v_c, 3d trial."
11. Average the three values of v_c for one RC and record in the column "v_c, average." Average also and record the value of v_c for two RCs, three RCs, etc. Draw a graph (in red pencil) of v_c (average) versus RC. Let RC be the horizontal axis.
12. Compute, and record in Table 46-1, the voltage v_c which should appear across C after one RC, two RCs, etc. For comparison with step 11, draw a graph (in green) of the computed values of v_c versus RC.

Charge Time Constant

13. Connect the circuit of Fig. 46-7*b*. Set the EVM on the 50-V or next higher range. Close S_1 and S_2 and set V at 30 V, as measured on the voltmeter.
14. Open S_2. Observe and record v_R, the voltage across the EVM, at every charge time-constant interval shown in Table 46-2.

NOTE: The charge time constant is the same as the discharge time constant, $R_{in}C$, recorded in Table 46-1.

Record the readings in the column labeled "v_R, 1st trial." Compute the voltage v_c across the capacitor at every time-constant interval and record in the column "v_c, 1st trial." Use the equation

$$v_c = V - v_R = 30 - v_R \qquad (46\text{-}7)$$

TABLE 46-1. Discharge Time Constant of a Capacitor

V, volts	R_{in}, Ω Computed	R_{in}, Ω Rated	*Discharge Time Constant,* $R_{in}C$ ______ s	*Volts across C*
				30

Discharge Time, RC	v_c, V *1st Trial*	*2d Trial*	*3d Trial*	*Average*	*Computed*
1					
2					
3					
4					
5					
10					

15. Close S_2, thus discharging capacitor C. V should be 30 V. Open S_2 and repeat step 14. Record the measured voltage in the column "v_R, 2d trial," and the computed values in the column labeled "v_c, 2d trial."
16. Repeat step 15. Record the measured voltage in the column "v_R, 3d trial," and the computed values in the column "v_c, 3d trial."
17. Average the three values of v_c at every time-constant interval and record in the column "v_c, average." Draw a graph (in pencil) of v_c average versus RC.
18. Compute, and record in Table 46-2, the voltage v_c which should appear across C at the end of one RC, two RCs, etc. For comparison with the graph in step 17, draw a graph of the computed values of v_c versus RC (in ink).

TABLE 46-2. Charge Time Constant of a Capacitor

Charge Time Constant, $R_{in}C$ ______ s	v_R, V			v_c, V				
	1st Trial	*2d Trial*	*3d Trial*	*1st Trial*	*2d Trial*	*3d Trial*	*Average*	*Computed*
1								
2								
3								
4								
5								
10								

QUESTIONS

1. What is meant by the term *time constant* for RC circuits?
2. How can we use the charging curve of a capacitor to measure time? Relate your answer to the universal time-constant chart.
3. What factors limit the accuracy of your measurements in this experiment?
4. What is the purpose of averaging the three trial values of v_c?
5. How do the average values of v_c compare with the computed values for:
   (*a*) The charge of a capacitor?
   (*b*) The discharge of a capacitor?
   Explain any discrepancies between computed and average values.
6. Do the graphs of v_c versus RC for charge, and v_c versus RC for discharge, agree with the universal time-constant chart? Comment on any variations from the normal.
7. Compute the time constant for the following values of R and C:
   (*a*) $R = 220{,}000\ \Omega$, $C = 0.003\ \mu F$
   (*b*) $R = 5600\ \Omega$, $C = 0.25\ \mu F$
   (*c*) $R = 2.2\ M\Omega$, $C = 250$ pF

Extra Credit

8. A periodic square wave (Fig. 46-8*b*) is applied across an RC circuit (Fig. 46-8*a*), and the voltages across R and C in turn are observed on an oscilloscope. Draw the waveforms which would appear across R and C. Explain how you determined what the waveforms would be.
9. Experimentally confirm your answer to question 8. Explain in detail the procedure you used and the results of your findings.

Answers to Self-Test

1. 31.6
2. 0.55
3. 1.1
4. 0.011
5. 5
6. 0.011
7. exponential

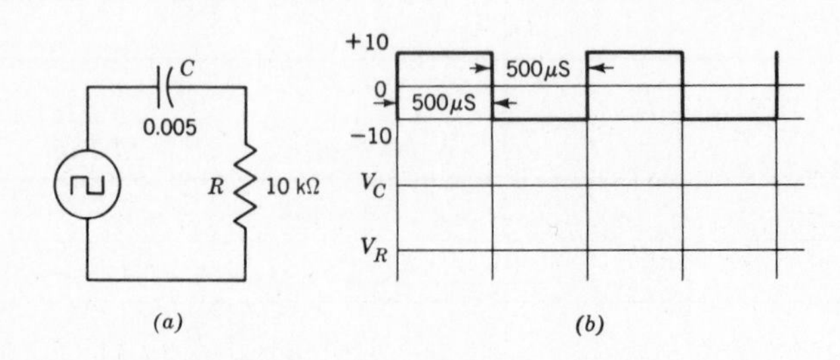

Fig. 46-8. Waveforms in an RC circuit.

47

EXPERIMENT

REACTANCE OF A CAPACITOR (X_C)

OBJECTIVE

To verify experimentally that the capacitive reactance X_C of a capacitor is given by the formula

$$X_C = \frac{1}{2\pi FC}$$

INTRODUCTORY INFORMATION

Reactance of a Capacitor

The capacitive reactance X_C of a capacitor is the amount of opposition it offers to current in an ac circuit. The unit of capacitive reactance is the ohm. However, like the X_L of a coil, the X_C of a capacitor cannot be measured with an ohmmeter. Rather, capacitive reactance must be measured indirectly from its effect on current in an ac circuit.

X_C is dependent on frequency and is given by the formula

$$X_C = \frac{1}{2\pi FC} \tag{47-1}$$

Here X_C is in ohms, C in farads, and F is frequency in hertz (cycles per second). The value of C in microfarads (μF) may be substituted directly in the formula

$$X_C = \frac{10^6}{2\pi FC} \tag{47-2}$$

For example, suppose it is required to find the reactance of a 0.1-μF capacitor at a frequency of 1000 Hz. Substituting in Eq. (47-2),

$$X_C = \frac{10^6}{(6.28)(1000)(0.1)} = 1592\ \Omega$$

It is apparent from Eqs. (47-1) and (47-2) that the higher the frequency, the smaller the reactance of a capacitor; and the lower the frequency, the higher the reactance. For direct current, where $F = 0$, X_C is infinite. Therefore, direct current will not flow through a capacitor.

The reactance of a capacitor may be determined by measurement. In the circuit of Fig. 47-1, a sinusoidal voltage V causes a current I to flow in the circuit. Ohm's law, extended to ac circuits, states that

$$I = \frac{V}{Z} \tag{47-3}$$

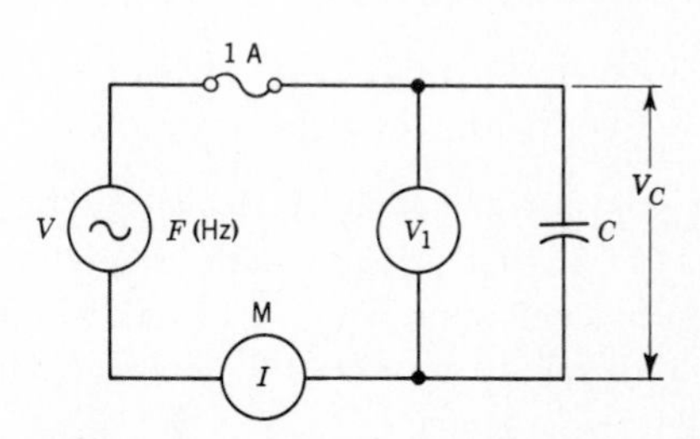

Fig. 47-1. A sinusoidal voltage V causes current I to flow in the circuit. We assume here that the resistance of the meter is zero. Therefore $Z = X_C$ and $V_C = V$.

where I is measured in amperes, V in volts, and Z in ohms. The term Z stands for impedance, which is the opposition to alternating current. In Fig. 47-1, the reactance of C is the impedance[1] of the circuit. That is,

$$X_C = Z \tag{47-4}$$

Hence, the capacitive current I is given by the equation

$$I = \frac{V}{X_C} = \frac{V_C}{X_C} \tag{47-5}$$

We see, therefore, that the amplitude of alternating current through a capacitor across which an ac voltage is connected is directly proportional to the amplitude of voltage (V_C) across the capacitor, and inversely proportional to the reactance (X_C) of the capacitor.

Equation (47-5) may be written in the form

$$X_C = \frac{V_C}{I} \tag{47-6}$$

We may employ Eq. (47-6) in determining the reactance of a capacitor at a given frequency F. We do this by measuring the voltage V_C across C with an ac voltmeter V and measuring the current I in the circuit with an ac ammeter. The values of V_C and I are then substituted in Eq. (47-6), and X_C is computed. Note in the experimental circuit of Fig. 47-1 that the voltmeter V_1 is connected directly across C and not across the voltage source V. Thus, we measure V_C and not V. The reason is that the ammeter M has resistance, across which there will be a voltage drop. If the resistance of M is high, its effect on impeding current would have to be considered if the

[1]We assume here that the resistance of the current meter is so small, compared to X_C, that we may ignore its effect on alternating current.

applied voltage V were measured. By measuring V_C, Eq. (47-6) can be used to find X_C.

An alternate method for measuring alternating current does not require the use of an ac ammeter.

Consider the circuit of Fig. 47-2. This is a series circuit. Thus, the current flowing through R and C is the same. With this circuit arrangement it is possible to determine the value of X_C experimentally by measuring the voltage V_C across C and V_R across R. To determine the current I in amperes in the circuit, the known value of R in ohms and the measured value V_R in volts are substituted in Ohm's law, Eq. (47-7).

$$I = \frac{V_R}{R} \qquad (47\text{-}7)$$

Knowing I, it is possible to find X_C by substituting for I and V_C in Eq. (47-6). It is also possible to find X_C without computing I by combining Eqs. (47-6) and (47-7) and solving for X_C. We get

$$X_C = \frac{V_C}{V_R} \times R \qquad (47\text{-}8)$$

By substituting the measured values V_C, V_R, and R in Eq. (47-8), the value of X_C is determined. Having found X_C experimentally, it is possible to verify approximately the validity of the formula for X_C. This is done by comparing the experimental value of X_C with the formula value. The capacitance C must be known or measured.

SUMMARY

1. The amount of opposition that a capacitor offers to current in an ac circuit is called capacitive reactance (X_C) and is measured in ohms.
2. The capacitive reactance of a capacitor is not constant but is inversely proportional to the ac frequency (F). That is, the higher F is, the smaller is X_C; the lower F is, the greater is X_C.
3. X_C is also inversely proportional to the capacitance C of a capacitor.
4. Capacitive reactance of a capacitor C, in a circuit of frequency F, may be computed from the formula

$$X_C = \frac{1}{2\pi FC}$$

where X_C is in ohms, F in hertz, and C in farads.

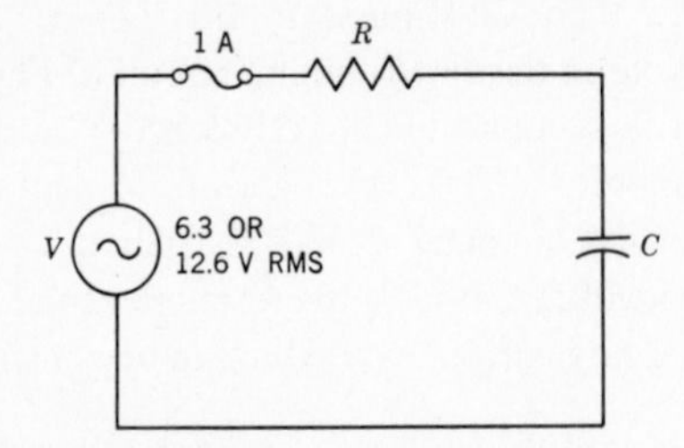

Fig. 47-2. Series *RC* circuit.

5. When C is given in microfarads (μF),

$$X_C = \frac{10^6}{2\pi FC}$$

6. X_C cannot be measured directly—only by its effect in an ac circuit. Thus in Fig. 47-1, we measure the voltage V_C across C and the current I in the circuit. Then, by extending Ohm's law to ac circuits, we can find X_C by substituting V_C and I in the formula

$$X_C = \frac{V_C}{I}$$

7. Another method of determining X_C is illustrated by the circuit of Fig. 47-2. Here the voltages V_R and V_C are measured across R and C, respectively. Then X_C is computed by substituting V_R and V_C in the formula

$$X_C = \frac{V_C}{V_R} \times R$$

SELF-TEST

Check your understanding by answering these questions:

1. The reactance of a capacitor is given in ________.
2. The reactance of a capacitor depends on the ________ of the applied voltage source and the ________ of ________.
3. The X_C of a capacitor whose value is 0.1 μF in a circuit where a 60-Hz signal is applied is ________ Ω.
4. If the frequency of the signal is changed to 6000 Hz, the X_C of the 0.1-μF capacitor is ________ Ω.
5. The voltage across a capacitor is 9.5 V, and the current in the capacitor is 0.005 A. The capacitive reactance of the capacitor is ________ Ω.
6. A 1000-Ω resistor and a 0.05-μF capacitor are connected in series. The voltage measured across the resistor is 3.5 V, and the voltage measured across the capacitor is 31.5 V. The X_C of the capacitor is ________ Ω.
7. The frequency of the applied voltage in question 6 is ________ Hz.

MATERIALS REQUIRED

- Power supply: Isolation transformer and variable autotransformer operating from the 120-V line
- Equipment: Electronic voltmeter, oscilloscope, 0–5-mA ac milliammeter, capacitance tester
- Resistors: ½-W, 5600-Ω
- Capacitors: 0.5 μF, 0.1 μF
- Miscellaneous: Fused line cord

PROCEDURE

Determining X_C by Voltmeter-Ammeter Method

1. With a capacitance tester, measure the values of C listed in Table 47-1 and record the figures in Table 47-1.
2. Connect the circuit (Fig. 47-1). Plug a variable auto-transformer (ac supply) into the isolated outlet of an isolation transformer connected to the ac line. V_1 is an EVM set on the 25-V ac voltage range. M is a 5-mA ac milliammeter. C is a 0.5-μF nonelectrolytic capacitor. Set the output of the ac supply at zero.
3. **Power on.** Increase the output of the ac supply until 2 mA of current is measured on the milliammeter. With the voltmeter, measure the voltage V_C cross the capacitor and record in Table 47-1. Compute, and record in Table 47-1, the X_C of the capacitor using the measured values of V_C and I. Show your computations. Increase the ac voltage until $I = 3$ mA. Record V_C and the computed value of X_C in Table 47-1.
4. Repeat step 3 for $I = 4$ mA. Compute the value of X_C, substituting the measured value of C in Eq. (47-2). F is the ac line frequency. Show your computation. Repeat steps 3 and 4 for a 0.1-μF capacitor, setting current to the levels shown in Table 47-1.

Determining X_C by Voltage Method

5. Measure the 5600-Ω resistor listed in Table 47-2 and record its value. Record also the measured values of the capacitors from Table 47-1. Record also the formula values of $X_C = 10^6/2\pi FC$ from Table 47-1.
6. Connect the circuit of Fig. 47-2. C is the 0.5-μF capacitor used previously. The line-isolated voltage supply V is set at 6.3 or 12.6 V rms 60 Hz or line frequency. Calibrate the oscilloscope for peak-to-peak ac voltage measurement. Measure, and record in Table 47-2, the peak-to-peak voltage V_C across C and the voltage V_R across R. Keep the hot lead of the scope at the junction of R and C. Compute X_C by using (1) V_C, V_R, and R and substituting these values in Eq. (47-8); and (2) the measured value of C and the line frequency. Show your computations.
7. Repeat step 6 for $C = 0.1$ μF.

TABLE 47-1. Voltampere Method to Determine X_C

C, μF				X_C, Ω	
Rated	Measured	I, mA	V_C, V	$X_C = \frac{V_C}{I}$	$X_C = \frac{10^6}{2\pi FC}$
0.5		2			
		3			
		4			
0.1		1			
		2			
		3			

TABLE 47-2. Voltage-Resistance Method to Determine X_C

C, μF		R, Ω					X_C, Ω	
Rated	Measured	Rated	Measured	V_C, V p-p	V_R, V p-p	I_{mA}	$X_C = \frac{V_C \times R}{V_R}$	$X_C = \frac{10^6}{2\pi FC}$
0.5			5600					
0.1			5600					

QUESTIONS

1. In your own words state Ohm's law for ac circuits.
2. (*a*) How would you apply Ohm's law to a pure capacitive circuit to determine the current in the capacitor?
   (*b*) Which parameters (variables) must you know to determine current in a capacitor?
3. Upon what parameters does the reactance X_C of a capacitor depend? Explain.
4. Refer to the data in Tables 47-1 and 47-2. Does the experimental value of $X_C (V_C/I)$ remain constant in this experiment? Should it? Comment on any unexpected results.
5. How does the measured value of $X_C (V_C/I)$ compare with the formula value $10^6/2\pi FC$? Comment on any discrepancy.
6. List the factors which limit the accuracy of the measurements in this experiment.

Answers to Self-Test

1. ohms
2. frequency; value; capacitance
3. 26,526
4. 265
5. 1900
6. 9000
7. 354

48

EXPERIMENT

MEASURING PHASE ANGLE BETWEEN VOLTAGE AND CURRENT IN A CAPACITIVE CIRCUIT

OBJECTIVES

1. To measure the phase angle between voltage and current in a capacitive circuit using a dual-trace oscilloscope
2. To measure the phase angle between voltage and current in a capacitive circuit directly by using a single-trace oscilloscope

INTRODUCTORY INFORMATION

Phase Relations between Voltage and Current in a Capacitor

In a circuit containing only capacitance, voltage and current are *not* in phase, but current *leads* voltage by 90°. A graph of voltage and current in a pure capacitive circuit, Fig. 48-1, shows this relationship. You will recall from Experiment 42 that this characteristic of a capacitor is just the opposite of that of an inductor, for in an inductor current *lags* voltage by 90°. It is apparent, then, that capacitors and inductors have opposite effects on voltage and current in an ac circuit. These effects will be considered in later experiments.

Measuring Phase Relations between Voltage and Current in a Purely Capacitive Circuit by Means of a Dual-trace Oscilloscope

In measuring phase relations between voltage and current in a purely capacitive circuit, using a dual-trace oscilloscope, we can use the same techniques as in Experiment 42, when phase relations between voltage and current in a pure inductive circuit were considered. It is suggested that students review Experiment 42 to refresh their memory.

It is possible to measure the phase difference between voltage and current in a capacitor by connecting an external resistor R in series with C, as in the circuit of Fig. 48-2. The value of the resistor chosen must be such that the ratio X_C/R is equal to or greater than 10 ($X_C \geqq 10$). This circuit will act almost like a pure capacitance, and the error in phase angle introduced by adding R will be less than 6° if $X_C/R = 10$. If $X_C/R > 10$, the angle of error will be smaller. In the case of Fig. 48-2, at 60 Hz, $X_C/R = 16.08$, and the error introduced is less than 4°.

Current i in the circuit of Fig. 48-2 is the same throughout the circuit. The voltage v_R across R is in phase with the current i in R. Hence by measuring the phase angle between the voltage v_R and the applied voltage v, we are in fact measuring the phase angle between current and voltage in the capacitor.

You will recall that this technique requires that the sine wave v applied to the series RC circuit, Fig. 48-2, be fed to the external triggering jack of the oscilloscope and that the trigger selector be set on Ext. The oscilloscope sweep circuit is then triggered at the same point in time regardless of the waveform we are viewing in the circuit. The applied voltage v fed to the vertical input of channel A acts as the reference

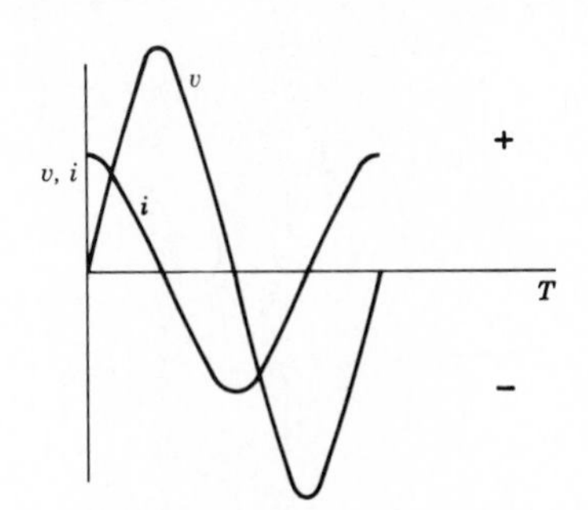

Fig. 48-1. In a pure capacitive circuit, voltage and current are *not* in phase, but current leads voltage by 90°.

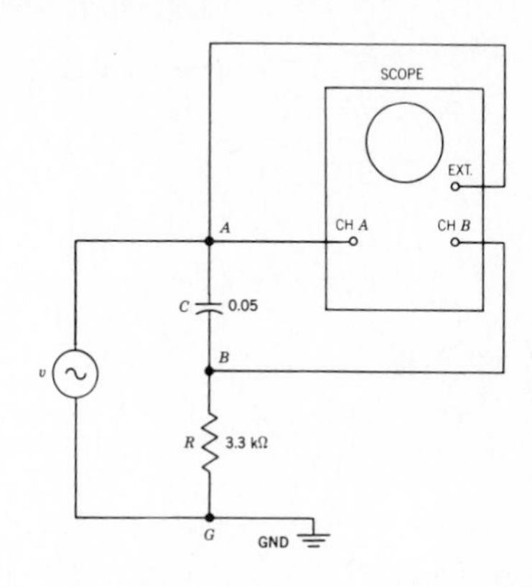

Fig. 48-2. Experimental circuit for demonstrating phase relationship between v and i in a capacitor using a dual-trace oscilloscope.

voltage for phase measurement. After proper adjustment of the channel *A* controls, the reference waveform is properly centered with respect to the vertical and horizontal axes of the graticule.

The waveform across *R* is applied to the vertical input of channel *B*, and after proper adjustment of the channel *B* vertical centering control, the *A and B* mode of the scope is selected. This mode places both waveforms on the same time axis, as in Fig. 48-1. We can then determine the phase difference by finding the difference *d* in baseline distance between the zero crossover points of the two waveforms. The distance *d* is then converted to degrees. The number of degrees is the required phase angle we are seeking. In the case of capacitor *C*, connected as in Fig. 48-2, *i* should *lead v* by 90°, approximately. A linear time base is assumed, or this technique cannot be used. If the sweep is linear, equal distances on the time base correspond to equal angular displacements.

Using a Single-trace Oscilloscope to Measure Phase between Voltage and Current in a Capacitor

A single-trace oscilloscope can also be used to measure the phase difference between two waveforms of the same frequency, in almost exactly the same way as a dual-trace scope is employed. The basic difference is that only one waveform appears on the scope at a time.

Suppose we wish to measure the phase relation between voltage and current in a capacitor *C*. A resistor *R* is again connected in series with *C*, as in Fig. 48-3, and the voltage source *v* is placed across this combination. As in the preceding analysis, if the ratio of $X_C/R \geqq 10$, the error in phase angle caused by adding *R* will be less than 6°. Also, as in the case of the dual-trace scope, the applied voltage *v* is used to trigger the sweep externally, as in Fig. 48-3. The voltage *v* fed to the vertical input of the scope serves as the reference for phase measurements. This reference sine wave is conveniently centered on the graticule, so that the start of the waveform coincides with a major vertical graph line, *A* in Fig. 48-4.

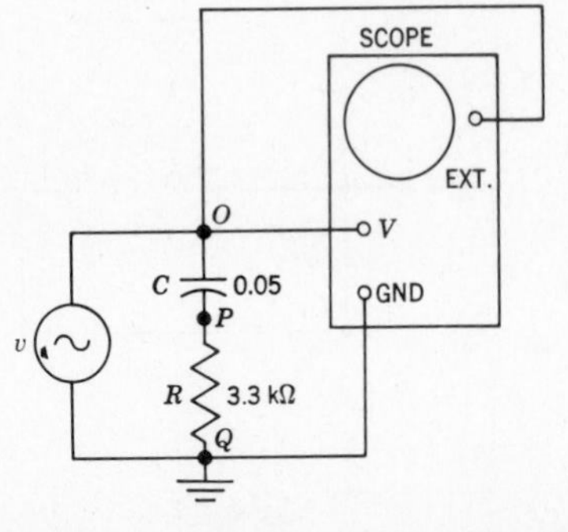

Fig. 48-3. Oscilloscope connections for demonstrating phase relations between the applied voltage *v* and current *i* in a capacitive circuit, using a single-trace scope.

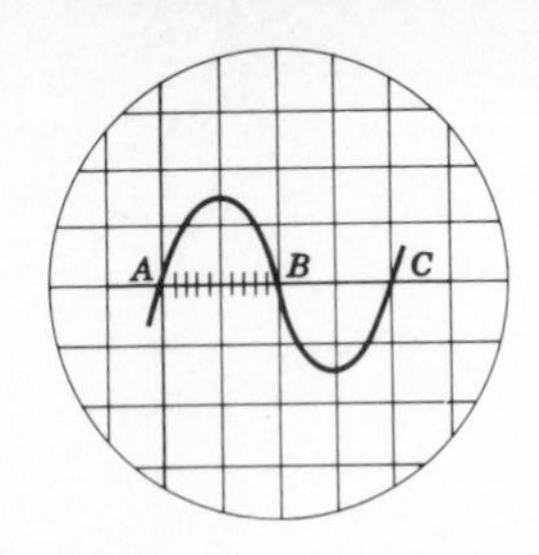

Fig. 48-4. Horizontal sweep used directly for phase measurements.

As before, it is assumed that the scope's sweep provides a linear time base, so that equal distances on the time axis correspond to equal angular displacements. Thus, in Fig. 48-4, one time-base division is divided into five graduations. Since one sine-wave alternation (180°) covers two graticule divisions (10 graduations), each graduation corresponds to 18°.

Once the reference waveform has been properly centered and its width in divisions noted, the number of degrees per graduation can be calculated. Then the vertical input (hot) lead of the oscilloscope is connected to point *P* (Fig. 48-3). The voltage waveform across *R* will now appear on the oscilloscope. The displacement (number of graduations) of the observed waveform, from the reference voltage, is measured on the time base. Multiplying the calculated *degrees per graduation* by the *number* of graduations of displacement, we get the phase difference between the applied voltage and current.

The position of the waveform across *R* relative to the reference waveform tells us whether the current *i* leads or lags the applied voltage *v*. Thus in Fig. 48-5, waveform 1 leads waveform 2 by 90°. Waveform 2 lags 1 by 90°.

SUMMARY

1. In a capacitive circuit alternating current and voltage are not in phase. If the circuit is a pure capacitance, that is, if it has no resistance, current leads voltage by 90°.
2. We can demonstrate this phase difference between the applied voltage and current in a capacitive circuit by using a dual-trace oscilloscope, connected as in Fig. 48-2. On the *A and B* mode of the oscilloscope, both the voltage and

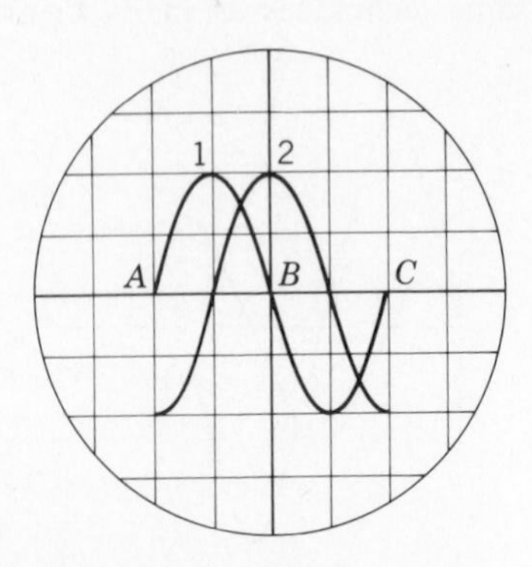

Fig. 48-5. Waveform 1 leads waveform 2 by 90°.

current waveforms appear on the screen, and their separation in degrees is the phase angle.
3. In order to use a dual- or single-trace oscilloscope to measure phase difference between two waveforms, the time base (sweep) of the oscilloscope must be linear.
4. It is also possible to measure the phase difference between voltage and current in a capacitor directly, by using a single-trace oscilloscope set on Ext. triggering, employing the circuit of Fig. 48-3.
5. The reference voltage waveform v is then centered on the graticule as in Fig. 48-4.
6. Once v is centered, the vertical hot lead of the oscilloscope is connected to the junction of C and R, point P in Fig. 48-3. The waveform across R now appears on the oscilloscope. This is the required current (i) waveform.
7. The horizontal displacement between v and i is measured on the oscilloscope graticule and changed into degrees.
8. Whether the current waveform leads or lags the reference waveform can be determined by noting whether the current waveform is positive or negative, respectively, at the start of the reference waveform (that is, at the time when the positive alternation of the reference waveform is just starting).

SELF-TEST

Check your understanding by answering these questions:

1. In a capacitor, current ____________ voltage by ____________ degrees.
2. If a dual-trace oscilloscope is used to measure phase difference between voltage and current in a capacitor C, the circuit of Fig. 48-2 may be used. The error introduced by connecting R in series with C is less than ____________ degrees as long as ____________ $\geqq 10$.
3. In the circuit of Fig. 48-2, the reference waveform is the one which is fed to ____________, and the current waveform is the one fed to ____________.
4. ____________ triggering is required in using a dual- or single-trace oscilloscope to measure the phase difference between two waveforms.
5. A single-trace oscilloscope can also be used to measure phase difference. Figure ____________ illustrates the circuit to achieve this.
6. In using a single-trace oscilloscope to measure phase difference, the width of a complete sine wave on the screen is two divisions. Each division is divided into 10 graduations. The displacement between the start of the waveform being measured and the start of the reference waveform is three graduations. The phase angle is ____________ degrees.
7. In question 6, the displaced waveform is in its negative alternation just at the time that the reference waveform is starting to go positive. The displaced waveform therefore ____________ the reference waveform.

MATERIALS REQUIRED

- Equipment: Dual- (and single-) trace oscilloscope; AF signal generator
- Resistors: ½-W 3300-Ω
- Capacitors: 0.05 μF

PROCEDURE

Measuring Phase Relations in a Capacitor Using a Single-trace Oscilloscope

1. Connect the circuit of Fig. 48-3. Set the trigger selector switch to Ext. Feed the AF generator output signal to the Ext. trigger jack on the oscilloscope. Adjust the generator output 60-Hz signal to 20 V peak-to-peak. Set the oscilloscope vertical-gain controls to two or three divisions of vertical deflection. Set the oscilloscope controls for a two- or three-cycle display.

2. Center the waveform vertically with respect to the horizontal graticule axis. Center the waveform horizontally so that the start of the positive alternation coincides with one of the major vertical graph lines on the graticule, as in Fig. 48-4. Determine the degree per (horizontal) graduation factor of the time base ____________.

3. Now connect the vertical input lead to point P, the junction of R and C in Fig. 48-3. Readjust the vertical-gain controls for two or three divisions of vertical deflection on the screen. Observe the waveform. Does it lead or lag the reference waveform? ____________

4. Measure the number of graduations of horizontal displacement of the waveform across R_1 from the reference and convert into degrees of phase difference. ____________ degrees

Measuring Phase Relations in a Capacitor Using a Dual-trace Oscilloscope

5. Connect the circuit of Fig. 48-2. The AF sine-wave generator remains set as in steps 1 to 4, that is, at 60 Hz, 20 V p-p.

6. Following the method in Experiment 42, measure the phase difference between voltage and current in capacitor C. ____________ degrees.

QUESTIONS

1. Explain how you measured the phase difference between voltage and current in a capacitor, using a dual-trace oscilloscope.
2. How do the results of your measurements using a single-trace and a dual-trace oscilloscope compare? Explain any difference.
3. What is the value of X_C/R in the experimental circuit? Explain how you computed this ratio.
4. Why are the applied voltage and the voltage across C assumed to be the same in Figs. 48-2 and 48-3?

Answers to Self-Test

1. leads; 90
2. 6; X_C/R
3. channel A; channel B
4. external
5. 48-3
6. 54
7. lags

49

EXPERIMENT

TOTAL CAPACITANCE OF CAPACITORS IN SERIES AND IN PARALLEL

OBJECTIVES

1. To verify experimentally that the total capacitance C_T of capacitors connected in parallel is $C_T = C_1 + C_2 + C_3 + \cdots$
2. To verify experimentally that the total capacitance C_T of capacitors connected in series is $1/C_T = 1/C_1 + 1/C_2 + 1/C_3 + \cdots$

INTRODUCTORY INFORMATION

Total Capacitance of Parallel-Connected Capacitors

Series- and parallel-connected capacitor combinations are frequently used in electronics. Consider the circuit of Fig. 49-1. In this circuit C_1 and C_2 are connected in parallel. The ac voltage V is applied equally across C_1 and C_2. There are two paths for current, namely, through C_1 and C_2. Current I_1 flows through C_1, and I_2 flows through C_2. I_1 and I_2 combine to form the total line current I_T in the circuit. As far as the ac generator V is concerned, it sees a capacitance C_T through which I_T is flowing. C_T is therefore the equivalent capacitance resulting from the parallel combination. The formula for C_T is

$$C_T = C_1 + C_2 \tag{49-1}$$

The formula may be extended for more than two capacitors connected in parallel, and is

$$C_T = C_1 + C_2 + C_3 + \cdots \tag{49-2}$$

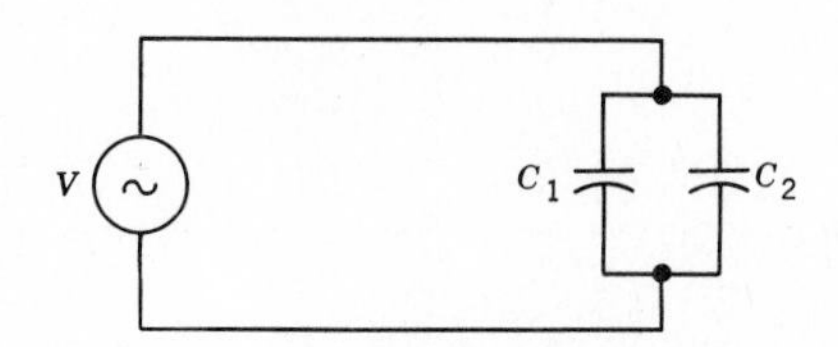

Fig. 49-1. Capacitors connected in parallel in an ac circuit.

Total Capacitance of Series-Connected Capacitors

In the series-connected circuit of Fig. 49-2, the same line current I flows through C_1 and C_2. The effect of connecting these capacitors in series is to increase the total capacitive reactance and thus reduce the line current which would flow if either capacitor were in the circuit alone.

Since the reactance of the series combination is greater than the reactance of either capacitor considered alone, the total capacitance C_T must be smaller than the capacitance of either C_1 or C_2. The formula for total capacitance C_T for two or more capacitors connected in series is

$$\frac{1}{C_T} = \frac{1}{C_1} + \frac{1}{C_2} + \frac{1}{C_3} + \cdots \tag{49-3}$$

The total capacitance C_T of capacitor combinations may be determined experimentally in a number of ways. The simplest is to connect the required combination and measure C_T with a capacitance tester. If a laboratory-standard capacitance bridge is used, C_T can be measured to a high degree of accuracy.

A second method, used in this experiment, is to determine the capacitive reactance X_{CT} of the combination. Once X_{CT} is determined, C_T may be found from the equation

$$C_T = \frac{1}{2\pi F X_{CT}} \tag{49-4}$$

The reactance of the capacitor combination may be found by measuring the current I_C and the voltage V_C across the combination. X_{CT} is then computed from the equation

$$X_{CT} = \frac{V_C}{I_C} \tag{49-5}$$

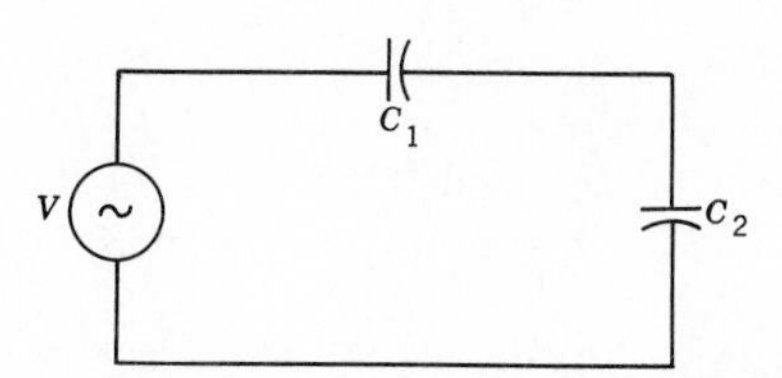

Fig. 49-2. Capacitors connected in series in an ac circuit.

The capacitive current may be measured with an ac ammeter. Or the ac voltage V_R across a series-connected resistor may be measured, and the series current, which is also the capacitive current I_C, may be computed from the equation

$$I_C = \frac{V_R}{R} \quad (49\text{-}6)$$

SUMMARY

1. In a circuit consisting of two or more capacitors connected in parallel, the total current I_T is equal to the sum of the branch currents. Therefore the total current is greater than the branch current in any one of the capacitors in the circuit.
2. In a circuit consisting of two or more parallel-connected capacitors, the total capacitance C_T is equal to the sum of each of the individual capacitors. That is,

$$C_T = C_1 + C_2 + C_3 + \cdots$$

3. A simple way to remember the above formula is to note that the method of computing C_T of *parallel-connected capacitors* is the same as the method of computing R_T of *series-connected resistors*.
4. In series-connected capacitors, the current is the same in each of the capacitors. However, the total reactance X_{CT} of series-connected capacitors is greater than the reactance of any one capacitor in the circuit. Therefore, the total current I_T in a circuit consisting of series-connected capacitors is less than the current would be if any one of the capacitors were in the circuit by itself.
5. In a circuit consisting of two or more series-connected capacitors, the total capacitance C_T of the combination is

$$\frac{1}{C_T} = \frac{1}{C_1} + \frac{1}{C_2} + \frac{1}{C_3} + \cdots$$

6. The total capacitance of *series-connected capacitors* is computed in the same way as the total resistance R_T of *parallel-connected resistors*.
7. One method of determining experimentally the total capacitance C_T of capacitor combinations is to measure the C_T of the combination with a capacitance bridge.
8. Another method is to measure the total alternating current I_T of the combination and the voltage V_C across the combination, and to determine X_{CT} using the formula

$$X_{CT} = \frac{V_C}{I_T}$$

9. Now, knowing X_{CT} and frequency F of the alternating current, find C_T by substituting X_{CT} and F in the formula

$$C_T = \frac{1}{2\pi F X_{CT}}$$

SELF-TEST

Check your understanding by answering these questions:

1. In Fig. 49-1, the current I_1 in C_1 is 0.05 A and the current I_2 in C_2 is 0.1 A. The total current I_T in the circuit is ______ A.
2. In the circuit of Fig. 49-1, C_1 = 0.2 μF and C_2 = 0.4 μF. The total capacitance C_T of the combination is C_T = ______ μF.
3. If the applied voltage in the circuit of question 1 is 10 V, the reactance
   (*a*) X_{C1} = ______ Ω
   (*b*) X_{C2} = ______ Ω
   (*c*) X_{CT} = ______ Ω
4. In the circuit of Fig. 49-2, C_1 = 0.25 μF and C_2 = 0.05 μF. The total capacitance C_T of this combination is C_T = ______ μF.
5. If the frequency F of the applied voltage V is 100 Hz, then in the circuit of question 4,
   (*a*) X_{C1} = ______ Ω
   (*b*) X_{C2} = ______ Ω
   (*c*) X_{CT} = ______ Ω
6. In the circuit of Fig. 49-3, I_T = 0.015 A, V_C = 6.3 V, and F = 60 Hz. Then
   (*a*) X_{CT} = ______ Ω
   (*b*) C_T = ______ μF

MATERIALS REQUIRED

- Power supply: Isolated 6.3-V rms (18-V peak-to-peak) ac source
- Equipment: Oscilloscope, capacitance tester, ac milliammeter (0–5 mA), EVM
- Capacitors: 0.5 μF, two 0.1 μF, 0.05 μF
- Resistor: ½-W 56,000-Ω
- Miscellaneous: One SPST switch; fused line cord

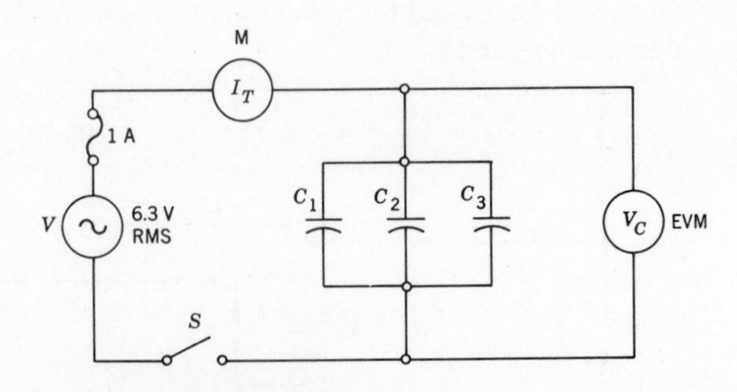

Fig. 49-3. Voltage-current method for determining X_{CT} and then C_T of parallel-connected capacitors.

PROCEDURE

Total Capacitance of Capacitors Connected in Parallel

Method 1

1. With a capacitance tester, measure the capacitance of each of the capacitors listed in Table 49-1 and record their values.
2. Measure also, and record in Table 49-1, the total capacitance C_T of each of these parallel combinations of capacitors:
   a. Capacitors C_1, C_2, C_3, C_4
   b. Capacitors C_2, C_3, C_4
   c. Capacitors C_1, C_2, C_3
   d. Capacitors C_1, C_2
   Compute, and record in Table 49-1, the C_T (formula value) of each of these combinations using the measured values of the individual capacitors for your computations. Show a sample computation.

Example. Assume that capacitor C_1 measures 0.47 μF and C_2, 0.11 μF. Then

$$C_T = C_1 + C_2 = 0.47 + 0.11 = 0.58\ \mu\text{F}$$

Method 2

3. Connect the circuit of Fig. 49-3. The capacitors are combination **c.** from step 2. *M* is an ac 0–5-mA meter. The EVM is set on the 10-V range to measure the voltage across the capacitors. *V* is an isolated 6.3-V rms, line-derived source. Measure, and record in Table 49-2, the total current I_T and the voltage V_C across the capacitors. Compute and record the value of X_{CT} applying Eq. (49-5) to the measured values of V_C and I_T.

   Substituting the computed value of X_{CT} in Eq. (49-4), compute and record the total capacitance C_T of combination **c.** Show your computations.

Sample Computation. Assume V_{CT} is 7.8 V, $I = 0.0021$ A, and $F = 60$ Hz. Then

$$X_{CT} = \frac{V_{CT}}{I_T} = \frac{7.8}{0.0021} = 3714\ \Omega$$

$$C_T = \frac{1}{2\pi F X_{CT}} = \frac{1}{6.28(60)(3714)}\ \text{F}$$

$$C_T = \frac{10^6}{6.28(60)(3714)}\ \mu\text{F}$$

$$C_T = 0.714\ \mu\text{F}$$

   From Table 49-1 transcribe the formula value of combination **c** to Table 49-2.

4. Repeat step 3 for combination **d.**

Total Capacitance of Capacitors Connected in Series

Method 1

5. With a capacitance tester, measure and record in Table 49-3 the total capacitance of each of the combinations of the following series-connected capacitors:

   NOTE: For rated value of capacitors, refer to Table 49-1.

   a. Capacitors C_1, C_2
   b. Capacitors C_1, C_2, C_3
   c. Capacitors C_1, C_2, C_3, C_4
   d. Capacitors C_1, C_4

TABLE 49-1. Measuring *C* with a Capacitance Tester

Capacitor No.	C_1	C_2	C_3	C_4	C_T: Parallel Combinations		
						Measured (Method 1)	Formula Value
Rated value, μF	0.5	0.1	0.1	0.05	**a.**		
					b.		
Measured value, μF					**c.**		
					d.		

TABLE 49-2. Determining C_T by the Voltmeter-Ammeter Method

Parallel Combination	I_T, mA	V_C, V	X_{CT}, Ω	C_T, μF Computed (Method 2)	C_T, μF Formula Value
c.					
d.					

TABLE 49-3. C_T of Series-Connected Capacitors

Series Combination	C_T, μF Method 1 Measured	Computed	V_R, V	V_{CT}, V	I_T, mA	X_{CT}, Ω	C_T, μF (method 2)
a.							
b.							
c.			X	X	X	X	X
d.			X	X	X	X	X

Compute, and record in Table 49-3, the C_T of each of these combinations, using the measured values of the individual capacitors listed in Table 49-1.

Example. Assume that capacitor 1 measures 0.47 μF and capacitor 2 measures 0.11 μF. Then for combination **a**:

$$\frac{1}{C_T} = \frac{1}{C_1} + \frac{1}{C_2}$$

and

$$\frac{1}{C_T} = \frac{1}{0.47} + \frac{1}{0.11} = 11.2$$

$$C_T = \frac{1}{11.2} = 0.089 \ \mu\text{F}$$

Method 2

6. Connect the circuit of Fig. 49-4. With an EVM measure and record in Table 49-3 the voltage V_R across R and the voltage V_{CT} across the series capacitors (points A to B). Verify these measurements with an oscilloscope calibrated to measure rms volts. Solve, and record in Table 49-3, the current I_T in this circuit by substituting the EVM measured values of V_R and R in the equation

$$I_T = \frac{V_R}{R}$$

Compute and record the value of X_{CT} by applying Eq. (49-5) to the measured values of V_{CT} and I_T. Substituting the computed value of X_{CT} in Eq. (49-4), compute and record the total capacitance C_T of series combination **a** shown in Fig. 49-4. (See sample computations, step 3.)

7. Add C_3, a 0.1-μF capacitor in series with C_1 and C_2, and repeat step 6. Record for combination **b**.

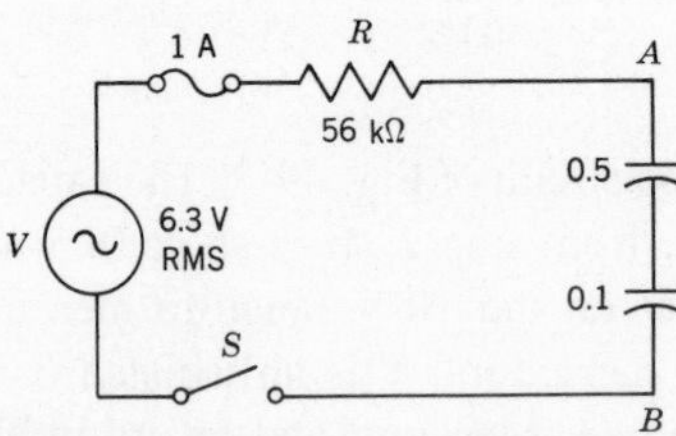

Fig. 49-4. Voltage-current method of determining X_{CT} and then C_T of series-connected capacitors.

QUESTIONS

1. (*a*) Refer to Fig. 49-1. What would be the effect on line current (total current I_T) of adding another capacitor C_3 in parallel with C_1 and C_2?
   (*b*) Explain why.
2. (*a*) Refer to Fig. 49-2. What would be the effect on line current of adding another capacitor C_3 in series with C_1 and C_2?
   (*b*) Explain why.
3. Refer to Table 49-1. How do the measured values of C_T (using method 1) compare with the formula values of C_T [computed using Eq. (49-1)]? Explain any discrepancies.
4. Refer to Table 49-2. How do the computed values of C_T (using method 2 for determining the total capacitance of parallel-connected capacitors) compare with the formula values? Explain any discrepancies.
5. In measuring C_T of parallel-connected capacitors, which method, 1 or 2, appears to give the more accurate results? Why?
6. State in your own words the formula for finding the total capacitance of capacitors connected in parallel.
7. Refer to Table 49-3. What verification, if any, do you have for the formula for total capacitance of series-connected capacitors? Explain.
8. Which of the two methods seems more accurate for measuring C_T of series-connected capacitors? Explain why?

Answers to Self-Test

1. 0.15
2. 0.6
3. (*a*) 200; (*b*) 100; (*c*) 66.7
4. 0.0417
5. (*a*) 6366; (*b*) 31,831; (*c*) 38,197
6. (*a*) 420; (*b*) 6.32

50

EXPERIMENT

THE CAPACITIVE VOLTAGE DIVIDER

OBJECTIVES

1. To derive analytically that the voltage V_1 across a capacitor C_1 in a series-connected capacitive voltage divider is given by the formula

$$V_1 = V \times \frac{C_T}{C_1}$$

where V = voltage applied across all the capacitors and C_T = total capacitance of the series-connected capacitors
2. To confirm experimentally the formula in objective 1

INTRODUCTORY INFORMATION

AC Voltage across a Capacitor

Ohm's law extended to ac circuits states that the current I in a circuit is the ratio of the applied voltage V to the total opposition to alternating current. We call this total opposition *impedance* and label it Z. Thus

$$I = \frac{V}{Z} \tag{50-1}$$

Determination of ac impedances can be quite complex. However, in a circuit containing only capacitance C, such as that in Fig. 50-1, the total opposition to alternating current is the reactance X_C of capacitor C. Therefore, in the circuit of Fig. 50-1, Z and X_C are the same and

$$I = \frac{V}{X_C} \tag{50-2}$$

From Eq. (50-2) it is quite evident that the voltage drop V_C across capacitor C is equal to the product of the current I in the capacitor and the reactance X_C of the capacitor.

$$V_C = I \times X_C \tag{50-3}$$

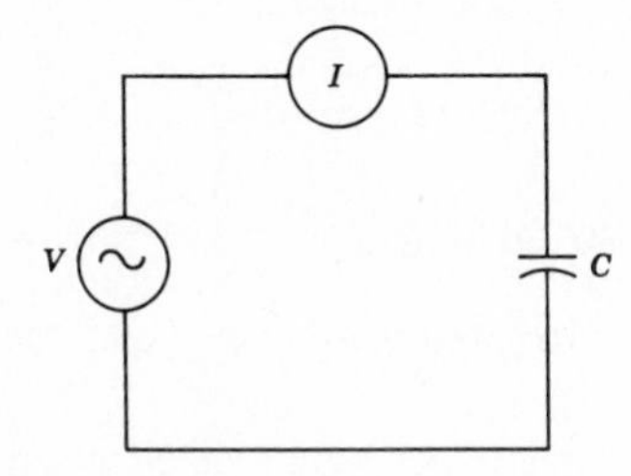

Fig. 50-1. The total opposition to alternating current in this circuit is the reactance X_C of C.

Capacitive Voltage Divider

If an ac voltage is applied across capacitors connected in series, as in Fig. 50-2, voltage division takes place across the capacitors. There is a voltage drop across each capacitor which is proportional to the reactance of the capacitor. Now since in a series circuit, the current I is the same everywhere in the circuit, the voltage V_1 across C_1 is

$$V_1 = I \times X_{C1}$$

Similarly

$$V_2 = I \times X_{C2} \tag{50-4}$$

Since capacitive reactance is inversely proportional to capacitance, the smaller the capacitor, the greater the voltage drop across it in a capacitive voltage divider. Moreover, since the reactance of a 0.05-μF capacitor is twice the reactance of a 0.1-μF capacitor, $V_1 = 2V_2$ in Fig. 50-2. If the applied voltage V were 18 V in that circuit, then V_1 = 12 V and V_2 = 6 V, with the sum of V_1 and V_2 being equal to the applied voltage V.

The relationships just discussed may be generalized. Thus in a three-capacitor divider (Fig. 50-3),

$$V_1 = I \times X_{C1} = I \times \frac{1}{2\pi FC_1}$$

$$V_2 = I \times \frac{1}{2\pi FC_2} \tag{50-5}$$

$$V_3 = I \times \frac{1}{2\pi FC_3}$$

Dividing V_1 by V_2 leads to

$$\frac{V_1}{V_2} = \frac{C_2}{C_1}$$

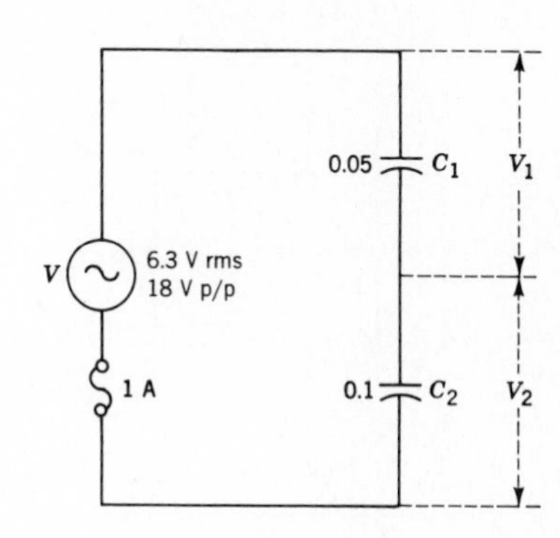

Fig. 50-2. Voltage distribution in a capacitive voltage divider.

Similarly $$\frac{V_2}{V_3} = \frac{C_3}{C_2} \tag{50-6}$$

and $$\frac{V_3}{V_1} = \frac{C_1}{C_3}$$

Now consider the total reactance X_{CT} of the series-connected capacitors in Fig. 50-3.

$$V = I \times X_{CT} \tag{50-7}$$

That is, the applied voltage V equals the product of I and the total reactance of the series-connected capacitors.

Since $V_1 = I \times X_{C1}$, dividing V_1 by V leads to

$$\frac{V_1}{V} = \frac{I \times X_{C1}}{I \times X_{CT}} = \frac{X_{C1}}{X_{CT}} = \frac{\frac{1}{2\pi FC_1}}{\frac{1}{2\pi FC_T}}$$

and $$\frac{V_1}{V} = \frac{C_T}{C_1}$$

Therefore $$V_1 = V \times \frac{C_T}{C_1} \tag{50-8}$$

Equation (50-8) states that the voltage drop across any capacitor C_1 in a series-connected capacitive voltage divider equals the product of the applied voltage and the ratio of the total capacitance C_T to the capacitance of C_1. Note that this is similar to the formula for voltage distribution in a series-connected resistive voltage divider, with the exception that in a resistive divider V_1 (across R_1) $= V \times (R_1/R_T)$.

That is, in a series capacitive divider the positions of C_1 and C_T [see Eq. (50-8)] are reversed, as compared with the positions of R_1 and R_T in a series-resistive divider.

It is now possible to apply Eq. (50-8) to the circuit of Fig. 50-2. In that circuit

$$\frac{1}{C_T} = \frac{1}{C_1} + \frac{1}{C_2} = \frac{1}{0.05} + \frac{1}{0.1}$$

and $$C_T = \frac{1}{30}\ \mu\text{F}$$

Substituting $C_T = 1/30$, $C_1 = 1/20$, and $V = 18$ in Eq. (50-8) gives

$$V_1 = 18 \times \frac{1/30}{1/20} = 12\ \text{V}$$

Similarly $$V_2 = 6\ \text{V}$$

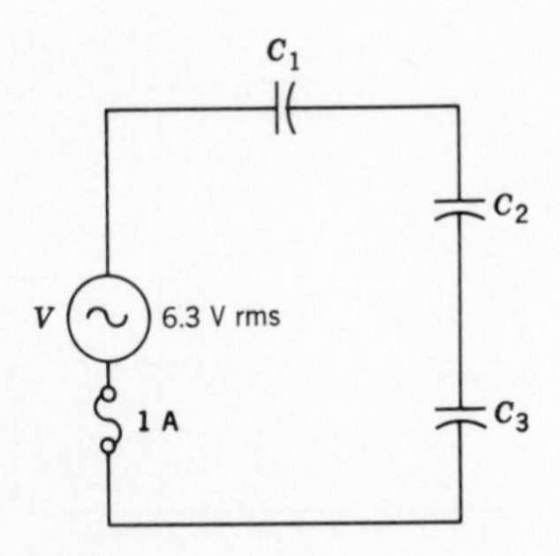

Fig. 50-3. Capacitive voltage divider.

These are the same results as were obtained in our earlier solution of this circuit.

We can now also verify Eqs. (50-6), for according to those equations,

$$\frac{V_1}{V_2} = \frac{C_2}{C_1}$$

and $$\frac{12}{6} = \frac{0.1}{0.05} = \frac{2}{1}$$

as expected.

SUMMARY

1. The voltage drop V_1 across a capacitor C_1 in a capacitive voltage divider is given by the formula $V_1 = I \times X_{C1}$, where I is the current in the capacitor and X_{C1} is the reactance of the capacitor.
2. In a series-connected capacitive voltage divider the ratio $V_1/V_2 = C_2/C_1$, where V_1 and V_2 are the voltage drops across C_1 and C_2, respectively, and C_1 and C_2 are measured in units of capacitance (that is, farads, microfarads, etc.). Similarly $V_2/V_3 = C_3/C_2$, $V_3/V_4 = C_4/C_3$, etc.
3. The total current I in a capacitive voltage divider where the applied voltage is V and the total capacitance is C_T is
$$I = \frac{V}{X_{CT}}$$
4. The voltage V_1 across a capacitor C_1 in a series-connected voltage divider with applied voltage V is
$$V_1 = V \times \frac{C_T}{C_1}$$

SELF-TEST

Check your understanding by answering these questions:

1. In the circuit of Fig. 50-1, $V = 10$ V and $X_C = 1000\ \Omega$. The current I in the circuit is ________ A.
2. In the circuit of Fig. 50-2, $X_{C1} = 1000\ \Omega$. Therefore
(a) X_{C2} = ________ Ω;
(b) X_{CT} = ________ Ω;
(c) I = ________ A.
3. In the capacitive voltage divider (Fig. 50-3), $C_1 = 0.1$ μF, $C_2 = 0.05$ μF, $C_3 = 0.5$ μF, and $V = 100$ V.
(a) V_1 = ________ V
(b) V_2 = ________ V
(c) V_3 = ________ V

MATERIALS REQUIRED

- Power supply: Isolated 6.3-V rms (18-V peak-to-peak) ac source
- Equipment: Capacitance tester, electronic voltmeter
- Capacitors: 0.5 μF, two 0.1 μF, 0.05 μF
- Miscellaneous: Fused line cord

PROCEDURE

1. With the capacitance tester check the value of each capacitor and record in Table 50-1.
2. Connect the circuit of Fig. 50-2. V is a 6.3-V rms (18-V peak-to-peak) source. Measure and record in Table 50-2 the source voltage V.
3. Measure also and record in Table 50-2 the voltages V_1 and V_2 across C_1 and C_2, respectively.
4. Using the measured values of C_1 and C_2 from Table 50-1, compute and record the voltages V_1 and V_2 which should appear across C_1 and C_2, respectively. Show your computations.
5. Add V_1 and V_2 (measured) and record in Table 50-2.
6. Connect the circuit of Fig. 50-3 using the values of C noted in Table 50-1. Measure and record V, V_1, V_2, and V_3. Add V_1, V_2, and V_3 (measured) and record in Table 50-2. Compute and record the voltages V_1, V_2, and V_3 which should appear in the circuit.
7. In Fig. 50-3, replace C_1 with C_4 and repeat step 6.

TABLE 50-1. Capacitor Measured Values

	C_1	C_2	C_3	C_4
Rated value (μF)	0.05	0.1	0.1	0.5
Measured value				

TABLE 50-2. Verifying Voltage Formula for Series-Connected Capacitors

	Measured Voltages					*Computed Voltages*				*Sum of Measured Voltages*
Step	V	V_1	V_2	V_3	V_4	V_1	V_2	V_3	V_4	
2, 3, 4, 5				X	X			X	X	
6					X				X	
7		X				X				

QUESTIONS

1. State in your own words the rule for computing the voltage V_A across a capacitor C_A in a series-connected capacitive voltage divider. Also write the formula for C_A.
2. Refer to Table 50-2. How do the measured voltage values compare with the computed values? Explain any differences.
3. Do the results of your measurements confirm the formula in question 1? Explain.
4. What conclusion can you reach about the sum of the voltages in a series-connected capacitive divider? Prove your answer by referring specifically to the data in Table 50-2.
5. Three capacitors are connected in series across a 100-V line. The capacitances of the capacitors are in the ratio 1:2:3. Find the voltage across each of the capacitors. Show your computations.

Answers to Self-Test

1. 0.01
2. (*a*) 500; (*b*) 1500; (*c*) 0.0042
3. (*a*) 31.25; (*b*) 62.5; (*c*) 6.25

51

EXPERIMENT

IMPEDANCE OF A SERIES *RL* CIRCUIT

OBJECTIVES

1. To verify experimentally that the impedance Z of a series RL circuit is given by the equation

$$Z = \sqrt{R^2 + X_L^2}$$

2. (Optional) To verify experimentally that the relationship between Z, R, and X_L is given by the equation

$$Z = \frac{R}{\cos\theta} = \frac{X_L}{\sin\theta}$$

where θ is the phase angle between R and Z

INTRODUCTORY INFORMATION

Impedance of a Series *RL* Circuit

The total opposition to alternating current in an ac circuit is called the *impedance* of the circuit. Consider the circuit of Fig. 51-1. If it is assumed that the choke coil (L) through which alternating current flows has zero resistance, the current is impeded only by the X_L of the choke. That is, $Z = X_L$. In this case, if $L = 8$ H and $F = 60$ Hz,

$$X_L = 2\pi FL = 6.28(60)(8) = 3014\ \Omega$$

How much current I will there be in the circuit if $V = 6.3$ V? Applying Ohm's law,

$$I = \frac{V}{X_L} \qquad (51\text{-}1)$$

where I is in amperes, V in volts, and X_L is in ohms. Therefore, $I = 6.3/3014 = 2.09$ mA.

If there is resistance R associated with the inductance L, or if L is in series with a resistor of, say, 3000 Ω (Fig. 51-2), there will be less than 2.09 mA of current. How much current will flow, assuming the same X_L as previously computed? Measurement shows that there is 1.485 mA in the circuit. It is evident that the total impedance of the resistor R connected in series with L is *not* simply the arithmetic sum of R and X_L. It can be demonstrated mathematically that Z is the result of the phasor sum of X_L and R. Recall from Experiment 44 that the term phasor is used to describe vectors that represent time-varying quantities. Since Z is related to the sinusoidal V and I, we will use the term phasor when discussing Z.

The phasor diagram of Fig. 51-3 shows that R and X_L are at right angles to each other and that X_L leads R by 90°. We say that there is a 90° phase difference between R and X_L. Phasor addition is achieved by drawing a line through A parallel to OB, and another line through B parallel to OA. C is the intersection of these two lines and is the terminus of the resultant phasor OC. In this case, the phasor AC is the same as X_L. Hence, OC or Z is the resultant of the phasor sum of X_L and R.

In Fig. 51-3, Z is seen to be the hypotenuse of the right triangle of which X_L and R are the legs. Applying the pythagorean theorem to this right triangle, we note that

$$Z = \sqrt{R^2 + X_L^2} \qquad (51\text{-}2)$$

Equation (51-2) makes it possible to determine the impedance of an RL circuit if R and X_L are known. For example, if $R = 40\ \Omega$ and $X_L = 25\ \Omega$, then

$$Z = \sqrt{40^2 + 25^2} = \sqrt{1600 + 625}$$
$$= \sqrt{2225} = 47.17\ \Omega$$

Equation (51-2) and the illustrative problem point up a very important difference between the mathematics of ac circuits and that of dc circuits. In a dc circuit consisting of series-connected resistors, the total opposition to current, R_T, is simply the arithmetic sum of R_1, R_2, etc. In an ac

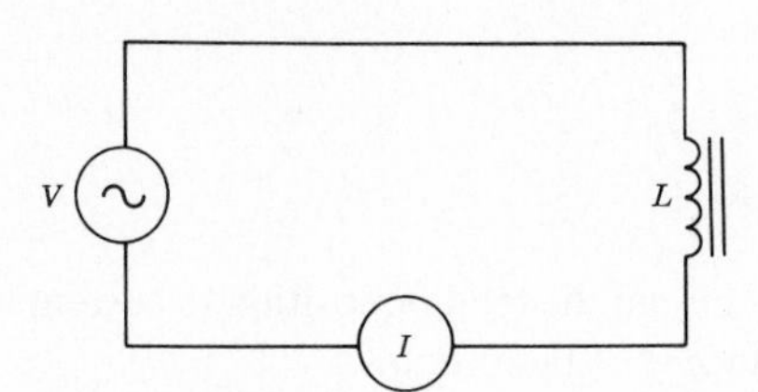

Fig. 51-1. The current I in a pure inductance is limited by the inductive reactance of the circuit.

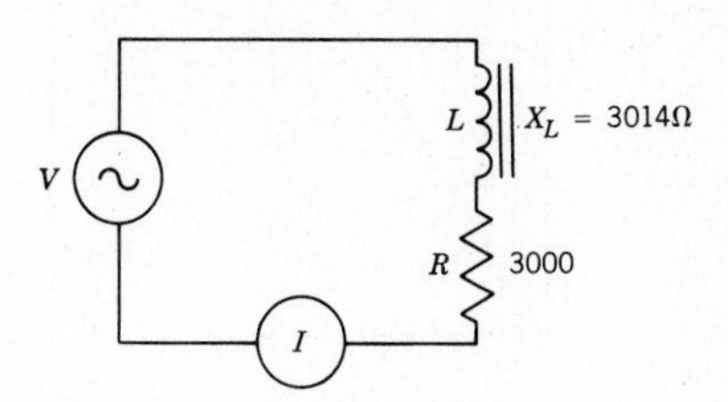

Fig. 51-2. The impedance of L in series with R is greater than that of L alone.

circuit consisting of a series-connected resistor and inductance, the total opposition to current, the impedance Z, is *not* the arithmetic sum of R and X_L, but *is the phasor sum of* R *and* X_L.

Impedance calculations based on Eq. (51-2) can become cumbersome. Trigonometry provides a more effective means of analyzing ac circuits, and trigonometry is therefore extensively employed in ac computations.

If R and X_L are known in Fig. 51-2, the angle θ between R and Z can be determined. For

$$\tan\theta = \frac{X_L}{R}$$

and
$$\theta = \arctan\frac{X_L}{R} \qquad (51\text{-}3)$$

We can now calculate Z. For

$$\frac{X_L}{Z} = \sin\theta$$

and
$$Z = \frac{X_L}{\sin\theta} \qquad (51\text{-}4)$$

Similarly
$$Z = \frac{R}{\cos\theta} \qquad (51\text{-}5)$$

The process of finding Z is resolved, therefore, to the following simple operations:

1. Find θ, the phase angle between Z and R, by applying Eq. (51-3).
2. Find Z by substituting the values of R, X_L, and θ in Eq. (51-4) or (51-5).

Example. If $R = 40\ \Omega$ and $X_L = 25\ \Omega$, find Z trigonometrically.

Solution. Draw the phasor triangle (see Fig. 51-3).

1. $$\tan\theta = \frac{X_L}{R} = \frac{25}{40} = 0.625$$

$$\theta = 32°$$

2. $$Z = \frac{X_L}{\sin\theta} = \frac{25}{\sin 32°} = 47.18\ \Omega$$

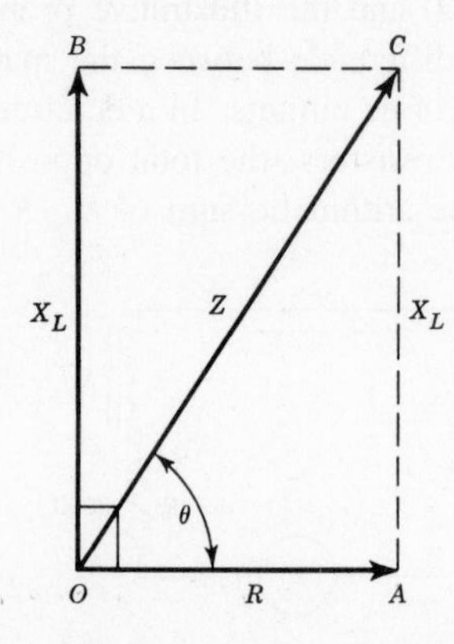

Fig. 51-3. Phasor relations in a series *RL* circuit. Impedance is a phasor quantity.

or
$$Z = \frac{R}{\cos\theta} = \frac{40}{\cos 32°} = 47.17\ \Omega$$

The results agree substantially with those we obtained when we used the right-triangle method.

Ohm's Law Applied to a Series *RL* Circuit

Ohm's law extended to ac circuits states that the current I in a circuit equals the ratio of the applied voltage V and impedance Z.

$$I = \frac{V}{Z} \qquad (51\text{-}6)$$

and
$$Z = \frac{V}{I} \qquad (51\text{-}7)$$

Equation (51-7) provides the means for verifying the relationships among Z, X_L, and R. Thus, in the circuit of Fig. 51-2 we can measure V and I and compute Z by substituting the measured values of V and I in Eq. (51-7). Z is then calculated by the right-triangle formula [Eq. (51-2)] or by the trigonometric formulas [Eqs. (51-4) and (51-5)]. If the results agree, i.e., if the value of Z is substantially the same for each of these methods, then we have verification of the theoretical formulas for Z.

Example. Assume that in the series circuit (Fig. 51-2), $X_L = 25\ \Omega$, $R = 40\ \Omega$, and the applied voltage $V = 47$ V. An ac ammeter is inserted in the circuit (the circuit must be broken and the ammeter inserted in series with R and L), and it measures 1 A. The impedance of the circuit can now be determined. For

$$Z = \frac{V}{I} = \frac{47}{1} = 47\ \Omega$$

The current measurement and subsequent computation for Z is one verification of Eqs. (51-2), (51-4), and (51-5).

Another technique for determining the current in the ac circuit of Fig. 51-2, without measuring current directly, is to measure the ac voltage V_R developed across the resistor R. The current in R may then be found from the formula

$$I = \frac{V_R}{R} \qquad (51\text{-}8)$$

Since current is everywhere the same in a series circuit, the computed value of I is the current in the circuit. We may then solve for the impedance of the circuit by substituting I in the equation $Z = V/I$, where V is the known measured value of the applied voltage and I is the computed current.

SUMMARY

1. In an ac circuit the total opposition to current is called the impedance Z of the circuit.
2. In a series *RL* circuit, the impedance Z is the phasor sum of R and X_L, where X_L *leads* R by an angle of 90° (Fig. 51-3).

3. From the impedance right triangle in Fig. 51-3, the numerical value of Z can be found from the equation

$$Z = \sqrt{X_L^2 + R^2}$$

4. In Fig. 51-3, if θ is the phase angle between Z and R, then

$$\theta = \arctan \frac{X_L}{R}$$

5. Z may be computed trigonometrically if X_L and R are known. Thus

$$Z = \frac{X_L}{\sin \theta}$$

or

$$Z = \frac{R}{\cos \theta}$$

where the angle θ is computed from the equation in summary step 4.

6. Ohm's law extended to ac circuits states that the current I equals the ratio of the applied voltage V and impedance Z. Thus

$$I = \frac{V}{Z}$$

SELF-TEST

Check your understanding by answering these questions. Students who have not studied trigonometry may skip questions 2, 3, and 4.

1. In the series RL circuit (Fig. 51-2), $R = 1000\ \Omega$ and $X_L = 2000\ \Omega$. $Z =$ __________ Ω.
2. In the same circuit as in question 1, $\theta =$ __________ degrees.
3. In the same circuit as in question 1,

   (a) $\dfrac{R}{\cos \theta} =$ __________ Ω

   (b) $\dfrac{X_L}{\sin \theta} =$ __________ Ω
4. The ratio $R/\cos \theta$ or $X_L/\sin \theta$, where θ is the phase angle between Z and R, also gives the __________ of the circuit of Fig. 51-2.
5. In a series RL circuit, $R = 500\ \Omega$, $X_L = 500\ \Omega$, and the applied voltage $V = 100$ V. The current I in the circuit is $I =$ __________ A.
6. In the circuit of Fig. 51-2, $V_R = 15$ V. The current in the inductor L is __________ A.

MATERIALS REQUIRED

- Power supply: Isolation transformer and variable autotransformer operating from the 120-V 60-Hz line
- Equipment: EVM, 0–25-mA ac meter
- Resistors: 2-W 5000-Ω; ½-W 2700-Ω
- Inductor: 8 H at 50 mA direct current, whose resistance is 250 Ω
- Miscellaneous: SPST switch; fused line cord

PROCEDURE

1. Connect the circuit of Fig. 51-4. S is open. R is a 2-W 5000-Ω resistor. L is an iron-core choke rated 8 H at 50 mA direct current. V_A is a variable autotransformer plugged into T, a 1:1 isolation transformer, which derives its power from the 120-V 60-Hz line. M is a 0–25-mA ac meter, and V is an EVM set on the ac voltage range.
2. Close S. **Power on.** Adjust the output of the ac supply until the meter measures 15 mA (0.015 A). Measure the rms voltage V_{AB} applied to L and R and record in Table 51-1. Also measure and record V_R, the voltage across R, and V_L, the voltage across L. Compute and record X_L and Z. ($X_L = V_L/I$; $Z = V_{AB}/I$.)
3. Substitute the computed value of X_L and R (5 kΩ) in the equation $Z = \sqrt{R^2 + X_L^2}$. Find Z and record in Table 51-1. Show computations.

   Confirm the measured value of I by computing I from the equation $I = V_R/R$. Record in Table 51-1. **Power off.**
4. Replace R with a 2700-Ω resistor. **Power on.** Repeat the measurements in steps 2 and 3 and record in Table 51-1.

Extra Credit

5. From Table 51-1 transcribe to Table 51-2 the values of $Z = V_{AB}/I$ and $X_L = V_L/I$ corresponding to $R = 5000\ \Omega$ and $R = 2700\ \Omega$. For each value of X_L and R, determine and record $\tan \theta$ and θ using Eq. (51-3). Next, compute and record the corresponding values of Z, using Eqs. (51-4) and (51-5). Show computations.

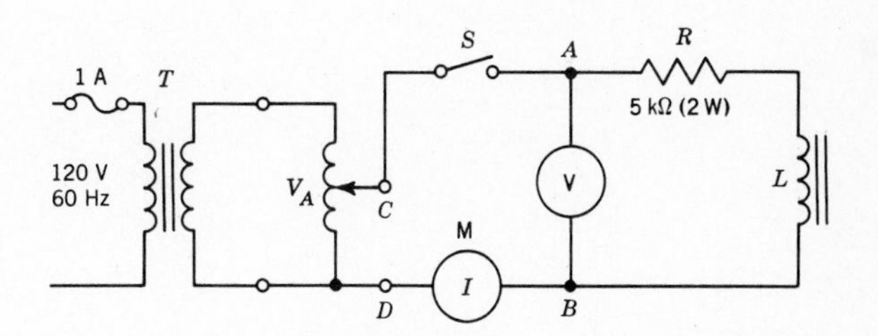

Fig. 51-4. Determining impedance Z of an RL circuit.

TABLE 51-1. Verifying the Impedance Formula for an *RL* Circuit

R, Ω	I, A	V_{AB}, V	V_R, V	V_L, V	$X_L = \frac{V_L}{I}$, Ω	$Z = \frac{V_{AB}}{I}$, Ω	$Z = \sqrt{R^2 + X_L^2}$ Ω	$I = \frac{V_R}{R}$, A
5000	0.015							
2700	0.015							

TABLE 51-2. Trigonometric Formulas for Impedance

R, Ω	$Z = \frac{V_{AB}}{I}$, Ω	$X_L = \frac{V_L}{I}$, Ω	$\tan \theta = \frac{X_L}{R}$	θ, degrees	Z (Ω)	
					$\frac{R}{\cos \theta}$	$\frac{X_L}{\sin \theta}$
5000						
2700						

QUESTIONS

1. Refer to Table 51-1. Are the two values of Z corresponding to R = 5000 Ω equal? Should they be? Are the two values of Z corresponding to R = 2700 Ω equal?
2. From the data in Table 51-1, what conclusion can you draw concerning the formula $Z = \sqrt{R^2 + X_L^2}$? Support your answer.
3. Refer to Table 51-1. Is the metered value of I (using the milliammeter) the same as the computed value $I = V_R/R$? Explain why it should be the same.
4. Define Z.
5. What are the possible sources of error in this experiment?
6. In Fig. 51-4, would it make any difference if the voltmeter V were connected across the points CD instead of AB? Explain why.

Extra Credit

7. Do the results of the data in Table 51-2 confirm Eqs. (51-4) and (51-5)? Explain.

Answers to Self-Test

1. 2236
2. 63.43
3. (*a*) 2236; (*b*) 2236
4. impedance
5. 0.1414
6. 0.005

52

EXPERIMENT

VOLTAGE RELATIONSHIPS IN A SERIES *RL* CIRCUIT

OBJECTIVES

1. To measure the phase angle θ between the applied voltage V and the current I in a series RL circuit
2. To verify experimentally that the relationships among the applied voltage V, the voltage V_R across R, and the voltage V_L across L are described by the equations

$$V = \sqrt{V_R^2 + V_L^2}$$

$$V_R = V \cos\theta = V \times \frac{R}{Z}$$

$$V_L = V \sin\theta = V \times \frac{X_L}{Z}$$

INTRODUCTORY INFORMATION

Phasors

A *phasor* is a quantity which is identified by two characteristics, *amplitude* and *direction*. This is the same definition as that for a *vector*. But where vectors are concerned with the *space* domain, phasors deal with the *time* domain. In electricity we will refer to sinusoidal voltages and currents as *phasor* quantities. Because it is related to sinusoidal voltages and currents we will also consider impedance a phasor quantity.

Sinusoidal voltages or currents of the same frequency may be added and subtracted readily when they are represented by rotating radius phasors. Phasor mathematics, as distinguished from scalar mathematics, is identical to vector mathematics. (Scalar math, that is, ordinary arithmetic, deals with quantities identified by just *one* dimension, namely amplitude. Phasor math deals with quantities identified by two dimensions, amplitude and direction.)

Phase Angle between Applied Voltage and Current in a Series *RL* Circuit

Since current is the same in every part of a series circuit, current I is shown as the reference phasor in considering the phase relations among V_R, V_L, and V in the series RL circuit (Fig. 52-1*a*). Current and voltage in a resistor are in phase. Hence the voltage V_R across R is drawn as an extension of the phasor I (Fig. 52-1*b*). Current in an inductance lags the voltage across the inductance by 90°. Therefore the phasor V_L is shown leading I by 90°. It is evident, therefore, that V_L leads V_R by 90°.

Phasor Sum of V_R and V_L Equals Applied Voltage V

V_R and V_L are, respectively, the voltages across R and L in the series RL voltage divider (Fig. 52-1*a*). Does the sum of V_R and V_L equal the applied voltage V? Obviously it *cannot* be the arithmetic sum, since V_R and V_L are 90° out of phase. But V *is* the *phasor* sum of V_R and V_L. This fact is shown in Fig. 52-1*b*, where V also appears as the hypotenuse of the right triangle of which V_R and V_L are the legs. Applying the pythagorean theorem,

$$V = \sqrt{V_R^2 + V_L^2} \tag{52-1}$$

what is the phase relationship between the applied voltage V and the current I in the inductive circuit? In Fig. 52-1*b*, I is seen to lag the applied voltage V by an angle θ. Is this angle θ the same as the angle θ between Z and R in the impedance phasor diagram in the previous experiment? They are the same, as will be evident from Fig. 52-2.

Figure 52-2*a* is a redrawing of Fig. 52-1*b*, with the composition of each of the voltages as shown. Thus, V_R is the

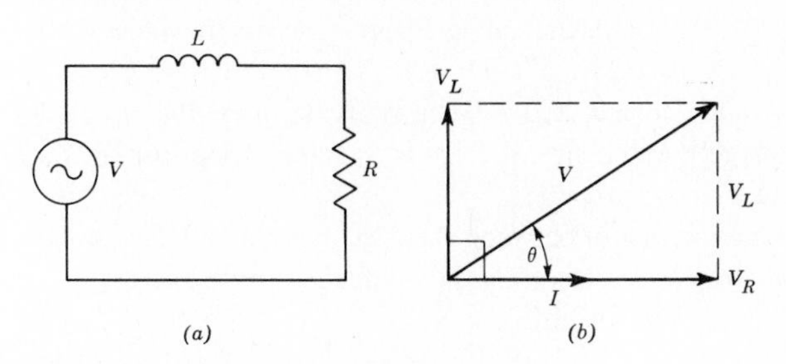

Fig. 52-1. Phase relations in a series *RL* circuit.

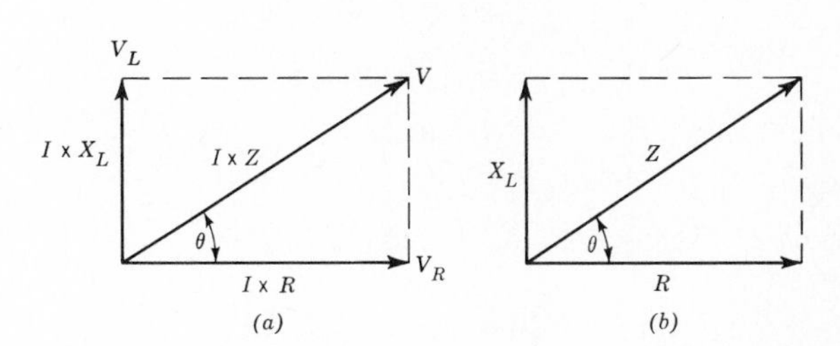

Fig. 52-2. The phase angle θ between *v* and *i* is the same as the angle between *Z* and *R*.

product of I and R, V_L is the product of I and X_L, and V is the product of I and Z. Since I is a common factor in each of these products, it may be eliminated, leaving the impedance diagram (Fig. 52-2*b*). It is evident, therefore, that the angle θ is the same in Figs. 52-1*b* and 52-2*b*.

Further analysis of Fig. 52-1*b* reveals the relationship between V, V_R, V_L, and the phase angle θ. Since

$$\frac{V_R}{V} = \cos\theta \qquad (52\text{-}2)$$

$$V_R = V\cos\theta \qquad (52\text{-}3)$$

Similarly it follows that

$$V_L = V\sin\theta \qquad (52\text{-}4)$$

From the impedance triangle we know that

$$\cos\theta = \frac{R}{Z}$$

and

$$\sin\theta = \frac{X_L}{Z}$$

Substituting these values in Eqs. (52-3) and (52-4), we find that

$$V_R = \mathrm{V} \times \frac{R}{Z} \qquad (52\text{-}5)$$

and

$$V_L = \mathrm{V} \times \frac{X_L}{Z} \qquad (52\text{-}6)$$

Equations (52-3), (52-4), (52-5), and (52-6) may be used to compute the voltages V_R and V_L in a series *RL* circuit where the applied voltage V, the resistance R, and the inductive reactance X_L are known.

Example. If 47 V is applied across a circuit consisting of a 40-Ω resistor in series with an inductance whose reactance is 25 Ω, what is the value of:

1. θ, the phase angle between V and I?
2. V_R, the voltage across R?
3. V_L, the voltage across L?

Solution 1. Using Eqs. (52-3) and (52-4),

1. $$\tan\theta = \frac{X_L}{R} = \frac{25}{40} = 0.625$$
$$\theta = 32^\circ$$

2. $$V_R = V\cos\theta = 47\cos 32^\circ = 40\text{ V}$$

3. $$V_L = V\sin\theta = 47\sin 32^\circ = 25\text{ V}$$

Solution 2. Using Eqs. (52-5) and (52-6),

1. $$\tan\theta = \frac{X_L}{R} = \frac{25}{40} = 0.625$$
$$\theta = 32^\circ$$

2. $$Z = \frac{R}{\cos\theta} = \frac{40}{\cos 32^\circ} = 47\ \Omega$$
$$V_R = V \times \frac{R}{Z} = 47 \times \left(\frac{40}{47}\right) = 40\text{ V}$$

3. $$V_L = V \times \frac{X_L}{Z} = 47 \times \frac{25}{47} = 25\text{ V}$$

We can check solutions 1 and 2 using Eq. (52-1). By substituting the computed values of V_R and V_L in Eq. (52-1), we find that

$$V = \sqrt{40^2 + 25^2} = 47\text{ V}$$

The computed value of V is substantially the same as the given applied voltage V, and our solutions are therefore correct.

SUMMARY

1. In a series *RL* (inductive) circuit, the current I lags the applied voltage V by some angle θ, called the phase angle.
2. The phase angle θ between V and I is the same as the angle θ between Z and R in the impedance phasor diagram of the *RL* circuit (Experiment 51). The angle θ is also the same as the angle between V and V_R in Fig. 52-1*b*.
3. The value of θ depends on the relative values of X_L, R, and Z, and may be computed from *any one* of the following formulas:
$$\theta = \arctan\frac{X_L}{R}$$
$$\theta = \arcsin\frac{X_L}{Z}$$
$$\theta = \arccos\frac{R}{Z}$$
4. In a series *RL* circuit the voltage drop V_L across the inductance leads the voltage drop V_R across the resistor by 90°.
5. V_R and V_L are added vectorially to give the applied voltage V in the circuit. That is, V is the phasor sum of V_R and V_L.
6. The relationship between the applied voltage V, the voltage drop V_R across R, and the voltage drop V_L across L is given by the formula
$$V = \sqrt{V_R^2 + V_L^2}$$

7. If the applied voltage V and the phase angle θ are known, then V_R and V_L may be calculated from these formulas:

$$V_R = V \cos \theta$$
$$V_L = V \sin \theta$$

8. If the applied voltage V and R, X_L, and Z are known in a series RL circuit, then V_R and V_L may be calculated from these formulas:

$$V_R = V \times \frac{R}{Z}$$

$$V_L = V \times \frac{X_L}{Z}$$

9. To verify that V is the phasor sum of V_R and V_L, we measure V, V_R, and V_L in a series RL circuit and substitute the measured values V_R and V_L in $\sqrt{V_R{}^2 + V_L{}^2}$. If the numerical result is equal to the measured voltage V, then we have the confirmation we require.

SELF-TEST

Check your understanding by answering these questions:

1. In the circuit of Fig. 52-1, the measured values of V_R and V_L are, respectively, 30 and 20 V. The applied voltage V must then equal ____________ V.
2. The phase angle between V and V_R is ____________ ____________ as the phase angle between the applied voltage V and the current I in a series RL circuit.
3. The phase angle θ between V and I in the problem of question 1 is ____________ degrees.
4. In the circuit of Fig. 52-1, $X_L = 600\ \Omega$, $R = 800\ \Omega$, and $V = 120$ V. In this circuit
   (a) $Z =$ ____________ Ω
   (b) $\theta =$ ____________ degrees
   (c) $V_R =$ ____________ V
   (d) $V_L =$ ____________ V
5. In a series RL circuit, arcsin X_L/Z = arcsin V_L/V. ____________ (true/false)

MATERIALS REQUIRED

- Power supply: Isolated 6.3-V rms source
- Equipment: EVM, dual-trace oscilloscope
- Resistors: ½-W 4700-Ω, 1200-Ω
- Inductor: 8 H at 50 mA direct current, whose resistance is 250 Ω
- Miscellaneous: SPST switch; fused line cord

PROCEDURE

Measuring the Phase Angle θ

In Fig. 52-1*a* the voltage V_R across R is in phase with the current I. The angle θ by which the applied voltage V leads the current I can therefore be found by measuring the angle between V and V_R by the method described in Experiment 42.

1. Connect the circuit of Fig. 52-3*a*. V is a 6.3-V rms line-isolated source. L is an iron-core choke rated 8 H at 50 mA direct current, whose resistance is 250 Ω. Since the circuit value of $R = 4700\ \Omega$, we can ignore the resistance of L. An insignificant error is introduced as a result. Connect the oscilloscope as shown. Note that the trigger switch is set on Ext.
2. Following the method used in Experiment 42, measure the phase angle between the applied voltage and current in the circuit of Fig. 52-3. Record the angle θ in Table 52-1.
3. To confirm that the measured angle θ above is the same as the angle between Z and R, proceed as follows: With an EVM, measure and record the rms voltage V_R across R. Compute I in the circuit by substituting the measured value V_R and the rated value of R in the equation

$$I = \frac{V_R}{R}$$

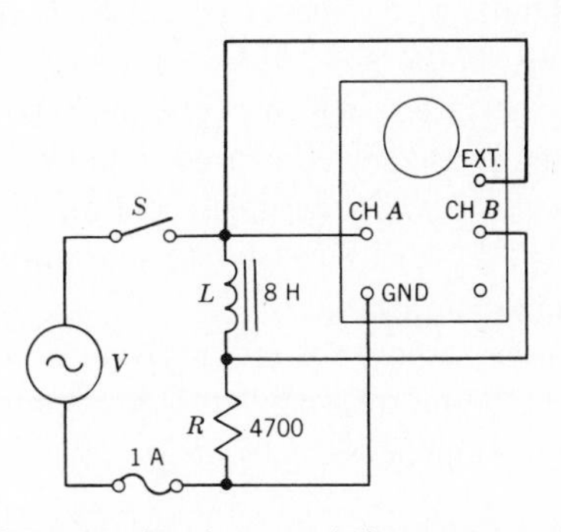

Fig. 52-3. Measuring the phase angle between v and i.

TABLE 52-1. Voltage and Phase Relationships in a Series *RL* Circuit

R, Ω	Phase Angle θ, degrees	V_R, V	$I = \frac{V_R}{R}$, A	V_L, V	$X_L = \frac{V_L}{I}$, Ω	θ (degrees) = arctan $\frac{X_L}{R}$
4700						
1200						

Show your computation. Now measure V_L rms across L. Compute and record the reactance X_L of L ($X_L = V_L/I$). Determine the angle θ by substituting the computed value X_L and the rated value of R in the equation

$$\theta = \arctan \frac{X_L}{R}$$

Show your computations. The two values of θ should be the same or reasonably close.

4. Replace R with a 1200-Ω resistor. Repeat the measurements and computations, steps 2 and 3.

Verifying the Formula $V = \sqrt{V_R^2 + V_L^2}$

5. Transcribe from Table 52-1 to Table 52-2 the measured rms voltages V_R and V_L for each value of resistance. For the corresponding values of V_R and V_L, evaluate $\sqrt{V_R^2 + V_L^2}$ and record in Table 52-2.

Optional: Extra Credit

6. From the data in Table 52-1 and 52-2, verify the relationships in Eqs. (52-3), (52-4), (52-5), and (52-6). List the transcribed data and computed data in tabular form. Explain the method used to verify and show your computations.

TABLE 52-2. Formulas for V_R, V_L, and Z

R, Ω	V Applied, V rms	V_R, V	V_L, V	$\sqrt{V_R^2 + V_L^2}$, V
4700	6.3			
1200	6.3			

TABLE 52-3. Applying Series RL Formulas

R, Ω	L, H	V Applied, V	F, Hz	V_L, V	V_R, V	θ, degrees	X_L, Ω	Z, Ω
10,000	0.03	100	100,000					
	0.05	80	50,000	60				

QUESTIONS

1. Explain why the two values of θ corresponding to R = 4700 Ω (Table 52-1) should be equal. If they are not, explain why. For R = 1200 Ω?
2. What are the factors which may contribute to errors in determining phase angle, as used in this experiment?
3. Do the results of your experiment confirm the relationship $V = \sqrt{V_R^2 + V_L^2}$? Substantiate your answer by referring to the proper data.
4. Determine and record the missing values in Table 52-3, which is based on measurements in a series RL circuit. Show your computations.

Extra Credit

5. What conclusions do you draw from procedural step 6? Substantiate your answer by referring to your computations.

Answers to Self-Test

1. 36.1
2. the same
3. 33.7
4. (*a*) 1000; (*b*) 36.9; (*c*) 96; (*d*) 72
5. true

53

EXPERIMENT

(OPTIONAL) THE J OPERATOR IN THE ANALYSIS OF RL CIRCUITS

OBJECTIVES

1. To define the meaning of the *j* operator as used in electric circuit analysis
2. To transform the phasor $a + jb$ into $\rho\angle\theta$, and $\rho\angle\theta$ into $a + jb$
3. To verify experimentally that the impedance Z of a series-connected *RL* circuit is

$$Z = R + jX_L = |Z|\angle\theta$$

4. To verify experimentally that the applied voltage V of a series-connected *RL* circuit is

$$V = V_R + jV_L = |V|\angle\theta$$

INTRODUCTORY INFORMATION

The *j* Operator

In Experiment 51 we learned that in a series *RL* circuit, inductive reactance X_L can be represented as leading the resistance R by 90° and that the impedance Z of the circuit is equal to the vector sum of X_L and R. Moreover, in Experiment 52 it was demonstrated that the voltage V_L across L leads by 90° the voltage V_R across R in an *LR* circuit, and that the applied voltage V is the phasor sum of V_R and V_L. To describe these relationships, the *j* operator is used in the mathematics of ac circuits. The symbol *j* represents the imaginary number $\sqrt{-1}$.

The symbol *j* as defined above is called a mathematical operator, because it represents a rotation of 90°. Consider the horizontal axis in Fig. 53-1 as being the axis of all *real* numbers, that is, all positive and negative numbers between $+\infty$ and $-\infty$. Then consider the *j* axis as the axis of all imaginary numbers between $+j\infty$ and $-j\infty$. The number $j5$ is the result of rotating a real number $+5$, 90° counterclockwise. The number $j5$, then, falls along the *j* axis, and is on the axis of imaginaries.

What would be the result of rotating the number $+j5$ another 90° counterclockwise? Graphically it is evident that the resultant number should be the real number -5. It can be shown that this graphical representation is also mathematically correct. For $j5$, rotated 90° CCW, is written as $j(j5)$, or simply $j^2 5$. Since $j = \sqrt{-1}$, $j^2 = -1$, and $j^2 5$ then becomes -5. Similarly, if the number -5 is rotated 90° CCW, it becomes $-j5$, and if $-j5$ is rotated 90° CCW, it becomes $+5$. That is, a number which has undergone four 90° rotations returns to its original starting position.

It should be quite evident, then, that the *j* operator can be treated as an algebraic symbol which obeys the laws of algebra. For

$$\begin{aligned} j &= \sqrt{-1} \\ j^2 &= -1 \\ j^3 &= -1j \\ j^4 &= +1 \\ j^5 &= j, \text{ etc.} \end{aligned}$$

The Complex Number *a* + *jb* (Rectangular Form)

Expressions involving *real* numbers and imaginaries are called *complex numbers* and are written in the form $a + jb$, where a and b are real numbers. A complex number $a + jb$ can be represented graphically as a point P in the complex plane (Fig. 53-1). Take, for example, the number $15 + j10$ ($a = 15$, $b = 10$). To find the point P that this represents, proceed to $+15$ on the axis of reals (Fig. 53-1) and erect a perpendicular to the horizontal axis at this point. Also, at $+j10$, on the axis of imaginaries, erect a perpendicular to the vertical axis. The intersection of these two perpendiculars is the point P, and it represents the expression $15 + j10$. To

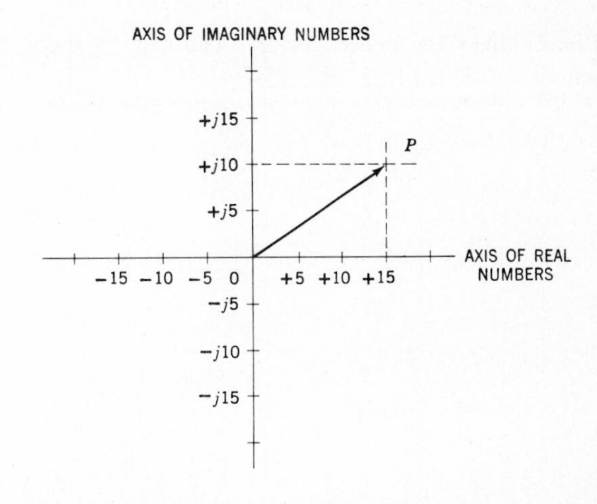

Fig. 53-1. Representation of numbers in the complex plane. *P* is the number $15 + j10$.

reach the point $15 + j10$ starting at the origin requires moving 15 units to the right along the horizontal axis (the axis of reals) and 10 units up along a perpendicular parallel to the j axis. Similarly, for *any* complex number $a + jb$, a represents the component along the horizontal axis, and b represents the component along the vertical axis. That is, a is the projection of point P on the horizontal axis, and b is the projection of point P on the vertical axis.

The complex number $a + jb$ is said to be in the *rectangular form* because a and b are the rectangular coordinates of point P. The complex number $a + jb$ can represent the point P or the radius vector drawn from the origin to P.

The Complex Number $\rho\angle\theta$ (Polar Form)

Polar coordinates can also be used to locate the point $15 + j10$. Refer to Fig. 53-2. We can find the point P if we know the length P of the radius vector OP and the angle θ. Since the triangle OQP is a right triangle, ρ can be computed by the pythagorean theorem. Thus $\rho = \sqrt{15^2 + 10^2} = 18.03$. Moreover, $\tan\theta = 10/15 = 0.667$, and therefore $\theta = 33.69°$. Thus the polar form of $15 + j10$ is $18.03 \angle 33.69°$.

In the form $\rho\angle\theta$ (Fig. 53-3*a*), ρ is the distance of the point P from the origin. Therefore ρ is the length of a radius, with center at the origin O describing a circular arc about the origin. The positive angle θ is the number of degrees of the arc, as the radius moves from the horizontal in a counterclockwise direction. The distance ρ is always positive. However, the angle θ may be negative. Thus the number $\rho \angle -\theta$ in Fig. 53-3*b* is obtained by rotating the radius $\theta°$ in a clockwise direction.

Transforming a Number in the Rectangular Form into Its Equivalent in Polar Form

To transform the complex number $a + jb$ into the polar form $\rho\angle\theta$, use the following general formula:

$$\rho = \sqrt{a^2 + b^2}$$
$$\theta = \arctan\frac{b}{a} \qquad (53\text{-}1)$$

The length of the radius vector ρ can also be found by trigonometry. Thus, from Fig. 53-2,

$$\rho = \frac{b}{\sin\theta} = \frac{a}{\cos\theta} \qquad (53\text{-}2)$$

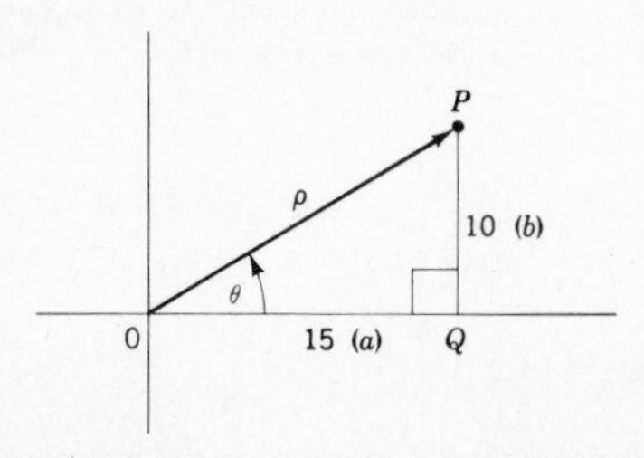

Fig. 53-2. Representation of a number in polar coordinates.

where $\theta = \arctan\dfrac{b}{a}$

Transforming from Polar Form into Rectangular Form

If a complex number is given in the polar form $\rho\angle\theta$, it can be transformed into the rectangular form $a + jb$. Consider the complex number $10 \angle 36.87°$ in Fig. 53-4. If a perpendicular is drawn from P to the horizontal axis, a right triangle is formed with the arms a and b.

Since $a = 10 \cos 36.87°$

$a = 8$

Moreover $b = 10 \sin 36.87°$

Therefore $b = 6$

The number $10\angle 36.87°$ can therefore be written as $8 + j6$. Similarly, any number in polar form $\rho\angle\theta$ can be written in the form $a + jb$, where

$$a = \rho\cos\theta \qquad (53\text{-}3)$$

and $b = \rho \sin\theta$

Therefore $\rho\angle\theta = \rho\cos\theta + j\rho\sin\theta \qquad (53\text{-}4)$

Impedance *Z* of an *RL* Circuit Expressed in Rectangular Form

Consider the series *RL* circuit in Fig. 53-5*a*. The impedance diagram for this circuit is shown in Fig. 53-5*b*. Since the impedance Z of this circuit is the phasor sum of X_L and R, and since X_L and R are at right angles, impedance Z may be written in the rectangular form, using the j operator. Thus if

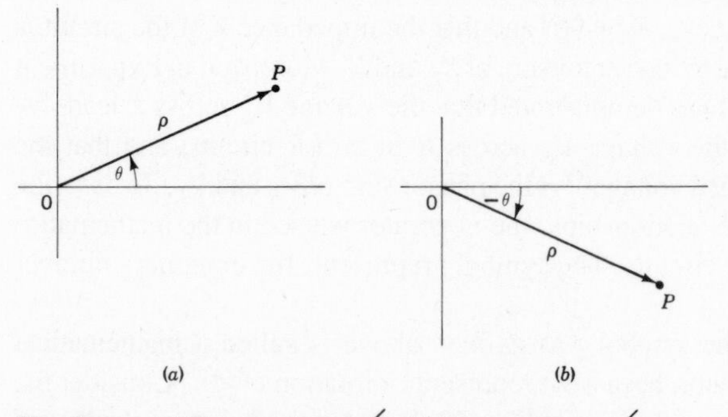

Fig. 53-3. (*a*) The number $\rho\angle\theta$; (*b*) the number $\rho\angle -\theta$.

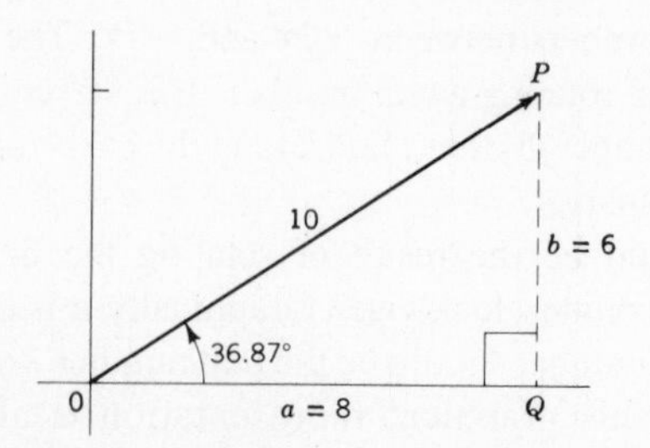

Fig. 53-4. The vector $10 \angle 36.87°$ can be transformed into the complex number $8 + j6$.

X_L is the inductive reactance of L, and R is the resistance of the resistor, then

$$Z = R + jX_L \qquad (53\text{-}5)$$

Equation (53-5) defines impedance as a phasor in the complex plane, with a horizontal component equal to R ohms and a vertical component equal to X_L ohms. This equation does not give the numerical value $|Z|$ of the impedance. If $|Z|$ is required, it may be computed from any one of the equations in (53-6):

$$|Z| = \sqrt{R^2 + X_L^2}$$

$$|Z| = \frac{R}{\cos\theta} \qquad (53\text{-}6)$$

$$|Z| = \frac{X_L}{\sin\theta}$$

where $\theta = \arctan \frac{X_L}{R}$

It is evident that Eqs. (53-6) are the same as those found in Experiment 51.

A numerical example will illustrate the manner in which Eqs. (53-5) and (53-6) are applied.

Example. In Fig. 53-5, $R = 80\ \Omega$ and $X_L = 60\ \Omega$.
(*a*) Write an equation for Z using the j operator.
(*b*) Find the numerical value $|Z|$.
(*c*) Find the phase angle θ.

Solution.
(*a*) $Z = R + jX_L = 80 + j60$
(*b*) $|Z| = \sqrt{R^2 + X_L^2} = \sqrt{80^2 + 60^2} = 100$
(*c*) $\theta = \arctan \frac{X_L}{R} = \arctan \frac{60}{80} = 36.87°$

Impedance *Z* of an *RL* Circuit Expressed in Polar Form

Since it is possible to express Z in rectangular form, it is also possible to express it in polar form. Thus in Fig. 53-5,

$$Z = |Z| \angle \theta$$

where $|Z| = \sqrt{X_L^2 + R^2} = \frac{R}{\cos\theta} = \frac{X_L}{\sin\theta} \qquad (53\text{-}7)$

and $\theta = \arctan \frac{X_L}{R}$

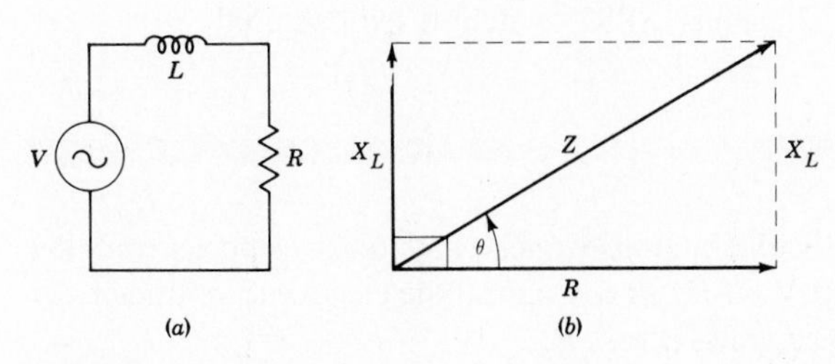

Fig. 53-5. (*a*) Series *RL* circuit; (*b*) impedance diagram of series *RL* circuit: $Z = R + jX_L$.

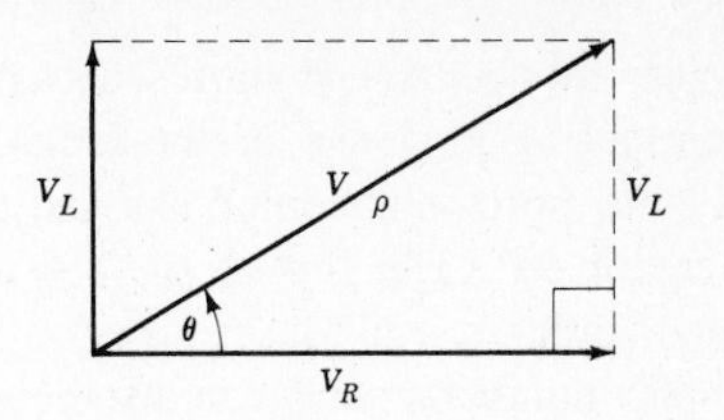

Fig. 53-6. $V = V_R + jV_L$, or $V = \rho\angle\theta$.

If $R = 80\ \Omega$ and $X_L = 60\ \Omega$, applying Eq. (53-7) we get

$$Z = 100 \angle 36.8°$$

V, V_R, and V_L in a Series *RL* Circuit Expressed in Rectangular Form; in Polar Form

In the series *LR* circuit of Fig. 53-5*a*, the relationship between the applied voltage V, the voltage drop V_R across R, and the voltage drop V_L across L can be represented by the phasor diagram in Fig. 53-6. Applying our definition of j, we can write the phasor V in rectangular form:

$$V = V_R + jV_L$$

The numerical value $|V|$ can be found from the equation

$$|V| = \sqrt{V_R^2 + V_L^2} = \frac{V_R}{\cos\theta} = \frac{V_L}{\sin\theta} \qquad (53\text{-}8)$$

where $\theta = \arctan \frac{V_L}{V_R}$

The phasor V can also be written in polar form:

$$V = |V| \angle \theta$$

where $|V| = \sqrt{V_R^2 + V_L^2} = \frac{V_R}{\cos\theta} = \frac{V_L}{\sin\theta} \qquad (53\text{-}9)$

and $\theta = \arctan \frac{V_L}{V_R}$

If we know the values of $|V|$ and θ, we can simply write V in the polar form:

$$V = |V| \angle \theta$$

We can then transform the polar into the rectangular form by writing

$$V = |V| \cos\theta + j|V| \sin\theta \qquad (53\text{-}10)$$

It is evident then that $|V| \cos\theta = V_R$, and $|V| \sin\theta = V_L$.

SUMMARY

1. The symbol j represents the imaginary number $\sqrt{-1}$. The symbol j is used as an operator in electric circuit analysis to indicate a counterclockwise rotation of 90°. The symbol $-j$ represents a clockwise rotation of 90°.
2. The sum of a real number a and an imaginary number jb is called a *complex* number and is written $a + jb$.

3. The complex number $a + jb$ represents a point P in the complex plane (Fig. 53-1), or the radius vector OP, where a is the horizontal component and b the vertical component of OP. The format $a + jb$ is called the rectangular form.
4. The complex number $a + jb$ can also be written in the polar form $\rho\angle\theta$, that is, $\rho\angle\theta = a + jb$, where

$$\rho = \sqrt{a^2 + b^2} = \frac{a}{\cos\theta} = \frac{b}{\sin\theta}$$

and $$\theta = \arctan\frac{b}{a}$$

5. If a vector is given in the polar form, that is, if ρ and θ are known, then the vector can be transformed into the rectangular form by the equation

$$\rho\angle\theta = \rho\cos\theta + j\rho\sin\theta$$

6. Rotating radius vectors used in the analysis of ac voltages and currents are called phasors.
7. Because it is related to voltage and current the impedance Z of a series RL circuit may also be considered a phasor quantity. X_L and R have a 90° phase difference. If the values of X_L and R are known, their impedance Z can be written in the rectangular form by the equation

$$Z = R + jX_L$$

8. Z can also be written in the polar form by the equation $Z = \rho\angle\theta$, where

$$\rho = \sqrt{R^2 + X_L^2} = \frac{R}{\cos\theta} = \frac{X_L}{\sin\theta}$$

and $$\theta = \arctan\frac{X_L}{R}$$

9. If the impedance of a series RL circuit is given in the polar form, that is, if the values of ρ and θ are known, then Z can be expressed in rectangular form by the equation

$$Z = \rho\cos\theta + j\rho\sin\theta$$

In this equation, $\rho\cos\theta = R$

and $\rho\sin\theta = X_L$

10. In a series RL circuit the phasor relationship between the applied voltage V, the voltage drop V_R across R, and the voltage drop V_L across L is given by the formula

$$V = V_R + jV_L$$

or $$V = |V|\angle\theta$$

where $$|V| = \sqrt{V_R^2 + V_L^2} = \frac{V_R}{\cos\theta} = \frac{V_L}{\sin\theta}$$

and $$\theta = \arctan\frac{V_L}{V_R}$$

SELF-TEST

Check your understanding by answering these questions:

1. In the circuit of Fig. 53-5a, $V = 20$ V, $R = 150\ \Omega$, and $X_L = 100\ \Omega$. The impedance Z, in rectangular form, is $Z =$ __________.
2. In the circuit of question 1, the phase angle θ between Z and R is __________ degrees.
3. In the circuit of question 1, impedance Z written in polar form is $Z =$ __________.
4. In the circuit of question 1, the applied voltage V, that is, the phasor V, written in polar form is $V =$ __________.
5. In the circuit of question 1, the phasor V written in rectangular form is $V =$ __________.
6. From the equation in question 5, we know that the voltage V_R across R is __________ V, while the voltage V_L across L is __________ V.
7. In the circuit of Fig. 53-5a, $Z = 100\ \Omega$, $V_R = 60$ V, and $V_L = 80$ V. The phasor V written in rectangular form is $V =$ __________.
8. In the circuit of question 7, the phase angle θ between V and V_R is __________ degrees.
9. In the circuit of question 7, the phasor V written in polar form is $V =$ __________.
10. In the circuit of question 7, the impedance Z written in polar form is $Z =$ __________.
11. In the circuit of question 7, the impedance Z written in rectangular form is $Z =$ __________.
12. From the equation in question 11, we know that $R =$ __________ Ω and $X_L =$ __________ Ω.

MATERIALS REQUIRED

- Power supply: Isolation transformer and variable autotransformer operating from the 120-V 60-Hz line
- Equipment: Electronic voltmeter
- Resistors: 2-W 5000 Ω
- Inductor: 8 H at 50 mA direct current whose resistance is 250 Ω
- Miscellaneous: SPST switch; fused line cord

PROCEDURE

Impedance *Z* of a Series *RL* Circuit

1. Connect the circuit of Fig. 53-7. S is open. R is a 2-W 5000-Ω resistor. L is an iron-core choke rated 8 H at 50 mA dc. V_A is a variable transformer plugged into T, a 1:1 isolation transformer which receives power from the 120-V 60-Hz line. Connect the electronic voltmeter, set on ac volts, across R.

2. Close S. **Power on.** Adjust the output of the ac supply until the voltage V_R across R measures 75 V.

3. Measure also and record in Table 53-1 the voltage V_L across L and the voltage V_{AB}, which is the numerical value $|V|$ of the voltage applied across the series-connected RL circuit.
4. Open switch S. **Power off.** Measure the resistance of R and record it in Table 53-1.
5. Using the measured values of V_R and R, compute and record in Table 53-1 the current I in the circuit.
6. Compute and record the value X_L of the reactance in the circuit, using the equation $X_L = V_L/I$, where V_L is the measured value.
7. Compute and record the phase angle θ from the equation

$$\theta = \arctan \frac{V_L}{V_R}$$

8. Compute and record in Table 53-1 the numerical value $|Z|$ using the relationship

$$|Z| = \sqrt{R^2 + X_L^2}$$

where R is the measured value and X_L is the computed value.
9. Using the computed values $|Z|$ and θ, write phasor Z in polar form in Table 53-1.
10. Transform the polar form of Z into its rectangular form, using the relationship

$$Z = |Z| \cos \theta + j|Z| \sin \theta$$

and record in Table 53-1.

Voltage Relationships in a Series *RL* Circuit

11. From the measured values V_R and V_L in Table 53-1, write the phasor V (applied voltage) in rectangular form using the equation $V = V_R + jV_L$ and record in Table 53-2.
12. Compute the numerical value $|V|$ of the phasor V using any one of the equations

$$|V| = \sqrt{V_R^2 + V_L^2} = \frac{V_R}{\cos \theta} = \frac{V_L}{\sin \theta}$$

and record in Table 53-2. Employ the values V_R, V_L, and θ from Table 53-1.
13. From Table 53-1 record in Table 53-2 the measured value $V_{AB} = |V|$.
14. Using the value of $|V|$ from step 12 and the value of θ from Table 53-1, write the phasor V in polar form in Table 53-2.

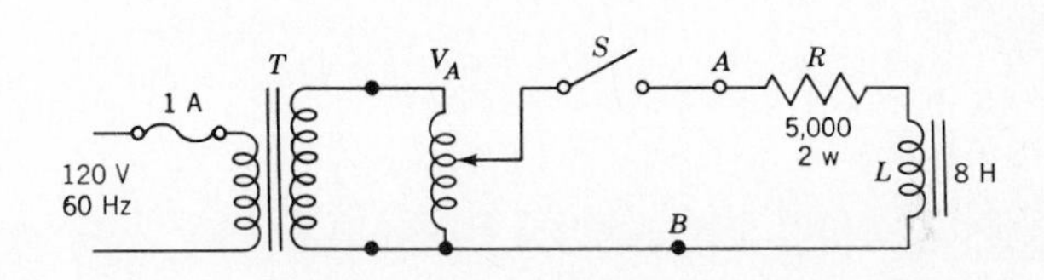

Fig. 53-7. Experimental circuit for determining the impedance Z of an RL circuit.

TABLE 53-1. Impedance Z in Polar and Rectangular Form

Measured Values				*Computed Values*			
V_R, V	V_L, V	$V_{AB} = \|V\|$, V	R, Ω	I, A	$\|Z\|$, Ω	θ, degrees	X_L, Ω
75							
Z in polar form				Z in rectangular form			
$Z =$				$Z =$			

TABLE 53-2. Voltage in Polar and Rectangular Form

V in rectangular form	*|V| computed*	*|V| measured*	*V in polar form*
$V =$			$V =$

QUESTIONS

1. Refer to the equation for Z in rectangular form, Table 53-1 and to Procedure step 10.
   (*a*) Does the measured value of $R = |Z| \cos \theta$?
   (*b*) Should it?
   (*c*) If the two values are not the same, explain why they are not.
2. Refer to the equation for Z in rectangular form, Table 53-1 and to Procedure step 10.
   (*a*) Does the computed value of $X_L = |Z| \sin \theta$?

(*b*) Should it?

(*c*) If the two values are not the same, explain why not.

3. Have you confirmed experimentally (approximately) the analytical statement that $Z = R + jX_L$? Refer specifically to your data to show such confirmation or lack of it.
4. If your data does not confirm (approximately) the statement $Z = R + jX_L$ explain why not.
5. Refer to Table 53-2 and to step 12. Does the computed value of $|V|$ equal the measured value of $|V|$? If not, why not?
6. Have you confirmed experimentally (approximately) the analytical statement that $V = V_R + jV_L$? Refer specifically to your data to show such confirmation or lack of it.
7. If your data does not confirm (approximately) that $V = V_R + jV_L$ explain why not.

Answers to Self-Test

1. $150 + j100$
2. 33.69
3. $180.28\angle 33.69°$
4. $20\angle 33.69°$
5. $16.64 + j11.1$
6. 16.64; 11.1
7. $60 + j80$
8. 53.13
9. $100\angle 53.13°$
10. $100\angle 53.13°$
11. $60 + j80$
12. 60; 80

54

EXPERIMENT

IMPEDANCE OF A SERIES *RC* CIRCUIT

OBJECTIVES

1. To verify experimentally that the impedance Z of a series RC circuit is given by the equation

$$Z = \sqrt{R^2 + X_C^2}$$

2. (Optional) To verify experimentally that the relationship between Z, R, and X_C is given by the equations

$$Z = \frac{R}{\cos\theta} = \frac{X_C}{\sin\theta}$$

where θ is the phase angle between Z and R.

INTRODUCTORY INFORMATION

Analysis of the impedance of RC circuits, Fig. 54-1, is very similar to analysis of the impedance of RL circuits (see Experiment 51). It can be demonstrated mathematically that the impedance Z of an RC circuit is a phasor quantity resulting from the phasor sum of X_C and R.

Figure 54-2 is a phasor diagram which shows that R and X_C are at right angles to each other. Note, however, that X_C *lags* R by 90°, in contradistinction to X_L, which leads R by 90°.

In Fig. 54-2, Z is shown as the phasor sum of R and X_C. Z is the hypotenuse of a right triangle of which R and X_C are the legs. Applying the pythagorean theorem to R, X_C, and Z, we see that

$$Z = \sqrt{R^2 + X_C^2} \tag{54-1}$$

Equation (54-1) makes it possible to determine the impedance of an RC circuit if R and X_C are known. For example, if $R = 30\ \Omega$ and $X_C = 40\ \Omega$, then $Z = \sqrt{30^2 + 40^2} = 50\ \Omega$.

Refer again to Fig. 54-2. If R and X_C are known, the angle θ between R and Z can be determined.

For

$$\tan\theta = \frac{X_C}{R} \tag{54-2}$$

and

$$\theta = \arctan\frac{X_C}{R} \tag{54-3}$$

We can now calculate Z. Thus

$$\frac{X_C}{Z} = \sin\theta \tag{54-4}$$

and

$$Z = \frac{X_C}{\sin\theta} \tag{54-5}$$

Similarly

$$Z = \frac{R}{\cos\theta} \tag{54-6}$$

The process of finding Z, therefore, is resolved to the following simple operations:

1. Find θ by applying Eqs. (54-2) and (54-3).
2. Find Z by applying either Eq. (54-5) or (54-6).

Example. If $R = 30\ \Omega$ and $X_C = 40\ \Omega$, find Z trigonometrically.

Solution. Draw the impedance triangle (see Fig. 54-2).

1. $$\tan\theta = \frac{X_C}{R} = \frac{40}{30} = 1.33$$

$$\theta = 53.1°$$

2. $$Z = \frac{R}{\cos\theta} = \frac{30}{\cos 53.1°} = 50\ \Omega$$

or

$$Z = \frac{X_C}{\sin\theta} = \frac{40}{\sin 53.1°} = 50\ \Omega$$

This result agrees with that obtained when we used the right-triangle formula.

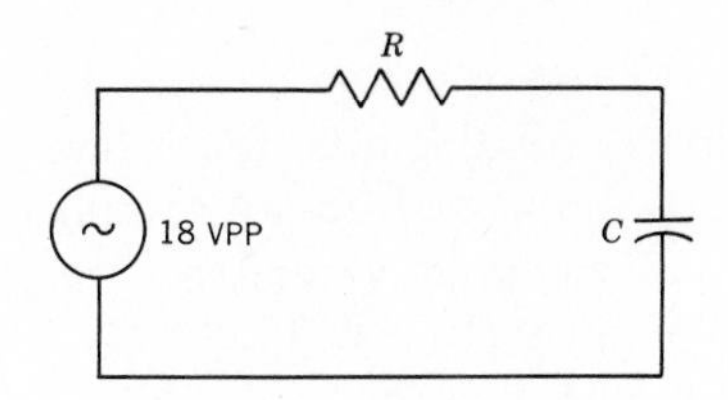

Fig. 54-1. Series *RC* circuit.

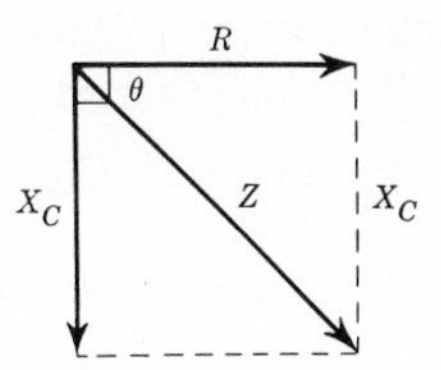

Fig. 54-2. Impedance of a series *RC* circuit.

Ohm's Law Applied to a Series *RC* Circuit

Ohm's law states that the current I in a circuit equals the ratio of the applied voltage V and impedance Z.

$$I = \frac{V}{Z} \tag{54-7}$$

and

$$Z = \frac{V}{I} \tag{54-8}$$

Equation (54-8) provides the means of verifying the relationship between Z, X_C, and R. Thus, in the circuit of Fig. 54-1, we can measure V and I and compute Z by substituting the measured values of V and I in Eq. (54-8). We can then calculate Z, using Eqs. (54-1), (54-5), or (54-6). If the results agree, that is, if the value of Z is substantially the same for each of these methods, then we have verification of the right-triangle relationship between Z, X_C, and R and also verification of Eqs. (54-1), (54-5), and (54-6).

Example. Assume that in the series circuit (Fig. 54-1), $X_C = 40\ \Omega$, $R = 30\ \Omega$, and the applied voltage $V = 10$ V. An ac milliammeter is inserted in the circuit, and it measures 0.2 A. We can now determine the impedance for

$$Z = \frac{V}{I} = \frac{10}{0.2} = 50\ \Omega$$

The current measurement and subsequent computation for Z is one verification of Eqs. (54-1), (54-5), and (54-6).

Another technique for determining the current in the ac circuit of Fig. 54-1 without measuring current directly is to measure the ac voltage V_R developed across the resistor R. The current in R may then be found from the equation

$$I = \frac{V_R}{R} \tag{54-9}$$

Since current is everywhere the same in a series circuit, the computed value of I *is* the current in the circuit. We may then solve for the impedance of the circuit in the manner described in the preceding example, that is, by substituting I in the equation

$$Z = \frac{V}{I}$$

where V is the known or measured value of applied voltage and I is the computed current.

SUMMARY

1. In a series-connected *RC* circuit, the impedance Z is the phasor sum of R and X_C, where X_C *lags* R by 90° (Fig. 54-2).
2. From the impedance right triangle in Fig. 54-2, the numerical value of Z can be found from the equation

$$Z = \sqrt{R^2 + X_C^2}$$

3. In Fig. 54-2, if θ is the phase angle between Z and R, then

$$\theta = \arctan \frac{X_C}{R}$$

4. Z may be computed trigonometrically if X_C and R are known. Thus

$$Z = \frac{X_C}{\sin \theta}$$

or

$$Z = \frac{R}{\cos \theta}$$

where θ is the angle computed from the equation in summary step 3.

5. The impedance Z of an *RC* circuit may also be computed, if the applied voltage V and the current I in the circuit are known, from the basic equation

$$Z = \frac{V}{I}$$

6. If the value of Z is substantially the same for each of the methods (equations), we have verification of the right-triangle relationship between Z, X_C, and R, and also verification of the equations which define Z.

SELF-TEST

Check your understanding by answering these questions. Students who have not studied trigonometry may skip questions 2, 3, and 4.

1. In the series *RC* circuit (Fig. 54-1), $R = 3000\ \Omega$, $X_C = 1200\ \Omega$, $Z =$ __________ Ω.
2. In the same circuit as in question 1, $\theta =$ __________ degrees.
3. In the same circuit as in question 1,

   (*a*) $\dfrac{R}{\cos \theta} =$ __________ Ω

   (*b*) $\dfrac{X_C}{\sin \theta} =$ __________ Ω

4. The ratio $R/\cos \theta$ or $X_C/\sin \theta$, where θ is the phase angle between Z and R, also gives the __________ of the circuit of Fig. 54-1.
5. In a series *RC* circuit, $R = 1200\ \Omega$, $X_C = 1500\ \Omega$, and the applied voltage $V = 50$ V. The current I is $I =$ __________ A.
6. In the circuit of question 1, $V_R = 45$ V. The current I in the capacitor C is __________ A.

MATERIALS REQUIRED

- Power supply: Line-isolation transformer and variable autotransformer operating from 120-V line
- Equipment: 0–5-mA and 0–25-mA ac milliammeters, or a multirange milliammeter covering these ranges; EVM
- Resistors: 1-W 5000-Ω; ½-W 22,000-Ω
- Capacitors: 0.5 μF, 0.1 μF
- Miscellaneous: SPST switch; fused line cord

PROCEDURE

1. Connect the circuit of Fig. 54-3. S is open. R is a 5000-Ω resistor; C is an 0.5-μF capacitor. V_A is a variable autotransformer plugged into T, a 1:1 isolation transformer. M is an ac milliammeter set on the 0–25-mA range. V is an EVM set to the 100-V ac range.
2. Close S. Adjust the output of the autotransformer until M measures 10 mA (0.01 A). With the voltmeter, measure the voltage V across the RC circuit and record in Table 54-1. Compute Z by substituting the measured values of V and I in Eq. (54-8). Record Z in the column V/I.
3. Compute X_C using the formula

$$X_C = \frac{1}{2\pi FC}$$

where F is the power-line frequency.
Record X_C in Table 54-1. Calculate and record Z using Eq. (54-1).
4. Measure and record V_R, the voltage across R. Compute I, using Eq. (54-9). Record I in the column V_R/R.
5. **Power off.** Replace C by a 0.1-μF capacitor and R by a 22,000-Ω resistor. Switch the milliammeter to the 0–5-mA range. Repeat steps 2, 3, and 4, but set the output of the variable autotransformer for 2 mA (0.002 A) of current.

Extra Credit

6. From Table 54-1 transcribe to Table 54-2 the values of $Z = V/I$ and X_C corresponding to $C = 0.5$ and 0.1 μF. For each C and R, determine tan θ and θ using Eqs. (54-2) and (54-3) and record in Table 54-2. Next compute the corresponding values of Z using Eqs. (54-5) and (54-6).

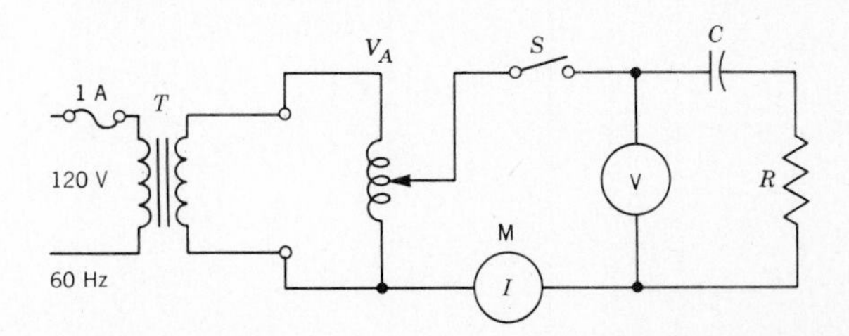

Fig. 54-3. Experimental circuit for determining the impedance Z of a series RC circuit.

TABLE 54-1. Impedance of a Series RC Circuit

R, Ω	C, μF	I, A	V, V	$Z = \frac{V}{I}$, Ω	X_C, Ω	$Z = \sqrt{R^2 + X_C^2}$, Ω	V_R, V	$I = \frac{V_R}{R}$, A
5,000	0.5	0.01						
22,000	0.1	0.002						

TABLE 54-2. Trigonometric Formulas for Impedance

R, Ω	C, μF	I, A	$Z = \frac{V}{I}$, Ω	X_C, Ω	$\tan\theta = \frac{X_C}{R}$	θ, degrees	Z (ohms) =	
							$\frac{R}{\cos\theta}$	$\frac{X_C}{\sin\theta}$
5,000	0.5	0.01						
22,000	0.1	0.002						

QUESTIONS

1. If the two values of Z in Table 54-1, corresponding to any one RC circuit, are equal, what conclusion can you draw concerning the formula $Z = \sqrt{R^2 + X_C^2}$? Are the values equal?
2. What are the possible sources of error in the data in Table 54-1?
3. For a specific Z in Table 54-1, is the value of I measured with a milliammeter the same as the value of I computed from the equation $I = V_R/R$? Should they be the same? Explain any discrepancies.
4. Define Z.

Extra Credit

5. Do the results of your data in Table 54-2 confirm Eqs. (54-5) and (54-6)? Explain.

Answers to Self-Test

1. 3231
2. 21.8
3. (*a*) 3231; (*b*) 3231
4. impedance
5. 0.026
6. 0.015

55

EXPERIMENT

VOLTAGE RELATIONSHIPS IN A SERIES *RC* CIRCUIT

OBJECTIVES

1. To measure the phase angle θ between the applied voltage V and the current I in a series RC circuit
2. To verify experimentally that the relationships among the applied voltage V, the voltage V_R across R, and the voltage V_C across C are described by the equations

$$V = \sqrt{V_R^2 + V_C^2}$$

$$V_R = V \cos \theta = V \times \frac{R}{Z}$$

$$V_C = V \sin \theta = V \times \frac{X_C}{Z}$$

INTRODUCTORY INFORMATION

There is a striking similarity between analysis of RC and RL circuits (see Experiment 52). The difference is that in an inductive circuit, current I lags the applied voltage V, whereas in a *capacitive* circuit, current I *leads* the applied voltage V.

Phase Relations between the Applied Voltage and Current in a Series *RC* Circuit

The current I is common in every part of a series circuit. Therefore, current is used as the reference phasor in considering the phase relations between V_R and V_C in the series RC circuit of Fig. 55-1*a*. Since current and voltage in a resistor are in phase, the voltage V_R is shown in Fig. 55-1*b* as an extension of the current phasor. But *current* in a *capacitor leads the voltage across* the capacitor by 90°. Therefore, the phasor V_C is shown lagging the current I by 90°. It is evident, therefore, that V_C lags V_R by 90°.

V_R and V_C are, respectively, the voltages across R and C in the series RC voltage divider (Fig. 55-1*a*). As in the case of an LR circuit, the applied voltage V is the phasor sum of V_R and V_C. This is shown in Fig. 55-1*b*. We also see that V is the hypotenuse of a right triangle of which V_R and V_C are the legs. Therefore, applying the pythagorean theorem,

$$V = \sqrt{V_R^2 + V_C^2} \tag{55-1}$$

Figure 55-1*b* also reveals the phase relationship between the applied voltage V and the current I in the series RC circuit. The current I is seen to lead the applied voltage V by the angle θ.

The question that suggests itself is: Is the angle θ by which current leads the applied voltage in a series RC circuit the same as the angle θ between the impedance phasor Z and the resistance R in the previous experiment? They are the same, as can be seen from Fig. 55-2. Part (*a*) is a redrawing of Fig. 55-1*b* that shows the phase relations between V, V_R, and V_C. But in addition, the composition of each of these voltages is indicated. Thus, V_R is the product of I and R, V_C the product of I and X_C, and V the product of I and Z. Since I is a common factor in each of these products, it may be eliminated, leaving the impedance diagram (Fig. 55-2*b*).

It is evident, therefore, that the angle θ is the same in Figs. 55-1*b* and 55-2*b*.

Analysis of Fig. 55-1*b* reveals the relationships among V, V_R, V_C, and the phase angle θ. Thus, since

$$\frac{V_R}{V} = \cos \theta \tag{55-2}$$

$$V_R = V \cos \theta \tag{55-3}$$

Similarly it follows that

$$V_C = V \sin \theta \tag{55-4}$$

Referring to the impedance triangle, it is also evident that

$$\cos \theta = \frac{R}{Z}$$

and

$$\sin \theta = \frac{X_C}{Z}$$

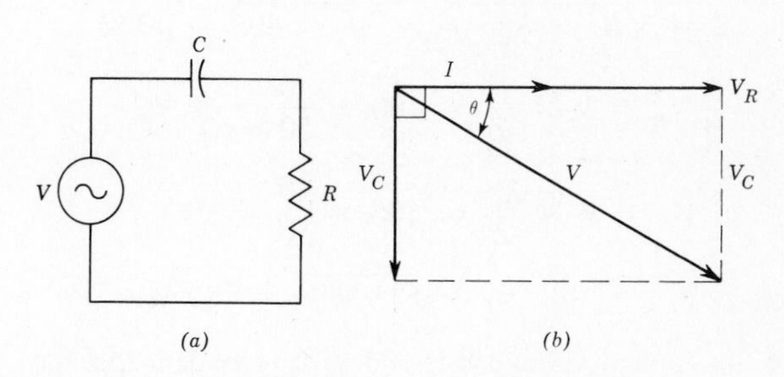

Fig. 55-1. Phase relations in a series *RC* circuit.

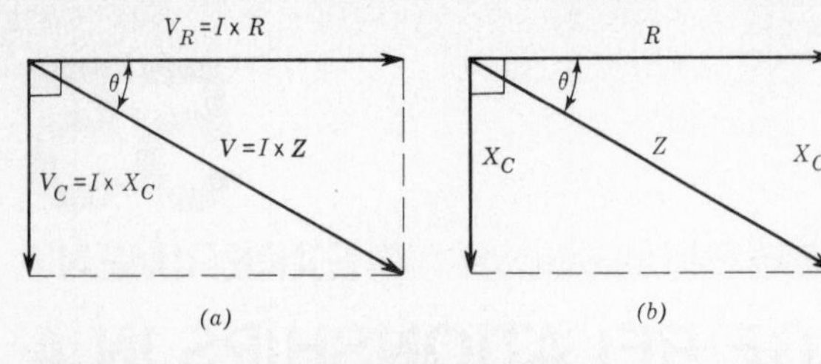

Fig. 55-2. The phase angle between *V* and *I* is the same as the angle between *Z* and *R* in a series *RC* circuit.

By substituting these values of the trigonometric functions in Eqs. (55-3) and (55-4), we find that

$$V_R = V \times \frac{R}{Z} \tag{55-5}$$

and

$$V_C = V \times \frac{X_C}{Z} \tag{55-6}$$

Equations (55-5) and (55-6) state that in a series *RC* circuit, the voltage V_R across *R* may be found by multiplying the applied voltage *V* by the ratio of the resistance and impedance (R/Z), and the voltage V_C across *C* is found by multiplying the applied voltage *V* by the ratio of capacitive reactance and impedance (X_C/Z).

Example. If 100 V is applied across a circuit consisting of a 30-Ω resistor in series with a capacitor whose reactance is 40 Ω, what is the value of:

1. θ, the phase angle between *V* and *I*?
2. V_R, the voltage across *R*?
3. V_C, the voltage across *C*?

Solution 1.

1.
$$\tan\theta = \frac{X_C}{R} = \frac{40}{30} = 1.33$$
$$\theta = 53.13^\circ$$

2.
$$V_R = V\cos\theta = 100\cos 53.13^\circ = 60\text{ V}$$

3.
$$V_C = V\sin\theta = 100\sin 53.13^\circ = 80\text{ V}$$

Solution 2.

1. As in the previous case, $\theta = 53.13^\circ$
2. Find *Z* by the right-triangle method.

$$Z = \sqrt{R^2 + X_C^2} = \sqrt{30^2 + 40^2} = 50\ \Omega$$

$$V_R = V \times \frac{R}{Z} = 100 \times \frac{30}{50} = 60\text{ V}$$

3.
$$V_C = V \times \frac{X_C}{Z} = 100 \times \frac{40}{50} = 80\text{ V}$$

Check: $\sqrt{V_R^2 + V_C^2} = \sqrt{60^2 + 80^2} = 100\text{ V} = V$

Since the applied voltage *V* is 100 V, it is evident that the solution to the problem is correct.

SUMMARY

1. In a series *RC* (capacitive) circuit, the current *I* *leads* the applied voltage *V* by some angle θ, called the phase angle.
2. The phase angle θ between *V* and *I* is the same as the angle θ between *Z* and *R* in the impedance diagram (Fig. 55-1*b* and 55-2*b*). The angle θ is also the same as the angle between *V* and V_R (Fig. 55-1*b*).
3. The value of θ depends on the relative values of X_C, *R*, and *Z*, and may be computed from any one of the following formulas:

$$\theta = \arctan\frac{X_C}{R}$$

$$\theta = \arcsin\frac{X_C}{Z}$$

$$\theta = \arccos\frac{R}{Z}$$

4. In a series *RC* circuit the voltage drop V_C across the capacitor *lags* the voltage drop V_R across the resistor by 90°.
5. V_C and V_R are phasors and are added to give the applied voltage *V*.
6. The relationship between the applied voltage *V* and the voltage drops V_R and V_C across *R* and *C*, respectively, is given by the formula

$$V = \sqrt{V_R^2 + V_C^2}$$

7. If the applied voltage value *V* and the phase angle θ are known, then V_R and V_C may be calculated from these formulas:

$$V_R = V\cos\theta$$
$$V_C = V\sin\theta$$

8. If the applied voltage *V* and *R*, X_C, and *Z* are known in a series *RC* circuit, then V_R and V_C may be calculated from these formulas:

$$V_R = V \times \frac{R}{Z}$$

$$V_C = V \times \frac{X_C}{Z}$$

9. To verify that *V* is the phasor sum of V_R and V_C in a series *RC* circuit, we measure *V*, V_R, and V_C and substitute the measured values of V_R and V_C in $\sqrt{V_R^2 + V_C^2}$. If the numerical result is equal to the measured value *V*, then we have the required confirmation.

SELF-TEST

Check your understanding by answering these questions:

1. In the circuit of Fig. 55-1*a*, the measured values of V_R and V_C are, respectively, 50 and 120 V. The applied voltage *V* must then equal __________ V.

2. The phase angle between voltages V and ____________ is the same as the phase angle between V and I in the circuit of Fig. 55-1*a*.
3. The phase angle θ between V and I in question 1 is ____________ degrees.
4. In the circuit of Fig. 55-1*a*, $X_C = 200\ \Omega$, $R = 300\ \Omega$, and $V = 60$ V. In this circuit,
   (*a*) $Z =$ ____________ Ω
   (*b*) $\theta =$ ____________ degrees
   (*c*) $V_R =$ ____________ V
   (*d*) $V_C =$ ____________ V
5. In a series *RC* circuit, $\arctan X_C/R = \arccos R/Z$. ____________ (true/false)

MATERIALS REQUIRED

- Power supply: Line-isolation transformer and variable autotransformer operating from 120-V ac line
- Equipment: Oscilloscope; EVM
- Resistors: ½-W 12,000-Ω; 5600-Ω
- Capacitor: 0.5 μF
- Miscellaneous: SPST switch; fused line cord

PROCEDURE

Measuring Phase Angle

In Fig. 55-1*a*, the voltage V_R across R is in phase with current I in the circuit. The phase angle θ, by which the applied voltage V lags the current I, can therefore be found by measuring the angle between V and V_R, using the method employed in Experiment 52.

1. Connect the circuit of Fig. 55-3*a*. V is the output of the variable autotransformer, which is plugged into a 1:1 isolation transformer in the manner of Fig. 55-3. Switch S is *off*. Set the output of V at 20 V, approximately. Switch the Trigger selector of the oscilloscope to Ext. **Power on.**
2. Following the method you used in Experiments 52 and 42, measure the phase angle between the applied voltage and current in the circuit of Fig. 55-3. Record the angle θ in Table 55-1.
3. Compute and record X_C of the capacitor C in Fig. 55-3*a*. For comparison, determine the phase angle θ by substituting the values of X_C and R in the equation

$$\theta = \arctan \frac{X_C}{R}$$

   Show your computation. The computed and measured values of θ should be very close to each other. Also compute and record the impedance Z of the circuit. Show your computation.

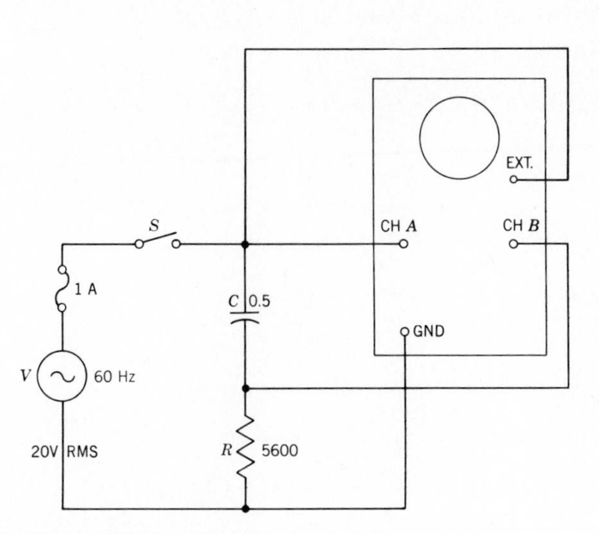

Fig. 55-3. Measuring the phase angle between *V* and *I*.

4. **Power off.** In the circuit of Fig. 55-3*a*, replace R with a 12,000-ohm resistor. As in the preceding steps, measure and record the new phase angle between V and I. Also compute and record the phase angle, as in step 3, for comparison. Also compute and record the impedance Z of the circuit.

Verifying the Formula $V = \sqrt{V_R^2 + V_C^2}$

5. Remove the oscilloscope connections from the circuit. Set the output of the isolated supply at 30 V, as measured on an EVM. With the voltmeter, measure the voltages V_R and V_C and record in Table 55-2. Calculate the value of $\sqrt{V_R^2 + V_C^2}$ and record in Table 55-2. Show your computations.
6. Repeat step 5 for each value of supply voltage V shown in Table 55-2.
7. **Power off.** Replace the 12,000-Ω resistor with a 5600-Ω resistor. Repeat, recording the measurements for V_R and V_C for each value of applied voltage in Table 55-2. In each case calculate and record the value of $\sqrt{V_R^2 + V_C^2}$.

Extra Credit

8. From the data in Tables 55-1 and 55-2, verify the relationships in Eqs. (55-3), (55-4), (55-5), and (55-6). Show your computations. Record the required data in tabular form.

TABLE 55-1. Phase Angle in a Series *RC* Circuit

R, Ω	*C*, μF	θ (*degrees*) Measured	X_C, Ω	θ (*degrees*) = arctan $\frac{X_C}{R}$	*Z*, Ω
5,600	0.5				
12,000	0.5				

TABLE 55-2. Voltages in a Series *RC* Circuit

R, Ω	*C*, μF	*V*, V	V_R, V	V_C, V	$\sqrt{V_R^2 + V_C^2}$ V
12,000	0.5	30			
		40			
		50			
5,600	0.5	30			
		40			
		50			

TABLE 55-3. Computation for Question 4

	R, Ω	*C*, μF	*V*, V	*F*, Hz	V_C, V	V_R, V	θ, degrees	X_C, Ω	*Z*, Ω
a	1200	0.10	80	1000					
b		0.25	100	500	50				
c				100	20	30		10,000	

QUESTIONS

1. What conclusion can you draw from the measured and computed results of θ in Table 55-1? Explain.
2. What are the sources of error in the measurements for θ (Table 55-1)?
3. What conclusion can you draw from the data in Table 55-2 concerning *V*, V_R, and V_C? Explain.
4. Find the missing values in Table 55-3, which is based on measurements in a series *RC* circuit. Show your computations.

Answers to Self-Test

1. 130
2. V_R
3. 67.4°
4. (*a*) 361; (*b*) 33.7; (*c*) 50; (*d*) 33.3
5. true

56

EXPERIMENT

(OPTIONAL) THE *j* OPERATOR IN THE ANALYSIS OF *RC* CIRCUITS

OBJECTIVES

1. To verify experimentally that the impedance Z of a series-connected RC circuit is

$$Z = R - jX_C = |Z| \angle{-\theta}$$

2. To verify experimentally that the applied voltage V of a series-connected RC circuit is

$$V = V_R - jV_C = |V| \angle{-\theta}$$

INTRODUCTORY INFORMATION

The use of the *j* operator (Experiment 53) facilitates analysis and computations of impedance, voltage, and current in RC circuits. In this experiment we will learn how to write Z of an RC circuit in rectangular form using the *j* operator. Moreover, the relationship between the applied voltage V and the voltage drops V_R across R and V_C across C will be written in rectangular form using the *j* operator. Z and V will also be written in the polar form.

Impedance *Z* of an *RC* Circuit Expressed in Rectangular Form

Consider the series RC circuit in Fig. 56-1*a*. The impedance diagram for this circuit is shown in Fig. 56-1*b*. Since the impedance Z is considered a phasor quantity whose components are X_C and R, and since X_C and R are at right angles, the impedance Z may be written in the rectangular form, using the *j* operator. It should be noted, however, that X_C *lags* R (considered as reference) by 90°. Therefore, since R is a positive real number lying on the axis of real numbers in the complex plane, X_C must lie on the axis of imaginary numbers (*j*) in the complex plane, and specifically on the *negative* side of the *j* axis. Thus if X_C is the capacitive reactance of C, and R is the resistance of the resistor in Fig. 56-1*a*, then

$$Z = R - jX_C \tag{56-1}$$

Equation (56-1) states that the impedance of a series-connected RC circuit is treated as a phasor, and when it is written in the rectangular form, the positive horizontal component is the value of resistance R, and the negative vertical component is the value of X_C. This equation does not give the *numerical* value $|Z|$ of the impedance. If $|Z|$ is required, it may be computed from the equation

$$|Z| = \sqrt{R^2 + X_C^2} \tag{56-2}$$

Refer again to Fig. 56-1*b*. Note that the phase angle is negative, $-\theta$, since Z lags R. The numerical value Z of the impedance can also be found from the equations

$$|Z| = \frac{R}{\cos(-\theta)} = \frac{R}{\cos\theta} \tag{56-3}$$

or

$$|Z| = \frac{-X_C}{\sin(-\theta)} = \frac{X_C}{\sin\theta}$$

A numerical example will illustrate the manner in which Eqs. (56-2) and (56-3) are applied.

Example. In Fig. 56-1*a*, $R = 30\ \Omega$ and $X_C = 40\ \Omega$.

1. Write an equation for Z using the *j* operator.
2. Find the numerical value $|Z|$.
3. Find the phase angle θ.

Solution.

1. $Z = R - jX_C = 30 - j40$
2. $|Z| = \sqrt{R^2 + X_C^2} = \sqrt{30^2 + 40^2} = 50$
3. $\theta = \arctan \dfrac{X_C}{R} = \arctan \dfrac{40}{30} = 53.1°$

Impedance *Z* of an *RC* Circuit in Polar Form

The polar form of the impedance Z of a series-connected RC circuit follows directly from the rectangular form. Thus in Fig. 56-1*b*

$$Z = |Z| \angle{-\theta}$$

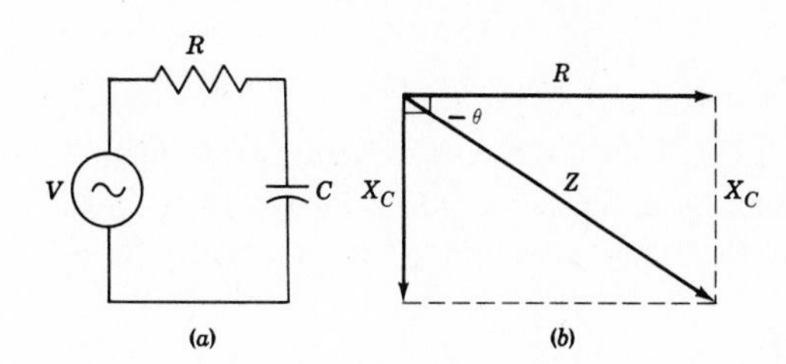

Fig. 56-1. (*a*) Series-connected *RC* circuit; (*b*) phasor diagram showing the relationship between *Z*, *R*, and X_C. Note that the phase angle is negative, $-\theta$, since *Z* lags *R*.

$$|Z| = \sqrt{R^2 + X_C^2} = \frac{R}{\cos\theta} = \frac{X_C}{\sin\theta} \quad (56\text{-}4)$$

where

$$\theta = \arctan \frac{X_C}{R}$$

and

Now if $R = 30\ \Omega$ and $X_C = 40\ \Omega$, applying Eqs. (56-4), we get

$$Z = 50\ \angle{-53.1°}$$

V, V_R, and V_C in a Series *RC* Circuit Expressed in Rectangular Form and in Polar Form

In the series *RC* circuit of Fig. 56-1*a*, the relationship between the applied voltage V and the voltage drop V_R across R and the voltage drop V_C across C can be represented by the phasor diagram in Fig. 56-2. The phasor V in rectangular form is written

$$V = V_R - jV_C \quad (56\text{-}5)$$

The numerical value V can be found from the equations

$$|V| = \sqrt{V_R^2 + V_C^2} = \frac{V_R}{\cos(-\theta)} = \frac{-V_C}{\sin(-\theta)}$$

or

$$|V| = \frac{V_R}{\cos\theta} = \frac{V_C}{\sin\theta} \quad (56\text{-}6)$$

where

$$\theta = \arctan \frac{V_C}{V_R}$$

In polar form, the phasor V is written

$$V = |V|\ \angle{-\theta} \quad (56\text{-}7)$$

where $|V|$ is defined in exactly the same way as it is in Eq. (56-6).

If we know the values of $|V|$ and θ for a series *RC* circuit, we can simply write V in polar form:

$$V = |V|\ \angle{-\theta}$$

We can then transform the polar into the rectangular form,

$$V = |V| \cos\theta - j|V| \sin\theta \quad (56\text{-}8)$$

Comparing Eqs. (56-5) and (56-8), we see that

$$V \cos\theta = V_R$$

and

$$V \sin\theta = V_C$$

SUMMARY

1. The impedance Z of a series *RC* circuit is considered a phasor quantity. X_C lags R by 90°. If the values of X_C and R are known, Z can be written in the rectangular form

$$Z = R - jX_C$$

2. Z can also be written in the polar form by the equation

$$Z = |Z|\ \angle{-\theta}$$

where

$$|Z| = \sqrt{R^2 + X_C^2} = \frac{R}{\cos\theta} = \frac{X_C}{\sin\theta}$$

and

$$\theta = \arctan \frac{X_C}{R}$$

3. If the impedance of a series *RC* circuit is given in the polar form, that is, if the numerical values $|Z|$ and θ are known, then Z can also be expressed in the rectangular form using the equation

$$Z = |Z| \cos\theta - j|Z| \sin\theta$$

In this equation $|Z| \cos\theta = R$, and $|Z| \sin\theta = X_C$.

4. In a series *RC* circuit the phasor relationship between the applied voltage V and the voltage drops V_R across R and V_C across C is given by the formulas

$$V = V_R - jV_C$$

or

$$V = |V|\ \angle{-\theta}$$

where

$$|V| = \sqrt{V_R^2 + V_C^2} = \frac{V_R}{\cos\theta} = \frac{V_C}{\sin\theta}$$

and

$$\theta = \arctan \frac{V_C}{V_R}$$

SELF-TEST

Check your understanding by answering these questions:

1. In the circuit of Fig. 56-1*a*, $V = 30$ V, $R = 220\ \Omega$, and $X_C = 150\ \Omega$. The impedance Z written in rectangular form is $Z =$ ____________.
2. In the circuit of question 1, the phase angle θ between Z and R is ____________ degrees.
3. In the circuit of question 1, the impedance Z written in polar form is $Z =$ ____________.
4. In the circuit of question 1, the phasor V written in polar form is $V =$ ____________.
5. In the circuit of question 1, the phasor V written in rectangular form is $V =$ ____________.
6. From the equation in question 5 we know that the voltage V_R across R is ____________ V, while the voltage V_C across C is ____________ V.
7. In the circuit of Fig. 56-1*a*, $Z = 1000\ \Omega$, $V_R = 60$ V, and $V_C = 80$ V. The phasor V written in rectangular form is $V =$ ____________.
8. In the circuit of question 7, the phase angle θ between V and V_R is ____________ degrees, and V ____________ (leads/lags) V_R by this angle.
9. In the circuit of question 7, the phasor V written in polar form is $V =$ ____________.
10. In the circuit of question 7, the phasor Z written in polar form is $Z =$ ____________.
11. In the circuit of question 7, the phasor Z written in rectangular form is $Z =$ ____________.
12. From the equation in question 11 we know that $R =$ ____________ Ω and $X_C =$ ____________ Ω.

MATERIALS REQUIRED

- Power supply: Isolation transformer and variable autotransformer operating from the 120-V ac line
- Equipment: Electronic voltmeter
- Resistors: ½-W 5600-Ω
- Capacitor: 0.5 μF
- Miscellaneous: SPST switch; fused line cord

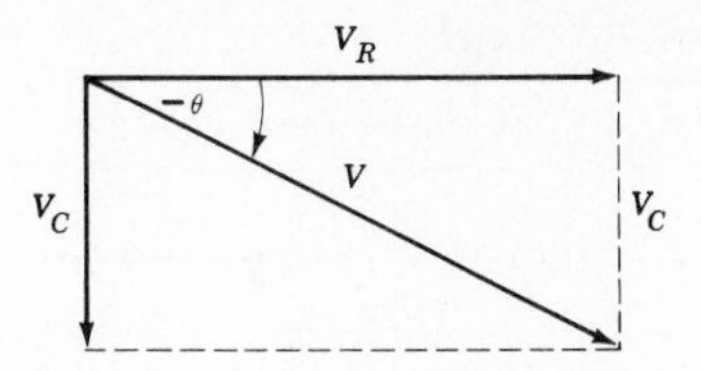

Fig. 56-2. $V = V_R - jV_C$ or $V = |V| \angle \theta$.

PROCEDURE

Impedance *Z* of a Series *RC* Circuit

1. Connect the circuit of Fig. 56-3. S is open. V_A is a variable autotransformer plugged into T, a 1:1 isolation transformer which receives power from the 120-V ac line. The voltmeter V_1, set on ac volts, is connected across R.
2. Close S. **Power on.** Adjust the output of the ac supply until the voltage V_R measured across $R = 36$ V.
3. Measure also and record in Table 56-1 the voltage V_C across C and the voltage V_{AB}, which is the numerical value $|V|$ of the voltage V applied across the series-connected RC circuit.
4. Open switch S. **Power off.** Measure the resistance of R and record in Table 56-1.
5. Using the measured values V_R and R, compute and record in Table 56-1 the current I in the circuit.
6. Compute and record the numerical value $|Z|$ of the impedance in the circuit using the equation $|Z| = |V|/I$, where $|V|$ is the measured value.
7. Compute and record the phase angle θ from the equation

$$\theta = \arctan \frac{V_C}{V_R}$$

8. Compute and record in Table 56-1 the value of X_C using the relationship

$$X_C = \frac{V_C}{I}$$

9. Using the computed value of Z and θ, write the phasor Z in polar form in Table 56-1.
10. Transform the polar form of Z into the rectangular form using the relationship

$$Z = |Z| \cos \theta - j|Z| \sin \theta$$

and record in Table 56-1.

Voltage Relationships in a Series *RC* Circuit

11. From the measured values V_R and V_C in Table 56-1, write the phasor V (applied voltage) in rectangular form and record in Table 56-2.
12. Compute the numerical value $|V|$ of the phasor V using any one of the equations

$$|V| = \sqrt{V_R^2 + V_C^2} = \frac{V_R}{\cos \theta} = \frac{V_C}{\sin \theta}$$

and record in Table 56-2. Use the values of V_R, V_C, and θ from Table 56-1.
13. From Table 56-1, record in Table 56-2 the measured value $V_{AB} = |V|$.
14. Using the value of $|V|$ from step 12 and the value of θ from Table 56-1, write the phasor V in polar form in Table 56-2.

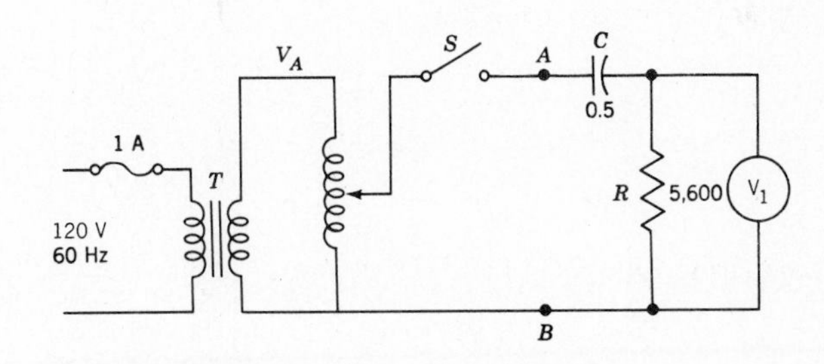

Fig. 56-3. Experimental circuit for determining the impedance Z of an RC circuit.

TABLE 56-1. Impedance in a Series *RC* Circuit

Measured Values				*Computed Values*			
V_R, V	V_C, V	$V_{AB} = \|V\|$, V	R, Ω	I, A	$\|Z\|$, Ω	θ, degrees	X_C, Ω
36							
Z in polar form				Z in rectangular form			
$Z =$				$Z =$			

TABLE 56-2. Voltages in a Series *RC* Circuit

V in Rectangular Form	$|V|$ *Computed,* V	$|V|$ *Measured,* V	*V in Polar Form*
$V =$			$V =$

QUESTIONS

1. Refer to the equation for Z in rectangular form, Table 56-1, and to procedural step 10.
   (*a*) Does the measured value of $R = |Z| \cos \theta$?
   (*b*) Should it?
   (*c*) If the two values are not the same, explain why they are not.
2. Refer to the equation for Z in rectangular form, Table 56-1, and to procedural step 10.
   (*a*) Does the computed value of $X_C = |Z| \sin \theta$?
   (*b*) Should it?
   (*c*) If the two values are not the same, explain why they are not.
3. Have you confirmed experimentally (approximately) the analytical statement that $Z = R - jX_C$? Refer specifically to your data to show that you have or have not confirmed it.
4. If your data does not confirm (approximately) that $Z = R - jX_C$, explain why not.
5. Refer to Table 56-2 and to step 12. Does the computed value of $|V|$ equal the measured value of $|V|$? If not, why not?
6. Have you confirmed experimentally (approximately) the analytical statement that $V = V_R - jV_C$? Refer specifically to your data to show that you have or have not confirmed it.
7. If your data do not confirm that $V = V_R - jV_C$, explain why not.

Answers to Self-Test

1. $220 - j150$
2. 34.3
3. $26.6 \angle -34.3°$
4. $30 \angle -34.3°$
5. $24.8 - j16.9$
6. 24.8; 16.9
7. $60 - j80$
8. 53.1; lags
9. $100 \angle -53.1°$
10. $1000 \angle -53.1°$
11. $600 - j800$
12. 600; 800

57

EXPERIMENT

POWER IN AC CIRCUITS

OBJECTIVES

1. To differentiate between "true" power and "apparent" power in ac circuits
2. To measure power in an ac circuit

INTRODUCTORY INFORMATION

Consumption of AC Power

Power dissipation in dc resistive circuits was defined as the product of the applied voltage and the current, that is, $W = V \times I$, where W is in watts, V in volts, and I in amperes. W can also be computed using the formulas $W = I^2R = V^2/R$. V and I, of course, are constant, not varying, values.

Consumption of power in ac circuits is more complex. One reason for this is that sinusoidal voltages and currents are continuously varying in amplitude, and may be in or out of phase. Another is that ac circuits contain both resistive and reactive components. Resistive components dissipate energy in ac circuits, just as they do in dc circuits. Reactive components, however, do not dissipate energy, but release to the power source in one alternation of the power cycle as much energy as they absorbed in the preceding alternation. The net result is that the total energy dissipated in an ac circuit is that associated with the resistive component, and with none of that associated with the reactive.

Apparent Power

Apparent power W_A is the input power to the ac circuit. W_A is defined as the product of the voltage V across and the current I in the ac circuit. Thus $W_A = V \times I$, just like the definition for power dissipation in a dc circuit. V, of course, is measured in volts and I in amperes when this formula is used. For this reason apparent power is sometimes referred to as volt-amperes. Electrical appliances are rated in voltamperes. It should be noted that the power consumed by an electrical appliance is not W_A but a quantity called *true power*, W_T.

True Power and Power Factor

The power consumed by an electrical device with both resistive and reactive components is defined as true power. In a circuit containing both resistive and reactive components, true power is a fractional part of apparent power. That fraction is called the *power factor* (PF) of the circuit. It is really the ratio of W_T to W_A. Thus

$$\text{PF} = \frac{W_T}{W_A}$$

and

$$W_T = W_A \times \text{PF} = V \times I \times \text{PF}$$

It can be shown that PF is cos θ, where θ is the phase angle between voltage and current in the ac circuit. PF is therefore a number between 0 and 1, depending on the circuit constants. Now, the definition for true power can be given as

$$W_T = V \times I \times \cos\theta \qquad (57\text{-}1)$$

From Eq. (57-1) it is evident that when $\theta = 0$, $\cos\theta = 1$ and $W_T = V \times I$. This condition occurs in a resistive circuit when voltage and current are in phase. So we can conclude that in a resistive circuit true power and apparent power are equal. In a reactive circuit, however, true power is always less than apparent power, because θ is an angle greater than 0, lying in the range 0 to 90°. For such angles, cos θ, the power factor, is less than 1. The most efficient use of ac power is when W_T is equal to or close to W_A, that is, when $\theta = 1$ or is close to 1.

We have defined θ as the angle by which current leads voltage in a capacitive circuit or lags voltage in an inductive circuit, and cos θ as the power factor in the circuit. In a series circuit we can also employ ac theory to calculate the power factor, using the following formulas:

$$\text{PF} = \cos\theta = \frac{R}{Z} = \frac{V_R}{V} \qquad (57\text{-}2)$$

In these formulas R is the total circuit resistance in ohms, Z is the circuit impedance in ohms, V_R is the voltage across R, and V is the voltage applied to the circuit.

There are other formulas besides (57-1) for calculating true power. Thus

$$W_T = I^2R = \frac{(V_R)^2}{R} \qquad (57\text{-}3)$$

In these formulas I is the circuit current in amperes, R is the total circuit resistance in ohms, V_R is the voltage across R, and W is in watts.

Example. Two examples will illustrate how to calculate power in ac circuits. In the circuit of Fig. 57-1, assume that the coil L has zero resistance; that $X_L = 1000\ \Omega$ at the

Fig. 57-1. Series RL circuit to illustrate computation of apparent power W_A and true power W_T.

frequency of the applied voltage; that $R = 250\ \Omega$; and that $V = 50$ V. We need to determine the apparent power, the true power, the power factor, and the phase angle.

Method 1

1.
$$W_A = V \times I$$

and

$$I = \frac{V}{Z}$$

$$Z = \sqrt{R^2 + X_L{}^2}$$

Therefore

$$Z = \sqrt{(250)^2 + (1000)^2} = 1031\ \Omega$$

and
$$I = \frac{50}{1031} = 0.0485 \text{ A}$$

Now,
$$W_A = 50 \times (0.0485) = 2.43 \text{ W}$$

2.
$$W_T = I^2R = (0.0485)^2\ (250) = 0.588 \text{ W}$$

3.
$$\text{PF} = \cos\theta = \frac{W_T}{W_A} = \frac{0.588}{2.43} = 0.2420$$

4.
$$\theta = \arccos\ (0.2420) = 76°$$

Method 2

1.
$$I = \frac{V}{Z}$$

and as above

$$Z = 1031\ \Omega$$

Therefore

$$I = \frac{50}{1031} = 0.0485 \text{ A}$$

Now
$$W_A = V \times I = 50(0.0485) = 2.43 \text{ W}$$

2.
$$\text{PF} = \cos\theta = \frac{R}{Z} = \frac{250}{1031} = 0.2425$$

3.
$$\theta = \arccos\ (0.2425) = 75.97°$$

4.
$$W_T = V \times I \times \cos\theta = 50(0.0485)(0.2425) = 0.588 \text{ W}$$

Measuring AC Power—Voltampere Method

Power in an ac circuit may be determined by a series of measurements, using instruments with which you are already familiar, namely the EVM and the oscilloscope. With the EVM we can measure the voltage V_R across R and the applied voltage V. We can then determine the power factor by substituting the measured values in the formula

$$\text{PF} = \frac{V_R}{V} = \cos\theta$$

Now, we can measure the resistance of R in the circuit and compute the current I, using the formula

$$I = \frac{V_R}{R}$$

Then, using the formula $W_T = V \times I \times \cos\theta$, we can calculate the true power.

Or, after measuring V_R and R, we can calculate W_T using the formula

$$W_T = \frac{(V_R)^2}{R}$$

The measurements are simple to make in a circuit like that of Fig. 57-1. The calculations are straightforward.

If we wish to measure the phase angle in the circuit directly, we can use an oscilloscope. By the method discussed in Experiment 42, we can determine θ by comparing the phase of the applied voltage with the phase of the voltage across R (which is the same as the phase of I).

Wattmeter

We can measure true power directly, eliminating all calculations, by using a wattmeter. For low-frequency power, such as 50 or 60 Hz, an instrument such as that shown schematically in Fig. 57-2 is used. It consists of a dynamometer-type movement using separate voltage and current coils. L_1, the voltage coil AB, plugs into the power line and is the movable coil of the meter to which a pointer is attached. The device whose power we wish to measure plugs into an outlet CD on the wattmeter. The current in the device flows through the current coil (L_2) and acts as the exciting magnetic field which interacts with the magnetic field about the voltage coil, causing the voltage coil to rotate. The pointer then moves along an arc calibrated in watts. The deflection of the pointer is proportional to the product of voltage, current, and the power factor in the circuit. Accordingly, the reading is in true power, that is $V \times I \times \cos\theta$.

Compensated Wattmeter

From Fig. 57-2 it is evident that the device whose power is being measured is in series with the current coil L_2. There is a voltage drop across L_2, so that the voltage applied to the load is somewhat lower than the power-line voltage. The wattmeter, however, "sees" the full power-line voltage which is applied to the voltage coil L_1. The effect is to cause a higher wattage reading on the meter than the load actually consumes. Since wattmeters are usually designed to measure large loads (hundreds of watts), the slightly higher reading of the meter is no problem. But for measurement of low-power loads (0 to 10 W), errors can be as high as 20 to 30 percent. The *compensated* wattmeter solves this problem for low-power loads. In the compensated wattmeter, an additional winding which carries a compensating current is back-wound around the current coil to counteract the error introduced by the meter. This meter provides accurate true power readings for low-power loads. The usual range for this type of meter is 0 to 10 W or 0 to 20 W.

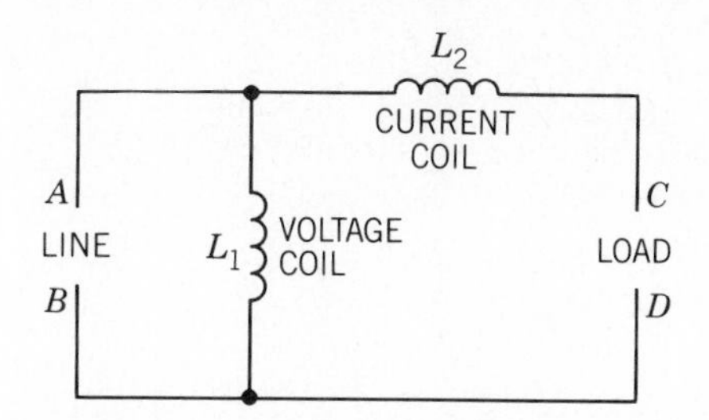

Fig. 57-2. Simplified circuit diagram of a low-frequency wattmeter.

SUMMARY

1. Power in ac circuits is consumed by the resistive, not the reactive, components.
2. Apparent power W_A in an ac circuit is the power which the power source presents to the ac circuit. $W_A = V \times I$, where V is the applied voltage and I is the current drawn by the circuit.
3. The true power dissipated by the circuit (or by the electrical appliance) is the product of V and I and the power factor (PF). PF = cos θ, where θ is the phase angle between voltage and current in the circuit, that is,

$$W_T = V \times I \times \cos\theta$$

4. Other formulas for true power are

$$W_T = I^2R = \frac{V_R^2}{R}$$

where I = current in the circuit, in amperes
R = total resistance of the circuit, in ohms
V_R = voltage measured across the total resistance of the circuit

5. The power factor (cos θ) of an ac circuit may be determined by measuring θ, the phase angle between the applied circuit voltage V and the current I, and computing cos θ.
6. Also, PF = $\cos\theta = R/Z = V_R/V$.
7. Power in an ac circuit may be determined by measuring the applied voltage V, the current I, and the phase angle θ, and substituting the measured values in the formula

$$W_T = V \times I \times \cos\theta$$

8. True power may be measured directly, using a wattmeter.
9. For accurate measurement of low-power devices, a compensated wattmeter is used. The range of this instrument is usually 0 to 10 W or 20 W.

SELF-TEST

Check your understanding by answering these questions. In the circuit of Fig. 57-1, L = 8 H, R = 1000 Ω, the frequency of the power source V is 60 Hz, and V = 100 V. Assume that L has zero resistance, and answer the following:

1. The reactance of L = ________ Ω.
2. The impedance of the circuit is ________ Ω.
3. The current I drawn by the circuit is ________ A.
4. W_A = ________ W.
5. The power factor of the circuit is ________.
6. The phase angle of the circuit is ________ degrees.
7. An accurate compensated wattmeter will read ________ W in the circuit.
8. I^2R = ________ W.
9. The ratio of true power to apparent power is ________.

MATERIALS REQUIRED

- Power supply: Variable autotransformer, isolation transformer
- Equipment: EVM, oscilloscope, 0–10-W compensated wattmeter equivalent to Simpson (Model 1379) catalog #10930
- Resistors: 5-W 100-Ω
- Capacitors: 5 μF/100 V, 10 μF/100 V
- Miscellaneous: Fused line cord; SPST switch

PROCEDURE

Measuring Power—Voltampere Method

1. Connect the circuit of Fig. 57-3. Close S. **Power on.** Adjust the output of the variable autotransformer so that V_{AB} = 75 V. Measure and record this voltage in Table 57-1.
2. Measure and record the voltage V_R across R.
3. Open switch S. **Power off.**
4. Measure and record the resistance of R.
5. Calculate and record the current I in the circuit. Calculate and record the apparent power W_A. Show your computations.
6. Calculate and record the true power W_T. Show your computations. Calculate also and record θ, the phase angle.

Measuring Power with a Wattmeter

7. Disconnect C and R from the circuit and insert the wattmeter as shown in Fig. 57-4. Note that the Line

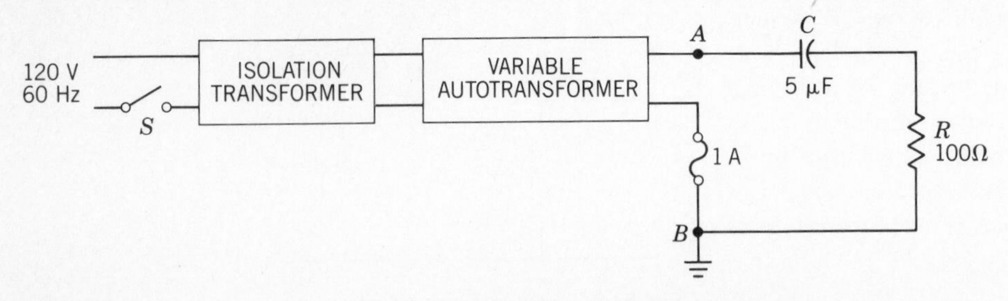

Fig. 57-3. Experimental circuit for measuring power by the voltampere method.

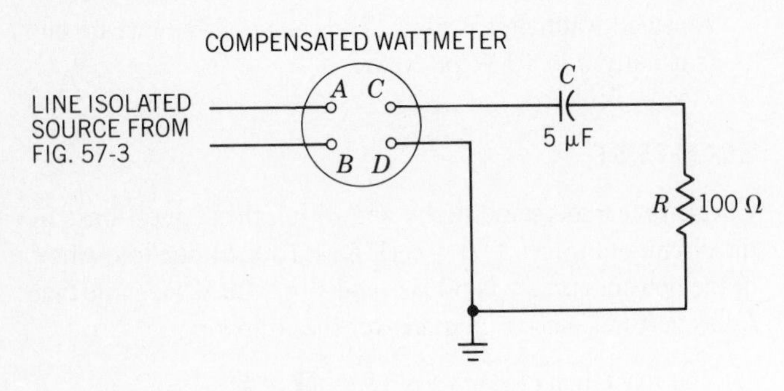

Fig. 57-4. Measuring power with a wattmeter.

TABLE 57-1. Power Measurement

Measured Values				
V_{AB}, V	V_R, V	R, Ω	W_T, W	θ, degrees

Calculated Values				
I, A	W_A, W	W_T, W	PF cos θ	θ, degrees

terminals on the meter are connected to the output of the variable autotransformer, set for 75 V. To the Load terminals on the meter connect the load, C, and R as shown.

NOTE: Be certain that the instructions for your wattmeter call for the same connections as those shown in Fig. 57-4. If they do not, follow the manufacturer's instructions for connecting your meter.

8. **Power on.** Close S. Adjust, if necessary, the variable autotransformer until the voltage measured across AB is 75 V. Record this voltage V_{AB} in Table 57-1. Measure also and record the wattage as read on the wattmeter.

Measuring Power Factor

9. With an oscilloscope measure the phase angle θ between voltage and current in the circuit, following the procedure outlined in Experiment 42. Record in Table 57-1. Indicate whether the current leads or lags the applied voltage. **Power off.**
10. From the measured value of θ, calculate and record in Table 57-1 the power factor in the circuit.

Measuring Power in Another Circuit

11. In the circuit of Fig. 57-3 replace C by a 10-μF capacitor. V_{AB} = 40 V.
12. Repeat steps 1 through 10 with the new circuit and record your measurements in Table 57-2.

TABLE 57-2. Power Measurement

Measured Values				
V_{AB}, V	V_R, V	R, Ω	W_T, W	θ, degrees

Calculated Values				
I, A	W_A, W	W_T, W	PF cos θ	θ, degrees

QUESTIONS

1. How does the calculated value of W_T compare with the measured value (using a wattmeter) in
   (a) Table 57-1?
   (b) Table 57-2?
   Explain any large discrepancy.
2. What does the wattmeter measure, W_T or W_A? Confirm your answer by referring to your experimental data.
3. In this experiment, which method appears more accurate for measuring W_T, the voltampere method or the wattmeter method? Refer to your data to substantiate your answer.

4. How does the measured value of θ compare with the calculated value in
   (*a*) Table 57-1?
   (*b*) Table 57-2?
   Explain any unusual discrepancy.
5. What is the significance of the power factor in an ac circuit?

Answers to Self-Test

1. 3016
2. 3177
3. 0.0315
4. 3.15
5. 0.3148
6. 71.65
7. 0.992
8. 0.992
9. PF

58

EXPERIMENT

FREQUENCY RESPONSE OF A REACTIVE CIRCUIT

OBJECTIVES

1. In a series-connected *RL* circuit, to measure the effect on impedance and on current of a change in frequency of the voltage source
2. In a series-connected *RC* circuit, to measure the effect on impedance and on current of a change in frequency of the voltage source

INTRODUCTORY INFORMATION

Impedance of a Series *RL* Circuit

The impedance of a series-connected *RL* circuit is given by

$$Z = \sqrt{R^2 + X_L^2} \tag{58-1}$$

It is apparent from Eq. (58-1) that if R remains constant in value, a change in X_L will affect Z. Thus, as X_L increases, Z also increases. As X_L decreases, Z decreases. Since

$$X_L = 2\pi FL \tag{58-2}$$

we can change X_L by either increasing or decreasing the size of L, with F remaining constant. Or we can change X_L by increasing or decreasing F, with L remaining constant.

Let us consider the effect on X_L of varying F, with L constant. It is evident from Eq. (58-2) that as F increases, X_L increases. As F decreases, X_L decreases. Z will therefore also increase or decrease, respectively, as F increases or decreases.

An example will illustrate this variation. Suppose the X_L of L in Fig. 58-1 changes with frequency as shown in Table 58-1. If $R = 30$, Z computed will take on the values shown in Table 58-1. Note that the effect of R on Z decreases as X_L increases. Figure 58-2 is a graph of Z versus F in this *RL* circuit.

Current versus Frequency in an *RL* Circuit

The current in an ac circuit is given by the equation

$$I = \frac{V}{Z} \tag{58-3}$$

Current varies inversely with Z. Since Z varies directly with F in a series *RL* circuit, current will also vary inversely with F. Thus, current in a series *RL* circuit will increase as F decreases, and will decrease as F increases.

If we assume that the voltage V applied to the circuit of Fig. 58-1 is 100 V, then the current takes on the values shown in Table 58-1. In Fig. 58-2 we also see a graph of I versus F. This is the *frequency response* of the circuit.

Analyzing the two graphs, it becomes evident that the impedance of an *RL* circuit varies directly with frequency, while the current varies inversely.

Impedance of a Series *RC* Circuit

The impedance of a series *RC* circuit is given by the equation

$$Z = \sqrt{R^2 + X_C^2} \tag{58-4}$$

Though Eqs. (58-4) and (58-1) are the same, except for the substitution of X_C for X_L, the variation of Z with frequency is very different for the *RC* circuit. The reason is that X_C varies inversely with frequency. Thus

$$X_C = \frac{1}{2\pi FC} \tag{58-5}$$

Hence, as F increases, X_C decreases. As F decreases, X_C increases. Therefore, the impedance of a series *RC* circuit increases with a decrease in frequency but decreases with an increase in frequency. That is, Z varies inversely with frequency.

Current versus Frequency in an *RC* Circuit

In a series *RC* circuit, as F decreases, X_C increases, Z increases, and I decreases. As F increases, X_C decreases, Z decreases, and I increases. It is evident, therefore, that in a series *RC* circuit, current varies directly with frequency F. Again this relationship is opposite to that of a series *RL* circuit.

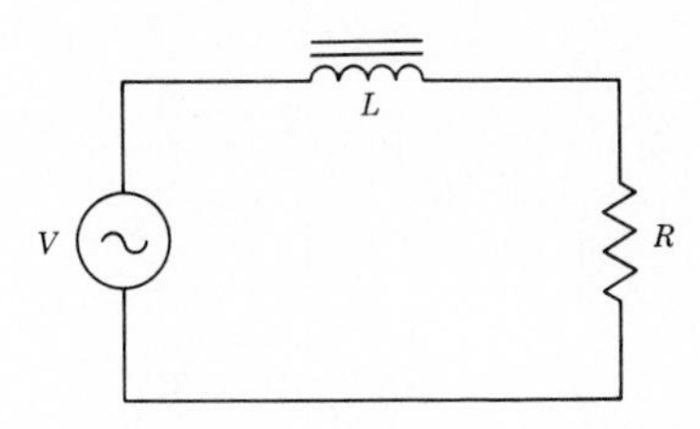

Fig. 58-1. Series *RL* circuit to determine the effect of change in frequency *F* on impedance *Z* and on current *I*.

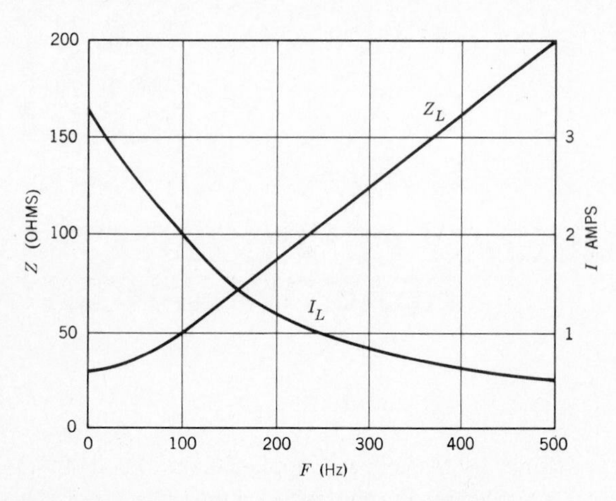

Fig. 58-2. *Z* and *I* versus *F* in a series *RL* circuit.

The foregoing analysis shows that the effects of a capacitor and an inductor on current in series *RC* and *RL* circuits are opposite. This fact is related to the phase relationship between voltage and current in a capacitor and an inductor, for in an inductor current lags voltage, while in a capacitor current leads the voltage.

SUMMARY

1. In a series *RL* circuit, as X_L increases, with *R* constant, *Z* increases.
2. X_L increases or decreases as the frequency *F* of the voltage applied to the circuit increases or decreases, respectively.
3. Therefore, as *F* increases in a series *RL* circuit, *Z* increases. As *F* decreases in a series *RL* circuit, *Z* decreases.
4. As *F* increases in a series *RL* circuit, *I* in the circuit decreases. As *F* decreases, *I* increases.
5. In a series *RC* circuit, X_C increases or decreases as the frequency *F* decreases or increases, respectively. That is, the effect of *F* on X_C is opposite to its effect on X_L.

TABLE 58-1. Variation of X_L and *I* with Frequency

F, Hz	X_L, Ω	$Z = \sqrt{R^2 + X_L^2}$, Ω	$I = \frac{V}{Z}$, A
0	0	30	3.33
100	40	50	2.00
150	60	67.1	1.49
200	80	85.4	1.171
250	100	104.4	0.958
300	120	123.7	0.808
400	160	162.8	0.614
500	200	202.2	0.495

6. In a series *RC* circuit, *Z* increases as *F* decreases, with *R* and *C* constant. Moreover, *Z* decreases as *F* increases.
7. In a series *RC* circuit, *I* decreases as *F* decreases, but *I* increases when *F* increases.
8. It is apparent, therefore, that the effects of a capactior and an inductor on impedance and current in a series-connected circuit are opposite.
9. The reactance of an inductor is $X_L = 2\pi FL$. The impedance of a series *RL* circuit is $Z = \sqrt{R^2 + X_L^2}$.
10. The reactance of a capacitor is $X_C = (1/2\pi FC)$. The impedance of a series *RC* circuit is $Z = \sqrt{R^2 + X_C^2}$.
11. The frequency response of a reactive circuit is the graph of *I* versus *F*.

SELF-TEST

Check your understanding by answering these questions:

1. The X_L of a 3.19-H choke at 100 Hz is ________ Ω.
2. The *Z* of a series *RL* circuit, where $R = 1000\ \Omega$ and $L = 3.19$ H, at 100 Hz is ________ Ω.
3. The *Z* of the same circuit as in question 2, at 200 Hz, is ________ Ω.
4. A generalization concerning a series *RL* circuit is that *Z* ________ (increases/decreases) with an increase in frequency, and vice versa.
5. In the circuit of question 2, if the applied voltage $V =$ 22.35 V, the current *I* in the circuit is ________ A.
6. In the circuit of question 3, if the applied voltage $V =$ 22.35 V, *I*, the current *I* in the circuit is ________ A.
7. A generalization concerning a series *RL* circuit is that *I* ________ (increases/decreases) with an increase in frequency, and vice versa.
8. In a series *RC* circuit, the X_C of the capacitor is 1000 Ω at 100 Hz. The X_C of the same capacitor at 200 Hz is ________ Ω, while at 50 Hz the X_C of the capacitor is ________ Ω.
9. The *Z* of a series *RC* circuit, with *R* and *C* constant, ________ (increases/decreases) with an increase in frequency, and vice versa.
10. If the *Z* of the *RC* circuit in question 8 at 100 Hz = 1410 Ω, the *Z* of the same circuit at 50 Hz is ________ (greater/less than) 1410 Ω.
11. The current *I* in the *RC* circuit of question 8 ________ (increases/decreases) when *F* increases from 50 Hz to 100 Hz.

MATERIALS REQUIRED

- Equipment: AF signal generator; electronic voltmeter
- Resistor: ½-W 10,000-Ω
- Capacitor: 0.1 μF
- Inductance: 8 H at 50 mA direct current

PROCEDURE

Impedance and Current versus Frequency in an *RL* Circuit

1. Connect the circuit of Fig. 58-1. L is an inductance rated 8 H at 50 mA direct current. R is a 10,000-Ω resistor. The voltage source is an audio-frequency signal generator.
2. Set the frequency of the generator at 50 Hz, and the output V at 10 V rms. With an EVM, measure, and record in Table 58-2, the output voltage V applied across the circuit. Measure also and record the voltage V_R across R. Compute and record the current I in the circuit ($I = V_R/R$, where V_R is in volts and R is the resistance in the circuit). Compute and record the impedance Z of the circuit ($Z = V/I$, where V is the voltage applied to the circuit and I is the calculated current). Show your computations.
3. Set the frequency of the generator at 100 Hz, with the output V at the same level as in the preceding step. Again measure the voltages V and V_R. Compute I and Z and record in Table 58-2.

NOTE: In the remaining checks, step 4, be certain that V is kept at the same level as in steps 2 and 3.

4. Repeat step 3 for each of the frequencies shown in the table.
5. From the data in Table 58-2, draw a graph of Z versus F, using the vertical axis for Z and the horizontal axis for F. Label the graph Z_L.
6. Using the same coordinate axes, but labeling the vertical axis I, draw a graph of I versus F. Label the graph I_L.

Impedance and Current versus Frequency in an *RC* Circuit

7. Connect the circuit of Fig. 58-3. Set the frequency of the generator at 500 Hz and output V at 10 V rms. With an EVM, measure the output voltage V applied across the circuit and record it in Table 58-3. Measure also and record the voltage V_R across R. Compute and record the current I in the circuit and the impedance Z of the circuit. Show your computations.
8. Repeat step 7 for the other frequencies shown in Table 58-3. Be certain that the voltage V is maintained at the same level as in step 7. You need not show your computations.
9. From the data in Table 58-3, draw a graph of Z versus F. Label the graph Z_C. Let Z be the vertical axis, F the horizontal.
10. From the data in Table 58-3, draw a graph of I versus F. Label the graph I_C. Use the same coordinate axes as in step 9, but let I be the vertical axis.

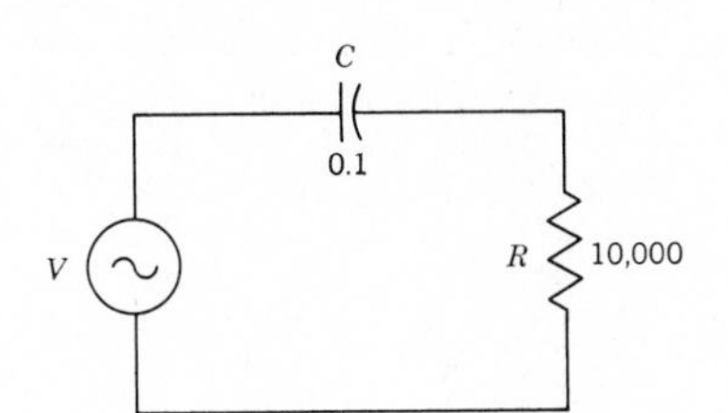

Fig. 58-3. Series *RC* circuit to determine the effect of change in frequency *F* on impedance *Z* and on current *I*.

TABLE 58-2. How *Z* and *I* Vary with Frequency in an *RL* Circuit

F, Hz	V, V	V_R, V	$I = \frac{V_R}{R}$, A	$Z = \frac{V}{I}$, Ω
50				
100				
150				
200				
250				
300				
350				
400				
450				
500				

TABLE 58-3. How *Z* and *I* Vary with Frequency in an *RC* Circuit

F, Hz	V, V	V_R, V	$I = \frac{V_R}{R}$, A	$Z = \frac{V}{I}$, Ω
500				
450				
400				
350				
300				
250				
200				
150				
100				
50				

QUESTIONS

1. Refer to your graph of Z_L versus F. How does Z_L vary with F? Why?
2. Is any portion of the graph Z_L versus F linear? Identify that portion, if any, and explain its linearity.
3. In the graph I_L versus F, how does I vary with respect to F? Why?
4. How does Z_C vary with respect to F? Why?
5. Compare the graphs of Z_C versus F and Z_L versus F. How are they different?
6. How does I_C vary with respect to F? Why?
7. Compare the graphs of I_L and I_C versus F. How are they different?

Extra Credit

8. At any specific frequency, what do you think would be the effect of connecting R, L, and C in series (in the same circuit) on:
   (*a*) Impedance
   (*b*) Current
   assuming that the source V used is the same as in this experiment? Why?
9. Suggest a formula for impedance in a series *RLC* circuit.
10. What is the minimum impedance of a series *RLC* circuit? Why?
11. What would be the frequency when the impedance is minimum in a series *LCR* circuit? Why?

Answers to Self-Test

1. 2004
2. 2.240
3. 4131
4. increases
5. 0.01
6. 0.0054
7. decreases
8. 500; 2000
9. decreases
10. greater
11. increases

59

EXPERIMENT

IMPEDANCE OF A SERIES *RLC* CIRCUIT

OBJECTIVE

To verify experimentally that the impedance Z of a series *RLC* circuit is $Z = \sqrt{R^2 + (X_L - X_C)^2}$

INTRODUCTORY INFORMATION

Impedance of an *RLC* Circuit

To understand the effects on alternating current of series-connected *RLC* circuits, it is necessary to recall the individual effect of each of these components.

The effects of a resistor in an ac circuit are the same as those in a dc circuit, since alternating current and voltage are in phase in a resistor. In a resistive divider, ac voltage drops across each of the resistors may be added arithmetically to get the value of applied voltage.

When a capacitor is connected in series with a resistor, the reactance of the capacitor together with the resistance of the resistor determines the effect on alternating current. Capacitive reactance depends on frequency. Hence, the effects of a capacitor are determined both by its size (capacitance) and by the frequency of the current in the circuit. In a totally capacitive circuit, i.e., one which contains only capacitance, alternating current leads voltage by 90°. In a series *RC* circuit, the alternating current leads the voltage by a phase angle less than 90°.

The effects of an inductor connected in series with a resistor in an ac circuit also depend on the frequency of the current and the size of the inductor. In a totally inductive circuit, i.e., one which contains only inductance, current lags voltage by 90°. In a series *RL* circuit, the current lags the voltage by an angle less than 90°.

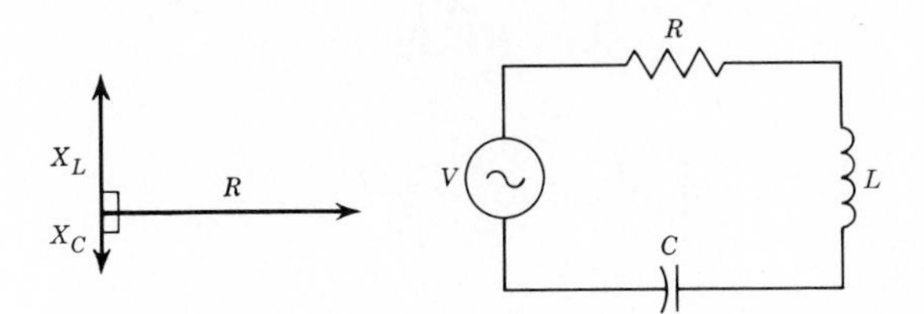

Fig. 59-1. Effect of an inductance and capacitance in a series *RLC* circuit.

It may be seen from the characteristics of an inductance and a capacitance that they have opposite effects on current and voltage in an ac circuit. This fact is demonstrated in the phasor representation of X_L and X_C in a series-connected *RLC* circuit (see Fig. 59-1).

The phasor sum of X_L and X_C is the same as their arithmetic difference. The resultant takes the direction of the larger reactance. Assume that X_L is larger than X_C in Fig. 59-1. The phasor diagram can then be redrawn as in Fig. 59-2*a*.

Note that Fig. 59-2*a* really represents an inductive circuit whose net reactance is the difference between X_L and X_C. The effect is as though the L and C of the series-connected *RLC* circuit had been replaced by an inductor whose reactance is equal to $X_L - X_C$. Figure 59-2*b* represents a capacitive circuit whose reactance is $X_C - X_L$.

Figure 59-2*a* indicates that the impedance Z of the *RLC* circuit is given by the formula

$$Z = \sqrt{R^2 + (X_L - X_C)^2} \qquad (59\text{-}1)$$

This formula holds whether X_L is greater or smaller than X_C because the reactive term is a squared term and is therefore always positive, or is zero when $X_L = X_C$.

Several conclusions may be drawn from the phasor diagram of Fig. 59-1 and from Eq. (59-1). One is that there must be some frequency, call it F_R, at which $X_L = X_C$, that is, at which X_L and X_C completely cancel each other, and the impedance Z of the series-connected *RLC* circuit is $Z = R$. At this frequency the impedance of the circuit is minimum, and

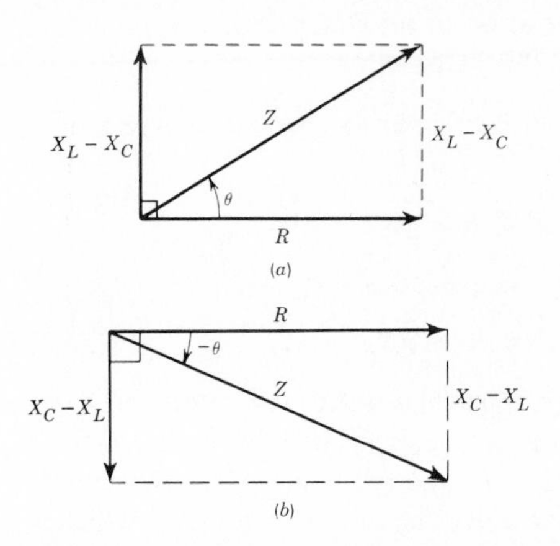

Fig. 59-2. Impedance of a series-connected *RLC* circuit.

the circuit is resistive. It is apparent, too, that for all frequencies F_H higher than F_R, X_L is larger than X_C, and the circuit is inductive. As F_H becomes increasingly higher, $X_L - X_C$ increases and Z increases. And finally, another conclusion may be drawn. For all frequencies F_L lower than F_R, X_C is larger than X_L. As F_L becomes ever lower, the value of $X_C - X_L$ becomes increasingly larger. So that Z again increases beyond the value $Z = R$, but now the *RLC* circuit is capacitive. A more complete discussion of the effect of frequency on an *RLC* circuit will be found in the next experiment.

(Optional) Impedance of an *RLC* Circuit Expressed Trigonometrically

Consider the case of a series *RLC* circuit which is inductive, that is, where $X_L > X_C$, as is shown in the phasor diagram in Fig. 59-2*a*. Here the phase angle θ between Z and R can be used to calculate Z, if $(X_L - X_C)$ and R are known. Thus

$$\theta = \arctan \frac{X_L - X_C}{R} \tag{59-2}$$

and

$$Z = \frac{R}{\cos \theta} = \frac{X_L - X_C}{\sin \theta}$$

Equations (59-2) are very similar to those in a simple *RL* circuit (see Experiment 51).

If $X_L < X_C$, the circuit acts like a capacitive circuit, whose vector diagram is Fig. 59-2*b*. Here the phase angle θ is negative, that is, Z lags R, which is the characteristic of an *RC* circuit. Equations (59-2) can still be used to define θ and Z. Note, however, that the quantity $X_L - X_C$ is negative, and therefore θ is negative.

Where $X_C > X_L$, it is possible to simplify calculations by using these equations:

$$\theta = \arctan \frac{X_C - X_L}{R} \tag{59-3}$$

$$Z = \frac{R}{\cos \theta} = \frac{X_C - X_L}{\sin \theta}$$

Equations (59-3) are similar to those that define θ and Z in an *RC* circuit (see Experiment 54).

(Optional) Impedance of an *RLC* Circuit Analyzed Using the *j* Operator

The impedance of an *RLC* circuit can be written in rectangular form. Thus

$$Z = R + jX_L - jX_C \tag{59-4}$$

and combining the reactive terms, we get

$$Z = R + j(X_L - X_C) \tag{59-5}$$

From Eq. (59-5) it is apparent that if $X_L > X_C$, the term $X_L - X_C$ is positive. The circuit then acts like an *RL* circuit, since Z has the form $Z = R + jX$ where $X = X_L - X_C$. However, if $X_L - X_C$ is negative, Z has the form $Z = R - jX$, which is the characteristic of an *RC* circuit.

(Optional) *Z* of an *RLC* Circuit Expressed in Polar Form

We can transform Eq. (59-5) into the polar form. Thus

$$Z = |Z| \angle \theta$$

where

$$Z = \sqrt{R^2 + (X_L - X_C)^2} = \frac{R}{\cos \theta} = \frac{X_L - X_C}{\sin \theta} \tag{59-6}$$

and

$$\theta = \arctan \frac{X_L - X_C}{R}$$

As we noted above, if $X_L > X_C$, θ is positive; if $X_L < X_C$, θ is negative.

SUMMARY

1. An inductor L and a capacitor C have opposite effects on current in an ac circuit. Thus in an inductor current *lags*, while in a capacitor current *leads* the applied voltage.
2. The total reactance X of a series-connected *RLC* circuit is the algebraic sum, that is, arithmetic difference, of X_L and X_C ($X = X_L - X_C$).
3. If $X_L > X_C$, that is, if the term $X_L - X_C$ is a positive number, the circuit acts like an *RL* circuit whose *net* inductive reactance $X = X_L - X_C$.
4. If $X_L < X_C$, that is, if the term $X_L - X_C$ is a negative number, the circuit acts like an *RC* circuit whose net capacitive reactance $X = X_C - X_L$.
5. The impedance Z of an *RLC* circuit is expressed by the formula
$$|Z| = \sqrt{R^2 + (X_L - X_C)^2}$$
6. The phase angle θ of a series *RLC* circuit is positive (that is, Z leads R) if $X_L > X_C$; θ is negative (that is, Z lags R) if $X_L < X_C$.
7. Z may also be expressed trigonometrically by the equations
$$|Z| = \frac{R}{\cos \theta} = \frac{X_L - X_C}{\sin \theta}$$
where
$$\theta = \arctan \frac{X_L - X_C}{R}$$
8. Using the *j* operator, Z is written
$$Z = R + j(X_L - X_C)$$
9. In polar form, Z of a series *RLC* circuit is written
$$Z = |Z| \angle \theta$$
where
$$\theta = \arctan \frac{X_L - X_C}{R}$$
and
$$|Z| = \sqrt{R^2 + (X_L - X_C)^2} = \frac{R}{\cos \theta} = \frac{X_L - X_C}{\sin \theta}$$

SELF-TEST

Check your understanding by answering these questions:

NOTE: You need not answer those questions requiring a knowledge of trigonometry, if you have not studied trigonometry.

1. In a series-connected *RLC* circuit, $X_L = 50\ \Omega$, $X_C = 20\ \Omega$, and $R = 40\ \Omega$. The net reactance X of this circuit is $X =$ ________ Ω. This reactance is ________ (inductive/capacitive).
2. The impedance Z of the circuit in question 1 is ________ Ω.
3. The circuit in question 1 acts like an ________ (*RC*/*RL*) circuit, whose ________ (X_C/X_L) = ________ Ω.
4. For the same values of applied voltage V, resistance R, inductance L, and capacitance C, if $X_L > X_C$, the current in an *RL* circuit is ________ than the current in an *RLC* circuit.
5. In a series-connected *RLC* circuit, $X_L = 100\ \Omega$, $X_C = 150\ \Omega$, and $R = 120\ \Omega$. In this circuit the net ________ (inductive/capacitive) reactance is ________ Ω, and the impedance phasor ________ (leads/lags) the resistance R.
6. In the circuit of question 5, the phase angle $\theta =$ ________ degrees and is ________ (negative/positive).
7. In the circuit of question 5, $|Z| =$ ________ Ω.
8. The impedance Z in question 5 expressed in rectangular form is $Z =$ ________.
9. The impedance Z in question 5 expressed in polar form is $Z =$ ________.
10. The reason θ is ________ (positive/negative) is that $X_L - X_C$ is ________ (positive/negative).

MATERIALS REQUIRED

- Power supply: Line-isolated source of 6.3-V rms 60-Hz
- Equipment: EVM, oscilloscope
- Resistor: ½-W 4700-Ω
- Capacitors: 0.5 μF
- Miscellaneous: SPST switch; choke: 8 H at 50 mA direct current; fused line cord

PROCEDURE

NOTE: In this experiment you will first determine experimentally the current and impedance in a series *RL* circuit. You will then connect a capacitor C in series with L and R, and again measure the current. There should be more current in the series *RLC* circuit than there was in the *RL* circuit because the impedance Z of the *RLC* circuit *in this experiment* is smaller than that of the *RL* circuit. The data will then be used to verify the impedance formula $Z = \sqrt{R^2 + (X_L - X_C)^2}$ for a series-connected *RLC* circuit.

1. Connect the circuit of Fig. 59-3. Use an isolated 60-Hz ac source approximately 18 V peak-to-peak. Measure this value of V with an EVM and record in Table 59-1.
2. Measure also the voltages V_R across R and V_L across L, and record in Table 59-1.

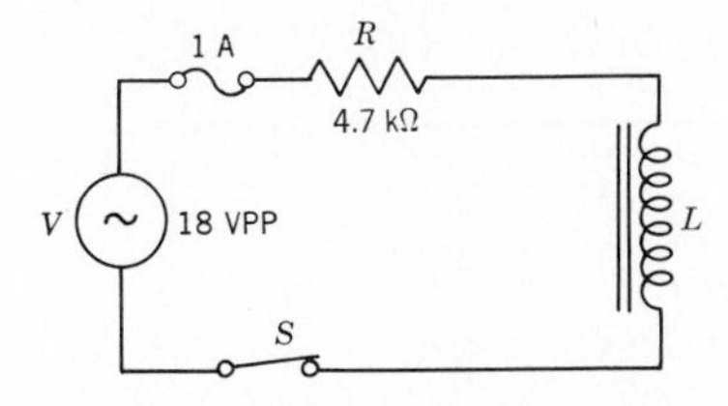

Fig. 59-3. Circuit to determine the value of Z.

TABLE 59-1. Impedance of a Series *RL* Circuit

V Applied, V	V_R, V	V_L, V	I_1, A	Z, Ω

3. Compute the current I_1 in the circuit and record in Table 59-1. Use the rated value of R and the measured value of V_R to determine I_1. Compute and record also the impedance Z of the circuit ($Z = V_{\text{applied}}/I_1$). Show your computations.
4. Connect a 0.5-μF capacitor in series with R and L as shown in Fig. 59-4.
5. Measure the applied voltages V, V_R, V_L, and V_C and record in Table 59-2.
6. From the measured values of V_R and the rated value of R, compute the current I_2. Also compute the values of X_L and X_C from the computed value I_2 and the measured values V_L and V_C. Record in Table 59-2. Show your computations.
7. Compute and record the impedance Z of the *RLC* circuit in Fig. 59-4, using the formula

$$Z = \frac{V_{\text{applied}}}{I_2}$$

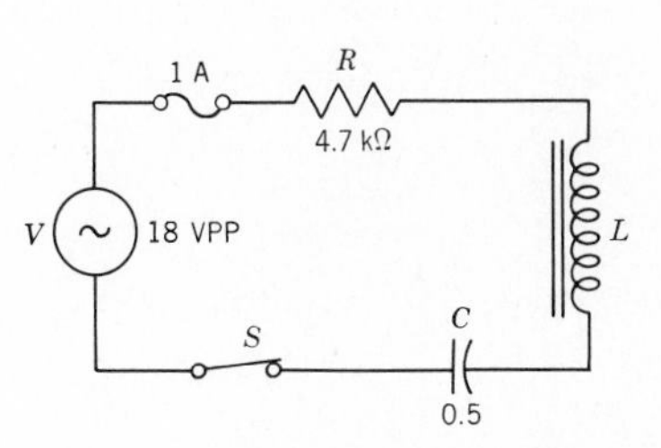

Fig. 59-4. Circuit to determine the impedance of *RLC* connected in series.

8. Compute and record Z using the formula

$$Z = \sqrt{R^2 + (X_L - X_C)^2}$$

Use the values X_L and X_C from Table 59-2.

(Optional) Extra Credit

9. Using the data in Table 59-2, record in a specially prepared Table 59-3 the impedance Z of the experimental *RLC* circuit in rectangular form.

10. Compute and record in Table 59-3 the phase angle θ. Show your computations.

11. Compute and record in Table 59-3 the impedance Z, in polar form, of the experimental *RLC* circuit. Show your computations.

TABLE 59-2. Impedance of a Series *RLC* Circuit

V Applied, V	V_R, V	V_L, V	V_C, V	I_2, A	X_C, Ω	X_L, Ω	Z (Ω) =	
							V/I_2	$\sqrt{R^2 + (X_L - X_C)^2}$

QUESTIONS

1. Refer to your data in Tables 59-1 and 59-2. How was the impedance of the circuit of Fig. 59-3 affected by adding a capacitor C, as in Fig. 59-4?
2. (*a*) State the condition for which the current I will be maximum and impedance Z minimum in the circuit of Fig. 59-4.
   (*b*) Compute the value of maximum current.
   (*c*) What is the value of minimum impedance?
3. How does the experimental value of Z in step 7 compare with the formula value in step 8? Explain any difference.
4. From your data in Table 59-2, is the circuit of Fig. 59-4 inductive or capacitive at line frequency? Explain why.
5. Refer to your data in Table 59-2. What is the net reactance of the circuit? Is it inductive or capacitive reactance?

(Optional) Extra Credit

6. Does the value of Z computed in procedural step 11 agree with the values of Z in Table 59-2? If not, why not?

Answers to Self-Test

1. 30; inductive
2. 50
3. *RL*; X_L; 30
4. less
5. capacitive; 50; lags
6. 22.6; negative
7. 130
8. $120 - j50$
9. $130 \angle -22.6°$
10. negative; negative

60

EXPERIMENT

VARIATION OF IMPEDANCE AND CURRENT OF A SERIES *RLC* CIRCUIT WITH CHANGES IN FREQUENCY

OBJECTIVE

In a series *RLC* circuit, to determine experimentally the effect on impedance and current of changes in frequency of the ac source

INTRODUCTORY INFORMATION

In Experiment 59 we verified that the impedance Z of a series-connected *RLC* circuit (Fig. 60-1) is given by the formula

$$Z = \sqrt{R^2 + (X_L - X_C)^2} \qquad (60\text{-}1)$$

In this experiment we will observe how a change in frequency of the voltage source affects circuit impedance and current.

Effect of Frequency on the Impedance of an *RLC* Circuit

Consider Eq. (60-1). The minimum value of impedance in an *RLC* circuit, for a given value of R, occurs when $X_L - X_C = 0$. At that point $Z = \sqrt{R^2} = R$, and the circuit acts like a pure resistance. The line current I is limited only by R. Hence I is maximum when $X_L - X_C = 0$.

Inductive reactance X_L and capacitive reactance X_C are dependent on frequency. It is obvious that there is a frequency F_R at which $X_L = X_C$. We will consider this fact at greater length in the next experiment. At this point let us merely note the effect on impedance of frequencies greater and smaller than F_R.

In the series *RLC* circuit we have been analyzing, as the frequency F of the sinusoidal voltage source increases beyond F_R, X_L increases and X_C decreases. Hence the circuit acts like an inductance whose reactance $(X_L - X_C)$ increases as F increases. As the frequency F decreases below F_R, X_C increases while X_L decreases. The circuit now acts like a capacitance whose reactance $(X_C - X_L)$ increases as F decreases.

A graph of impedance versus frequency for a series *RLC* circuit is shown in Fig. 60-2. This curve shows minimum Z at F_R. It shows also that Z is capacitive (Z_C) for frequencies less than F_R and inductive (Z_L) for frequencies greater than F_R.

The variation of current versus frequency is also sketched graphically in Fig. 60-2.

SUMMARY

1. In a series *RLC* circuit there is a frequency F_R of the voltage source V at which $X_L = X_C$. At this frequency the impedance Z of the *RLC* circuit is minimum and $Z = R$, the resistance in the circuit.
2. For frequencies higher than F_R, the X_L of the inductor increases while X_C decreases and the circuit acts like an *RL* circuit, with $Z = \sqrt{R^2 + (X_L - X_C)^2}$. As F increases beyond F_R, Z increases beyond R.
3. For frequencies lower than F_R, the X_C of the *capacitor* increases while X_L decreases. The circuit acts like an *RC* circuit, with $Z = \sqrt{R^2 + (X_L - X_C)^2}$. As F decreases below F_R, Z increases beyond R.
4. In a series *RLC* circuit the current $I = V/Z$, where V is the applied voltage.

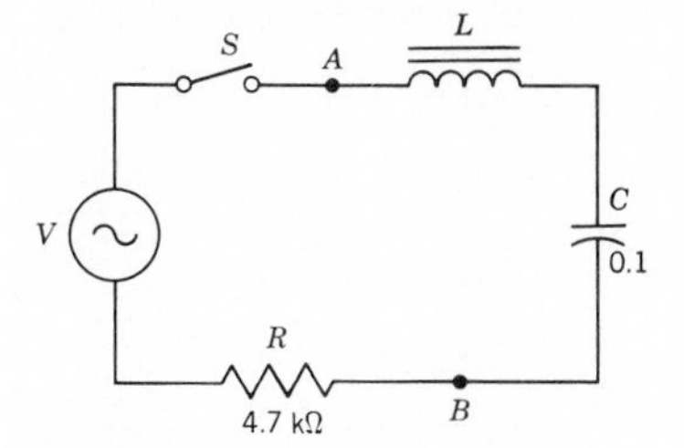

Fig. 60-1. Circuit to determine the effect of frequency on the impedance of *RLC* connected in series.

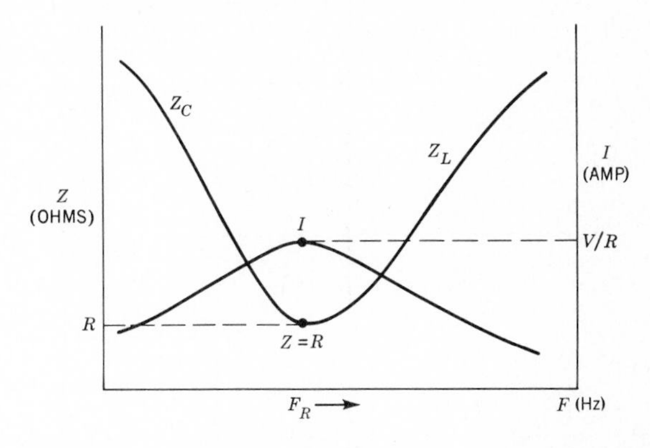

Fig. 60-2. Variation of Z and I in a series-connected *RLC* circuit.

SELF-TEST

Check your understanding by answering these questions:

1. In the circuit of Fig. 60-1, for a particular frequency F_R, $X_L = X_C = 1000\ \Omega$. The impedance Z of the circuit is ________ Ω.
2. The circuit in question 1 acts like a pure ________.
3. For the frequency $F_R + 100$, the impedance of the circuit in Fig. 60-1 is ________ (greater/less than) 4700 Ω and the circuit is ________ (inductive/capacitive).
4. For the frequency $F_R - 100$, the impedance of the circuit in Fig. 60-1 is ________ (greater/less than) 4700 Ω and the circuit is ________ (capacitive/inductive).
5. As F increases above F_R or decreases below F_R, the current I in the circuit ________ (increases/decreases/remains the same).

MATERIALS REQUIRED

- Equipment: Electronic voltmeter, oscilloscope, audio-frequency sine-wave generator
- Resistor: ½-W 4700-Ω
- Capacitor: 0.1 μF
- Miscellaneous: SPST switch; choke, 8 H at 50 mA direct current

PROCEDURE

Effect of Frequency on Impedance

1. Connect the circuit of Fig. 60-1. The ac voltage source is an audio-frequency signal generator. Set the generator output at three-quarters of maximum voltage and the frequency at 150 Hz.

2. With the EVM connected across R and an oscilloscope connected across points AB, increase or decrease the frequency of the signal generator, as required, until the voltage V_{AB} is minimum. At this point, the voltage V_R across R should be maximum.

3. Note the frequency F_R at which there is a minimum V_{AB} and record it in Table 60-1. With the EVM, measure and record the applied voltage V and the voltages V_R, V_L, V_C, and V_{AB} (the combined voltage across the inductance and capacitor). Compute and record the difference between V_L and V_C. Also compute and record the current $I = V_R/R$ and the impedance $Z = V_{\text{applied}}/I$. Show your computations.

4. Decrease the frequency of the generator by 20 Hz and record the new frequency, $F_R - 20$. Set the output V of the generator at the same level as for F_R in step 3. Repeat the measurements and computations of step 3 at the frequency $F_R - 20$ and record in Table 60-1.

5. Repeat step 4 for each of the frequencies shown in Table 60-1. Be *certain* that for each frequency the output of the generator is kept at the same voltage level as in step 3.

6. From the data in Table 60-1, draw a graph of:

a. Z versus F

b. I versus F

Extra Credit

7. For each value of F in Table 60-1, determine the values of X_L, X_C, and Z of the circuit, using the formulas $X_L = V_L/I$, $X_C = V_C/I$, and $Z = \sqrt{R^2 + (X_L - X_C)^2}$. Record your results in Table 60-2, which you will draw. For comparison, transcribe to Table 60-2 the computed values of Z from Table 60-1.

TABLE 60-1. Measurements to Show How Z and I Vary with F

Frequency, Hz	*V Applied*, V	V_R, V	V_L, V	V_C, V	V_{AB}, V	$V_L - V_C$ *or* $V_C - V_L$, V	I, A	Z, Ω
$F_R - 100$								
$F_R - 80$								
$F_R - 60$								
$F_R - 40$								
$F_R - 20$								
F_R								
$F_R + 20$								
$F_R + 40$								
$F_R + 60$								
$F_R + 80$								
$F_R + 100$								

QUESTIONS

1. Refer to the data in Table 60-1 and to the graph of Z versus F. Explain in your own words the effect on Z of a change in F.
2. Refer to the data in Table 60-1 and to the graph of I versus F. Explain in your own words the effect on I of a change in F.
3. In Fig. 60-1, what would be the effect on Z, if any, of interchanging L and C? Why?
4. In Table 60-1, comment on the realtionship, if any, between the measured voltage V_{AB} and the voltage $V_L - V_C$ at any specific frequency. Explain any unexpected results.

Extra Credit

5. Discuss the results of your computations in step 7 on verification of the impedance formula for an *RLC* circuit [Eq. (60-1)]. Explain any unexpected results.
6. Derive a general formula for the frequency at which $X_L = X_C$ in a series *RLC* circuit.
7. Assume that the external resistor R in Fig. 60-1 is short-circuited. What will limit the value of current I, when $X_L = X_C$?

Answers to Self-Test

1. 4700
2. resistance
3. greater; inductive
4. greater; capacitive
5. decreases

61

EXPERIMENT

IMPEDANCE OF A PARALLEL *RL* AND OF A PARALLEL *RC* CIRCUIT

OBJECTIVES

1. To determine experimentally the impedance of a parallel *RL* circuit
2. To determine experimentally the impedance of a parallel *RC* circuit

INTRODUCTORY INFORMATION

Impedance of a Parallel *RL* Circuit

The circuit of Fig. 61-1 consists of two parallel branches, a resistive branch R and an inductive branch L. The same voltage V appears across each branch. V causes a current I_R in R, whose value is $I_R = V/R$. V also causes a current I_L in L, whose value is $I_L = V/X_L$.

What are the relationships among V, I_R, I_L, and I_T (the total line current) in the circuit? Figure 61-2 is a phasor diagram of the relationship among these quantities. Since voltage is common to each branch of the circuit, V is shown as the reference phasor. In the resistive branch, voltage and current are in phase. I_R is therefore shown in phase with V. In a pure inductance, current lags the applied voltage by 90°. I_L is therefore shown lagging V by 90°.

The phasor sum of currents I_R and I_L forms the line current I_T, as shown in Fig. 61-2. Since I_L and I_R are legs of a right triangle and I_T is the hypotenuse of this triangle,

$$I_T = \sqrt{I_R^2 + I_L^2} \tag{61-1}$$

Figure 61-2 shows also that the line current I_T lags the applied voltage V by some angle θ. Theta may be determined from the equation

$$\theta = \arctan \frac{I_L}{I_R} \tag{61-2}$$

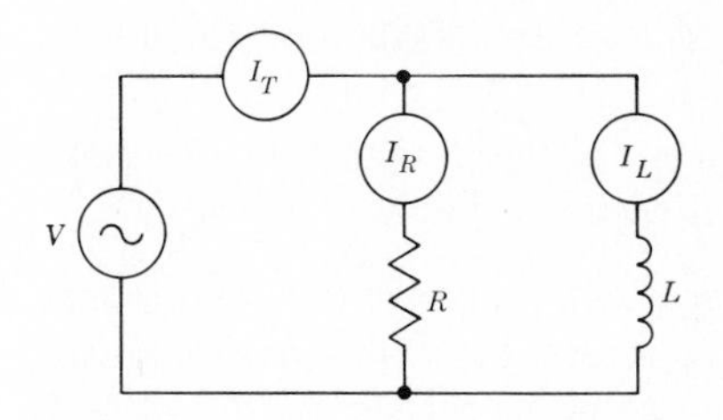

Fig. 61-1. Total current in a parallel *LR* circuit.

The impedance of the circuit (Fig. 61-1) may be found by substituting the measured values of applied voltage and line current in the equation

$$Z = \frac{V}{I_T} \tag{61-3}$$

Impedance of a Parallel *RC* Circuit

A resistive branch R and a capacitive branch C constitute the paths for current in the parallel RC circuit of Fig. 61-3. Analysis of this circuit is similar to that of the parallel RL circuit.

In Fig. 61-3 the applied voltage V is common to both branches. The current I_R in R is given by the equation $I_R = V/R$. The current I_C in C is given by the equation $I_C = V/X_C$. The relationships among V, I_C, I_R, and I_T (the line current) are shown in the phasor diagram (Fig. 61-4). Again V is used as the reference phasor. The current I_R in R is shown in phase with V. The current I_C in C is shown *leading* V by 90°. The line current I_T is the phasor sum of I_R and I_C. Since I_R and I_C are the legs of a right triangle and I_T is the hypotenuse of this triangle,

$$I_T = \sqrt{I_R^2 + I_C^2} \tag{61-4}$$

Figure 61-4 also shows that the line current I_T *leads* the applied voltage by an angle θ, whose value may be determined from the equation

$$\theta = \arctan \frac{I_C}{I_R} \tag{61-5}$$

The impedance Z of the circuit in Fig. 61-3 may be found by substituting the measured values of applied voltage and line current in the equation

$$Z = \frac{V}{I_T} \tag{61-6}$$

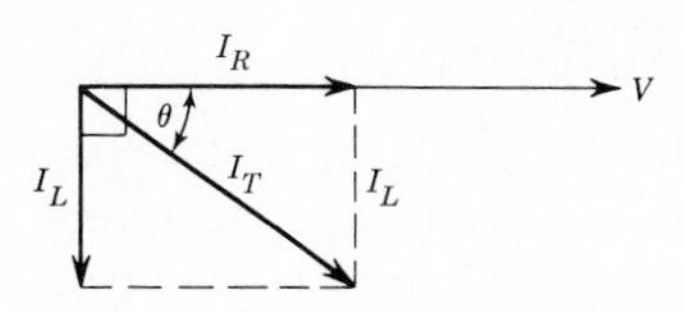

Fig. 61-2. Total current I_T in a parllel *LR* circuit is the phasor sum of I_R and I_L.

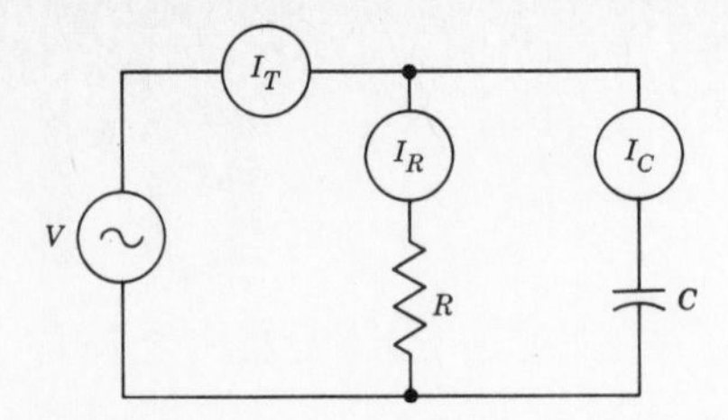

Fig. 61-3. Total current in a parallel *RC* circuit.

It is interesting to note that the line current in a parallel *RC* circuit is a leading current (with reference to the applied voltage), whereas the line current in a parallel *RL* circuit is a lagging one.

SUMMARY

1. In a parallel *RL* or *RC* circuit, the voltage across each branch of the circuit is the same and is therefore used as the reference phasor in an analysis of the relationships among the currents in the circuit.
2. The branch currents I_R, I_C, and I_L are phasors. The total (line) current I_T is the phasor sum of I_R and I_L in an *RL* circuit, and of I_R and I_C in an *RC* circuit.
3. In a parallel *RL* circuit, the current in *L lags* the current in *R* by 90°. Therefore, the numerical value of I_T can be computed from the formula

$$I_T = \sqrt{I_R^2 + I_L^2}$$

4. In a parallel *RL* circuit, the total current *lags* the applied voltage *V* by some angle θ, where

$$\theta = \arctan \frac{I_L}{I_R}$$

5. In a parallel *RC* circuit, the current in *C leads* the current in *R* by 90°. Therefore, the numerical value of I_T can be computed from the formula

$$I_T = \sqrt{I_R^2 + I_C^2}$$

6. In a parallel *RC* circuit, the total current *leads* the applied voltage *V* by some angle θ, where

$$\theta = \arctan \frac{I_C}{I_R}$$

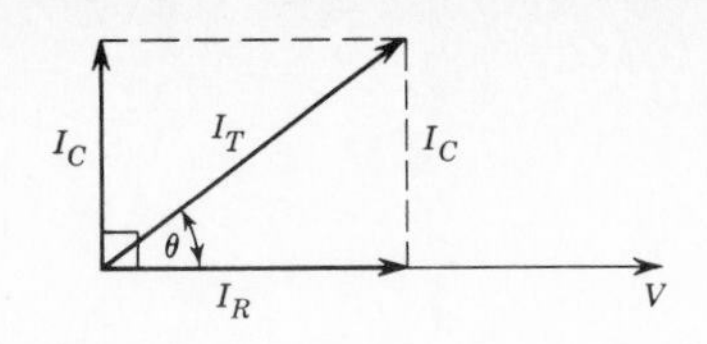

Fig. 61-4. Total current I_T in a parallel *RC* circuit is the phasor sum of I_R and I_C.

7. The impedance *Z* of an *RL* or an *RC* circuit may be computed from the equation

$$Z = \frac{V}{I_T}$$

where *V* is the applied voltage and I_T is the value of total or line current.

SELF-TEST

Check your understanding by answering these questions:

1. In a parallel *RL* circuit, the applied voltage $V = 100$ V; the measured value of $I_L = 0.2$ A and of $I_R = 0.5$ A. The numerical value of I_T = ____________ A.
2. The angle θ by which total current I_T (in question 1) ____________ (leads/lags) the applied voltage *V* is ____________ degrees.
3. The impedance of the circuit in question 1 is ____________ Ω.
4. In a parallel *RC* circuit, the applied voltage $V = 30$ V, and the measured values of $I_C = 0.1$ A and $I_R = 0.2$ A. The numerical value of I_T = ____________ A.
5. The angle θ by which the total current I_T (in question 4) ____________ (leads/lags) the applied voltage *V* is ____________ degrees.
6. The impedance of the circuit in question 4 is ____________ Ω.

MATERIALS REQUIRED

- Power supply: Source of line-isolated variable ac voltage (variable autotransformer plugged into line-isolation transformer)
- Equipment: 0–25-mA ac meter; EVM
- Resistor: 1-W 5000-Ω
- Capacitor: 0.5 μF
- Miscellaneous: Two SPST switches; iron-core choke rated 8 H at 50 mA direct current; fused line cord

PROCEDURE

Impedance of a Parallel *RL* Circuit

1. Connect the circuit of Fig. 61-5. Switches S_1 and S_2 are open. *V* is the output of a variable autotransformer plugged into a line-isolation transformer (1:1) which derives its power from the 120-V ac line. *M* is a 0–25-mA ac meter or the equivalent of a multirange ac milliammeter. V_1 is an EVM set on ac voltage. *L* is an iron-core choke rated at 8 H, with resistance of 250 Ω.

2. Close S_1. Slowly increase the output of the power source until the voltage V_{AB} across the 5000-Ω resistor reads 50 V rms. *M* now measures the current I_R. Record I_R in Table 61-1.

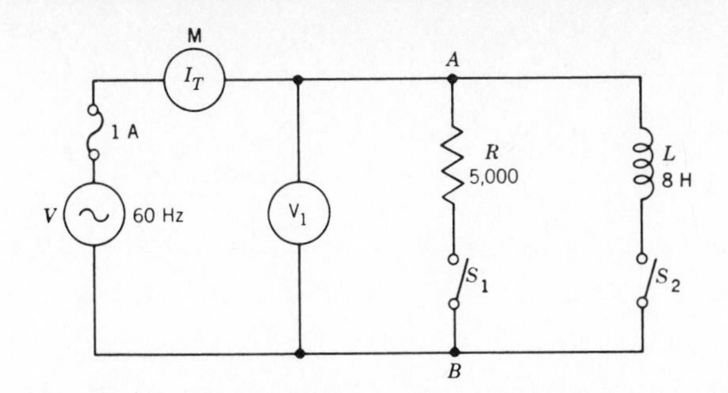

Fig. 61-5. Experimental circuit for measuring the total current of a resistor *R* in parallel with an inductor *L*.

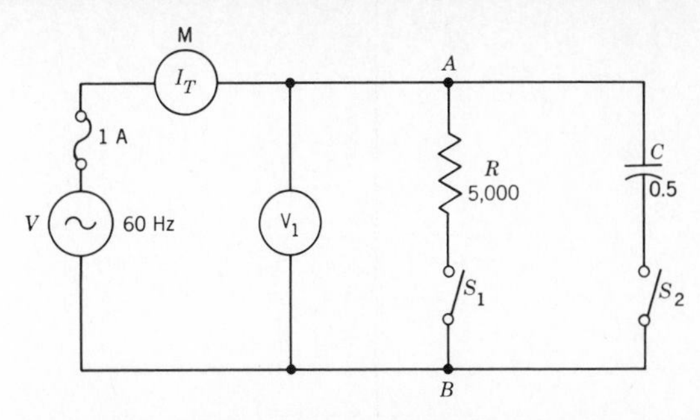

Fig. 61-6. Experimental circuit for measuring the total current of a resistor *R* in parallel with a capacitor *C*.

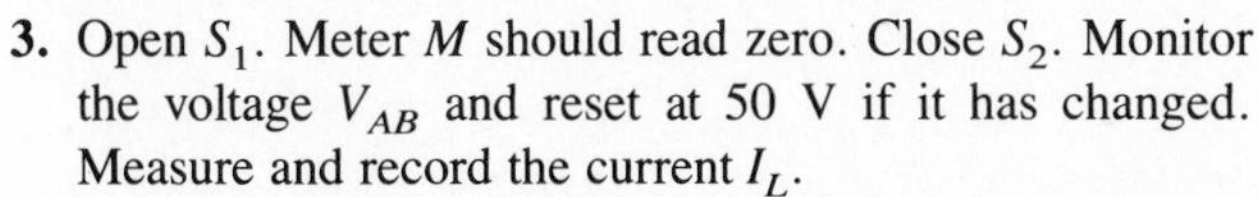

3. Open S_1. Meter *M* should read zero. Close S_2. Monitor the voltage V_{AB} and reset at 50 V if it has changed. Measure and record the current I_L.
4. Close S_1. Both switches are now closed. *L* and *R* now constitute a parallel circuit. Reset the voltage V_{AB} at 50 V, if necessary. Measure and record the line current I_T. **Power off.** Open S_1 and S_2. Compute and record the impedance *Z* of the circuit by substituting the measured values *V* and I_T in $Z = V/I_T$.
5. Substitute the measured values of I_R and I_L in the equation

$$I_T = \sqrt{I_R^2 + I_L^2}$$

Compute this value of I_T and record in Table 61-1. Show your computations.

Impedance of a Parallel *RC* Circuit

6. Replace *L* by a 0.5-μF capacitor as in Fig. 61-6. Close S_1. Only *R* is in the circuit. Set the voltage V_{AB} at 60 V. *M* now reads the current I_R. Record I_R in Table 61-2.
7. Open S_1. Close S_2. Only *C* is now in the circuit. Monitor the voltage V_{AB} and reset at 60 V if it has changed. Measure and record the current I_C.
8. Close S_1. Both switches are now closed. *R* and *C* are in the circuit. Reset the voltage V_{AB} at 60 V, if necessary. Measure and record the line current I_T. **Power off.** Compute and record the impedance *Z* of the circuit by substituting the measured values of *V* and I_T in the equation $Z = V/I_T$. Substitute the measured values of I_R and I_C in the equation $I_T = \sqrt{I_R^2 + I_C^2}$. Compute this value of I_T and record in Table 61-2. Show your computations.

TABLE 61-1. Impedance of a Parallel *RL* Circuit

V_{AB}, V	I_R, A	I_L, A	I_T, A	$Z = \frac{V}{I_T}$, Ω	$\sqrt{I_R^2 + I_L^2}$, A
50					

TABLE 61-2. Impedance of a Parallel *RC* Circuit

V_{AB}, V	I_R, A	I_C, A	I_T, A	$Z = \frac{V}{I_T}$, Ω	$\sqrt{I_R^2 + I_C^2}$, A
60					

QUESTIONS

1. Refer to Table 61-1. Does the measured value of I_T equal the computed value $\sqrt{I_R^2 + I_L^2}$? Should it? Explain any unexpected results.
2. Refer to Table 61-2. Does the measured value of I_T equal the computed value $\sqrt{I_R^2 + I_C^2}$? Should it? Comment on any unexpected results.
3. What are the sources of error in this experiment?
4. In computing the impedance of a parallel *RL* or *RC* circuit by the equation $Z = V/I_T$, you found only the arithmetic value of *Z*. Compute the phase angle of *Z*, and in each case indicate whether it is a leading or lagging angle.
5. Draw a phasor diagram of the branch currents in each circuit with both switches closed. Show total current and the phase angle between applied voltage and total current. Draw to scale.

Answers to Self-Test

1. 0.539
2. lags; 21.8°
3. 186
4. 0.224
5. leads; 26.6°
6. 134

62

EXPERIMENT

IMPEDANCE OF A PARALLEL *RLC* CIRCUIT

OBJECTIVE

To determine experimentally the impedance of a circuit containing a resistance R in parallel with an inductance L, in parallel with a capacitance C

INTRODUCTORY INFORMATION

The simplest type of parallel *RLC* circuit is illustrated in Fig. 62-1. The ac voltage V is common to each leg in the parallel network. The current I_R in R may be calculated from the equation

$$I_R = \frac{V}{R} \qquad (62\text{-}1)$$

where I is in amperes, V is the rms voltage, and R is the resistance in ohms of the resistor. Similarly the currents I_C in C and I_L in L may be calculated, respectively, from the equations

$$I_C = \frac{V}{X_C} \qquad (62\text{-}2)$$

$$I_L = \frac{V}{X_L} \qquad (62\text{-}3)$$

Note that the voltage across each leg is the applied voltage V.

The total line current I_{line} is equal to the phasor sum of I_R, I_C, and I_L. This is consistent with the analysis and results of Experiment 61, where parallel *RL* and *RC* circuits were considered.

Figure 62-2 is a phasor diagram showing the phase relationship between the applied voltage V and the currents in each leg. Thus, I_R is in phase with V, I_C leads V by 90°, and I_L lags V by 90°. Since I_C and I_L are 180° out of phase, they may be subtracted arithmetically to give the reactive current in the circuit, as in Fig. 62-3. If I_L is larger than I_C, the circuit is inductive. If I_C is larger than I_L, the circuit is capacitive, as in Fig. 62-3. The line current may be obtained by finding the phasor sum of I_R and $I_C - I_L$, as in Fig. 62-3.

The line current may also be calculated by substituting the values of I_R, I_C, and I_L in the equation

$$I_{\text{line}} = \sqrt{I_R^2 + (I_C - I_L)^2} \qquad (62\text{-}4)$$

This formula holds as long as there is no resistance in series with L or C in the circuit of Fig. 62-1.

The line current I_{line} is the sum of the instantaneous values of the resistive and reactive currents and is not in phase with either current. It differs from I_R, the resistive current taken as reference, by the angle θ. This angle may be computed from the formula

$$\tan\theta = \frac{I_C - I_L}{I_R}$$

$$\theta = \arctan\frac{I_C - I_L}{I_R} \qquad (62\text{-}5)$$

Having found the line current I_{line}, we can now compute the value of impedance Z in ohms of the circuit of Fig. 62-1 by substituting the values of I in amperes and V in volts in the equation

$$Z = \frac{V}{I_{\text{line}}} \qquad (62\text{-}6)$$

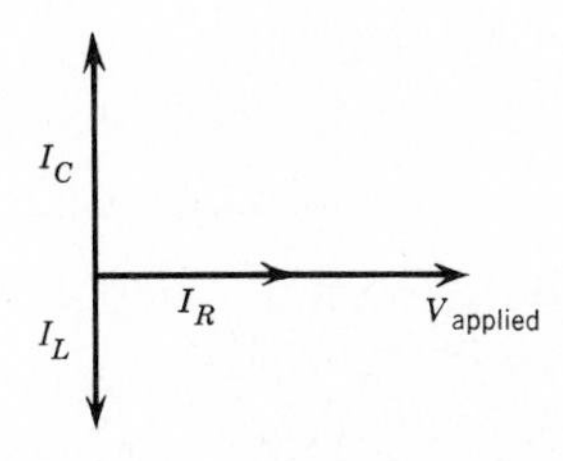

Fig. 62-2. Phase relations between applied voltage and current in each leg of a parallel *RLC* circuit.

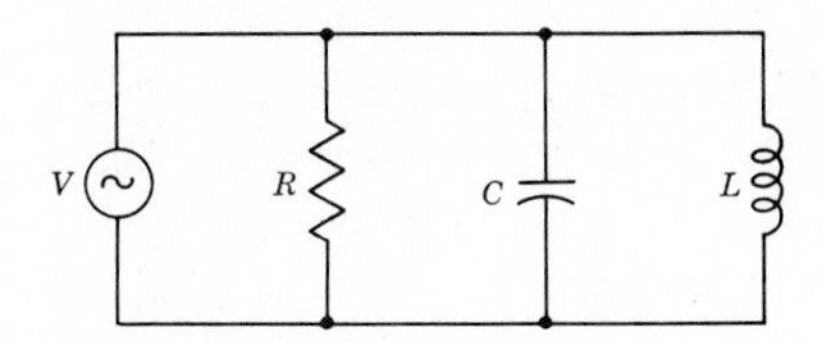

Fig. 62-1. Parallel *RLC* circuit.

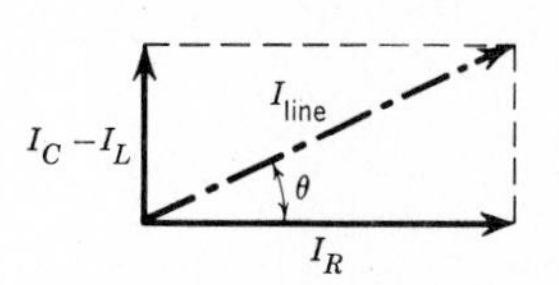

Fig. 62-3. Solving for line current in a capacitive *RLC* circuit.

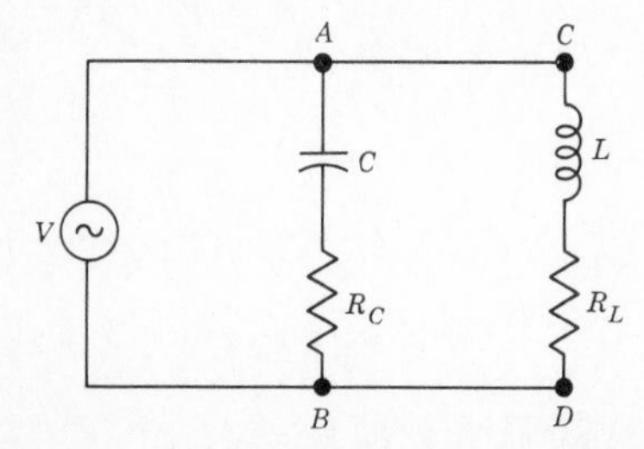

Fig. 62-4. A parallel *RLC* network consisting of two series-connected branches.

Since I_L and I_C are dependent on X_L and X_C, respectively, it is apparent that Z is a function of the frequency of the voltage V applied to the circuit.

NOTE: One important difference between parallel dc and ac circuits should be pointed out. In parallel dc circuits the total line current I_T increases with every parallel branch added. In parallel ac circuits the line current does not always increase as more branches are added. For example, if Fig. 62-1 initially consisted of a resistive branch in parallel with a capacitive branch, the effect of adding a third inductive branch *might* be to *reduce* the line current, a fact which is evident from Eq. (62-4). Adding an inductive branch, as in Fig. 62-1, reduces line current if $|I_C - I_L| < |I_C|$. This occurs whenever $|I_L| < |2I_C|$.

A corollary to this is that the total impedance of parallel ac circuits is not always less than the smallest branch impedance, but is sometimes *greater* than this branch impedance.

A more complex circuit is shown in Fig. 62-4. The ac generator applies a voltage V to two parallel branches AB and CD. Each branch is a series network. Thus the capacitive branch AB contains R_C and C connected in series. The inductive branch CD contains R_L and L connected in series. The impedance of each branch may be found by the methods outlined in the previous experiments dealing with series-connected RL and RC circuits. Thus, the impedance of branch AB is

$$Z_{AB} = \sqrt{R_C^2 + X_C^2} \tag{62-7}$$

The current I_{AB} in this branch may be computed as follows:

$$I_{AB} = \frac{V}{Z_{AB}} \tag{62-8}$$

where the value of Z_{AB} is taken from Eq. (62-7).

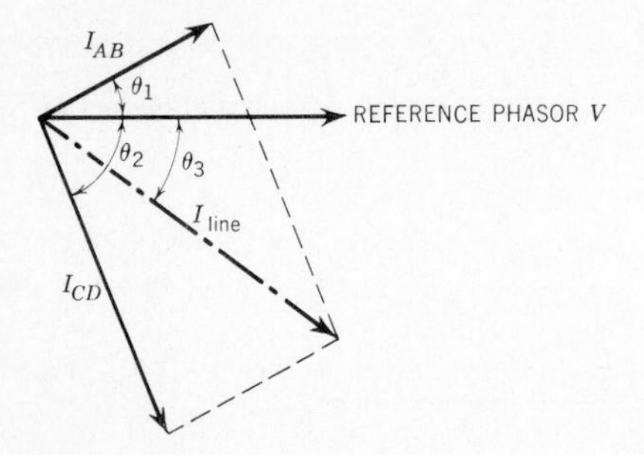

Fig. 62-5. Finding line current in a parallel *RLC* network.

The current I_{AB} is a leading current with relation to the applied voltage V, and the phase angle θ is given by the formula

$$\theta_1 = \arctan \frac{X_C}{R_C}$$

or

$$\theta_1 = \arctan \frac{V_C}{V_{R_C}} \tag{62-9}$$

The impedance of branch CD is

$$Z_{CD} = \sqrt{R_L^2 + X_L^2} \tag{62-10}$$

The current I_{CD} in this branch is

$$I_{CD} = \frac{V}{Z_{CD}} \tag{62-11}$$

where the value of Z_{CD} is taken from Eq. (62-10). The current in branch CD is a lagging current whose phase angle θ_2 may be found from the formula

$$\theta_2 = \arctan \frac{X_L}{R_L}$$

or

$$\theta_2 = \arctan \frac{V_L}{V_{R_L}} \tag{62-12}$$

After the currents in each branch have been found, it is possible to determine line current (I_{line}) by finding the phasor sum of I_{AB} and I_{CD} as in Fig. 62-5. Note that the phase angle of line current, θ_3, is different from θ_1 or θ_2.

The line current in Fig. 62-5 is shown as a lagging (inductive) current because in our illustration we assumed that Z_{CD} was smaller than Z_{AB}. Hence, in this case, the inductive current in branch CD is larger than the capacitive current in branch AB. Since the inductive current predominates, the total current may be considered inductive.

The value of total impedance Z that a generator sees "looking" into the circuit of Fig. 62-4 may be computed from the formula

$$Z = \frac{V}{I_{line}} \tag{62-13}$$

It is evident that Eqs. (62-6) and (62-13) are identical and that the method of finding the impedance of a parallel *RLC* circuit, regardless of the number or type of branches involved, is the same.

SUMMARY

A method for finding the impedance of a parallel *RLC* circuit requires the following steps:

1. Calculate the impedance of each branch or leg of the parallel network by the methods already discussed in previous experiments. Determine both the numerical value $|Z|$ and the phase angle θ.
2. Find the current in each branch and the phase angle between the applied voltage V and the branch current.

Branch current is computed using the equation

$$I = \frac{V}{|Z|}$$

Note that the phase angle (in degrees) is the same as the angle θ in summary statement 1, but it is opposite in sign. Thus the phase angle of a capacitive current is positive (+), and that of an inductive current is negative (−).

3. Obtain the phasor sum of all branch currents. The resultant is the line current, whose amplitude and phase angle are determined from the phasor diagram which has been drawn to scale.
4. If the parallel *RLC* circuit has no resistance in the capacitive or inductive branch, that is, if the circuit is exactly as in Fig. 62-1, then the net reactive current is $I_C - I_L$ if the capacitive current is larger than the inductive current, or $I_L - I_C$ if the inductive current is larger than the capacitive current.
5. The total current I_{line} for the circuit in summary statement 4 is found by using the formula

$$I_{\text{line}} = \sqrt{I_R^2 + (I_C - I_L)^2}$$

and the phase angle θ between the applied voltage *V* and the total current is

$$\theta = \arctan \frac{I_C - I_L}{I_R}$$

6. Find the impedance |*Z*| of the parallel *RLC* circuit using the formula $|Z| = V/I_{\text{line}}$, where *V* is the voltage across the parallel network and I_{line} is the line current.

SELF-TEST

Check your understanding by answering these questions:

1. In the circuit of Fig. 62-1, $R = 120\ \Omega$, $X_C = 150\ \Omega$, and $X_L = 180\ \Omega$. The applied voltage $V = 18$ V.
   (*a*) The current I_R in R = ________ A.
   (*b*) The phase angle between *V* and I_R is ________ degrees.
2. For the circuit in question 1,
   (*a*) The current I_C in C = ________ A.
   (*b*) The phase angle between *V* and I_C = ________ degrees.
3. For the circuit in question 1,
   (*a*) The current I_L in L = ________ A.
   (*b*) The phase angle between *V* and I_L is ________ degrees.
4. The net reactive current in the circuit in question 1 is ________ A, and it is ________ (capacitive/inductive).
5. The total current (I_{line}) in the circuit of question 1 = ________ A.
6. The phase angle between the applied voltage *V* and I_{line} in the circuit of question 1 is ________ degrees and its sign is ________ (+ or −).
7. The total impedance in the circuit of question 1 is ________ Ω.

MATERIALS REQUIRED

- Power supply: Source of line-isolated variable ac voltage (variable autotransformer plugged into line-isolation transformer)
- Equipment: 0–25-mA ac meter; EVM
- Resistor: 1-W 5000-Ω
- Capacitor: 0.5 μF
- Miscellaneous: Three SPST switches; iron-core choke rated 8 H at 50 mA direct current; fused line cord

PROCEDURE

1. Connect the circuit Fig. 62-6. All switches are open. *V* is the output of a variable autotransformer plugged into a line-isolation transformer which derives its power from the 120-V ac line. *V* is set at 0 V. *M* is a 0–25-mA ac meter, or the equivalent of a multirange milliammeter. V_1 is an EVM set on ac voltage. *L* is an iron-core choke rated 8 H at 50 mA direct current.
2. Close S_1. Slowly increase the voltage until $V_{AB} = 60$ V. Measure the current I_R and record it in Table 62-1.
3. Open S_1. Close S_2. Readjust *V* if necessary until $V_{AB} = 60$ V. Measure and record the current I_L.
4. Open S_2. Close S_3. Readjust *V* if necessary until $V_{AB} = 60$ V. Measure and record the current I_C.
5. Now close S_1. At this point S_1 and S_3 are closed. *R* and *C* are in the circuit. Readjust *V* if necessary until $V_{AB} = 60$ V. Measure and record the current I_{RC}.
6. Now close S_2. At this point all switches are closed. Readjust *V* if necessary until $V_{AB} = 60$ V. Measure and record the line current I_{RLC}. Note whether I_{RLC} is larger or smaller than I_{RC}. **Power off.**

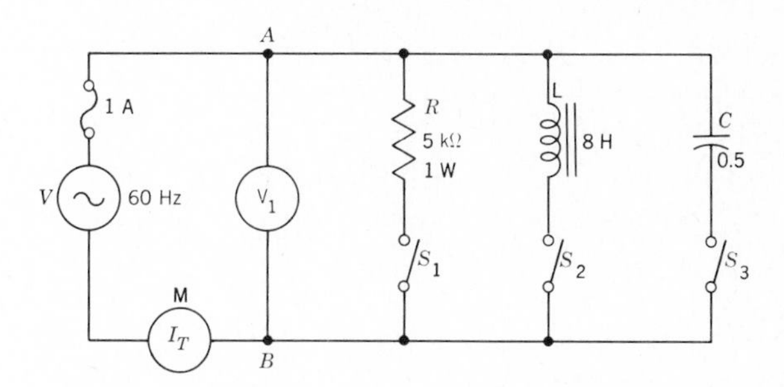

Fig. 62-6. Experimental circuit for determining the impedance of *R*, *L*, and *C* connected in parallel.

7. Find I_{line} by substituting the measured values I_R, I_L, and I_C in the formula

$$I_{\text{line}} = \sqrt{I_R^2 + (I_C - I_L)^2}$$

Record in Table 62-1.

8. Compute Z by substituting in the formula

$$Z = \frac{V_{AB}}{I_{RLC}}$$

the measured values of I_{RLC} and V_{AB}. Record in Table 62-1.

9. Compute and record the angle θ, the phase angle between V_{AB} and I_{line}, by substituting in the equation

$$\theta = \arctan \frac{I_C - I_L}{I_R}$$

the measured values of I_C, I_L, and I_R.

TABLE 62-1. Measurements in a Parallel *RLC* Circuit

V_{AB}, V	I_R, A	I_L, A	I_C, A	I_{RL}, A	I_{RC}, A	I_{RLC}, A	I_{line} (A) = $\sqrt{I_R^2 + (I_C - I_L)^2}$	Z, Ω	θ, degrees
60									

QUESTIONS

1. In Table 62-1, how does the measured value I_{RLC} compare with the formula value $\sqrt{I_R^2 + (I_C - I_L)^2}$? Should they be equal? Explain any unexpected results.
2. Which current is larger, I_{RC} or I_{RLC}? Why?
3. Is the impedance of the parallel *RLC* circuit higher or lower than the impedance of the parallel *RC* circuit? *RL* circuit? Why?
4. Is the line current in Fig. 62-6 capacitive or inductive? Why?
5. Draw a series equivalent circuit for Fig. 62-6. Compute the value of resistance and reactance of the series equivalent circuit.

Answers to Self-Test

1. (*a*) 0.15; (*b*) zero
2. (*a*) 0.12; (*b*) +90
3. (*a*) 0.10; (*b*) −90
4. 0.02; capacitive
5. 0.151
6. 7.6; +
7. 119

63

EXPERIMENT

RESONANT FREQUENCY AND FREQUENCY RESPONSE OF A SERIES RESONANT CIRCUIT

OBJECTIVES

1. To determine experimentally that the resonant frequency F_R of a series resonant LC circuit is given by the formula

$$F_R = \frac{1}{2\pi\sqrt{LC}}$$

2. To measure the frequency response of a series resonant LC circuit

INTRODUCTORY INFORMATION

Resonant Frequency of Series *LC* Circuit

Consider the circuit of Fig. 63-1. Assume V is an ac generator whose frequency and output voltage are manually adjustable. For a particular frequency F and output voltage V, there will be a current I such that $I = V/Z$, where Z is the impedance of the circuit. The voltage drops across R, L, and C will be given by IR, IX_L, and IX_C, respectively.

If the generator frequency is changed but V remains constant, the current I and the voltage drops across R, L, and C will change.

There is a frequency F_R called the *resonant* frequency at which

$$X_L = X_C \tag{63-1}$$

It is a simple matter to determine, analytically, the resonant frequency F_R. Consider the following:

$$X_L = 2\pi FL$$

and

$$X_C = \frac{1}{2\pi FC} \tag{63-2}$$

At F_R, when $X_L = X_C$,

$$2\pi F_R L = \frac{1}{2\pi F_R C} \tag{63-3}$$

Equation (63-3) can be put into the form

$$F_R^2 = \frac{1}{(2\pi)^2 LC} \tag{63-4}$$

Taking the square root of both sides of Eq. (63-4) leads to the formula for the resonant frequency of a series resonant circuit, namely

$$F_R = \frac{1}{2\pi\sqrt{LC}} \tag{63-5}$$

Here F_R is given in hertz, L in henrys, and C in farads. Equation (63-5) is used to find the resonant frequency of a series circuit containing a given value of L and of C.

Example. Find the resonant frequency of a series LC circuit where $L = 8$ H and $C = 0.01$ μF.

Solution.

1. $$F_R = \frac{1}{2\pi\sqrt{LC}}$$

Substituting the given values in 1 results in

2. $$F_R = \frac{1}{2\pi\sqrt{8 \times 0.01 \times 10^{-6}}}$$

3. $$F_R = \frac{1}{2\pi\sqrt{8 \times 1.0 \times 10^{-8}}}$$

4. $$F_R = \frac{10^4}{2\pi\sqrt{8}}$$

5. $$F_R = 563 \text{ Hz}$$

Characteristics of a Series Resonant Circuit

Some characteristics of a series-connected RLC circuit were investigated in a previous experiment. The facts observed were:

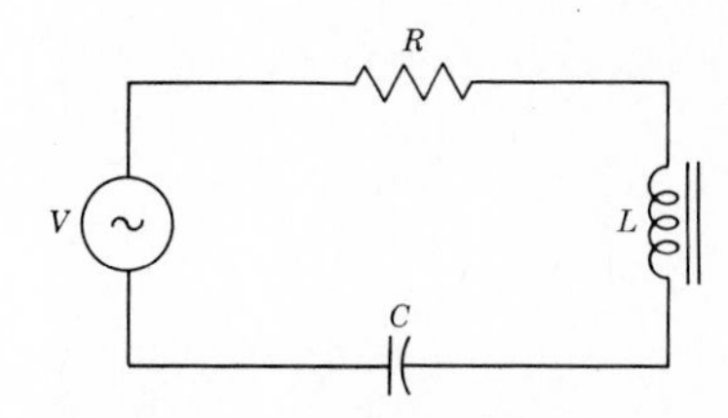

Fig. 63-1. Series *RLC* circuit.

1. The voltage drop across a reactive component is equal to the product of the current I in the circuit and the reactance X of the component.
2. Capacitive reactance X_C tends to cancel the effects of inductive reactance X_L.
3. The impedance Z of a series RLC circuit is

$$Z = \sqrt{R^2 + (X_L - X_C)^2}$$

4. The impedance Z of the circuit is minimum when $X_L = X_C$, and circuit I is maximum at this point.

Fact 4 states that *at resonance, the impedance of a series LC circuit is minimum* and the *current I* in the *circuit is maximum*. This fact, which was established experimentally in a preceding experiment, can also be determined analytically. For at F_R

$$Z = \sqrt{R^2 + (0)^2} = R$$

Therefore, a *minimum impedance* $Z = R$ exists at F_R. Also, since $I = V/Z = V/R$, and V is assumed to be fixed, there will be *maximum current* I in the circuit at resonance. Note that at F_R the applied voltage V appears across R, and resonant current I may be computed from

$$I = \frac{V}{R} \tag{63-6}$$

Since at resonance I is limited only by the value of R, the circuit is said to be *resistive*. For all frequencies higher than F_R, X_L is larger than X_C and the circuit is *inductive*. For all frequencies lower than F_R, X_C is larger than X_L and the circuit is *capacitive*.

At *resonance* the voltage V_L across L and the voltage V_C across C are *maximum* and equal. Theoretically, at the resonant frequency, when R becomes zero, the current I becomes infinite and the voltages V_L and V_C become infinitely large. Practically, this condition is never realized because L has some resistance R_L which may be considered in series with it.

Frequency Response of a Series Resonant Circuit

The frequency-response characteristic can be determined experimentally by injecting a constant-amplitude signal voltage into the circuit at the resonant frequency and at frequencies on both sides of resonance. The voltage across L or C is measured, and a graph of V_L or V_C versus F is plotted. This is one form of the frequency-response characteristic of the circuit.

The circuit I can also be measured at F_R and at frequencies on both sides of F_R. A graph of I versus F is another form of the frequency-response characteristic of the circuit.

In this experiment we will make a response check of V_C versus F.

Notes on Procedure

In this experiment the student will determine the resonant frequency F_R of a series LC circuit. There will be no resistance in the circuit other than the internal resistance of the coil. Resonance will therefore be determined by tuning the signal-generator frequency for maximum voltage V_L across L, or V_C across C. An electronic voltmeter connected across C will be used as the voltage indicator. One fact must be borne in mind. The input impedance of the voltmeter may affect the resonant frequency of the circuit when the meter is connected across L or C in the circuit.

The ground connections of the electronic voltmeter and the signal generator should always be at the same common point, to avoid extraneous effects which may arise from internal connections within the instruments. The component across which a voltage is to be measured is connected so that one leg is grounded. The order in which the components are connected in the series circuit will not affect circuit operation.

When loaded by a frequency-selective circuit, the voltage output of most signal generators will not remain constant as the frequency is varied. In order to compensate for this characteristic of the generator, the generator-output level must be monitored and maintained at the same voltage output, if necessary, at each new frequency setting of the instrument. Thus, a constant signal input to the circuit under test will be maintained. In this experiment the generator-output level will be monitored with an oscilloscope, and the generator-output cables will be terminated by a 100-Ω resistor to minimize generator-signal voltage variations.

SUMMARY

1. In a series RLC circuit there is a frequency F_R called the *resonant frequency*, which may be computed from the formula $F_R = 1/(2\pi\sqrt{LC})$. At resonance $X_L = X_C$.
2. At F_R, the impedance of the circuit is minimum and $Z = R$.
3. At F_R, the current I in the circuit is maximum and $I = V/R$, where V is the voltage applied across the circuit.
4. At resonance the voltages V_L across L and V_C across C are maximum and equal.
5. At resonance a series RLC circuit acts as a resistance; above resonance it is inductive, and below resonance, capacitive.

SELF-TEST

Check your understanding by answering these questions:

1. In a series resonant circuit, $X_L = 1200\ \Omega$ at resonance. The value of X_C at resonance is __________ Ω.
2. When the frequency F of the signal source applied to a series RLC circuit is higher than the resonant frequency F_R, then X_L __________ (equals/>/<) X_C and the circuit is __________.
3. In a series RLC circuit $R = 10\ \Omega$, $L = 100\ \mu H$, $C = 0.001\ \mu F$, and the applied voltage $V = 1.2$ V. The resonant frequency F_R of this circuit is __________ kHz.

4. In the circuit of question 3, the circuit current I = ____________ A at resonance.
5. In the circuit of question 3, the voltage across L, V_L = ____________ V.
6. In the circuit of question 3, the impedance Z at resonance is ____________ Ω.

MATERIALS REQUIRED

- Equipment: AF sine-wave generator, oscilloscope, EVM
- Capacitors: 0.005 μF, 0.01 μF, 0.05 μF
- Miscellaneous: 30-mH (approximately) choke
- Resistor: ½-W 100-Ω

PROCEDURE

Resonant Frequency

1. Adjust the oscilloscope for proper viewing and calibrate it for voltage measurements.
2. Set the frequency of the AF generator at 30,000 Hz and connect the oscilloscope across the output of the signal generator. Terminate the generator-output cables with a 100-Ω carbon resistor.
3. Connect the circuit of Fig. 63-2. Leave the scope connected across the generator output. Set the output of the generator at 2 V peak-to-peak. *Maintain generator-signal output voltage at this level for the remainder of the experiment.*
4. Now connect the electronic voltmeter across C as in Fig. 63-2 and observe the rms voltage V_C across C. Vary the generator frequency first on one side of 30,000 Hz and then on the other side of 30,000 Hz until the maximum voltage V_C appears across C. Read the generator frequency. This is the resonant frequency F_R of the circuit. Record this frequency in Table 63-1 in the "Measured" column.
5. Replace the 0.001-μF capacitor in Fig. 63-2 with a 0.05-μF capacitor. Following the method in steps 3 and 4, find the new resonant frequency of the circuit and record in Table 63-1.
6. Repeat step 5, using a 0.01-μF capacitor.
7. For each value of L and C used, compute the resonant frequency of the circuit using Eq. (63-5). Record each value of F_R in the column labeled "Computed."

Frequency Response

8. Leave the circuit connected as in step 6 and again find the resonant frequency F_R. (It should be the same frequency as in step 6.) Record in Table 63-2.
9. With the oscilloscope measure the generator (still connected across the circuit) output and readjust the generator level, if necessary, for 2 V peak-to-peak. *Maintain generator output at this level throughout.*
10. Now measure the voltage V_C across C and record in Table 63-2.
11. Fill in the "Frequency" column in Table 63-2, performing the required frequency additions and subtractions.
12. Make a frequency-response check of the circuit by setting the generator frequency where possible to the frequencies listed in Table 63-2, and noting and recording the voltage V_C across C for each frequency setting.

TABLE 63-1. Resonant Frequency of a Series LC Circuit

Step	L, mH	C, μF	*Resonant Frequency* (Hz) *Measured*	*Computed* $F_R = \frac{1}{2\pi\sqrt{LC}}$
4/7	30	0.001		
5/7	30	0.05		
6/7	30	0.01		

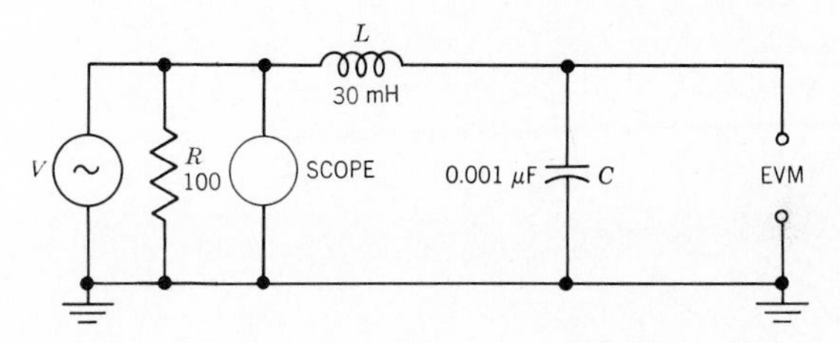

Fig. 63-2. Determining the resonant frequency of an LC circuit.

TABLE 63-2. Frequency Response Data

Frequency Deviation, Hz	*Frequency*, Hz	V_C, V rms
$F_R - 6000$		
$F_R - 5000$		
$F_R - 4000$		
$F_R - 3000$		
$F_R - 2000$		
$F_R - 1000$		
$F_R - 500$		
Resonance F_R		
$F_R + 500$		
$F_R + 1000$		
$F_R + 2000$		
$F_R + 3000$		
$F_R + 4000$		
$F_R + 5000$		
$F_R + 6000$		
v, volts peak-to-peak		2

QUESTIONS

1. In a series *RLC* circuit, under what conditions is it possible for the voltages V_L and V_C to be higher than the applied generator voltage *V*?
2. Under the conditions in question 1, what limits V_L and V_C?
3. What is the impedance of a series *LC* circuit at resonance?
4. Outline a method for computing the inductance *L* of the coil from the experimental data in this experiment, using Eq. (63-5) for the resonant frequency of a series circuit.
5. Find the inductance *L* by the method described in question 4. Use the data from Table 63-1.
6. Using the data in Table 63-2, draw a frequency-response curve showing V_C as a function of *F*. What type of curve is this?
7. In making a frequency-response check, why is it necessary to maintain a fixed generator output over the range of frequencies used?
8. How do the measured values of F_R compare with the computed values in Table 63-1? Explain any discrepancies.

Answers to Self-Test

1. 1200
2. >; inductive
3. 503,292 Hz
4. 0.12
5. 37.9
6. 10

64

EXPERIMENT

EFFECT OF Q ON FREQUENCY RESPONSE AND BANDWIDTH OF A SERIES RESONANT CIRCUIT

OBJECTIVES

1. To measure the effect of circuit Q on frequency response
2. To measure the effect of circuit Q on bandwidth at the half-power points

INTRODUCTORY INFORMATION

Circuit Q and Frequency Response

In Experiment 63 we studied the response curve of a series resonant LC circuit. In that experiment the only resistance in the circuit was that associated with the coil. Theoretically, at resonance $X_L = X_C$ and the impedance $Z = R_L$, where R_L equals the resistance of the coil.

The amount of coil resistance R_L, therefore, determines the current flow through the circuit at resonance if there is no external resistance. The R_L and X_L of the coil (assuming no other resistance) determine the quality or Q of the circuit, which is given by the formula

$$Q = \frac{X_L}{R_L} \tag{64-1}$$

The Q of the circuit also determines the rise in voltage across L and C at the resonant frequency F_R. The voltage developed across L is given by the equations

$$V_L = IX_L = \frac{V}{R} \times X_L \tag{64-2}$$

If the circuit resistance R is the resistance R_L associated with the coil, then

$$V_L = V \times \frac{X_L}{R_L} = VQ \tag{64-3}$$

Also, since $X_L = X_C$ at resonance,

$$IX_L = IX_C$$

and $\quad V_L = V_C$

Therefore $\quad V_C = VQ \qquad (64-4)$

Equations (64-3) and (64-4) become significant for values of $Q > 1$. For such values V_C and V_L are greater than the applied voltage V. Obviously, also, the higher the value of Q, the greater the voltage gain in the circuit. This is the *first* example of voltage gain.

Q is also significant when we consider the frequency response of a series resonant circuit. The frequency-response characteristic can be determined by injecting a constant-amplitude signal voltage V into the circuit at the resonant frequency and at frequencies on either side of resonance. The voltage across L or C is measured, and a graph of V_L or V_C versus F is plotted. This is one form of the frequency-response characteristic of the circuit.

The circuit current I can also be determined. A graph of I versus F is another form of the frequency-response characteristic of the circuit.

Circuit Q and Bandwidth

Figure 64-1 is a graph of the response of a series resonant circuit. Three significant points have been marked on the curve. These are F_R, the resonant frequency, and F_1 and F_2. The latter points are located at 70.7 percent of maximum (maximum is at F_R) on the curve. They are called the *half-power points*, and the frequency separation between them is $F_2 - F_1$. This frequency separation is called the bandwidth B of the circuit. Bandwidth, therefore, is given by the equation

$$B = F_2 - F_1 \tag{64-5}$$

Bandwidth is related to Q. It can be shown that B is given by the equation

$$B = \frac{F_R}{Q} \tag{64-6}$$

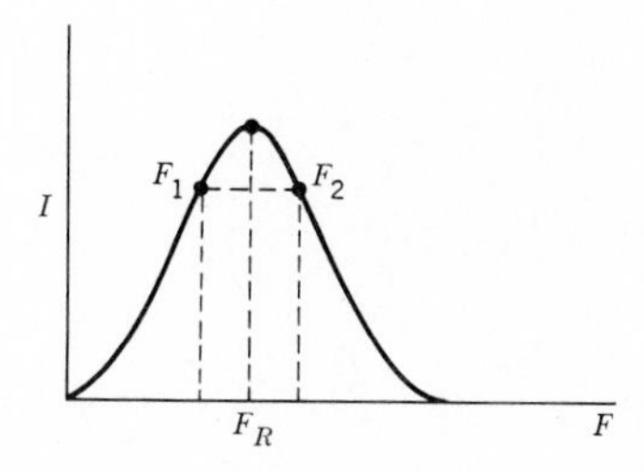

Fig. 64-1. Response curve of a series-resonant circuit.

As was noted in the preceding experiment, the resonant frequency F_R of a series *RLC* circuit can be computed from the formula

$$F_R = \frac{1}{2\pi\sqrt{LC}} \qquad (64\text{-}7)$$

where F_R is in hertz, L in henrys, and C in farads. Since the formula does not include R, it is apparent that the resonant frequency of the series *RLC* circuit is not affected by the size of the resistance R. What R does affect, however, is the bandwidth and amplitude of the response curve. The larger the value of R, the lower is the value of Q [see Eq. (64-1)], and the wider is the bandwidth. Moreover, the lower the value of Q, the lower is the value of current I in the circuit, and the lower are the voltages V_L across L and V_C across C.

Series resonant circuits are used in radio, television, and industrial electronics as frequency-selective circuits and as traps to eliminate unwanted signals. Normally such circuits require a highly peaked response with narrow bandwidth. To achieve such a response, the Q of the circuit must be high. Hence high-Q coils are employed. In such circuits the Q of the circuit is mainly determined by the Q of the coil.

However, there are some applications in electronics where wideband frequency-selective circuits are required. In such cases coil "loading" is achieved by the use of external resistors.

Notes on Experimental Procedure

In this experiment you will determine the resonant frequency F_R of a series *RLC* circuit. An external resistor R will be inserted in series with L and C to simulate the coil resistance. The effect of the value of R on the amplitude of current I will be investigated. The effect of the value of R on frequency response and bandwidth will also be determined.

You will determine resonance by tuning the signal-generator frequency for maximum voltage V_R across R, or V_C across C. You will again use the oscilloscope as a voltage indicator to monitor the generator output. You will use an electronic voltmeter to measure the ac voltage across C or R. As was noted in the previous experiment, the input impedance of the voltmeter may affect the resonant frequency of the circuit when the scope is placed across C or R.

Also noted in the preceding experiment, *the ground connections of the meter, the scope, and the signal generator must always be at the same common point, to avoid extraneous effects which may arise from internal connections within the instruments.*

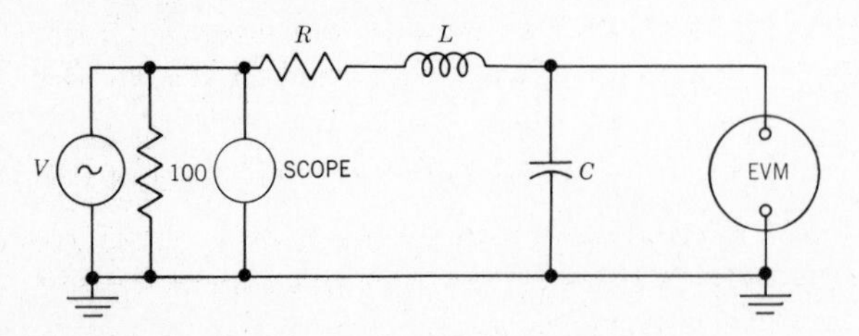

Fig. 64-2. Effect of Q on frequency response.

Also, as in the preceding experiment, the signal-generator output must be maintained at a constant level. To accomplish this, the output voltage of the generator will be monitored with the oscilloscope. As the generator is turned to each new frequency, manual adjustments will be made in the output level, if necessary. The generator-output cables will also be terminated by a 100-Ω carbon resistor.

SUMMARY

1. The Q or quality of a coil is defined by the formula $Q = X_L/R$.
2. In a series resonant circuit, where the only resistance in the circuit is coil resistance R_L, the Q of the circuit is determined by the Q of the coil.
3. In a series resonant circuit the voltages across L and C are equal and are affected by the Q of the circuit. These voltages may be computed from the formulas

$$V_L = V_C = V \times Q$$

where V is the applied voltage and V_L and V_C are the voltages across L and C, respectively.
4. If circuit Q is > 1, then the voltages V_L and V_C are larger than the applied voltage. Hence, in a series resonant circuit we find the first example of *amplification*.
5. Bandwidth B of a response curve is defined as the difference between the two frequencies F_2 and F_1 at the half-power points (Fig. 64-1). The half-power points are those points which appear at 70.7 percent of maximum on the response curve. Maximum output occurs at the resonant frequency F_R.
6. Bandwidth is related to circuit Q and is given by the formula

$$B = \frac{F_R}{Q}$$

7. The lower the Q of the circuit, the wider the bandwidth and the "flatter" the response curve.
8. The lower the Q of the circuit, the lower the amplitude of the response curve, and therefore the circuit gain is lower.

SELF-TEST

Check your understanding by answering these questions.

In the circuit of Fig. 64-2, $L = 50$ mH, $C = 0.01$ μF, and $R = 20$ Ω. Assume that the resistance of the coil $R_L = 0$. The applied voltage $V = 2.5$ V. The output voltage is V_C. Answer the following questions concerning this circuit.

1. The resonant frequency F_R = ____________ kHz.
2. The X_L of the coil at resonance is ____________ Ω.
3. The Q of the circuit is ____________ at resonance.
4. The voltage V_L across L is ____________ V at resonance.
5. The bandwidth of the circuit is ____________ Hz.
6. The voltage V_C across C at the half-power points is ____________ V.

MATERIALS REQUIRED

- Equipment: AF signal generator, the same as in Experiment 63, oscilloscope, EVM
- Resistors: ½-W 10- and 47-Ω, and two 100-Ω
- Capacitors: 0.01 μF
- Miscellaneous: 30 mH (approximate) choke, the same as in Experiment 63

PROCEDURE

Effects of Circuit Q on the Frequency Response of a Series Resonant Circuit

1. Connect the circuit of Fig. 64-2. $R = 10\ \Omega, L = 30$ mH (approximately), and $C = 0.01\ \mu$F.
2. Set the AF generator output at 1 V rms and vary the generator frequency until the maximum voltage V_C (as measured by the EVM) appears across C. Note the generator frequency F_R at this point and record it in Table 64-1. If necessary, reset the generator output v_G at 1 V rms at F_R.

NOTE: In the response check which follows, it is necessary to maintain the output of the generator at this constant level, 1 V rms, over the range of frequencies $F_R \pm 6$ kHz. *The voltage v_G (1 V rms) will be used as the standard voltage for setting the signal-generator attenuator for the remainder of this experiment. With the scope across the output cable of the generator (terminated by a 100-Ω carbon resistor), which is connected across the circuit under test, the generator attenuator is adjusted for the standard voltage v_G each time the frequency of the generator is changed.*

3. The generator frequency is still set at F_R output at 1 V rms. The EVM is still connected across C. Measure, and record in Table 64-1, the rms voltage V_C across C.
4. Fill in Table 64-1 with the proper information in the column labeled "Frequency," performing the additions and subtractions noted in the column labeled "Frequency deviation, Hz."
5. Make a frequency-response check of the circuit by setting the generator frequency, where possible, to the frequencies listed in Table 64-1. Measure and record in Table 64-1 the voltage V_C across C for each frequency setting.
6. Replace R by a 47-Ω resistor and repeat steps 3 through 5. Record your data in Table 64-2.
7. Replace R by a 100-Ω resistor and repeat steps 3 through 5. Record your data in Table 64-3.

Effect of R on Resonant Frequency

8. Rearrange the circuit as in Fig. 64-3. $R = 10\ \Omega$.
9. Keeping the generator output at 1 V rms, vary the generator frequency until circuit resonance is indicated

TABLE 64-1. Frequency Response, $R = 10\ \Omega$

Frequency Deviation, Hz	Frequency, Hz	V_C, V rms
$F_R - 6000$		
$F_R - 5000$		
$F_R - 4000$		
$F_R - 3000$		
$F_R - 2000$		
$F_R - 1000$		
$F_R - 500$		
Resonance F_R		
$F_R + 500$		
$F_R + 1000$		
$F_R + 2000$		
$F_R + 3000$		
$F_R + 4000$		
$F_R + 5000$		
$F_R + 6000$		
v_G, V rms:		

TABLE 64-2. Frequency Response, $R = 47\ \Omega$

Frequency Deviation, Hz	Frequency, Hz	V_C, V rms
$F_R - 6000$		
$F_R - 5000$		
$F_R - 4000$		
$F_R - 3000$		
$F_R - 2000$		
$F_R - 1000$		
$F_R - 500$		
Resonance F_R		
$F_R + 500$		
$F_R + 1000$		
$F_R + 2000$		
$F_R + 3000$		
$F_R + 4000$		
$F_R + 5000$		
$F_R + 6000$		
v_G, V rms:		

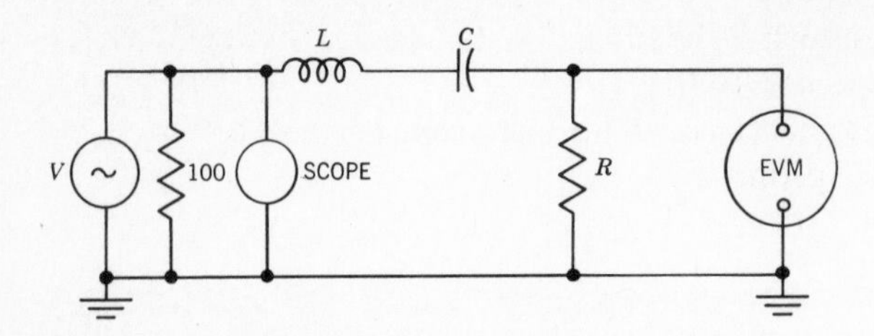

Fig. 64-3. Impedance of a series-resonant circuit.

by a maximum voltage reading across R. Note the resonant frequency and V_R at resonance, and record them in Table 64-4.

10. Repeat steps 8 and 9 for $R = 47\ \Omega$ and $R = 100\ \Omega$.
11. Measure and record in Table 64-4 the resistance of the coil.

Graphs of Response Curves

12. Using the data in Tables 64-1 to 64-3, graph the response curves of the circuits under test. Identify each curve. These graphs indicate the effects of Q on the frequency-response characteristics of the series LC circuit in the range of resonance.
13. In the graphs you have drawn, determine and mark the half-power points. Determine the bandwidth of each graph, and compute the circuit Q in each case. Show your computations.

Extra Credit

14. Refer to Fig. 64-3. State how you could determine experimentally the phase relationship between applied voltage v_G and current I at resonance. Perform this measurement and show your results.

TABLE 64-3. Frequency Response, $R = 100\ \Omega$

Frequency Deviation, Hz	*Frequency,* Hz	V_C, V rms
$F_R - 6000$		
$F_R - 5000$		
$F_R - 4000$		
$F_R - 3000$		
$F_R - 2000$		
$F_R - 1000$		
$F_R - 500$		
Resonance F_R		
$F_R + 500$		
$F_R + 1000$		
$F_R + 2000$		
$F_R + 3000$		
$F_R + 4000$		
$F_R + 5000$		
$F_R + 6000$		
v_G, V rms:		

TABLE 64-4. Effect of R on F_R

Resistance of Coil	R, Ω		
	10	47	100
Resonant frequency, Hz			
V_R at resonance			

QUESTIONS

1. At resonance, what is the theoretical relationship between V_R and the applied generator voltage, assuming that L has zero resistance? Is this relationship confirmed by the measurements in Table 64-4? If not, explain why.
2. What is meant by the Q of a series resonant circuit?
3. What effect does Q have on sharpness of frequency response of a series LC circuit?
4. Using the measured value of F_R in Tables 64-1 to 64-3, what is the Q of the series resonant circuit when the external resistance of R is
   (*a*) 10 Ω?
   (*b*) 47 Ω?
   (*c*) 100 Ω?
   Note that the approximate value of circuit resistance R_L is $R_L = R_{coil} + R$, where R is the external resistor, and R_{coil} is the measured coil resistance.
5. The half-power points are defined by the frequency at which the output voltage is 70.7 percent of the output voltage at resonance. What is the frequency difference at the half-power points for $R = 10, 47,$ and $100\ \Omega$? Use the measured value of F_R in Tables 64-1 to 64-3.
6. What effect, if any, did R have on the resonant frequency in steps 8 to 10?
7. At resonance what is the phase relationship between applied voltage and current?

Answers to Self-Test

1. 7.118
2. 2236
3. 111.8
4. 279.5
5. 63.7
6. 197.6

65

EXPERIMENT

CHARACTERISTICS OF PARALLEL RESONANT CIRCUITS

OBJECTIVES

1. To determine experimentally the resonant frequency of a parallel *RLC* circuit
2. To measure the line current and impedance at the resonant frequency
3. To measure the effect of variations in circuit frequency on the impedance of a parallel *RLC* circuit

INTRODUCTORY INFORMATION

Characteristics of a Parallel Resonant Circuit

Resonant Frequency of a High-*Q* Circuit

Consider the circuit of Fig. 65-1*a*, consisting of *C* and *L* in parallel. This theoretical circuit must be modified as in Fig. 65-1*b* to reflect the resistance R_L of coil *L*. It is assumed here that the *Q* of this circuit is high (that is, that R_L is small compared with X_L) and that the resistance of *C* and the wiring resistance of the circuit are negligible and may be ignored.

There is a particular frequency at which $X_L = X_C$. This frequency may be defined as the condition for parallel resonance in a high-*Q* circuit and is similar to the condition for series resonance.

There are other definitions for parallel resonance. Thus parallel resonance may be considered as the frequency at which the impedance of the parallel circuit is maximum. Also, parallel resonance may be considered as the frequency at which the parallel impedance of the circuit has unity power factor. These three definitions may lead to three different frequencies, each of which may be considered as the resonant frequency. In circuits whose *Q* is greater than 10, however, the three conditions lead to the same resonant frequency.

In a high-*Q* circuit, the formula for the resonant frequency F_R is the same as in the case of series resonance and is given by

$$F_R = \frac{1}{2\pi\sqrt{LC}} \qquad (65\text{-}1)$$

Line Current

Consider again the circuit of Fig. 65-1*b*. If the resistance R_L of the inductance *L* is small, then at resonance the impedance X_C of the capacitive branch is practically equal to the impedance

$$\sqrt{X_L^2 + R_L^2}$$

of the inductive branch. The current in each branch may therefore be considered equal. These currents, however, are practically 180° out of phase. Hence, their phasor sum, which is the line current, is very small. Moreover, since the impedance which this parallel resonant circuit presents to the circuit is defined as $Z = V/I_{\text{line}}$, it is apparent that the impedance is very high (since line current is low). As opposed to line current, however, the circulating current in the parallel resonant circuit is high at resonance.

Frequency Response

The characteristics of a parallel resonant circuit to frequencies on either side of the resonant frequency may now be analyzed. Thus, for a frequency F_a higher than F_R, the reactance X_C of the capacitive branch is lower than the reactance X_L of the inductive branch. There is therefore a greater capacitive than inductive current, and the circuit may be considered capacitive. Similarly, it may be shown that for a frequency F_b lower than F_R, the circuit is inductive. The

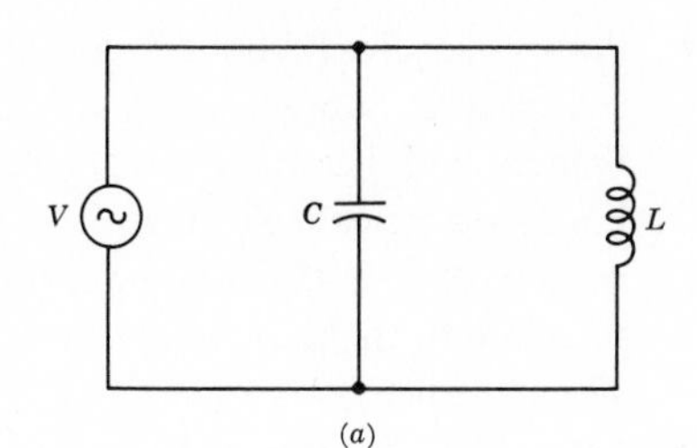

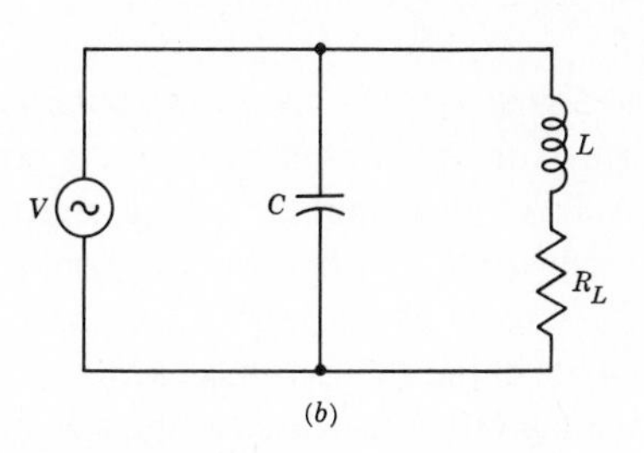

Fig. 65-1. Parallel-resonant circuits.

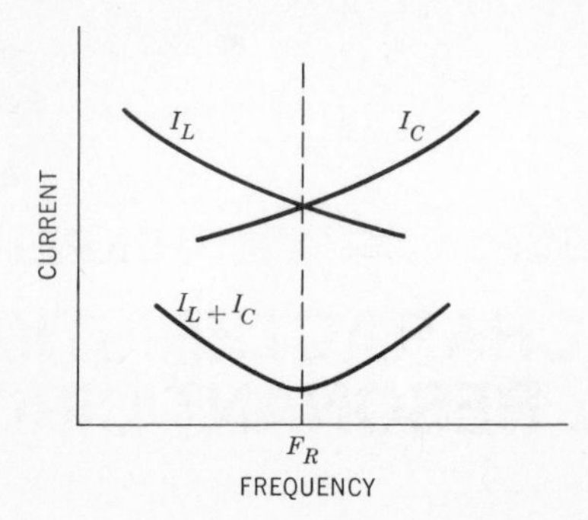

Fig. 65-2. Current versus frequency in a parallel-resonant circuit.

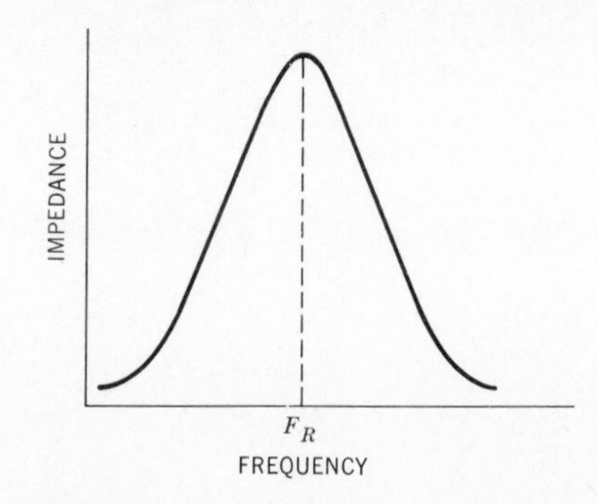

Fig. 65-3. Impedance versus frequency in a parallel-resonant circuit.

graph of Fig. 65-2 illustrates these relationships. Individual graphs for I_L, I_C, and I_{line} are shown. It is evident that at resonance line current is minimum, that I_L equals I_C, and that I_L and I_C are each greater than line current.

Impedance

Figure 65-3 is a graph of impedance versus frequency in a parallel resonant circuit. This graph resembles the frequency-response characteristic of a series resonant circuit. It shows that circuit impedance is maximum at resonance and falls off on either side of resonance.

In this experiment you will determine the resonant frequency F_R of a parallel LC circuit (see Fig. 65-4). A resistor R is inserted in series with the parallel LC circuit. You will measure the signal voltage V_R across R. You can then compute the line current from the relationship

$$I_R = \frac{V_R}{R} = I_{\text{line}} \qquad (65\text{-}2)$$

You will determine resonance by tuning for minimum voltage V_R across R, and hence for minimum line current. You will then vary the signal generator on either side of resonance, monitoring and maintaining a constant generator output. By measuring V_R, computing I_{line}, and plotting the values of I_{line}, you will secure a graph of line current versus frequency.

You will measure the signal voltage V_T across the tank circuit at resonance and for frequencies on both sides of resonance. You will then calculate the value of the impedance Z of the tank circuit from the relationship

$$Z = \frac{V_T}{I_{\text{line}}} \qquad (65\text{-}3)$$

and will plot a graph for the Z of the tank circuit from the calculated values of Z.

You will find that the signal voltage across the parallel resonant (tank) circuit is maximum at resonance. It falls off on either side of resonance. The amplitude of the signal across the tank circuit is directly proportional to the Q of the circuit. If we "load" the circuit by placing a resistor in parallel with it, the amplitude of the signal voltage across the circuit decreases. The loading effect increases as the size of the parallel resistor decreases.

You will also observe the reactive characteristic of the tank circuit as the signal-generator frequency is varied on both sides of resonance; that is, you will observe whether the circuit is inductive or capacitive for frequencies on both sides of resonance.

In this experiment an audio oscillator will be used as the signal source, since the values of L and C selected provide a circuit whose resonant frequency is in the AF range.

SUMMARY

1. In a parallel LC circuit, if the Q of the circuit is greater than 10, the resonant frequency of the circuit may be computed from the formula

$$F_R = \frac{1}{2\pi\sqrt{LC}}$$

At this frequency $X_L = X_C$.

2. In a high-Q circuit the impedance of the inductive branch equals (approximately) the impedance of the capacitive branch.
3. Since the voltage across each of the parallel branches in a high-Q circuit is the same, and the impedances of both branches are equal, the currents in both branches are equal. That is, $I_L = I_C$.
4. The currents, however, are approximately 180° out of phase, since one is capacitive and the other inductive. They therefore tend to cancel each other, and the line current, which is the phasor sum of I_C and I_L, is therefore very low, approximately zero in a very high-Q circuit.
5. The impedance of a high-Q parallel resonant LC circuit is therefore very high, since $Z = V_{\text{applied}}/I_{\text{line}}$ and I_{line} is approximately zero.
6. For frequencies higher than F_R, the current in the capacitive branch is higher than the current in the inductive branch, and the circuit acts like a capacitive circuit.
7. For frequencies lower than F_R, current in the inductive branch is higher than current in the capacitive branch, and the circuit acts like an inductance.
8. The impedance-response curve of a high-Q parallel resonant circuit resembles the frequency-response curve of a series resonant circuit.

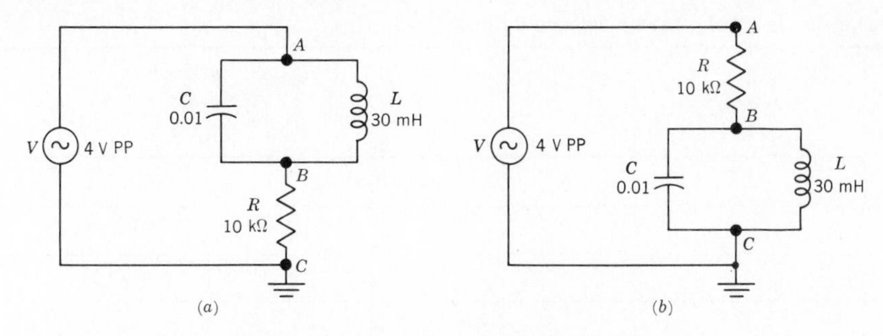

Fig. 65-4. Experimental parallel-resonant circuits.

SELF-TEST

Check your understanding by answering these questions:

1. The Q of a parallel resonant circuit in Fig. 65-1*a* is 100. At resonance the current in the inductive branch is 50 mA and the voltage across the tank circuit is 100 V. The current in the capacitive branch is ____________ mA.
2. In the parallel resonant circuit of Fig. 65-4, the line current at resonance is 0.05 mA. The voltage across the tank circuit is 2.5 V. The impedance of the tank circuit is ____________ Ω.
3. For the circuit in question 2, at frequencies higher than F_R, the impedance of the parallel resonant circuit is ____________ (higher/lower) than the impedance at F_R.
4. In the circuit of Fig. 65-4, the resistance R_L of L is 12 Ω. The resonant frequency of the circuit F_R = ____________ Hz.
5. The Q of the coil in the circuit of question 4 is ____________.
6. The voltage across R, in Fig. 65-4, is 1.5 V at resonance. The line current in the circuit is ____________ mA.

MATERIALS REQUIRED

- Equipment: Audio generator, oscilloscope
- Resistors: Two ½-W, 33-Ω, and 10,000-Ω
- Capacitor: 0.01 μF
- Miscellaneous: 30-mH (approximately) choke

PROCEDURE

1. Resonant Frequency of a Parallel *LC* Circuit

a. Connect the circuit of Fig. 65-4*a*.

b. Adjust the scope controls for proper viewing and the AF generator attenuator controls for a 4-V peak-to-peak signal output. Connect the scope across the 10,000-Ω resistor, points B and C, with the ground of the scope going to the ground of the circuit.

c. Set the generator frequency at 10,000 Hz. Observe the response on the scope.

d. Increase or decrease the generator frequency until a frequency F_R is reached where minimum signal voltage V_R appears across R. This point may be identified by the fact that V_R increases on either side of F_R. F_R is the resonant frequency of the circuit. Record in Table 65-1.

2. Line Current and Impedance of a Parallel Resonant Circuit

a. Check the generator output v and maintain it at 4 V peak-to-peak during the remainder of this exercise.

b. Measure the voltage v_R across R (BC) at resonance, and record it in Table 65-1.

c. Adjust the generator frequency in turn to each of the frequencies listed in Table 65-1, maintaining a 4-V peak-to-peak output signal v. At each frequency, measure and record v_R.

d. Connect the circuit of Fig. 65-4*b* and repeat steps **a**, **b**, and **c** for voltage measurements v_T across the tank circuit (BC).

e. Convert each peak-to-peak value of v_R into its rms value. Compute the value of line current, $I_{\text{line}} = V_R(\text{rms})/R$, and record in Table 65-1.

f. From the computed values of I_{line}, draw a graph of line current versus frequency for the parallel resonant circuit.

g. Convert each peak-to-peak value of v_T into its rms value. From the rms value of V_T and the computed value of line current, compute the value of impedance Z of the parallel resonant circuit for every frequency in Table 65-1. Record in Table 65-1.

h. Draw a graph of Z versus frequency.

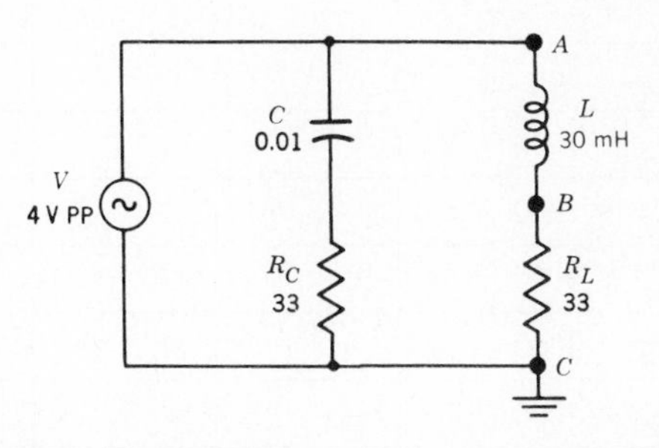

Fig. 65-5. Circuit to determine current in the inductive and capacitive legs of a parallel-resonant circuit.

TABLE 65-1. Response Characteristics of a Parallel Resonant Circuit

Frequency Deviation, Hz	*Frequency,* Hz	v_R	v_T	I_{line} *(computed)* = $\frac{V_R\ (rms)}{R}$	*Z (computed)* = $\frac{V_T\ (rms)}{I_{\text{line}}}$
$F_R - 6000$					
$F_R - 5000$					
$F_R - 4000$					
$F_R - 3000$					
$F_R - 2000$					
$F_R - 1000$					
$F_R - 500$					
F_R					
$F_R + 500$					
$F_R + 1000$					
$F_R + 2000$					
$F_R + 3000$					
$F_R + 4000$					
$F_R + 5000$					
$F_R + 6000$					

TABLE 65-2. Reactance of a Parallel Resonant Circuit

Frequency Deviation, Hz	*Frequency,* Hz	v_{R_L}	v_{R_C}	*Current in L,* mA	*Current in C,* mA
$F_R - 6000$					
$F_R - 5000$					
$F_R - 4000$					
$F_R - 3000$					
$F_R - 2000$					
$F_R - 1000$					
$F_R - 500$					
F_R					
$F_R + 500$					
$F_R + 1000$					
$F_R + 2000$					
$F_R + 3000$					
$F_R + 4000$					
$F_R + 5000$					
$F_R + 6000$					

3. Reactance of a Parallel Resonant Circuit at Frequencies Above and Below Resonance

a. Connect the circuit of Fig. 65-5. Use the same values of L and C as in parts 1 and 2. Set the generator output at 4 V peak-to-peak and maintain it at this value.

NOTE: The resonant frequency of this circuit may be assumed to be the same as in Fig. 65-4.

b. Measure the voltages v_{R_L} across R_L and v_{R_C} across R_C at resonance, and record in Table 65-2.

c. Measure and record the voltages across R_L and R_C for every frequency shown.

d. Compute the current in the inductive branch from the relationship $I_L = V_{R_L}(\text{rms})/R_L$ for every frequency shown, and record in Table 65-2.

e. Compute the current in the capacitive branch from the relationship $I_C = V_{R_C}(\text{rms})/R_C$ for every frequency shown, and record in Table 65-2.

f. Draw graphs of I_L and I_C versus frequency from the computed values in Table 65-2.

QUESTIONS

1. What determines the Q of a parallel RLC circuit?
2. What is meant by a high-Q parallel RLC circuit?
3. What is the formula for resonance of a high-Q parallel RLC circuit?
4. How does line current vary with frequency in a parallel RLC circuit?
5. How does impedance vary with frequency in a parallel RLC circuit?
6. What is the nature of the current (that is, is it inductive, capacitive, or resistive) of a parallel RLC circuit
   (*a*) at resonance?
   (*b*) for frequencies higher than resonance?
   (*c*) for frequencies lower than resonance?

Answers to Self-Test

1. 50
2. 50,000
3. lower
4. 9189
5. 144
6. 0.15

66

EXPERIMENT

LOW-PASS AND HIGH-PASS FILTERS

OBJECTIVES

1. To determine experimentally the frequency response of a low-pass filter
2. To determine experimentally the frequency response of a high-pass filter

INTRODUCTORY INFORMATION

Frequency Filters

Frequently electronic signal currents consist of two or more frequency components. For example, the signal of a broadcast radio station consists of a radio-frequency carrier and audio-frequency intelligence which is impressed on the carrier by a process called *modulation*. The modulated signal, consisting of the high-frequency carrier and the sidebands which form a modulated wave, together carrying the intelligence contained in the low-frequency audio components, is broadcast by the radio transmitter and selected by a radio receiver. The high-frequency carrier, a unique frequency assigned to a specific station, makes it possible for the receiver to tune in or select the broadcast station. But once in the receiver, the high-frequency carrier is no longer required. Only the amplitude variations of the modulated wave at the audio-frequency rate are needed. In one of the stages of the receiver the carrier frequency is eliminated; the audio variations are obtained, amplified, and left to actuate the loudspeaker. Processes called *detection* and frequency filtering are used to accomplish this.

There are many other examples of frequency complex signals in electronics. A television-station signal is one example. An FM station signal is another. In all cases frequency-selective filters are used to separate the wanted and unwanted signal currents, passing on the required frequency components to the circuits which process them.

There are different types of filters. Some are highly selective and will permit a single frequency or a very narrow band of frequencies through, attenuating all others. These are called *narrowband filters*. Others, called *wideband filters*, will pass a wide band of frequencies while attenuating all others. Filters have other classifications, including those classified as *low-pass* and *high-pass* filters. We will examine the characteristics of the latter two types in this experiment.

High-Pass Filter

Filtering is accomplished by the action of capacitors, inductors, and resistors, in combination. *LC* resonant circuits are one example of frequency filters, for these highly selective circuits favor one frequency over all others.

A capacitor, theoretically, offers an infinite reactance to a zero-frequency signal, that is, to direct current. Therefore, if a capacitor is placed in series with a load resistance R_L in Fig. 66-1*b*, it will pass the ac component of a complex signal but block the dc component of that signal. Figure 66-1*a* shows a 6-V peak-to-peak ac signal combined with 5 V dc. The resulting combined signal varies between +8 and +2 V. Figure 66-1*b* shows how a coupling capacitor *C* is connected in the circuit to block the dc but permit the ac component through. Figure 66-1*c* shows the pure ac sine wave across *R*. Note that the +5 V dc has been filtered by capacitor *C*.

The reactance of a capacitor varies inversely with frequency. This characteristic is used to pass certain frequencies but reject others. Consider the circuit of Fig. 66-2*a*. The level of ac voltage v_R coupled by capacitor *C* to resistor *R* is

$$v_R = V \cos\theta = V \times \frac{R}{Z}$$

where $$\theta = \arctan \frac{X_C}{R} \qquad (66\text{-}1)$$

The angle θ depends on the relative values of *C* and *R*, and on the frequency *F* of the applied signal source. Assume there are three frequencies in the complex signal *V*, namely 159 Hz, 1590 Hz, and 15,900 Hz. The reactance X_C of *C* at each of these frequencies is, respectively, 10,000, 1000, and 100 Ω. The voltage V_R delivered to *R* at 159 Hz is less than 10 percent of *V*. At 1590 Hz (the frequency at which $X_C = R$) over 70 percent of *V* is delivered to *R*, and at 15,900 Hz over 99 percent of *V* is delivered to *R*. It is evident, therefore, that *C* will permit the higher frequencies to reach *R* with a minimum of attenuation, but will attenuate the lower frequencies. This is one type of high-pass filter.

The circuit of Fig. 66-2*a* can be modified by adding an inductor *L* in parallel with the load resistor *R* to form another high-pass filter (Fig. 66-2*b*). *L* has a low reactance at the low frequencies and effectively reduces the output voltage across *R* at frequencies for which $X_L < R/10$ (approximately). At higher frequencies the value of X_L increases. For frequencies

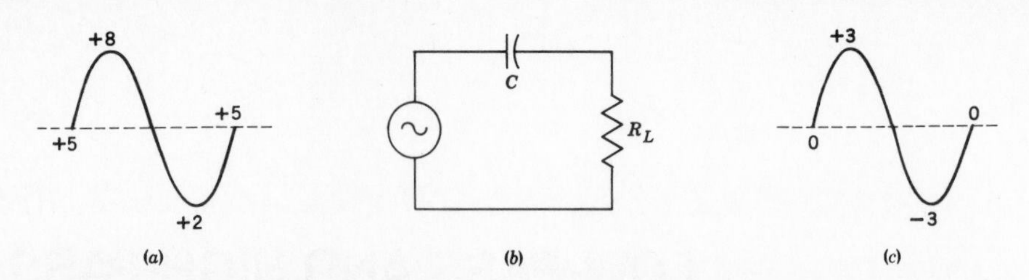

Fig. 66-1. (*a*) A 6-V signal imposed on a +5-V dc axis. The resulting combined signal varies between +8 and +2 V. (*b*) The combined dc and ac signal is connected in series with a capacitor *C* and a resistor R_L. (*c*) Capacitor *C* blocks the dc component and permits only the ac component through. Signal variation across R_L is between +3 and −3 V, a pure ac voltage.

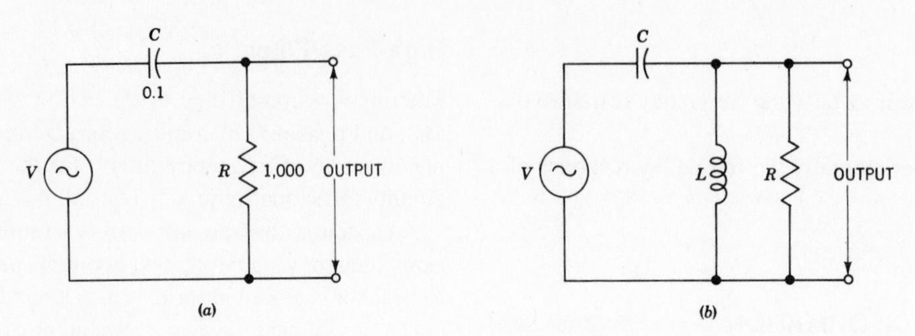

Fig. 66-2. High-pass filters. (*a*) *C* blocks the lower- and passes the higher-frequency components of *V* to *R*. (*b*) Adding *L* modifies the action of the filter.

at which $X_L > 10R$, the parallel impedance of X_L and *R* approaches *R*, stabilizing the load at the value *R*. The inclusion of inductor *L* modifies the response characteristics of the circuit, which is still a high-pass filter.

Low-Pass Filter

The reactance of an inductor varies directly with frequency. In Fig. 66-3 this characteristic of *L* is used to pass the low frequencies to *R*, but reject the high frequencies if the input contains high and low frequencies. As in the case of the *RC* circuit in Fig. 66-2, the *LR* circuit in Fig. 66-3 constitutes a voltage divider. The amplitude of the voltage delivered by *V* to *R* will depend on the inductance of *L*, the resistance of *R*, and the frequency of *V*. As frequency increases, X_L increases and the voltage across *R* decreases. As frequency decreases, X_L decreases and the voltage across *R* increases. This is an example of a low-pass filter.

An example of another low-pass filter is shown in Fig. 66-4*a*. The addition of capacitor *C* in this circuit increases the filtering of higher frequencies. For as the frequency *F* of the input signal increases, not only does X_L increase, reducing the output voltage (V_{out}), but X_C decreases, reducing the impedance of the output circuit, thus further reducing V_{out}.

If *L* is replaced by a resistor R_1, as in Fig. 66-4*b*, the result will still be a low-pass filter because of the bypass (attenuating) action of capacitor *C*. Where the cut-off high frequency starts depends on the value of the two resistors and of the capacitor *C*. We will use here the definition of cutoff frequency as that frequency whose output voltage is just 70.7 percent of maximum output, that is, one whose attenuation is at least 29.3 percent of maximum.

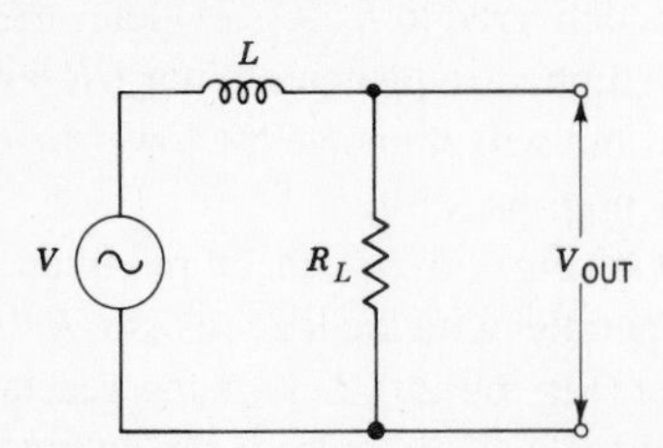

Fig. 66-3. Low-pass filter. *L* blocks the higher and passes the lower frequencies to R_L.

SUMMARY

1. Complex electronic currents contain many frequency components. Electric filters are used to separate these frequency components, attenuating those that are unwanted and passing on those that are required.
2. Electric filters consist of combinations of capacitors, inductors, and resistors. The reactance of a capacitor is inversely proportional to the frequency of the signal in the circuit. The reactance of an inductor is directly proportional to the frequency.
3. Figures 66-2*a* and *b* show examples of high-pass filters. The circuits are basically voltage dividers. With increasing frequency, X_C decreases, and the output voltage across *R* is accordingly higher.
4. Figures 66-3 and 66-4 are low-pass filters. Here the voltage-divider action is affected by the increase in X_L with an increase in frequency. Accordingly, the output voltage across *R* is greater for low frequencies than it is for high

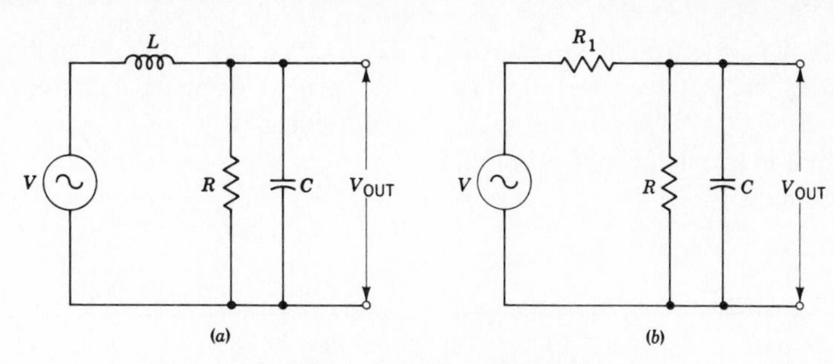

Fig. 66-4. Another low-pass filter.

frequencies. High frequencies are attenuated, and low frequencies are passed on to the output.

5. The terms low and high frequencies are relative. The frequency-response characteristic of the circuits depends on the circuit parameters, that is, on the values of L, C, and R, and on the circuit configuration. In all cases, the value of output voltage for each frequency component can be computed, using the methods of ac circuit analysis.

SELF-TEST

Check your understanding by answering these questions:

1. Circuits which attenuate some frequencies and pass others are called __________ __________.
2. DC can be separated from ac signals by means of a __________ __________.
3. In the high-pass circuit of Fig. 66-2*a*, the frequency at which the output voltage across R is 70.7 percent of maximum is the frequency at which X_C = __________ Ω.
4. The cutoff frequency in the filter of Fig. 66-2*a* is __________ Hz.
5. In the low-pass filter of Fig. 66-3, the output voltage across $R_L = V \times$ __________ $= V \times$ __________.
6. In the circuit of Fig. 66-3, if R = 500 Ω and L = 2 H, all frequencies higher than __________ Hz will be attenuated more than 29.3 percent; that is, the cutoff frequency is __________ Hz.

MATERIALS REQUIRED

- Equipment: AF sine-wave generator, EVM
- Resistor: ½-W 10,000-Ω
- Capacitor: 0.1 μF
- Miscellaneous: 8-H choke

PROCEDURE

High-Pass Filter

1. Connect the circuit of Fig. 66-5. An ac voltmeter is connected across the AF generator to measure its output.

2. Set the generator frequency at 60 Hz and adjust the generator output for 7 V rms. With the same voltmeter, measure the voltage V_R across R and record it in Table 66-1.

3. Repeat step 2 for each of the frequencies in Table 66-1.

CAUTION. Be certain that *generator output* is checked at each frequency setting and adjusted for 7 V rms.

4. Draw a graph (in pencil) of V_R (vertical axis) versus F, using the data in Table 66-1. Identify the cutoff frequency.

5. Connect an 8-H choke across R and repeat steps 2 and 3.

6. Draw a graph (in ink) on the same coordinate axes as in step 4, using the data from step 5. Identify the cutoff frequency.

Low-Pass Filter

7. Connect the circuit of Fig. 66-3. L is an 8-H choke, and R = 10,000 Ω. An ac voltmeter is connected across the generator.

8. Set generator frequency at 20 Hz, and adjust generator output for 7 V rms. With the same voltmeter, measure the voltage V_R across R and record it in Table 66-2.

9. Repeat step 8 for each of the frequencies in Table 66-2.

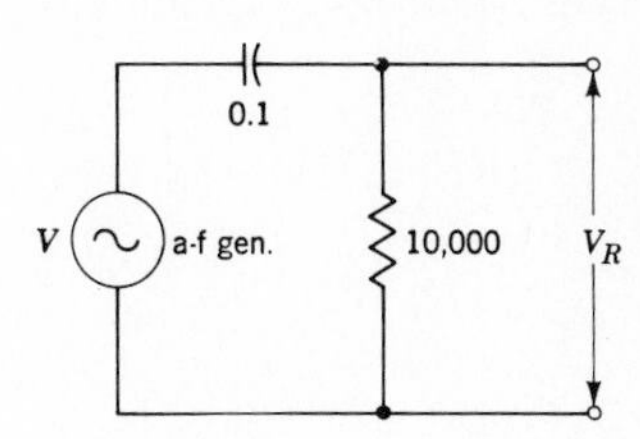

Fig. 66-5. Experimental circuit to determine the characteristics of a high-pass filter.

10. Draw a graph (in pencil) of V_R (vertical axis) versus F (horizontal axis), using the data in Table 66-2. Identify the cutoff frequency.
11. Connect a 0.1-μF capacitor in parallel with R as in Fig. 66-4*a* and repeat steps 8 and 9.
12. Draw a graph (in ink) on the same coordinate axes as in step 10, using the data from step 11. Identify the cutoff frequency.

TABLE 66-1. High-Pass Filter

Frequency, Hz	*Gen. Output,* V rms	V_R *(across R),* V rms	
		Without L	*L in parallel with R*
60	7		
80	7		
100	7		
120	7		
140	7		
160	7		
180	7		
200	7		
300	7		
500	7		
1000	7		
2000	7		
3000	7		

TABLE 66-2. Low-Pass Filter

Frequency, Hz	*Gen. Output,* V rms	V_R *(across R),* V rms	
		Without C	*C in parallel with R*
20			
30	7		
40	7		
50	7		
60	7		
80	7		
100	7		
150	7		
200	7		
300	7		
500	7		
1000	7		
2000	7		

QUESTIONS

1. Refer to Fig. 66-5. What is the calculated frequency F at which $X_C = R$? (Show your computations.)
2. Is the computed frequency in question 1 the same as the cutoff frequency in graph 1? Should it be? If the two frequencies are not the same, explain why they are not.
3. Refer to your data in Table 66-1 and to the two graphs. What was the effect of adding an 8-H choke in parallel with R on the frequency response of the high-pass filter?
4. Refer to procedural step 7. At what frequency does $X_L = R$? Show your calculations.
5. Is the frequency calculated in step 7 the same as the cutoff frequency on the graph in step 10? If not, explain why not.
6. Refer to your data in Table 66-2 and to the two graphs drawn from these data. What was the effect of adding a 0.1-μF capacitor in parallel with R on the frequency response of the low-pass filter?

Answers to Self-Test

1. frequency filters
2. blocking capacitor
3. 1000
4. 1592
5. R_L/Z; $\cos\theta$
6. 39.8; 39.8

67

EXPERIMENT

PHASE-SHIFTING NETWORKS

OBJECTIVES

1. To determine experimentally the range of phase variation in a network containing a fixed C in series with a variable R
2. To determine experimentally the range of variation in a phase-shift bridge circuit

INTRODUCTORY INFORMATION

Series *RC* Phase-shift Circuit

Phase-shifting networks used for electronic control are an interesting application of ac circuit theory. In its simplest form, a phase-shifting network may consist of a capacitor in series with a variable resistor, as in Fig. 67-1. The input voltage V is applied across the series circuit points FH, and the output voltage V_C is taken from the capacitor across points GH. As the size of R is varied, the phase of V_C varies with respect to V applied.

A phasor analysis of the circuit will clarify the phase relationships and range of phase variation as R is varied from maximum to minimum resistance. Figure 67-2*a* shows the phase relationships in the circuit for R greater than X_C. V applied (FH) is the reference phasor. Since this is a capacitive circuit, current I leads the applied voltage by some angle θ. In the resistor R, voltage and current are in phase. Hence V_R, the voltage across R, is shown lying along the phasor I. The voltage V_C across C lags I by 90°. If from H a line HG is drawn perpendicular to V_R, HG_1 will represent the phasor V_C, and FG_1 will be the phasor V_R. From the right triangle FG_1H we see that

$$V = \sqrt{V_R^2 + V_C^2} \qquad (67\text{-}1)$$

This is consistent with our knowledge of the voltage relationships in an RC circuit.

Angle ϕ_1 is the angle by which the voltage V_C lags the applied voltage. From triangle FG_1H it is evident that

$$\tan \phi_1 = \frac{V_R}{V_C} = \frac{R}{X_C} \qquad (67\text{-}2)$$

Since R is greater than X_C, V_R is greater than V_C. Therefore the ratio $V_R/V_C > 1$, and ϕ_1 is an angle greater than 45°. As R increases, ϕ_1 increases. If we set a limit on the maximum value of R at $R = 10\,X_C$, then

$$\tan \phi_1 = \frac{V_R}{V_C} = \frac{10}{1}$$

and $\phi_1 = 84.3°$. The maximum value of ϕ_1 is 90° when R becomes infinitely large.

What is the value of ϕ_2 if $R = X_C$? Figure 67-2*b* shows this condition. Here $V_R = V_C$ and $\tan \phi_2 = V_R/V_C = 1$. Therefore, $\phi_2 = 45°$.

Figure 67-2*c* shows that ϕ_3 becomes smaller than 45° when R becomes less than X_C. The limiting value of ϕ_3 is zero degrees, which occurs when $R = 0$. What this means, simply, is that when $R = 0$, the applied voltage and the voltage across C are the same, and are in phase.

The maximum range of phase delay with the RC network of Fig. 67-1 is from zero up to but not including 90°.

The amplitude of the voltage across C decreases as the phase angle between V_C and V increases.

Another characteristic is apparent from Fig. 67-2. Since the angle FGH is always a right angle, point G must lie along a semicircle whose diameter is FH. As R varies continuously from infinity to the value $R = X_C$, and finally to the value $R = 0$, G moves in a CCW direction along the semicircle from point H to G_2, and finally to point F.

Phase-shift Bridge Circuit

The range of variation of the angle ϕ can be increased to almost 180° by means of a bridge-type phase-shift circuit, Fig. 67-3. The applied voltage V appears across points FH, with point P as the center tap of the power source. That is, the voltage FP = voltage $PH = V/2$. The output voltage V_0 is taken from the points PG. As R is varied from maximum

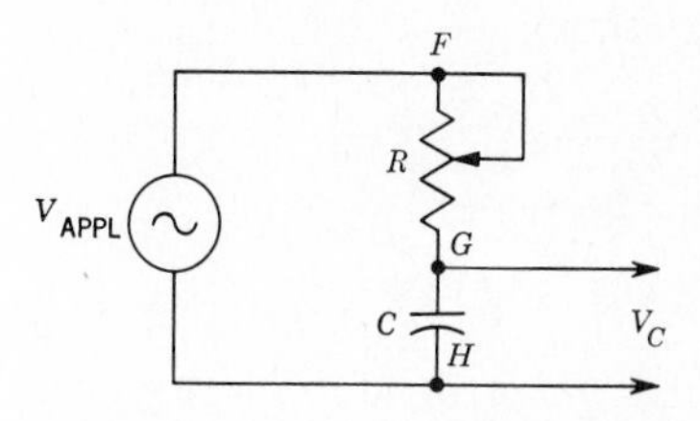

Fig. 67-1. *RC* phase-shifting circuit.

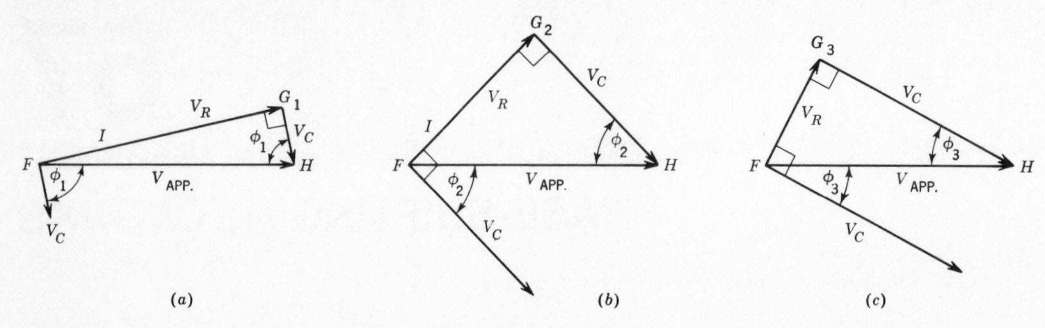

Fig. 67-2. Phasor diagrams for *RC* phase-shift circuits.

resistance to zero ohms, the phase angle ϕ of V_0 varies from some small value up to 180°. The angle ϕ is a leading angle relative to the voltage *FH*.

The phasor diagram (Fig. 67-4) illustrates these conditions. *FH* is the reference phasor (*V* applied), with *P* as center. Since current leads voltage in a capacitive circuit, the phasor *I* is shown leading the applied voltage *FH*, by some angle ϕ. Current *I* and voltage V_R in *R* are in phase. Hence the resistance voltage phasor *FG* lies along *I*. The voltage V_C across the capacitor lags the current *I* (and hence the voltage across *R*) by 90°. Therefore, the phasor *GH* perpendicular to *FG* represents the voltage V_C across the capacitor. *P* is the center of *FH* (since voltage FP = voltage $PH = V/2$). Hence, *P* is the center of the semicircle along which point *G* moves as the resistance of *R* is varied.

The phasor *PG* represents the output voltage V_0 from *P* to *G*, and ϕ is the angle by which V_0 leads the applied voltage *FH*. Since *P* is the center of the semicircle *FGH*, the radii *PG* and *PF* are equal, and angle PFG = angle $PGF = \theta$. In triangle *PFG* the exterior angle ϕ is equal to the sum of the two remote interior angles, or

$$\measuredangle\phi = 2\measuredangle\theta \tag{67-3}$$

and

$$\measuredangle\theta = \measuredangle\frac{\phi}{2} \tag{67-4}$$

From right triangle *FGH*,

$$\tan\theta = \frac{V_C}{V_R} = \frac{X_C}{R} \tag{67-5}$$

Therefore

$$\tan\frac{\phi}{2} = \frac{X_C}{R} \tag{67-6}$$

It is evident from Eq. (67-6) that when $R = 0$, $X_C/R = \infty$, $\phi/2 = 90°$, and $\phi = 180°$. Moreover, when $X_C = R$, $X_C/R = 1$, $\phi/2 = 45°$, and $\phi = 90°$. When *R* becomes very large, say $R = 10\,X_C$, $X_C/R = 1/10$, $\tan\phi/2 = 0.1000$, $\phi/2 = 5.75°$, and $\phi = 11.50°$.

Relative to the applied voltage *V* taken from *F* to *H*, the output voltage V_0 taken from *P* to *G* is a leading voltage and the angle of lead ϕ varies from 180° to 11.5° as *R* increases from zero ohms to $R = 10\,X_C$.

If the output voltage is taken from *G* to *P*, V_0 is a lagging voltage and the angle of lag equals $180° + \phi$.

SUMMARY

1. In the simple *RC* circuit of Fig. 67-1, as *R* is varied from zero ohms to its maximum value, the phase of the output voltage V_C with reference to the applied voltage $V_{applied}$ varies from zero degrees up to some maximum value determined by the size of *R* and the value of X_C. The value in degrees of this phase angle θ is given by the formula

$$\tan\theta = \frac{R}{X_C}$$

2. The angle θ is the angle by which the output voltage V_C *lags* the applied voltage $V_{applied}$.
3. In the simple *RC* phase-shifting circuit of Fig. 67-1, $\theta = 0°$ when $R = 0\,\Omega$; when $R = 10X_C$, $\theta = 84.3°$. The

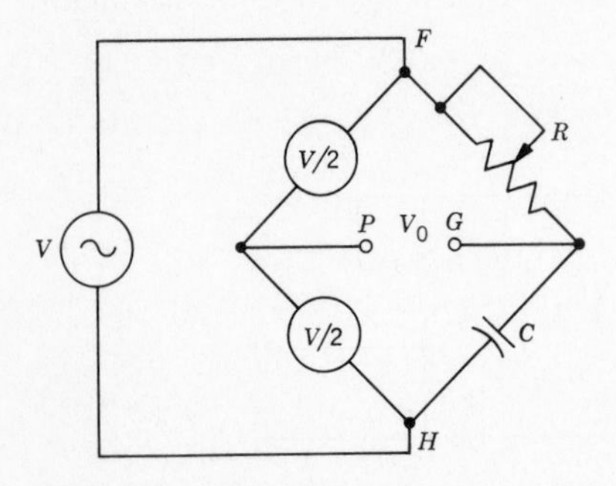

Fig. 67-3. Bridge-type *RC* phase-shift network.

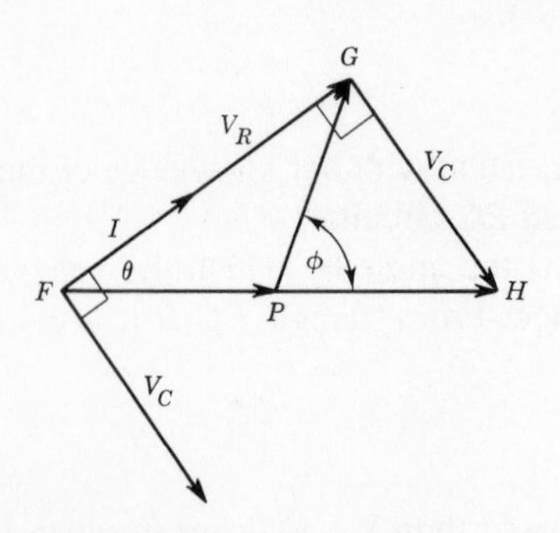

Fig. 67-4. Phasor diagram showing phase relations in bridge-type *RC* phase-shift circuit.

angle θ approaches, but never equals, 90°, because R cannot be made infinitely large. We can therefore say that the range of phase variation in the simple RC circuit containing a fixed value C and a variable R is approximately 90°.

4. The amplitude of the output voltage V_C in the simple RC phase-shifting circuit of Fig. 67-1 *decreases* as R increases in resistance.
5. In the phase-shift bridge circuit of Fig. 67-3, the range of variation of the phase-shift angle is increased up to but not including 180°.
6. In the phase-shift bridge circuit, the output voltage V_0 is taken from the junction of R and C (point G) and from the center point P of a low-impedance network across the applied voltage V.
7. The phase-shift angle ϕ in Fig. 67-3 is computed from the formula

$$\tan\frac{\phi}{2} = \frac{X_C}{R}$$

8. When $R = 0$, $\phi/2 = 90°$ and $\phi = 180°$. When $R = 10X_C$, $\phi = 11.5°$. The larger the value of R with relation to X_C, the smaller is the phase-shift angle ϕ.
9. The amplitude of the output voltage V_0 in the phase-shift bridge circuit remains constant over the range of variation of R. The output voltage $V_0 = V_{\text{applied}}/2$.
10. Whether the output phase angle in a phase-shift bridge circuit is leading or lagging depends on the voltages under comparison. Thus, in Fig. 67-3, if V_0 is taken from P to G (V_{PG}), relative to the voltage V_{FH} taken from F to H, V_0 is a *leading* voltage. However, if the output voltage V_{GP} is taken from G to P, V_0 is a *lagging* voltage relative to the applied voltage V_{FH} taken from F to H.

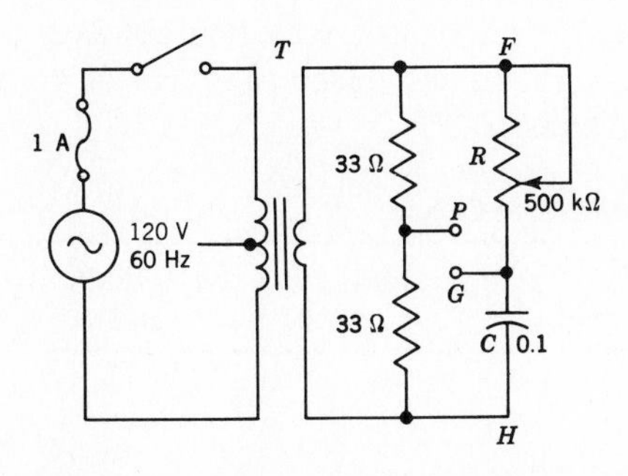

Fig. 67-5. Experimental phase-shift circuit.

SELF-TEST

Check your understanding by answering these questions.

Answer questions 1 through 4 relative to Fig. 67-1:

1. When $R = X_C$, the phase angle between V_C and V_{FG} is __________ degrees.
2. For the circuit in question 1, V_C (V_{GH}) __________ (leads/lags) V_{FH} by __________ degrees.
3. The maximum range of phase variation between V_C and V_{FH} approaches but never equals __________ degrees.
4. The amplitude of the output voltage V_C is constant. __________ (true/false)

Answer questions 5 through 8 relative to Fig. 67-5:

5. When $R = X_C$, the phase angle between V_{GP} relative to V_{FH} is __________ degrees.
6. For the circuit in question 5, V_{PG} __________ (leads/lags) V_{FH} by __________ degrees.
7. The maximum range of phase variation between V_{GP} and V_{FH} is approximately __________ degrees.
8. The output voltage V_{GP} = __________ V.

MATERIALS REQUIRED

- Power supply: Source of 120-V 60-Hz
- Equipment: Oscilloscope
- Capacitor: 0.1 μF
- Miscellaneous: 500,000-Ω 2-W potentiometer; AF output transformer (6K6 to speaker); two 33-Ω ½-W resistors; SPST switch; fused line cord

PROCEDURE

Series *RC* Circuit

1. Connect the circuit of Fig. 67-5. T is a voltage stepdown transformer (AF output) whose primary is connected to the 120-V 60-Hz line. (The center tap in the primary is not used.) The secondary is an untapped low-impedance winding. To obtain a center tap for P, two 33-Ω resistors are connected as shown in Fig. 67-5. R is a 500,000-Ω 2-W potentiometer connected as a rheostat, and C is a 0.1-μF capacitor. *Do not apply power.*

2. Connect the vertical "hot" lead of the oscilloscope to point F, the top of the secondary, and the ground lead to point H. Calibrate the oscilloscope for voltage measurements.

3. **Power on.** The applied-voltage sine wave should be visible on the screen. Set the oscilloscope on "Line" trigger. If Line trigger is not available, set the trigger selector on "External." Ext. trigger may be provided from point F on the secondary of T, connected to the jack marked "External Trigger." Adjust the oscilloscope's sweep and trigger controls until three cycles appear stationary on the screen. Center the waveform vertically with respect to the X axis. Center the waveform horizontally so that the middle (approximately) sine wave is centered with respect to the Y axis, with the positive alternation on the left and the negative alternation on the right, as in Table 67-1. This sine wave is the reference waveform 1, and phase measurements will be

made relative to this reference. Since a complete cycle is 360°, equal subdivisions of the time base represent equal angles. Thus, if each division is divided into 10 equal lengths, each length measures 18°. In the procedure which follows, movement of the waveform to the right of this position indicates a lagging angle, and movement of the waveform to the left, a leading angle. Once you have obtained the proper reference waveform, *do not* readjust the sweep, trigger, centering, or gain controls of the oscilloscope. Measure and record the peak-to-peak amplitude of the reference waveform.

4. Keep the ground lead of the oscilloscope at point *H*. Connect the vertical input lead to point *G*. Adjust the phase-shift control to minimum ($R = 0$). Observe and record this waveform 2 in proper time phase with the reference voltage, waveform 1. Measure and record the amplitude of waveform 2 and the number of degrees (if any) of phase shift.
5. Adjust *R* until there is a 45° phase shift. Draw this waveform 3 in proper time phase with reference waveform 1. Measure and record the amplitude of waveform 3.
6. Adjust *R* to maximum resistance. Draw waveform 4 in proper time phase with reference waveform 1. Measure and record the amplitude of waveform 4, and the number of degrees of phase shift.

Phase-shift Bridge Circuit

7. Remove the ground lead of the oscilloscope from point *H* and connect it to point *P*. The "hot" lead is connected to point *F*. Trigger connection and all controls remain set as previously. There should be a stable presentation of three sine waves on the screen. If there is not, readjust the sweep and trigger controls until the three sine waves "hold." Set horizontal-centering control as in step 3 until the middle sine wave is properly centered as in Table 67-2. This is the reference waveform 1. Measure and record the peak-to-peak amplitude of the waveform.
8. Keep the ground lead at point *P*. Move the hot lead of the oscilloscope to point *G*. The output appearing on the screen is the voltage V_{GP} (from *G* to *P*). Adjust the phase-shift control to minimum resistance ($R = 0$). Draw this waveform 2 in proper time phase with reference waveform 1. Measure and record the peak-to-peak amplitude of the waveform and the number of degrees (if any) of phase shift.
9. Adjust *R* until there is a 90° phase shift between V_{GP} and the reference waveform. Draw this waveform 3 in proper time phase with the reference. Measure and record the peak-to-peak amplitude of the waveform.
10. Adjust *R* to maximum resistance. Draw this waveform 4 in proper time phase with reference waveform 1. Measure and record the peak-to-peak amplitude of the waveform and the number of degrees of phase shift.
11. Reverse the vertical input leads of the oscilloscope so that the hot lead is connected to point *P* and the ground lead to point *G*. The output voltage on the screen is V_{PG}. Repeat steps 8, 9, and 10. The corresponding waveforms are 5, 6, and 7. Draw 5 on the same time axis as 2; 6 on the same axis as 3; and 7 on the same axis as 4. Identify the waveforms.

TABLE 67-1. Series *RC* Phase-shift Circuit

Waveform Number	*Waveform*	*Volts, Peak-to-Peak*	*Phase-shift, Degrees*
1	+ 0 −		
2	+ 0 −		
3	+ 0 −		45°
4	+ 0 −		

TABLE 67-2. Phase-shift Bridge Circuit

Waveform Number	*Waveform*	*Volts, Peak-to-Peak*	*Phase-shift, Degrees*
1			
2, 5			
3, 6			90°
4, 7			

QUESTIONS

1. Refer to Table 67-1. Compare the experimental range of variation of the phase-shift angle with that predicted by Eq. (67-2) for the minimum and maximum values of R. Show your computations.
2. Refer to the waveforms in Table 67-1. Is the phase-shift angle leading or lagging? How do you know? Is this consistent with the theory of this circuit?
3. Refer to Table 67-2. Compare the experimental range of variation of the phase-shift angle with that predicted by Eq. (67-6) for the minimum and maximum values of R. Show your computations.
4. Refer to the phase-shift waveforms 2, 3, and 4 in Table 67-2. Is the phase-shift angle leading or lagging? Which should it be theoretically?
5. What is the phase relationship between waveforms (*a*) 2 and 5? (*b*) 3 and 6? (*c*) 4 and 7?
6. Explain the basic differences between the output of the simple *RC* phase-shift circuit and the *RC* bridge-type phase-shift circuit.

Answers to Self-Test

1. 45
2. lags; 45
3. 90
4. false
5. 90
6. leads; 90
7. 180
8. $V_{FH}/2$

68

EXPERIMENT

NONLINEAR RESISTORS—THERMISTORS

OBJECTIVES

1. To observe the self-heating effect of current on the resistance of a thermistor
2. To determine experimentally the variation in resistance with time of current in a thermistor

INTRODUCTORY INFORMATION

We have been concerned to this point with the laws governing the flow of current in dc and ac circuits. Thus we verified Ohm's law, which shows the relationship between resistance (R), voltage (V), and current (I) in dc and ac circuits. Next we studied Kirchhoff's laws for more complex networks, and the various network theorems. Throughout our discussion and in our experimental work, we dealt with resistors whose resistance remained relatively constant despite variations in current flow and operating temperature.

There is a class of components known as *thermistors* whose resistance changes with changes in operating conditions, particularly with changes in operating temperature. As its name implies, a thermistor is a thermally sensitive resistor. It is part of a large family of semiconductors. At this time we need merely mention that a semiconductor is a substance whose resistivity lies somewhere between that of conductors and that of insulators. We will study the characteristics of semiconductors in greater detail in electronics.

The thermistor is a two-terminal component which may be used in either ac or dc circuits. It is manufactured in a number of shapes, such as beads, rods, disks, washers, and flakes (see Fig. 68-1).

Temperature Characteristic

A fundamental characteristic of the thermistor is its change of resistance with changes in temperature. The manner in which the resistance varies with temperature is given by the approximate formulas

$$R = R_0 \times e^k$$

and

$$k = B\left(\frac{1}{T} - \frac{1}{T_0}\right) \qquad (68\text{-}1)$$

where R = resistance at any temperature T, in kelvins (K)
R_0 = resistance at reference temperature T_0 (K)
B = a constant whose value depends on the thermistor material (determined from measurements at 0° and at +50°)
e = 2.7183 (base of natural logarithms)

Formula (68-1) shows that the resistance variation is nonlinear. Thermistors are frequently referred to as *nonlinear resistors*.

Thermistors exhibit negative temperature coefficients (NTC) and positive temperature coefficients (PTC). The resistance of NTC thermistors *decreases* as their temperature *rises* and *increases* as their temperature *falls*. The resistance of *PTC* thermistors *increases* as their temperature *rises*, and *decreases* as their temperature *falls*. In this experiment we will be concerned mainly with NTC thermistors, and the following discussion refers to them.

Figure 68-2 shows the variation for three NTC thermistors, compared with the resistivity of platinum under similar temperature changes. Note the tremendous variation of resistance in thermistor 1, a change of 10,000,000 to 1 over a temperature range of 500°C, while that of platinum changes by a factor of less than 10 to 1.

Figure 68-3 shows the variation in resistance of another NTC thermistor. Here a range of 200°C causes a variation in resistance from 1000 Ω to approximately 2 Ω.

The resistance-temperature characteristic of thermistors has been utilized in electrical and electronic applications. In particular, thermistors are used as protective devices for

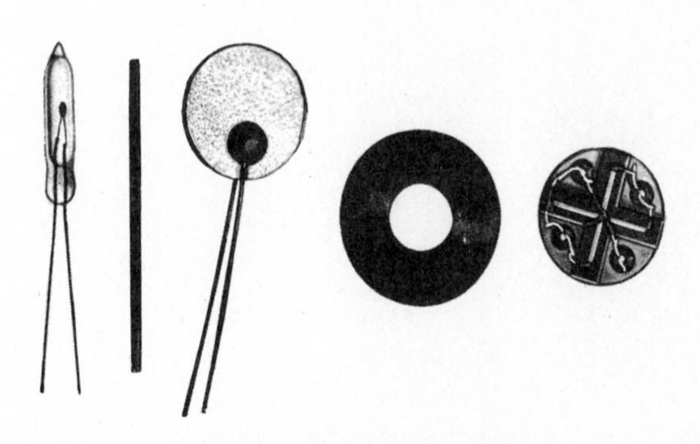

Fig. 68-1. Thermistor types (*Western Electric Co.*).

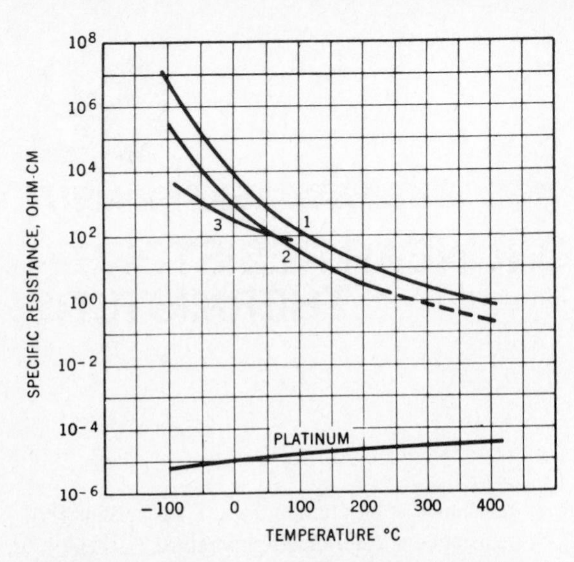

Fig. 68-2. Temperature-resistance values of three thermistors compared with a representative metal (*Western Electric Co.*).

electric bulb filaments during the warmup period. They are used for temperature measurement, control, and compensation. Thermistors are also widely used in television circuits, such as the degaussing circuits for color picture tubes. They may also be used in time-delay relay circuits.

Static Voltampere Characteristic

There are secondary characteristics which arise from the relationship between temperature and resistance that thermistors exhibit. Among these is the static relationship between

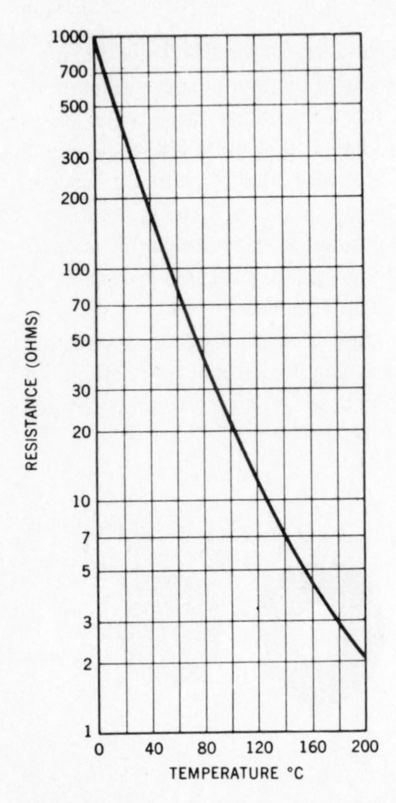

Fig. 68-3. Resistance versus temperature in a thermistor (*Western Electric Co.*).

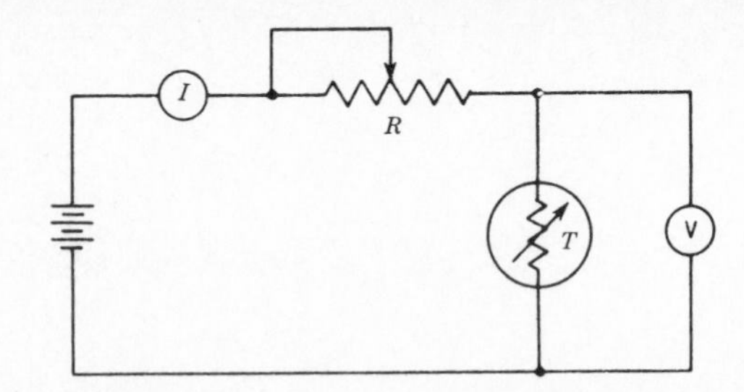

Fig. 68-4. Test circuit for determining the static *VI* characteristic of a thermistor.

voltage and current. This secondary characteristic results from the self-heating that occurs when current flows through a thermistor. The static characteristic may be determined using the test circuit shown in Fig. 68-4. In this circuit a rheostat R is used as a limiting resistor to vary the amount of current through the thermistor. As the resistance of R is decreased, the current I increases. As the resistance of R is increased, the current I decreases. A voltmeter is used to monitor the voltage across the thermistor for every value of current. Each time R is varied and a change in current is effected, enough time must be allowed for the voltage across the thermistor to reach a steady value. The resultant graph, therefore, reflects steady-state conditions of voltage versus current and is therefore called the *static voltampere characteristic*.

The thermistor whose resistance-temperature characteristic is shown in Fig. 68-3 has the static voltampere characteristic shown in Fig. 68-5. A temperature of 25°C was the starting temperature for the tests from which the data for the graph were secured. The numbers on the graph, namely, 53, 73, 102, etc., indicate the thermistor temperature, measured after current had stabilized at the level shown on the graph. Thus, at 0.05 A, the voltage across the thermistor is 11 V and the temperature is 53°.

The graph for this thermistor shows that up to a temperature of 53° there is a linear relationship between voltage and current. That is, Ohm's law holds during this interval, when only small current flow is involved. The reason is that there is insufficient heat dissipated at this low level of current to affect the cold resistance of the thermistor. As current increases, however, the heat dissipated increases, thermistor temperature increases, and its resistance drops nonlinearly. The voltage across the thermistor now increases nonlinearly to a peak value V_M, designated as the "self-heat" voltage. Beyond this value, the voltage across the thermistor drops nonlinearly with increasing current, as shown in Fig. 68-5.

The static voltampere characteristic of thermistors is utilized in the following applications:

- Temperature alarm devices
- Pyrometer
- Flow meter
- Switching devices
- Voltage surge devices
- Gas detectors

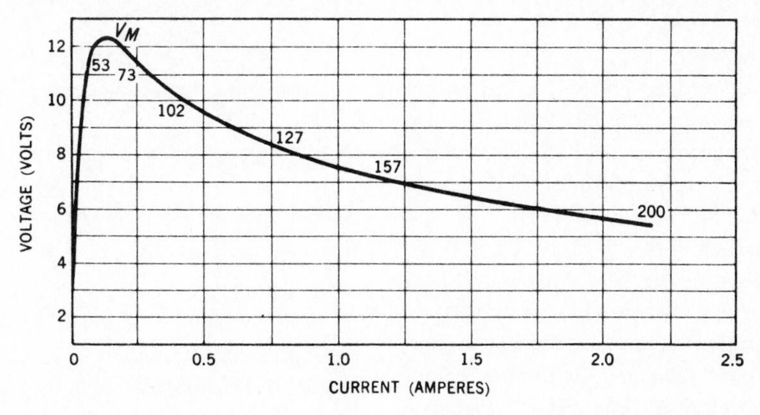

Fig. 68-5. Thermistor static voltampere characteristic (*General Electric Co.*).

Dynamic Characteristic

A thermistor requires a certain time interval to react to a change in the external circuit. A change in current through a thermistor from one value to another does not result in an instantaneous change in temperature, and hence there is no instantaneous change in resistance. These changes do occur, but they require a definite time. The thermal mass of the thermistor element will determine the time interval. A small element will change temperature more rapidly than an element with larger mass.

Figure 68-6 shows a test circuit for determining the dynamic characteristics of a thermistor. V is an ac power supply whose output is adjustable, and R_L is the load resistance. An EVM set on ac volts is connected across the load R_L to monitor the voltage across R_L. When switch S is open, there is no voltage applied to the circuit. At the instant S is closed, the cold resistance of the thermistor and the value of the load resistor R_L constitute the voltage divider which determines the voltage across R_L. As current starts to flow, the temperature of the NTC thermistor increases, its resistance decreases, and more of the source voltage appears across the load. Within a certain time interval, the circuit is stabilized. That is, the thermistor resistance is stabilized and the voltage across the load becomes constant.

Figure 68-7 is a graph showing the dynamic characteristic of a thermistor for two different values of supply voltage and load resistance. Note that with 48 V applied and a load resistance of 17 Ω, steady state is reached in approximately 75 s. With 115 V applied and a load of 50 Ω, steady state is reached in about 15 s.

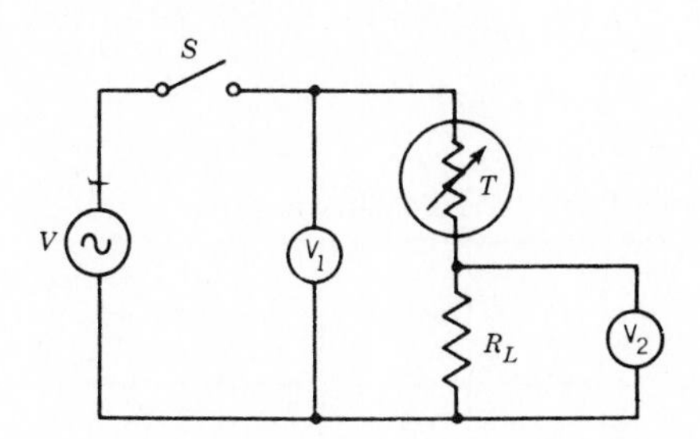

Fig. 68-6. Test circuit for determining the dynamic characteristic of a thermistor.

The dynamic characteristic of a thermistor is utilized in expandors and compressors (audio devices), switching devices, voltage-surge absorbers, and low-frequency negative-resistance devices.

In this experiment you will first observe the effect on the resistance of a thermistor of an increase in operating temperature due to self-heating. You will then determine experimentally the time required to stabilize the resistance of a thermistor due to self-heating.

SUMMARY

1. Thermistors are semiconductor devices whose resistance changes with changes in operating temperature.
2. Thermistors are called nonlinear resistors because their resistance varies exponentially, not linearly, as a function of temperature.

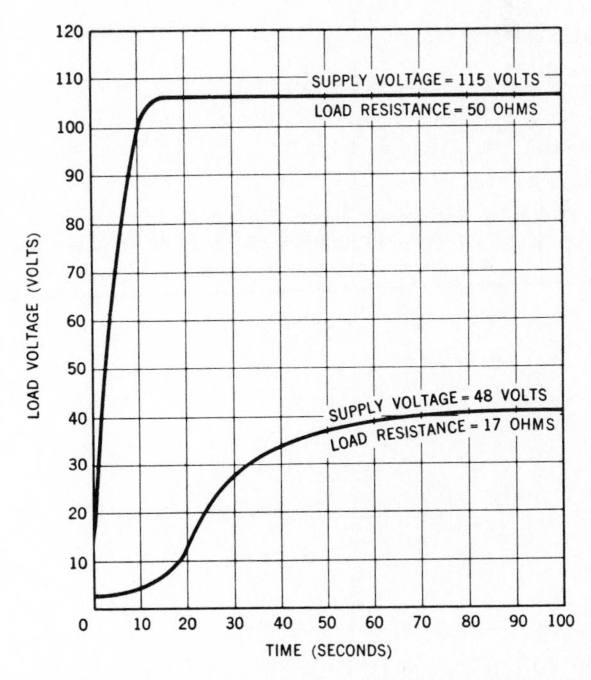

Fig. 68-7. Dynamic characteristic of a thermistor (*General Electric Co.*).

3. The *resistance* of negative temperature coefficient (NTC) thermistors *decreases* as their temperature *increases*, and *increases* as their temperature *decreases*. The resistance of positive temperature coefficient (PTC) thermistors *increases* as their temperature *increases*, and *decreases* as their temperature *decreases*.
4. Compared with the variation in resistance of a thermistor, the resistance of a carbon resistor remains relatively constant despite changes in operating temperature.
5. Thermistors exhibit unique temperature-resistance characteristics, depending on their size and construction. Thus, one particular thermistor changes in resistance from 1000 to 2 Ω over a temperature change of 200°C. Another changes from 10,000,000 to 1 Ω over a temperature variation from −100°C to +400°C.
6. Because of the self-heating effect of current in a thermistor, this device changes resistance with changes in current.
7. A secondary characteristic of a thermistor is the change in current it exhibits in an electric circuit. A graph of current versus voltage (Fig. 68-5) of a thermistor, called a static voltampere characteristic, shows that current does *not* vary linearly with applied voltage.
8. A thermistor does not undergo instantaneous changes in resistance with changes in temperature. A certain time interval, determined by the thermal mass of the thermistor, is required to accomplish the resistance change. A thermistor with a smaller mass will change more rapidly than one with a larger mass.
9. The dynamic characteristic of a thermistor is a graph of the time it takes a thermistor to stabilize its resistance as a function of the applied voltage (Fig. 68-7) and current.
10. Thermistors are used in temperature alarm devices, pyrometers, switching devices, voltage surge devices, etc.

SELF-TEST

Check your understanding by answering these questions:

1. A thermistor is a ____________ (linear/nonlinear) resistor.
2. The resistance of an NTC thermistor ____________ (increases/decreases) with a decrease in temperature.
3. An NTC thermistor has a ____________ (positive/negative/zero) temperature coefficient.
4. It takes ____________ for a thermistor to undergo a change in resistance with a change in temperature.
5. The larger the thermal mass of a thermistor the ____________ (more/less) time it will take to change its resistance.
6. A(n) ____________ (increase/decrease) in current in an NTC thermistor will cause a decrease in its resistance.
7. Some applications of thermistors are: (*a*) ____________, (*b*) ____________, and (*c*) ____________.
8. Current in a thermistor varies linearly with voltage across it. ____________ (true/false)

MATERIALS REQUIRED

- Power supply: Manually variable dc voltage source
- Equipment: Digital VOM; VOM or EVM
- Resistors: 5-W 500-Ω; 10-W 300-Ω
- Miscellaneous: Thermistor with the following characteristics: 50 Ω at 100 mA, 300 Ω cold (Carborundum A1204J-16 or the equivalent); SPST switch; SPDT switch

PROCEDURE

NOTE: It will be necessary for two students to work together as a team in this experiment. One will note time and will call out the voltages measured across the thermistor T and across the fixed resistor R_2, Fig. 68-8. To make the readings at the required time intervals, this student will need a watch with a seconds hand. The second student will record the readings in Tables 68-1 and 68-2, as required. Since the measurements change rapidly with time, a digital VOM will expedite measurements.

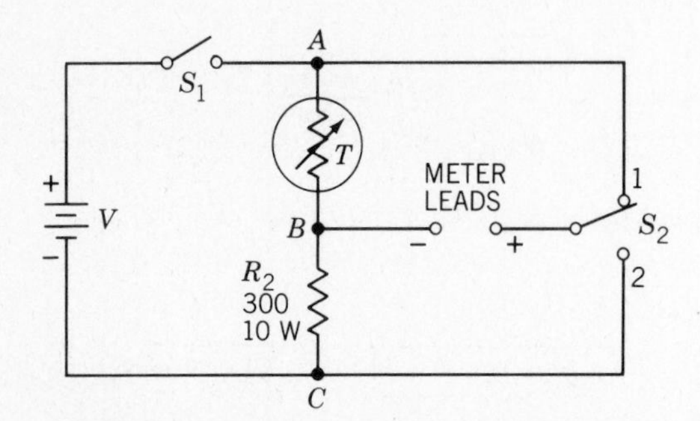

Fig. 68-8. Experimental circuit for determining the dynamic characteristic.

The voltage readings across the fixed resistor R must be made *immediately* after the reading across the thermistor T. A fast way to switch the voltmeter from T to R_2 is to use an SPDT switch connected as in Fig. 68-8. The negative meter lead is attached to point B, the junction of T and R; the positive lead is attached to the arm of S_2. In position 1 of S_2, the voltage V_{AB} is measured across T. In position 2, the voltage V_{BC} is measured across R. In recording the voltages we are not concerned with the polarity, only with the magnitude.

TABLE 68-1. Control Circuit Measurements

		Voltage, V	
Resistance, Ω		$t = 0$	$t = 5$ min
R_1	V_1		
R_2	V_2		

The voltage across R_2 will increase with time, up to 80 percent of the applied voltage V. The voltage across the thermistor will decrease with time. The sum of the voltages across R_2 and T should equal the applied voltage V.

Control Experiment

1. Connect the circuit of Fig. 68-9. Switch S is *off*. Measure and record in Table 68-1 the resistance of R_1 and R_2.
2. Set the adjustable supply V at 30 V. **Power on.** Monitor, and maintain the supply voltage at 30 V.
3. Measure and record in Table 68-1 the voltages V_1 across R_1, and V_2 across R_2.
4. Allow power to remain on for 5 min. Repeat the voltage measurements across R_1 and R_2, and record in Table 68-1.
5. **Power off.**

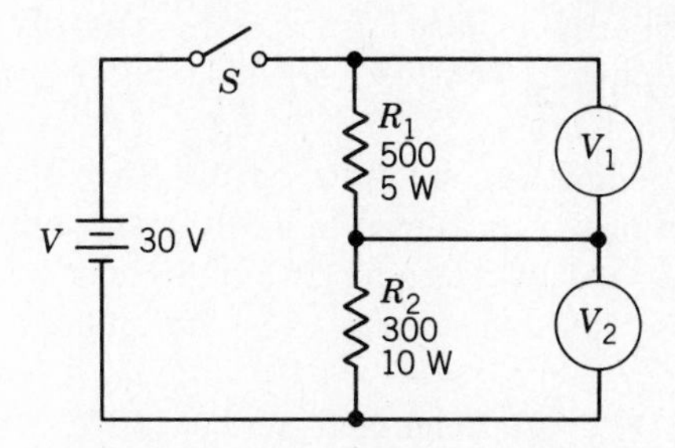

Fig. 68-9. Experimental control circuit.

Dynamic Characteristics of a Thermistor

NOTE: Before you apply power to the experimental circuit, Fig. 68-8, be certain that you understand the remaining checks which must be made. The timing factor is critical. Read carefully the following steps, the Note prior to Procedural step 1, and refer to Table 68-2.

6. Measure and record in Table 68-2 the cold resistance of the thermistor.
7. Connect the circuit of Fig. 68-8. S_1 is *off*. S_2 is in position 1. The output of the variable supply is still set at 30 V. Maintain it at this level during step 8.
8. **Power on** (close S_1). Measure and record in Table 68-2 the voltage V_T across the thermistor, and *immediately* afterward the voltage V_{R2} across R_2 at $t = 0$ (that is, as soon as power is applied), at $t = 15$ s, etc., following the time schedule in Table 68-2. The series of checks for this step is complete at $t = 5$ min, that is, 5 min after power is applied.
9. At this point you must measure the hot resistance of the thermistor when power is turned off. To accomplish this, leave power **on**, but remove the hot lead of the meter from the arm of S_2. The negative lead of the meter remains connected, as before. Now set the meter on the Ohms function, low range.
10. *Now* turn **power off** (open S_1). Immediately connect the positive lead of the ohmmeter to point A. Measure and record the hot resistance of T. *Let the thermistor cool for about 5 min.*
11. Set the output of the variable supply at 18 V. Maintain it at this value during step 12.
12. Repeat step 8 with the supply set at 18 V.
13. Repeat step 9.

TABLE 68-2. Thermistor Characteristic Data

Thermistor Resistance		*Cold*			*Hot, V = 30 V*			*Hot, V = 18 V*		
V applied	t	0	15 s	30 s	45 s	1 min	2 min	3 min	4 min	5 min
30	V_T									
	V_2									
18	V_T									
	V_2									

QUESTIONS

1. What is meant by negative temperature coefficient? Positive temperature coefficient?
2. What is the temperature coefficient of an ordinary constant-value resistor? Substantiate your answer by referring to the data in Table 68-1.
3. What is the temperature coefficient of our experimental thermistor?
4. How can you determine the temperature-versus-resistance characteristic of a thermistor? Is time involved?
5. From the data in Table 68-1, how would you compute the hot resistance of a thermistor without actually measuring it?
6. Compute the hot resistance of the thermistor after 5 min operation in the circuit of Fig. 68-9, with (*a*) 28 V applied; (*b*) 56 V applied. Why is there a difference, if any, in these two values?

7. How do the computed hot resistance values of the thermistor compare with the measured values? Explain any discrepancies.
8. What limitations are there on the amount of current which may be permitted to flow through a thermistor?
9. What is the relationship between the voltage across the thermistor and the voltage across the limiting resistor R_2 at any instant of time?
10. From the data in Table 68-2, for $V = 30$ V, compute the
    (*a*) Maximum current through the thermistor.
    (*b*) Minimum current. Show all computations.

Answers to Self-Test

1. nonlinear
2. increases
3. negative
4. time
5. more
6. increase
7. (*a*) temperature alarm devices; (*b*) pyrometers; (*c*) voltage surge devices (or switching devices)
8. false

69

EXPERIMENT

NONLINEAR RESISTORS—VARISTORS (VDRs)

OBJECTIVES

1. To determine experimentally the voltampere characteristic of a varistor
2. To determine experimentally the relationship between voltage and resistance

INTRODUCTORY INFORMATION

In the previous experiment you discovered that there is a class of resistors, called thermistors, whose resistance is dependent on temperature. There is another group of nonlinear resistors whose resistance is voltage-dependent. These voltage-sensitive resistors are called *varistors* or VDRs. The current in a varistor varies as a power of the applied voltage, and for a particular varistor it may increase as much as 64 times when the applied voltage is doubled.

Varistors are made of silicon carbide mixed with a ceramic binder, then fired at a high temperature. A metallic coating is sprayed on, and electrical contact is made to this coating. Typical shapes and sizes of varistors include rods, disks, and cylinders. The smaller sizes are usually supplied with tinned copper leads. This two-terminal device can be connected in a circuit like any linear resistor. Varistors have wide application in voltage regulation circuits.

Voltampere Characteristic

The voltampere characteristic of a silicon-carbide varistor is given by the approximate equation

$$I = KV^n \qquad (69\text{-}1)$$

where I is the current in amperes
K is a constant (amperes at 1 V)
n is an exponent

The value of K depends on the resistivity and the dimensions of the varistor, while n depends upon certain factors in the manufacturing process.

Values of n may range from 2 to 6. The higher the value of n, the more nonlinear is the current and resistance characteristic. For example, doubling the voltage of a varistor will increase the current by a ratio of 8 if $n = 3$, and by a ratio of 64 if $n = 6$. The value of n is not constant over the entire voltage range. Therefore it is usually specified between an upper and lower voltage limit.

Other relationships between voltage, current, and resistance which may be derived by application of Ohm's law are:

$$V = CI^b \qquad (69\text{-}2)$$

$$RI^a = C \qquad (69\text{-}3)$$

and

$$R = \frac{1}{KV^{n-1}} \qquad (69\text{-}4)$$

where $b = 1 - a$, $n = 1/b$, and $K = 1/C^n$.

The values of n and K are the same as those in Eq. (69-1).

Equations (69-1), (69-2), and (69-3) are straight lines when graphed on logarithm paper.

Electrical Specifications

Varistors are rated for the continuous power they can dissipate at a specific temperature (e.g., 1 W at 40°C) and for the maximum voltage which may be applied. In addition, other electrical characteristics may be specified in one or more of the following ways:

1. Voltage-Current Ratio. The constant current level at a standard voltage and tolerance is given, together with a specified voltage ratio. The ratio replaces the n in the equation $I = KV^n$.

Example. 1 mA at 68 V direct current ± 15 percent,

$$\text{Ratio } \frac{V \text{ at } 1.10 \text{ mA}}{V \text{ at } 0.1 \text{ mA}} = 2.20 \text{ max}$$

2. Voltage-Current n is specified instead of the maximum voltage ratio, in addition to the constant current level.

Example. For 1 mA at 10 V dc ± 15 percent,

$$n = 2.88 \text{ min between 5 and 10 V direct current}$$

or

$$n = 3.20 \pm 10 \text{ percent between 5 and 10 V}$$

3. Resistance-Voltage n. The resistance plus tolerance is given for a specified constant voltage plus the n value in a specified voltage range.

Example. 10,000 Ω ± 20 percent at 10 V direct current,

n = 2.88 min between 5 and 10 V direct current

or

n = 3.20 ± 10 percent between 5 and 10 V

4. *Resistance at Two Voltages*. Resistance plus tolerance is specified at two voltages.

Example.

10,000 Ω ± 20 percent at 10 V direct current,

45,000 Ω ± 30 percent at 5 V direct current

Other facts which the technician should know about a varistor are:

1. It possesses a negative temperature coefficient.
2. It is sensitive to high humidity. A standard varnish coating reduces the humidity effect appreciably.
3. The equivalent circuit for a varistor is a resistor and capacitor in parallel.
4. Though the standard varnish coating provides good insulation, the varistor should be mounted away from all conducting surfaces.
5. The varistor is a nonpolarized device.
6. The varistor is unaffected by pressure or vibration.
7. Varistors are used as in Fig. 69-1, to protect circuits and components from voltage surges when an ac or dc magnetic or inductive circuit current is suddenly interrupted. Varistors used in lightning arrestors serve a similar protective function. They may also be used as shunts across switch contacts to prevent sparking. In combination with other resistors, varistors may be used to provide voltage and current regulation.

SUMMARY

1. Varistors are voltage-dependent resistors (VDRs). They are nonlinear devices whose resistance is dependent on the voltage across them.
2. The current in a varistor is an exponential function of the voltage across it. Thus

$$I = K \times V^n$$

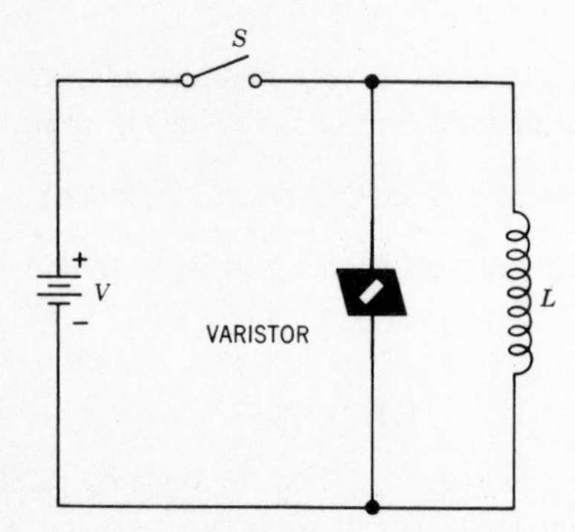

Fig. 69-1. Protective circuit using a varistor.

where K is a constant, depending on the type of varistor, and n is the exponent whose range is from 2 to 6.
3. Varistors are rated for their continuous power dissipation at a specified temperature and for the maximum voltage they can tolerate.
4. Other electrical specifications may include one of the following:
   (*a*) Voltage-current ratio, together with the constant current level at a standard voltage level and tolerance
   (*b*) Voltage-current n and the constant current level at a standard voltage level and tolerance
   (*c*) Resistance-voltage n, between a specified voltage range, together with the resistance plus tolerance for a specified constant voltage
   (*d*) Resistance plus tolerance at two specified voltages
5. A varistor has a negative temperature coefficient.
6. A varistor is sensitive to high humidity.
7. A varistor is a nonpolarized device which is unaffected by pressure or vibration.
8. Varistors are used as protective devices in lightning arrestors, and in ac and dc circuits to protect components against voltage surges.

SELF-TEST

Check your understanding by answering these questions:

1. The resistance of a varistor __________ (increases/decreases) as the voltage across it increases.
2. The equation which defines the voltampere characteristic of a varistor is $I = K \times V^n$. __________ (true/false)
3. Referring to the equation which defines the voltampere characteristic of a varistor, if $n = 2$ and the voltage across the varistor is tripled, the current in the varistor will __________ (increase/decrease) __________ (how many) times.
4. If the temperature of a varistor increases, its resistance __________.
5. In a protective circuit, varistors are most reliable in a __________ (humid/dry) climate.
6. Polarity must be observed in connecting a varistor in a dc circuit. __________ (true/false)

MATERIALS REQUIRED

- Power supply: Variable regulated dc source
- Equipment: Multirange dc micromilliammeter (the current ranges of a 20,000-Ω/V VOM will suffice); electronic voltmeter
- Resistors: 5-W 500-Ω
- Miscellaneous: Varistor (Carborundum, Electronics Division, Globar Plant, Niagara Falls, N.Y.) Part #463BNR-32, or the equivalent; SPST switch

PROCEDURE

1. Connect the circuit of Fig. 69-2. I is a multirange dc micromilliammeter (the current ranges on a 20,000-Ω/V VOM may be used), and V_1 is an EVM set on dc voltage to measure V_{AB}. V is set at zero volts.
2. Close S. Increase the output of power supply V so that V_{AB} takes, in turn, each of the voltages shown in Table 69-1. Measure and record the current I at each of these voltages. Measure also and record the voltage V_{AB} for which $I = 0.1 \times 10^{-3}$ A (0.1 mA) and $I = 1.0 \times 10^{-3}$ A (1 mA).
3. Compute and record the resistance of the varistor at each voltage setting, using Ohm's law.
4. Plot a graph of V versus I, with V as the vertical and I as the horizontal axis. Use log-log graph paper.
5. Compute and record the ratio

$$\frac{V \text{ at } 1 \times 10^{-3} \text{ A}}{V \text{ at } 0.1 \times 10^{-3} \text{ A}}$$

Show your computation.

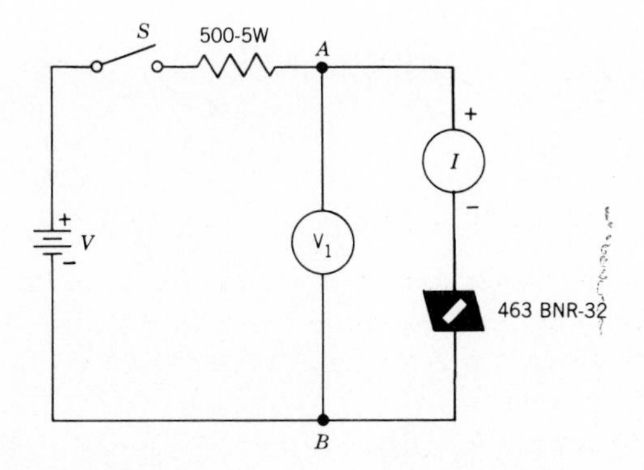

Fig. 69-2. Experimental circuit for determining the voltampere characteristic of a varistor.

Extra Credit

6. Compute also and record the values of n and K (within the range 0.1×10^{-3} A to 1.0×10^{-3} A). Show your computations [see Eq. (69-1)].
7. Compute also and record the values b, a, and C [see Eqs. (69-1) to (69-3)].

TABLE 69-1. Varistor Characteristics

V_{AB}, V	I, A	$R = \frac{V}{I}$, Ω
0		
2		
4		
6		
8		
10		
15		
20		
25		
30		
35		
40		
	(0.1 mA) 0.1×10^{-3}	
	(1.0 mA) 1.0×10^{-3}	

QUESTIONS

1. What is the difference between a linear and nonlinear resistor?
2. Comment on and explain the significance of your graph of V versus I. Why was the graph made on log paper?
3. From the data in Table 69-1, comment on the relationship between V and R.

Extra Credit

4. Accepting the empirical equation $I = KV^n$, derive Eqs. (69-2), (69-3), and (69-4).

Answers to Self-Test

1. decreases
2. true
3. increase; 9
4. decreases
5. dry
6. false

A1

APPENDIX

ELECTRONIC COMPONENTS AND THEIR SYMBOLS

Though electronic devices are fairly complex, they are basically made up of a small number of easily identifiable components. The technician must be able to identify these components readily. Photographs of a number of these parts are shown.

Resistors are the components most frequently used in electronic circuits. They are available in different values, shapes, and sizes. Manufacturers have adopted a standard EIA (Electronic Industries Association) color-coding system for determining resistance or ohmic values of low-power carbon resistors. You must learn this coding system. Higher-power resistors usually have the resistance value imprinted on their bodies.

There are fixed-value resistors, such as those shown in Fig. A1-1, and variable resistors such as those in Fig. A1-2.

Capacitors are also found in a wide variety of values, shapes, and sizes. Capacitors are classified as to their physical and electrical characteristics. Figure A1-3 shows some representative types. Capacitance values are given in units called microfarads (μF) or picofarads (pF). These values are either identified by EIA color code or imprinted on the body of the capacitor.

Inductors and transformers make up another classification of components. Wirewound coils and chokes are classified as inductors. These are wound on different forms and cores. There are air-core coils, iron-core, powdered iron-core, etc. (Fig. A1-4). Transformers are coupled inductors. They serve many purposes. There are power transformers, filament

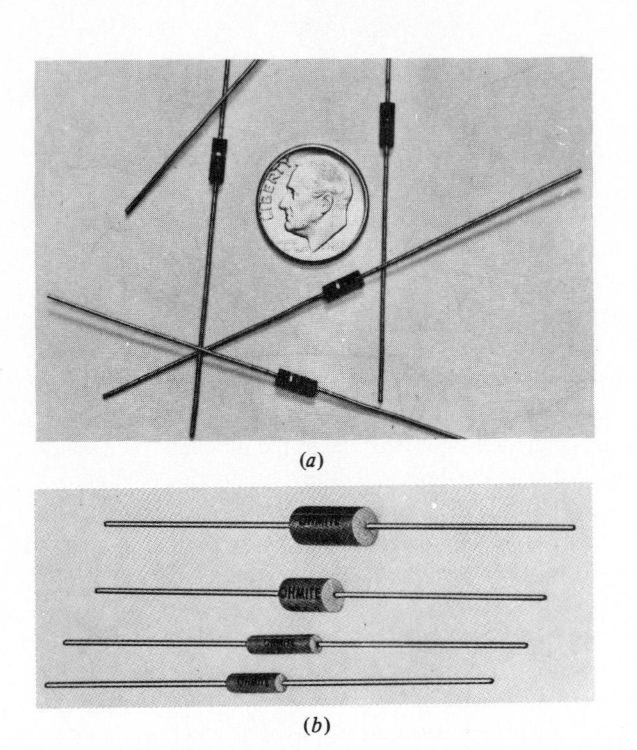

(*a*)

(*b*)

Fig. A1-1. (*a*) Quarter-watt resistors (*Speer Carbon Co.*); (*b*) molded silicone-ceramic wirewound resistors (*Ohmite Mfg. Co.*).

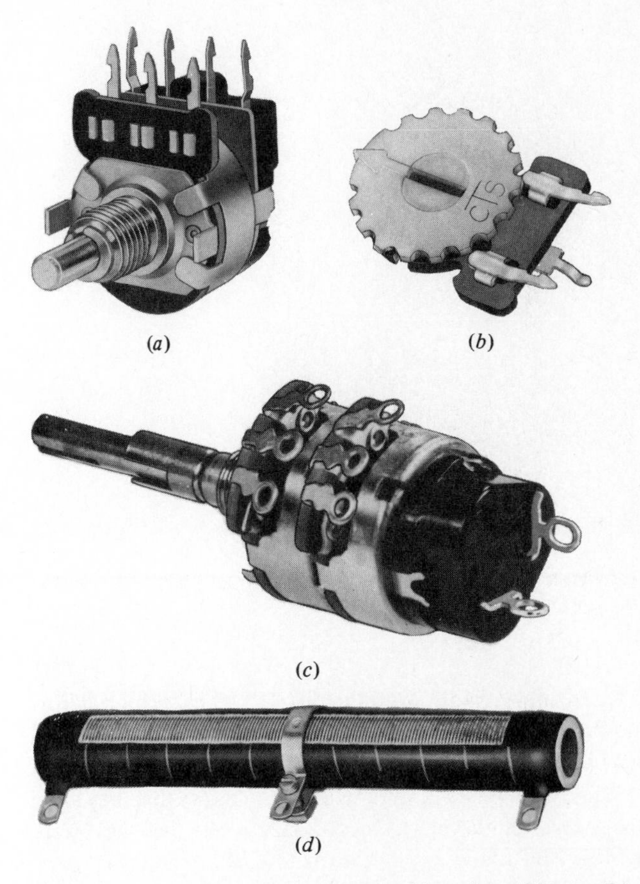

(*a*) (*b*)

(*c*)

(*d*)

Fig. A1-2. (*a*) Miniature composition variable resistor (*Chicago Telephone Supply Corp.*); (*b*) knob-operated composition trimmer (*Chicago Telephone Supply Corp.*); (*c*) dual-section composition variable resistor with switch (*Chicago Telephone Supply Corp.*); (*d*) vitreous enamel adustable resistor (*Ohmite Mfg. Co.*).

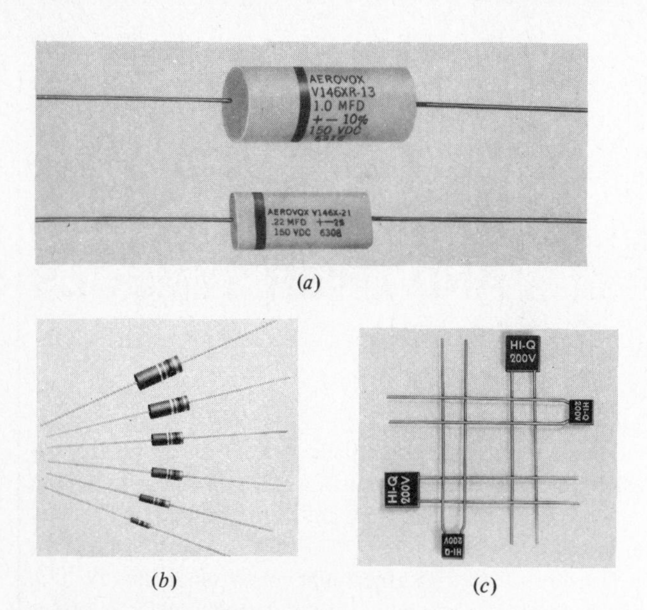

Fig. A1-3. (*a*) Mylar capacitors (*Aerovox Corp.*); (*b*) molded cerafil capacitors (*Aerovox Corp.*); molded ceramic capacitors for transistor circuits (*Aerovox Corp.*).

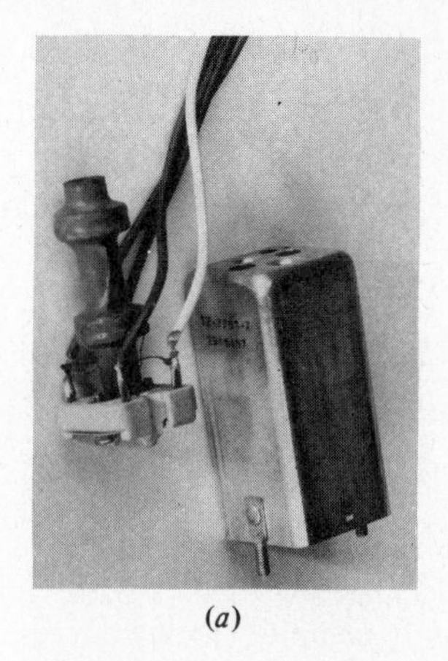
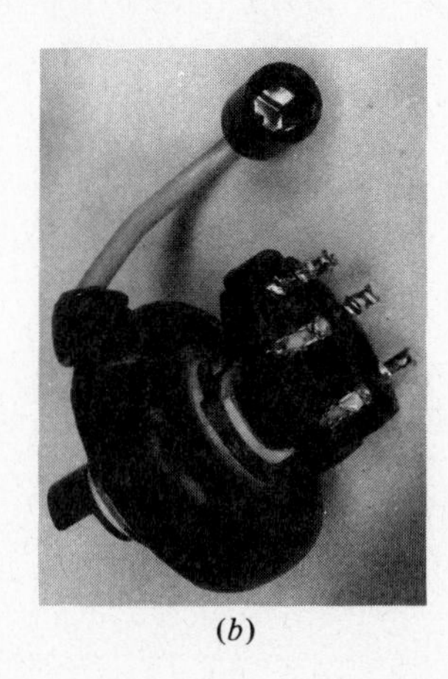

Fig. A1-5. Transformers: (*a*) typical *if* transformer (*F. W. Sickles Div. of General Instrument Corp.*); (*b*) horizontal sweep transformer (*General Electric Company*).

Fig. A1-4. Typical oscillator coil for broadcast band (*F. W. Sickles Div. of General Instrument Corp.*).

Fig. A1-6. Representative tube type, 9-pin miniature (*RCA*).

transformers, radio-frequency (RF) transformers (Fig. A1-5*a*), horizontal-sweep transformers (Fig. A1-5*b*), and many others.

Almost everyone familiar with tubes knows that they were the sinews of the electronics industry, but that they were replaced by transistors. Nevertheless, the technician must be familiar with them because some tube devices are still in use and must be serviced. Figure A1-6 shows a representative tube type, a 9-pin miniature.

Semiconductor devices are now the life blood of the industry. These include, among other devices, transistors, integrated circuits (ICs), and microprocessors, Fig. A1-7. ICs and microprocessors now occupy center stage. Within these tiny chips are contained complete electronic systems and circuits. These two devices have reduced the size and cost of digital computers and other electronic products. The microprocessor is on the threshold of revolutionizing communications, industrial processes, and control. It is predicted that in the next 20 years, with the growth in the use of microprocessors, the electronics industry will rank in size with the automotive industry.

Many other components are also used in electronics. You will become acquainted with these as you use them in the course of your work.

You should examine some of the basic electronic components and become familiar with their physical appearance and circuit symbols.

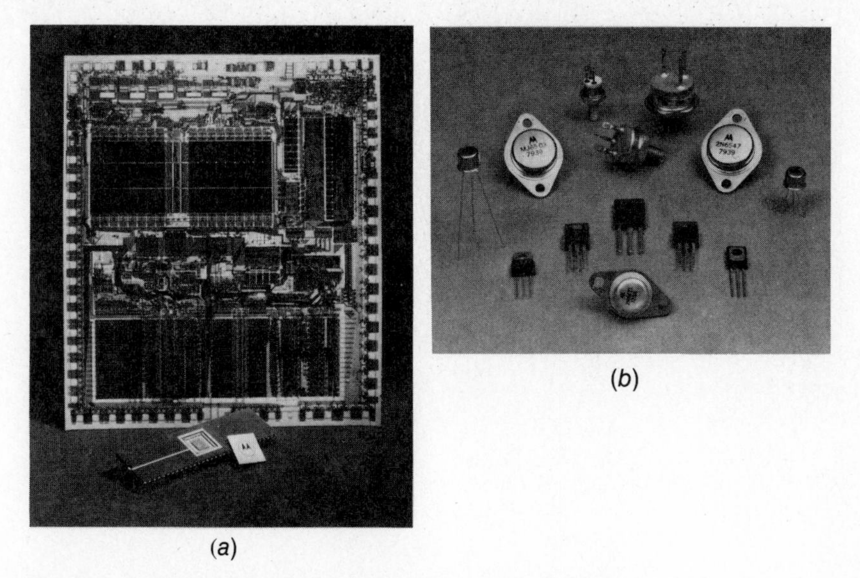

Fig. A1-7. (*a*) Representative transistors and other semiconductor control devices; (*b*) microprocessor, with an enlarged view of the chip's structure. (*Motorola Semiconductor Products, Inc.*)

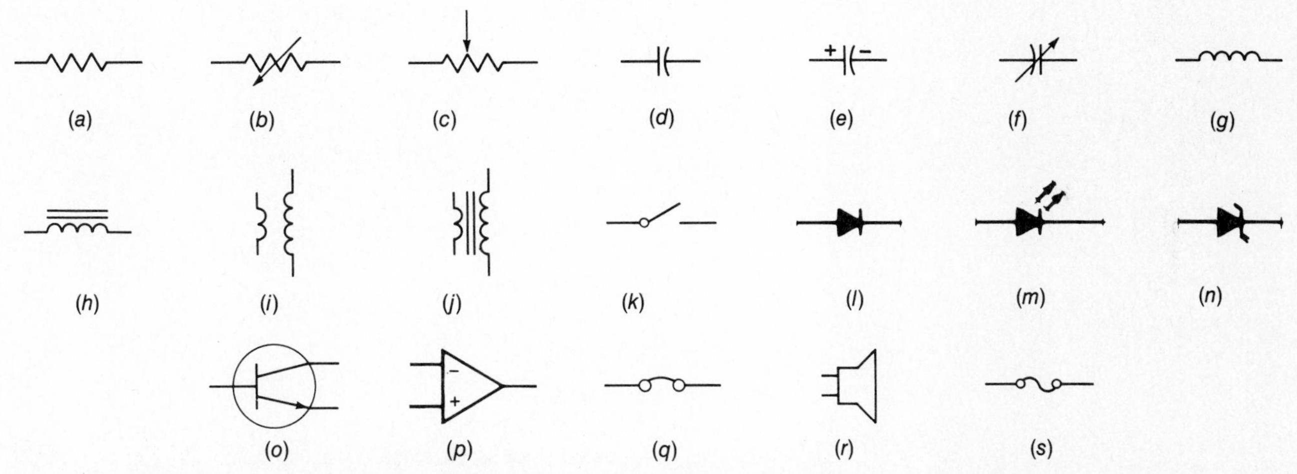

Fig. A1-8. Standard electronic component circuit symbols: (*a*) fixed resistor; (*b*) variable resistor; (*c*) potentiometer (resistor with adjustable tap); (*d*) capacitor (fixed); (*e*) polarized (electrolytic) capacitor; (*f*) variable capacitor; (*g*) inductor (air-core); (*h*) inductor (iron-core); (*i*) transformer (air-core); (*j*) transformer (iron-core); (*k*) switch; (*l*) diode (solid state); (*m*) light-emitting diode (LED); (*n*) zener diode; (*o*) transistor; (*p*) operational amplifier (op amp); (*q*) earphones; (*r*) speaker (loudspeaker); (*s*) fuse.

Some standard electronic component symbols are shown in Fig. A1-8. They specify components which are used in a particular electronic device. The manner in which these components are connected in the device is illustrated in a drawing called a circuit or schematic diagram. In this diagram each component is represented by its unique symbol. We will next discuss how to interpret circuit diagrams.

A2
APPENDIX
THE SCHEMATIC DIAGRAM

The Schematic Diagram

The schematic diagram is an electronic blueprint. It tells the technician where and how the electrical parts are connected in a specific circuit, such as a radio circuit. There are a few simple rules to learn:

1. Parts are shown by their electrical symbols.
2. The leads which extend from a part are shown by straight lines.
3. (*a*) If two or more parts are joined physically at a common point (terminal), this point is indicated by a heavy dot on the schematic. For three resistors which have a lead of one connected to leads of the others, we draw a simple schematic (Fig. A2-1). We can, if we wish, identify the junction point by some letter, such as *A* (Fig. A2-2).
   (*b*) If two lines in a schematic diagram cross one another and no dot appears at the point of crossing, they are *not* connected at this point.

Breadboarding

The schematic diagram, then, is a set of instructions to the technician for assembling and connecting the parts of a circuit. Schematics are used in the experimental laboratory as well as in the production and servicing processes. The physical product created on the production line is in permanent form. The experimental circuit in the laboratory is usually assembled in temporary form so that circuit changes may be easily made until the product design is final.

Experimental circuits are assembled on a "breadboard." This is a device which ensures simple, rapid circuit connection and disassembly. Moreover, the breadboarding technique allows component parts to be used over and over again, permitting a wide variety of circuit configurations for study and experimentation. The experiments in this manual will be facilitated by, and are intended for, breadboarding devices.

Wiring Methods

Handwiring

Components in electronic circuits may be connected by handwiring or printed wiring techniques. In the handwiring process, component leads are secured to terminal posts. The pigtails (leads) of parts which must be electrically connected are secured at a common terminal post. If the parts are too far apart, their terminal leads may not be long enough to reach a common post. In that case, the pigtails of these remote parts (which must be joined electrically) are secured to separate terminal posts. A wire is then used to connect these terminal posts. This is the handwiring technique of interconnecting components in a chassis or on a breadboard.

Etched Wiring

An advanced means of interconnecting parts in mass-produced electronic devices uses etched (commonly called "printed") wiring connections. This method requires the use of a plastic, ceramic, glass, or phenolic board on which the components are mounted. Interconnection is made by means of conductive surfaces, which are etched on the board.

One technique for making etched circuit boards is as follows: The conductive circuit pattern is designed on paper. A clear acetate is then made of the original art. A silk screen of the circuit pattern is made. A protective material is placed on the copper foil using the silk screen of the circuit pattern. The unprotected portion of the copper foil is etched away in an acid bath. The protective coating is also removed from the circuit pattern remaining on the board. By another silk-screen process, solder resist is applied to the pattern except in the areas where solder is to be applied. In some instances, silk screen processes apply a "road map" to the top and bottom of the board. Then holes are punched into the boards.

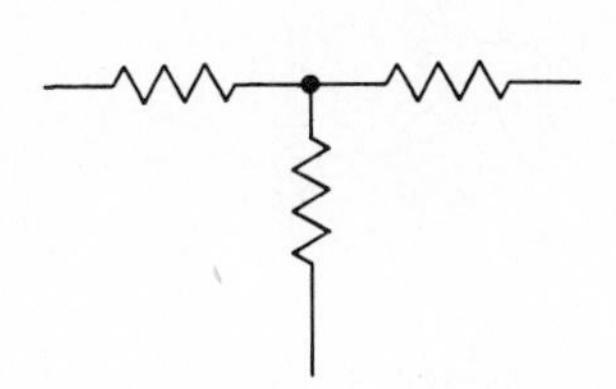

Fig. A2-1. Three resistors meeting at a common point.

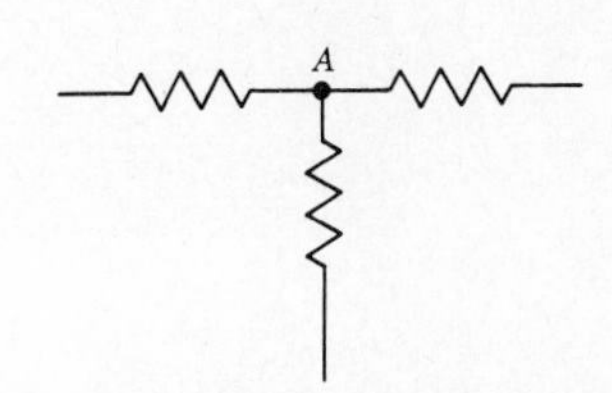

Fig. A2-2. Common point identified by the letter *A*.

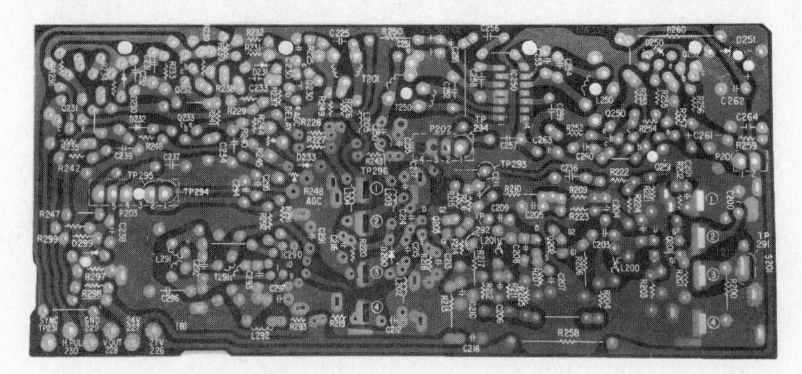

Fig. A2-3. Etched board ready for component insertion.

This requires tooling, which is more accurate than drilling. A precise hole size which will produce a good solder fillet around the lead is mandatory.

Figure A2-3 shows an etched board ready for component insertion. The components are mounted on the board by inserting their leads through the predrilled holes. The leads are then soldered to the conductive surfaces which "form" the pattern of interconnecting "printed" wiring between components. Figure A2-4*a* is a top view of a finished etched board. The components are on the other side of the board (Fig. A2-4*b*). The circular blobs are points at which component leads have been soldered.

(*a*)

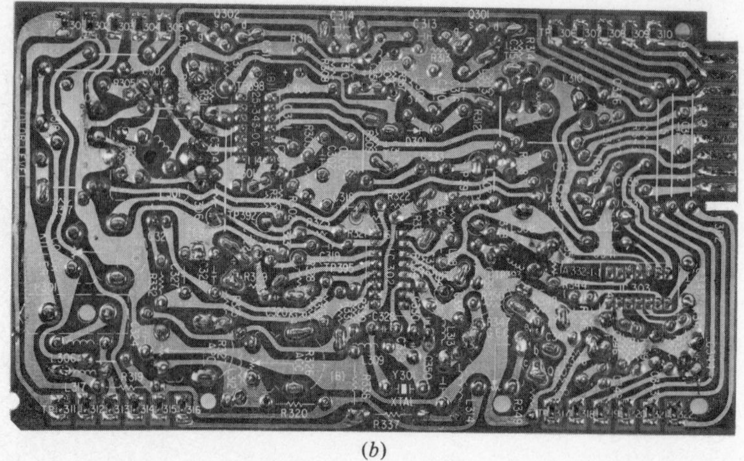

(*b*)

Fig. A2-4. (*a*) Top and (*b*) bottom views of completed etched board with components soldered.

A3 APPENDIX
FAMILIARIZATION WITH HAND TOOLS USED IN ELECTRONICS

After an experimental electronic circuit has been breadboarded, its operating characteristics are tested by the technician. Design changes may be indicated and are effected as required. When the technician and engineer are satisfied that the circuit is performing as it should, it is ready for prototype assembly.

The proper chassis layout of circuit components is determined, the chassis is prepared, and the technician mounts the major components. In this phase of chassis preparation and component assembly, the technician may use chassis forming, drilling, and cutting power tools. This phase will also require mechanical hand tools such as the scribe, punch, hammer, screwdriver, wrench, hacksaw, and file.

Hand Tools Used in Electronics

Electrical assembly follows the preparation of the chassis and mounting of parts. This is the stage which requires electrical interconnection of components. It is concerned with the preparation and soldering of conductive wiring between parts. The common hand tools used in electrical assembly include diagonal pliers, long-nose pliers, soldering aid, wire stripper, soldering iron, soldering pencil, soldering gun, knife, and heat sink.

Diagonal pliers, commonly called dykes, cutters, or diagonals, are used for cutting soft wire and component leads. They should not be used for cutting hard metals such as iron or steel. Some diagonals have a small, notched cutting surface for stripping the insulation from a wire. This stripping hole will accommodate #22 wire, which is the standard hookup wire used in electronics.

Long-nose pliers are used to hold wire so that the stripped end may be twisted around a terminal post or pushed through a terminal eye. Long-nose pliers sometimes have cutting edges so that the pliers can serve for both gripping and cutting wires. The 5-in nose pliers is a popular size, as is the 5-in diagonal.

The needle-nose pliers is a variation of the long-nose. Its long, thin jaws can get at difficult-to-reach spots.

The soldering aid is a useful tool which simplifies soldering jobs. A standard aid has a sharp metal pointer at one end and a slotted V-type grip at the other. One function of the pointer end is to help clear solder out of terminal eyes on solder lugs. The gripping end is useful in unwrapping wire and component leads when these are being unsoldered from terminal posts.

The wire stripper removes insulation from hook-up wires. There are different types of strippers, ranging from the simple type found on diagonal pliers to automatic multisized strippers which can handle wires of different diameters. The automatic trip-type stripper is popular with electronics technicians. In addition to mechanical, there are also thermal strippers.

The soldering iron is still a standard electrical hand tool in the industry, although many technicians favor the soldering gun. A heating element inside the iron utilizes power from the power line. The heat is channeled to the tip. The tip is applied to and heats the area to be soldered.

Soldering irons are rated by the amount of power they dissipate, and thus indirectly by the amount of heat they can develop. One-hundred-watt irons are standard in handwired operations, where wires and leads are soldered to terminals. Heavier irons (250-W) are used to solder connections to large metallic surfaces (such as a metal chassis) requiring more intense heat.

The low-wattage soldering pencil is a soldering iron with a smaller heating capacity. It is used for soldering or unsoldering components from a printed wiring board, or in delicate soldering applications requiring low heat levels. Interchangeable tips in a wide variety of shapes and sizes add flexibility to the soldering pencil.

The soldering gun has gained wide popularity because of its fast heating characteristic. A trigger switch on the gun handle applies power to the heating and soldering tip and heats it in 30 s (approximately). Unlike the soldering iron, which is slow heating and must be left on as long as it is in use, the soldering gun is heated only at the moment it is to be used. Between soldering operations it is left off. Popular sizes are 100- and 125-W guns.

A pocket knife is useful in many small tasks requiring cutting and scraping. It can be used for scraping and cleaning the terminal ends of wires and components, in preparation for soldering.

The heat sink is a small metal clip used to prevent overheating during soldering or unsoldering of heat-sensitive electronic parts. The heat sink is clipped onto the lead between the body of the part and the terminal point at which heat from a soldering iron or gun is applied. It absorbs heat and reduces the amount of heat conducted to the component.

Desoldering tools simplify the job of cleaning solder from etched (pc) board solder holes while component leads are being removed from the holes. The holes must be free of solder before the terminals of a new component may be inserted. Desoldering tools used by the technician take two forms. The first is a spring-loaded vacuum suction tool, the second a suction device in the form of a hollow rubber ball attached to a stainless steel or plastic tube. In operation the open end of the tube is placed on a heated solder joint or solder hole. With the spring-loaded device, the spring is released and the molten solder is sucked into the tube clearing the joint or hole of solder. With the ball-type tool, the ball is squeezed, and as it is released the solder is again sucked into the tube.

CAUTION: A word of caution about the use of hand tools. They should never be used in a live circuit, that is, in a circuit to which electric power is applied. These tools are made of metal, and metal is a conductor of electricity. Failure to observe this safety precaution may result in electrical shock or damage to the circuitry.

Wires

Hookup wire may come as a solid or stranded conductor. Stranded wire is the accepted standard for hookup wire. However, solid hookup wire is often used in circuits which are not subjected to movement. Solid wire is used for house wiring, bell wiring, transformer windings, etc. Component terminal leads are usually made of solid wire.

Stranded wire is made of wire threads twisted together. The number and diameter of the threads depend on the application and on the quantity of current the wire will carry.

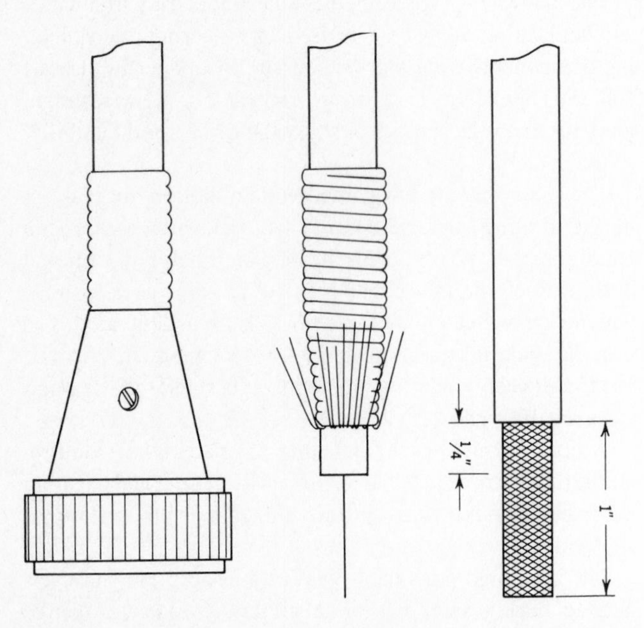

Fig. A3-1. Preparation of coaxial cable for attaching microphone connector.

The larger the current, the larger must be the diameter of the wire. Stranded wire is flexible and may be bent and twisted without danger of breaking it.

Wires conduct electricity. To prevent adjacent wires from making electrical contact, or from touching metallic or conductive surfaces, wires are covered with an insulated coating. The industry now makes extensive use of FR-1 (fire retarding insulation) as wire covering. Older methods of insulating wires included solid-wire insulation, which used shellac or enamel into which the wire was dipped. Stranded wire was insulated by a plastic, rubber, or cloth coating.

Multiwire cables are frequently used to interconnect electronic units or devices. For example, the several units of a radar set are interconnected by multiwire cables. A cable is made up of two or more insulated wires inside a common housing. The housing is usually a plastic or rubber sleeve, but the sleeve may also be a wire braid. Cables are designed and constructed for special applications requiring two or more separate insulated electrical conductors.

Coaxial cable is a very special type of shielded cable. It has an inner and an outer conductor. The inner conductor may be solid or stranded. Around the inner conductor is a plastic or phenolic insulating sleeve. A braided shield is fitted around the insulating sleeve of the inner conductor. The braid acts as a second conductor around the inner conductor and is insulated from it. A rubber or plastic sleeve covers the braid and provides insulation for the coaxial assembly. Shielded cable is also manufactured without any external insulating cover.

Wire Preparation

Two or more wires that provide a conductive path for electricity must be electrically connected. This means that an uninsulated surface on one wire must be mechanically connected to an uninsulated surface on the other wire. To ensure that the wires will not separate or the connection corrode, they are soldered at the junction.

Before wires may be connected and soldered, they must be properly prepared. This involves stripping away the insulation at the ends of the wire, thus providing terminal leads which may be connected to each other or to a terminal post. It also involves cleaning insulation or oxidation from the bare ends of the wire.

Mechanical wire strippers are usually used by the technician, though thermal strippers may also be employed. The technician selects the stripping hole which will fit the wire, inserts the wire in the tool, and squeezes the handles. The insulation is automatically removed and the stripper resets in one operation.

After the insulation is removed, the technician examines the wire. If it has a shiny look, no further preparation is needed. However, a dull- or dark-appearing wire must be cleaned before it is connected and soldered. The wire may be cleaned by scraping it gently with a knife or with emery cloth. This same process is employed in removing the insulation from shellac- or enamel-covered solid wires. The technician must be careful not to "nick" the wire.

The technician may also be required to prepare coaxial or shielded wire. This type of wire is used in circuit connections where a shielded conductor is required. Coaxial cable is used for intercabling two assemblies of an electronic system and for instrument leads. The braid around the coaxial wire acts as the grounded shield for the inner conductor, which is the "hot" wire.

In preparing an uncovered shielded wire for use, the braid is combed away with a soldering aid, awl, or other pointed tool. The braid is then twisted together to form the ground lead. The insulated covering is stripped from the inner conductor. The inner and outer conductors are scraped if necessary before connection to the circuit.

The processes for preparing a coaxial cable for use as an instrument lead are illustrated in Fig. A3-1. The two ends of the cable require individual treatment because one end must be terminated in a microphone connector (other connectors are also used), and the other in a probe or in two leads, depending on the use for which the cable is intended.

The preparation for mounting a microphone connector is shown in Fig. A3-1. One inch of cable insulation is first stripped away, exposing the braid. The spring shield of the connector is then placed over the braid. The braid is folded back and cut to fit the narrow neck of the spring. The insulation is stripped from the inner conductor. The knurled holding ring is placed over the spring so that the inner conductor protrudes through the eyelet. The ring is pushed back as far as it will go. The set screw is tightened. Care must be taken not to "short" the inner conductor to the braid or the wire shield.

The center conductor protruding through the eyelet is cut flush with the eyelet. This completes preparation of the cable, and the inner conductor is ready to be soldered to the eyelet.

The insulation from the other end of the coaxial cable is stripped and the braid trimmed back. The insulation is stripped from the inner conductor. The braid is now combed and used directly, or a short piece of flexible ground wire may be soldered to it.

A4
APPENDIX
SOLDERING TECHNIQUES

Soldering is generally required to assure permanent electrical connections. Wires, or wires and terminals, are wrapped or twisted together, then solder is melted into the heated joint. When the heat is removed, the solder and wire cool, making the soldered joint look like a solid piece of metal. It is not possible, after proper soldering, to separate the wires at a joint except by breaking them or by unsoldering them.

Solder is an alloy of lead and tin. It has a low melting point and comes in wire form for electronics use. Electronics solder is made up of 60 percent tin and 40 percent lead, though the composition may vary for certain applications. Rosin-core solder is used for soldering electronic components. The rosin is a flux which flows onto the surface to be soldered, assuring a more perfect union. Acid and soldering paste should *not* be used in electronics.

Proper soldering requries the following:

- Clean metallic surfaces
- Sufficient heat applied to the joint to melt solder when solder is applied to the heated wire surface

The solder must not come in direct contact with the tip of the soldering iron or soldering gun because it will melt readily. If the wires or terminals to be soldered have not been preheated sufficiently, the molten solder will *not* adhere to their surface. The joint may look well soldered, but the chances are that it is not. *Cold*-solder joints provide poor electrical contact. Defects arising from poor soldering are difficult to discover when troubleshooting.

Soldering is not always used in the manufacturing process. There are many instances of solderless wire-wrap connections in use today.

Tinning

To ensure maximum transfer of heat from the iron to the surfaces at the joint, the tip of the iron or gun must be tinned. A new soldering tip or a tip that has been used for a long period of time must be cleaned by scraping it with a knife or emery paper, steel wool, a wire brush, or fine sandpaper. If the tip is badly pitted, it may be necessary to file it. This technique applies to *copper* tips.

Many modern tips are gold-plated or iron-bearing. A gold-plated tip should be cleaned by wiping it against a wet sponge. Iron-bearing tips can be cleaned with a wire brush. These tips should never be filed or cleaned with sandpaper or emery cloth.

After the tip is cleaned, it is heated. Solder is permitted to melt onto the tip, tinning it. If the iron is overheated and permitted to discolor before solder is applied, it will be difficult to tin.

It is not only necessary to tin the iron; the surfaces to be soldered should also be cleaned and tinned. Tinned surfaces assure good electrical connections and proper bonding when soldered. Wire may be tinned by placing it on the tip of the iron and heating it sufficiently so that the wire will melt solder. Stranded wire should be twisted together before it is tinned. Terminals should also be tinned.

The tip of a soldering iron tends to become dark and dirty when in use. To keep the iron at maximum-heat-transfer efficiency, the tip should be cleaned periodically by wiping it on a damp sponge. A wire brush will also clean the soldering tip.

Mechanical Connections

Wire and component leads should be secured by a three-quarter to a full turn around a terminal post, as in Fig. A4-1*a*. The tendency of the novice is to wind several turns around the post. This is wasteful of terminal area and creates difficulties if it is ever necessary to unsolder and remove the lead. In Fig. A4-1*b* wire is secured to a terminal strip.

If two wires are to be joined together, a hook splice is used (Fig. A4-2). It is unnecessary to twist the wires together before soldering.

Heat Sinks

Heat sinks are used to avoid heat damage to heat-sensitive components during soldering. Solid-state devices such as transistors and diodes are extremely sensitive to heat and can be permanently damaged if they are overheated. Any small metal clip, an alligator clip for example, may be used as a heat sink. Figure A4-3 illustrates the use of a heat sink. It is clipped onto the lead between the component and the point at which the soldering iron is applied. The clip acts as a heat load and reduces heat transfer to the component.

After the iron is removed, the heat sink should be kept in place until the joint has cooled.

Wherever they can be used, heat sinks are a good practice. However, they are not practical in many physical applications, where solid-state devices have been soldered into circuit boards in such a way that they do not allow space for

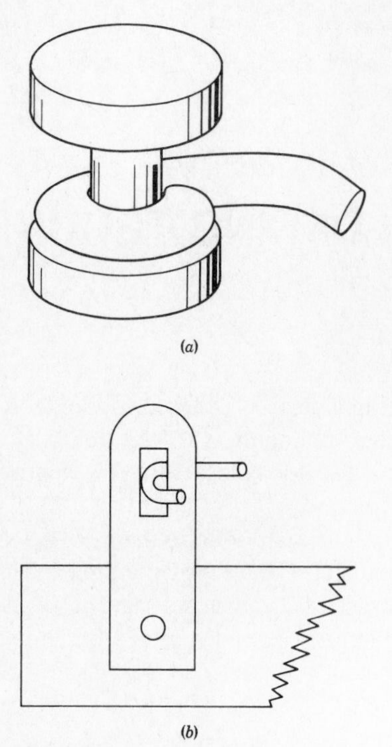

Fig. A4-1. Component lead secured to (*a*) terminal post before soldering, wrap is ¾ to 1 full turn; (*b*) terminal strip.

heat sinks. The present trend in the manufacture of solid-state devices provides for plug-in components, for example, plug-in transistors and plug-in integrated circuits.

Soldering Aid

The pointed end of a soldering aid is used to clean terminal eyes from which wires or leads have been unsoldered and removed. Heat is applied to the terminal, and the sharp tip is worked into the terminal eye until it is free of solder and ready to receive another wire.

The slotted end of a soldering aid is used to grasp wires while they are being soldered or unsoldered.

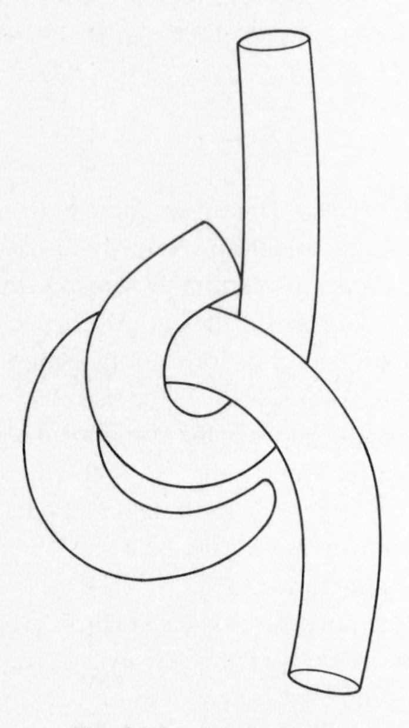

Fig. A4-2. Hook splice.

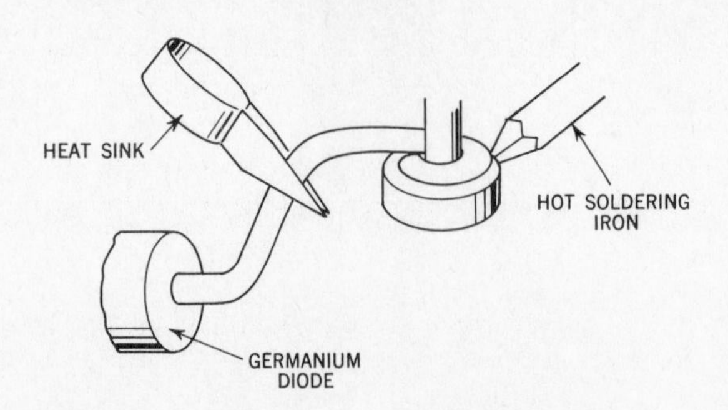

Fig. A4-3. Heat sink protecting germanium diode.

Soldering on Etched (Printed) Wiring Circuit Boards

The network of interconnecting conductive paths on a "printed" wiring board consists of thin copper strips and pads bonded to the plastic board. Leads of components mounted on the board are inserted through holes in the board and the conductive copper. These leads are soldered to the copper at the terminus of the hole through which they emerge. If excessive heat is applied to the copper, it may lift from the board, or the miniature components mounted on the board may be damaged.

A 30-W (approximately) soldering pencil is therefore used to heat the junction. This low-wattage iron provides an effective means of controlling heat. As in handwired circuits, component leads should be cleaned and tinned before soldering. Sixty-forty rosin-core solder or eutectic solder should be used. The surface of the copper bonded to the board should also be properly prepared before soldering. The copper can be cleaned by scraping it gently in the vicinity of the terminal hole, or an ink eraser can be used to clean the surface at the solder point.

Another factor which must be considered in working on "printed" boards is the proximity of the conductive strips and pads to each other. It is essential to avoid excess solder to prevent bridging the gap between two paths. Application of excessive solder should be avoided in handwired circuits also. Just enough solder should be melted into a junction to effect a proper bond. If tiny solder globules form in the junction area, remove them by cleaning the soldering tip on a sponge or cloth, then apply the cleaned tip to the solder globules at the junction.

Unsoldering Components

Unsoldering is the reverse of the soldering process. The terminal point is heated with a properly tinned iron. The lead to be unsoldered is grasped with a soldering aid or with a pair of long-nose pliers. When the solder melts, the lead is gently unwrapped and removed from the post. If the lead was inserted through a terminal eye, the eye is cleaned so that it will be ready to accept a new lead.

Cleaning a terminal eye, or circuit board terminal "hole," may be accomplished by manipulating a soldering aid in the heated terminal hole, or by use of a desoldering suction tool as previously described.

Excess solder or solder globules which may have separated from the post are shaken out of the chassis or board. Wire braid and special solder "sucker" tools are especially useful in retrieving these solder globules.

SOME SAFETY PRECAUTIONS: The soldering gun or iron operates at temperatures high enough to cause serious burns. Observe these safety precautions:

(*a*) Do not permit hot solder to be sprayed into the air by shaking a hot gun or iron or a hot-soldered joint.
(*b*) Always grasp a soldering gun or iron by its insulated handle. Do *not* grasp the bare metal part.
(*c*) Do not permit the metal part of a soldering gun or iron to rest on combustible material. An iron should always rest on a soldering stand.

A5

APPENDIX

USE AND CARE OF THE VOM

The commercial VOM combines within one instrument the functions of dc and ac voltage, direct current, and resistance measurement. Additional functions such as decibel, or dB, measurement (with which you are not yet familiar) are sometimes included.

The design of commercial VOMs follows the techniques discussed in the experiments dealing with meter movements. Circuit arrangements, however, are more sophisticated, including ring shunts for current ranges and series and series-parallel circuits for resistance ranges. Complex switching circuits are employed for range and function selection.

The general-purpose commercial VOM popularly uses a 50-μA meter movement, resulting in a 20,000-Ω/V dc voltmeter. The ohms/volt characteristic on ac volts is lower.

AC Voltmeter Function

A variety of meters and circuit arrangements are used for measuring ac volts. A common arrangement is one in which a "rectifier" in the meter changes alternating into direct current. The meter then measures this direct current. However, the scale of the meter is calibrated in terms of ac volts. The calibrations are generally in rms or "effective" values, which are 0.707 of peak value.

Figure A5-1 is a simple ac voltmeter circuit. Input ac voltage is applied across the meter terminals. The voltage is rectified by X_1, a solid-state device with which you will become more familiar in your study of basic electronics. The resulting direct current actuates the meter movement. The rectified current is limited by R_1, which therefore acts as a voltage-range multiplier. With the exception of X_1, this circuit is similar to a dc voltmeter circuit.

The circuit in Fig. A5-1 was simplified to explain the theory of operation. A more practical arrangement for a multirange ac voltmeter is shown in Fig. A5-2. A switch permits selection of five different ranges, 2.5 to 1000 V, for measurement between the terminals labeled Neg. and Pos. Note that polarity need not be observed in ac measurements. The polarity designations at the meter terminals (which are used for both direct and alternating current) apply to dc measurements.

A separate 5000-V ac jack may be used with the Neg. jack for measuring up to 5000 V of alternating current.

The rectifier circuit in Fig. A5-2 is more complex than that shown in Fig. A5-1. Moreover, a shunt resistor R_{24} is placed across the meter movement. This arrangement improves the

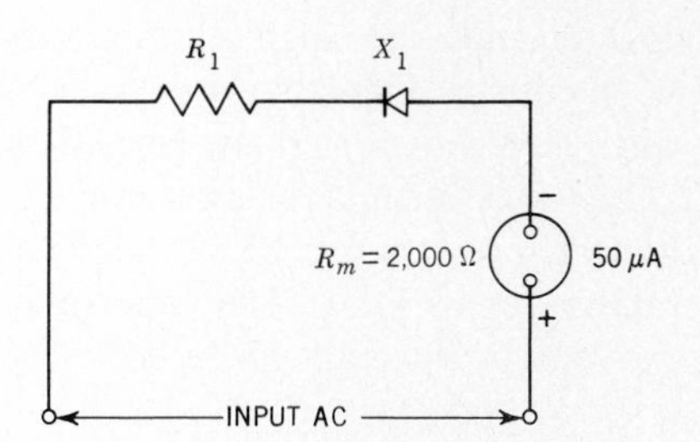

Fig. A5-1. Simplified ac voltmeter.

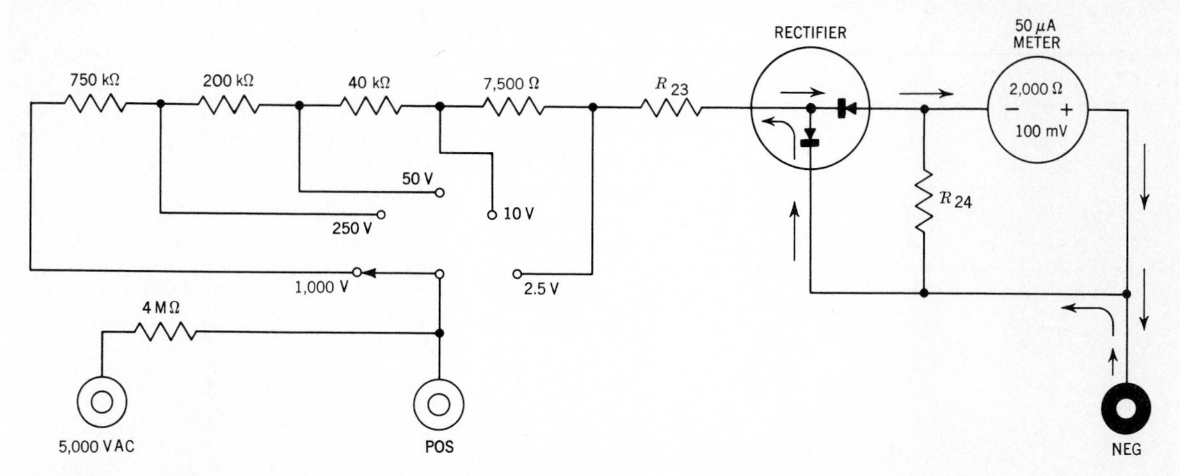

Fig. A5-2. Multirange ac voltmeter circuit (*Simpson Electric Co., Div. of American Gage & Machine Co.*).

linearity of the ac voltage scale. An improvement in linearity is achieved at the expense of meter sensitivity. The shunting effect of R_{24} reduces the ac ohms/volt characteristic of the voltmeter. Common ratings for ac voltmeters using a 50-μA movement vary between 1000 and 10,000 Ω/V. Therefore, *ac voltmeters using the same movement load the circuit more than dc voltmeters do*. As in dc voltmeters, the higher the ohms/volt rating, the smaller the loading effect of the meter on the circuit. Obviously, also, the higher voltage ranges of an ac voltmeter cause less loading than the lower ranges.

The ac voltmeter is used in the same manner as the dc voltmeter except that measurements are made in ac circuits and polarity need not be observed. In electronics we frequently want to measure the ac voltage at a point where dc voltage is also present. To accomplish this, the technician must insert a blocking capacitor, usually 0.1 μF, in series with the ac voltmeter leads. This will prevent the meter from reading the dc voltage but will permit ac voltage readings.

This blocking capacitor is frequently included in the design of the meter in the form of an additional meter jack labeled Output. One of the ac voltmeter leads is inserted into the output jack. The other is inserted into Common. Figure A5-3 shows this output function. Observe that with the exception of the 0.1-μF capacitor, which was added, the circuit is identical with that of Fig. A5-2. The voltage-measuring circuit is affected by the 0.1-μF capacitor. Hence, readings are relative, not absolute.

Function and Range Selection

The methods of function and range selection vary with the type of meter and design. Figure A5-4 shows a commercial VOM. A switch selects the functions, labeled AC, +DC, and −DC. The AC position is for measuring ac volts and decibels. The switch is set to the DC position to measure dc volts, direct current, or resistance. The output jack is used for measuring ac volts in a circuit where dc voltage is also present.

The range switch is used in conjunction with the function switch. The ranges inscribed on the panel are available at the + and − (common) terminal jacks of the meter. Observe that additional ranges are provided by jacks on the meter.

Zero Ohms is a manual control whose purpose is to zero the ohmmeter. Zero Ohms has the same function here as in the experimental ohmmeter, namely, to compensate for changes in the internal resistance of the instrument-contained ohmmeter battery.

Meter Scales

Figure A5-5 shows five scales on the meter pictured in Fig. A5-4. The top scale is for ohms. Zero ohms is on the right-hand side. Center-scale reading is approximately 12.5 Ω. The scale is nonlinear and crowds in the region of high resistance. When the range switch is set on $R \times 1$, the readings are made directly from the ohm scale. On the $R \times 100$ or $R \times 10{,}000$ range, scale values are multiplied by 100 and 10,000, respectively.

The dc scale is found immediately below the ohm scale. On this scale, read dc volts when the range switch is set on any of the voltage ranges. When the range switch is set on any of the current ranges, the dc scale serves also for current readings. The range designation 10 V, 100 mA, etc., corresponds to full-scale deflection of the pointer.

There are two separate ac scales on this meter immediately below the dc scales. The upper scale is for ac voltages greater than 2.5 V. The lower ac scale is for voltages up to 2.5 V. Both scales are somewhat nonlinear, crowding at the zero end.

Decibels are read on the fifth and last scale of this instrument. This is a nonlinear, logarithmic scale.

Accuracy

The accuracy of the general-purpose VOM is usually rated at full-scale deflection and is 2 to 3 percent on dc function, and 5 percent on ac function. Because these ratings are at full-

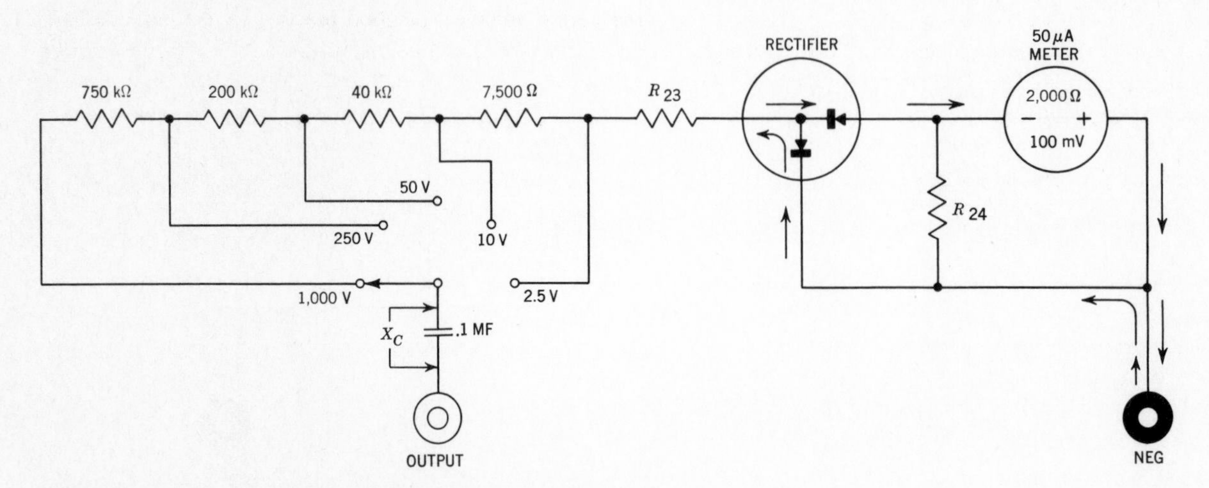

Fig. A5-3. "Output" meter circuit (*Simpson Electric Co., Div. of American Gage & Machine Co.*).

Fig. A5-4. Commercial VOM—Simpson 260.

scale deflection, the percentage of accuracy of the meter may be lower on the lower levels of the range. For example, 2 percent on the 250-V range is 5 V. This means that a 5-V inaccuracy can occur anywhere on the scale. At the 50-V mark on the 250-V range, the voltage may therefore be 50 ± 5 V. This represents a possible error of 10 percent at this level.

Care in the Handling and Use of a VOM

The technician must use common sense and care in using a multitester. Though it is ruggedly constructed and capable of giving many years of satisfactory performance if properly handled, a VOM may still be easily damaged if it is used carelessly. Read the operating instructions before using the meter.

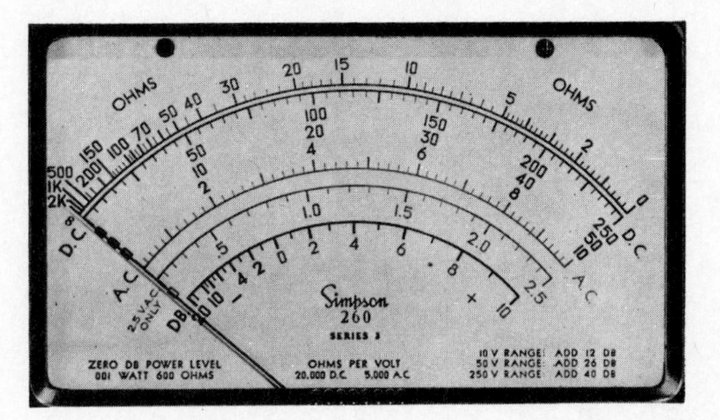

Fig. A5-5. VOM scale.

Following these simple rules will help secure trouble-free operation of the multitester. Before inserting the meter into the circuit be certain that:

1. The meter switch or switches have been set to the proper measurement function. If voltage is to be measured, the meter must be set for reading voltage. If current is to be measured, the meter switch or switches must be set for reading current, etc.
2. The meter switch or switches have been first set to the highest voltage or current range in measuring an unknown voltage or current. This will reduce the possibility of meter overload and damage.
3. The meter test leads have been plugged into the proper test jacks.
4. Polarity is being observed in measuring dc voltage and current.
5. Power is disconnected before resistance is measured in a circuit.
6. In measuring current, the circuit has been broken so that the meter may be inserted in series with the circuit.
7. The voltage or current to be measured does not exceed the range or capabilities of the meter.
8. When work with the VOM is completed, it should be left on the highest ac voltage range or on "transit."

CAUTION—DANGER: Improper handling of the multimeter or any instrument may result in personal injury. Therefore, in measuring voltage it is good practice to use just one hand. Keep the other hand behind your back. Hold the *insulated* part of the test prod and not the metal tip. In measuring current, shut **power off** before breaking the circuit. Insert the milliammeter before turning **power on.** Always inform your instructor of any personal injury, damage to equipment, or any possible hazards.

A6
APPENDIX
PARTS REQUIREMENTS

Components with asterisk (*) are added new parts for this edition.

Resistors, ½ W

Resistance, Ω	*Quantity*	*Resistance, Ω*	*Quantity*	*Resistance, Ω*	*Quantity*
10	1	1,800	1		
33	2	2,200	1	22 kΩ	1
47	1	2,700	1		
100	2	3,300	1		
330	1	4,700	1	56 kΩ	1
		5,100 (5%)	2	100 kΩ	2
470	1	5,600	1	120 kΩ	1
560	1	6,800	1		
		8,200	1	1 MΩ	2
1,000	3	10 kΩ	1	2.2 MΩ	1
1,200	3	12 kΩ	1		
1,500	1	15 kΩ	1		

Resistors, Higher Wattage

Resistance, Ω	*Watts*	*Quantity*
15	25	1
250	4	1
300	10	1
500	5	1
5,000	5	1
*100	5	1

- *Potentiometers* (2-W): One of each unless otherwise specified.

NOTE: Potentiometers with symbol ‡ will be required only if resistance decade box is not available.

- ‡100-Ω, 500-Ω, 10 kΩ (two), 100 kΩ, 500 kΩ
- *Solid State Diodes:* 1N2615
- *Thermistor:* 50-Ω/100-mA, 300-Ω cold, Carborundum-*1204J or equivalent

Capacitors

Capacitance, μF	*Quantity*	*Voltage*
0.005	1	400
0.01	1	400
0.05	1	400
0.1	2	400
0.5	1	400
1.0	1	400
*5.0	1	100
*10.0	1	100

- *Varistor:* Carborundum* #463-BNR-32 or equivalent
- *Transformers, Chokes, and Coils:*
  Choke: Two, 8-H @ 50 mA dc
  Audio Output Transformer: 6K6 GT, push-pull to 3.2 Ω
  RF Coil: 30 mH
  Filament Transformer: 117 V primary, 6.3 V secondary

NOTE: Filament Transformer is required only if 6.3-V/60-Hz source is not otherwise available.

■ *Miscellaneous:*

(1) Breadboarding device with terminals and connectors
*(2) *Power-line isolation transformer*
*(3) *Variable autotransformer*
(4) *Wire:*
  (*a*) #18 and #22 stranded hookup wire
  (*b*) 20 ft of #18 varnish insulated copper wire
  (*c*) *Nichrome wire:* #40 Standard (60% nickel, 16% chrome, 24% iron), 2 ft
  *(*d*) Instrument leads alligator type
(5) *Solenoid:* 100 turns of #18 varnished insulated copper wire wound in three layers on a 3-in long × 1-in diameter hollow cardboard or plastic cylindrical form
(6) *Coil Form and Core:*
  (*a*) 2-in long × ½-in diameter, hollow cardboard or plastic cylinder, or copper tubing
  (*b*) 2-in soft-iron core to fit inside coil form
(7) *Switches:* SPDT (one), SPST (two)
*(8) *Relay:* dc, SPDT, 400-Ω field, 7 mA pickup, 1A contacts RBM type 10730-8 or equivalent
*(9) 60-W wired *test lamp and socket*
(10) *Magnets:* two bar magnets, one horseshoe magnet
(11) *Iron filings in salt shaker type dispenser*
(12) *Dry cells and cell holders:* four 1½-V
(13) *Alligator clips:* three
(14) *Defective resistors:* for Experiment 19 (These may be made by changing color code on good resistors.)
(15) *Cardboard or glass plane:* 8½ × 11 in (one)
(16) *Insulators for Experiment 5:*
  (*a*) Rubber strip, 2 × ¼ in
  (*b*) Plastic strip, 2 × ¼ in
  (*c*) Wooden strip, 2 × ¼ in
(17) *Fused (1 A)* line cord
(18) *Meter movement:* 0–1 mA, 50-Ω, 50-mV
(19) *Resistance decade box (2 W):* Adjustable in 1-Ω steps from 1 to 99,999-Ω
(20) *Magnetic compass*